Student's Solutions Manual
to Accompany

Calculus
with Analytic Geometry
FOURTH EDITION

HOWARD ANTON
DREXEL UNIVERSITY

Prepared by

Albert Herr
Drexel University

JOHN WILEY & SONS, INC.
New York   Chichester   Brisbane   Toronto   Singapore

ISBN 0-471-55140-6

Printed in the United States of America

Printed and bound by Port City Press, Inc.

10 9 8 7 6 5 4 3 2

Preface

This supplement to the *fourth edition* of **Calculus with Analytic Geometry** by Howard Anton is designed to serve as a handy reference for students. It contains answers and, where appropriate, detailed solutions to the more than 3500 odd-numbered exercises in the text. The solutions show intermediate algebraic steps and manipulations that are a frequent cause of trouble for students, and also show details in the use of the various formulas that are encountered in calculus. Diagrams are used throughout the manual to assist in the visualization of problems and to suggest strategies for their solutions. High quality computer-generated figures ensure accuracy where a graph is needed in a solution or answer.

Special thanks are due to everyone at **Techsetters Incorporated** for the outstanding technical assistance that was provided for the production of this manual – typing, generation of art, and many other services that facilitated its preparation.

Contents

CHAPTER 1
Coordinates, Graphs, Lines

EXERCISE SET 1.1

1. **(a)** rational
    (d) rational
    (g) rational

    (b) integer, rational
    (e) integer, rational
    (h) integer, rational

    (c) integer, rational
    (f) irrational

3. **(a)**
    $$x = 0.123123123\cdots$$
    $$1000x = 123.123123123\cdots$$
    $$999x = 123$$
    $$x = \frac{123}{999} = \frac{41}{333}$$

    (b)
    $$x = 12.7777\cdots$$
    $$10x = 127.7777\cdots$$
    $$9x = 115$$
    $$x = \frac{115}{9}$$

    (c)
    $$x = 38.07818181\cdots$$
    $$100x = 3807.81818181\cdots$$
    $$99x = 3769.74 = \frac{3769.74}{99}$$
    $$= \frac{376974}{9900} = \frac{20943}{550}$$

    (d) $0.4296000\cdots = 0.4296$
    $$= \frac{4296}{10000}$$
    $$= \frac{537}{1250}$$

5. **(a)** If r is the radius, then $D = 2r$ so $\left(\frac{8}{9}D\right)^2 = \left(\frac{16}{9}r\right)^2 = \frac{256}{81}r^2$. The area of a circle of radius r is πr^2 so $256/81$ was the approximation used for π.
    (b) $256/81 \approx 3.16049$, $22/7 \approx 3.14268$, and $\pi \approx 3.14159$ so $256/81$ is worse than $22/7$.

7. line 2: blocks 3, 4; line 3: blocks 1, 2
    line 4: blocks 3, 4; line 5: blocks 2, 4, 5
    line 6: blocks 1, 2; line 7: blocks 3, 4

9. **(a)** always correct (add -3 to both sides of $a \le b$)
    (b) not always correct (correct only if $a = b = 0$)
    (c) not always correct (correct only if $a = b = 0$)
    (d) always correct (multiply both sides of $a \le b$ by 6)
    (e) not always correct (correct only if $a \ge 0$)
    (f) always correct (multiply both sides of $a \le b$ by the nonnegative quantity a^2)

11. **(a)** all values because $a = a$ is always valid   **(b)** none

1

13. (a) yes, because $a \leq b$ is true if $a < b$ (b) no, because $a < b$ is false if $a = b$ is true

15. (a) $\{x : x$ is a positive odd integer$\}$ (b) $\{x : x$ is an even integer$\}$
   (c) $\{x : x$ is irrational$\}$ (d) $\{x : x$ is an integer and $7 \leq x \leq 10\}$

17. (a) false, there are points inside the triangle that are not inside the circle
   (b) true, all points inside the triangle are also inside the square
   (c) true   (d) false   (e) true
   (f) true, a is inside the circle   (g) true

19. (a)    (b)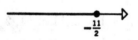

    (c)   (d)

    (e)   (f)

21. (a) $[-2, 2]$ (b) $(-\infty, -2) \cup (2, +\infty)$

23. $3x - 2 < 8$ **25.** $4 + 5x \leq 3x - 7$
        $3x < 10$ $2x \leq -11$
         $x < \dfrac{10}{3}$ $x \leq -\dfrac{11}{2}$
   $S = (-\infty, 10/3)$ $S = (-\infty, -11/2]$

27. $3 \leq 4 - 2x < 7$
    $-1 \leq -2x < 3$
    $\dfrac{1}{2} \geq x > -\dfrac{3}{2}$
   $S = (-3/2, 1/2]$

29.

$$\frac{x}{x-3} < 4$$

$$\frac{x}{x-3} - 4 < 0$$

$$\frac{x - 4(x-3)}{x-3} < 0$$

$$\frac{12 - 3x}{x-3} < 0$$

$$\frac{4-x}{x-3} < 0$$

$$S = (-\infty, 3) \cup (4, +\infty)$$

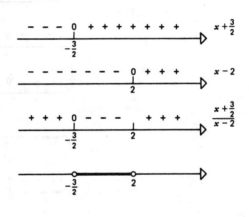

31.

$$\frac{3x+1}{x-2} < 1$$

$$\frac{3x+1}{x-2} - 1 < 0$$

$$\frac{3x+1-(x-2)}{x-2} < 0$$

$$\frac{2x+3}{x-2} < 0$$

$$\frac{x+3/2}{x-2} < 0$$

$$S = (-3/2, 2)$$

33.

$$\frac{4}{2-x} \le 1$$

$$\frac{4-(2-x)}{2-x} \le 0$$

$$\frac{x+2}{2-x} \le 0$$

$$S = (-\infty, -2] \cup (2, +\infty)$$

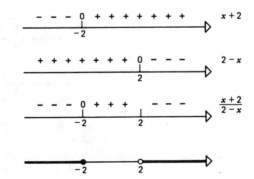

35.
$$x^2 > 9$$
$$x^2 - 9 > 0$$
$$(x+3)(x-3) > 0$$
$$S = (-\infty, -3) \cup (3, +\infty)$$

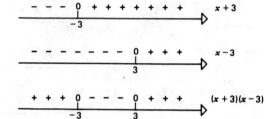

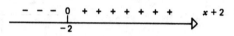

37. $(x-4)(x+2) > 0$
$$S = (-\infty, -2) \cup (4, +\infty)$$

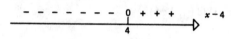

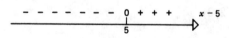

39. $x^2 - 9x + 20 \le 0$
$$(x-4)(x-5) \le 0$$
$$S = [4, 5]$$

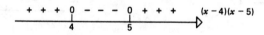

41.

$$\frac{2}{x} < \frac{3}{x-4}$$

$$\frac{2}{x} - \frac{3}{x-4} < 0$$

$$\frac{2(x-4) - 3x}{x(x-4)} < 0$$

$$\frac{-x-8}{x(x-4)} < 0$$

$$\frac{x+8}{x(x-4)} > 0$$

$$S = (-8, 0) \cup (4, +\infty)$$

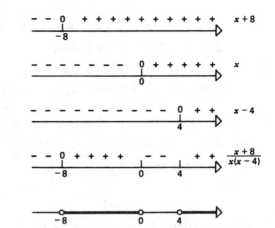

43. By trial-and-error we find that $x = 2$ is a root of the equation $x^3 - x^2 - x - 2 = 0$ so $x - 2$ is a factor of $x^3 - x^2 - x - 2$. By long division we find that $x^2 + x + 1$ is another factor so $x^3 - x^2 - x - 2 = (x-2)(x^2 + x + 1)$. The linear factors of $x^2 + x + 1$ can be determined by first finding the roots of $x^2 + x + 1 = 0$ by the quadratic formula. These roots are complex numbers so $x^2 + x + 1 \neq 0$ for all real x; thus $x^2 + x + 1$ must be always positive or always negative. Since $x^2 + x + 1$ is positive when $x = 0$, it follows that $x^2 + x + 1 > 0$ for all real x. Hence $x^3 - x^2 - x - 2 > 0$, $(x-2)(x^2 + x + 1) > 0$, $x - 2 > 0$, $x > 2$, so $S = (2, +\infty)$.

45. $\sqrt{x^2 + x - 6}$ is real if $x^2 + x - 6 \geq 0$. Factor to get $(x+3)(x-2) \geq 0$ which has as its solution $x \leq -3$ or $x \geq 2$.

47. $25 \leq \frac{5}{9}(F - 32) \leq 40$, $45 \leq F - 32 \leq 72$, $77 \leq F \leq 104$.

49. **(a)** Assume m and n are rational, then $m = \frac{p}{q}$ and $n = \frac{r}{s}$ where p, q, r, and s are integers so $m + n = \frac{p}{q} + \frac{r}{s} = \frac{ps + rq}{qs}$ which is rational because $ps + rq$ and qs are integers.

(b) (proof by contradiction)   Assume m is rational and n is irrational, then $m = \frac{p}{q}$ where p and q are integers. Suppose that $m + n$ is rational, then $m + n = \frac{r}{s}$ where r and s are integers so $n = \frac{r}{s} - m = \frac{r}{s} - \frac{p}{q} = \frac{rq - ps}{sq}$. But $rq - ps$ and sq are integers, so n is rational which contradicts the assumption that n is irrational.

51. (by example)   Given that $\sqrt{2}$ is irrational, then
$$(-1)\sqrt{2} = -\sqrt{2} \text{ is irrational (Exercise 50, part (b))}$$
but $\sqrt{2} + (-\sqrt{2}) = 0$, which is rational;

$\sqrt{2} + \sqrt{2} = 2\sqrt{2}$, which is irrational (Exercise 50, part (b));

$(\sqrt{2})(\sqrt{2}) = 2$, which is rational;

$1 + \sqrt{2}$ is irrational (Exercise 49, part (b))

and so $\sqrt{2}(1 + \sqrt{2}) = \sqrt{2} + 2$, which is irrational (Exercise 49, part (b)).

53. **(I)** The average of two rational numbers is rational:

Assume m and n are rational; then $m = \dfrac{p}{q}$ and $n = \dfrac{r}{s}$ where p, q, r, and s are integers. The average of m and n is

$$\frac{1}{2}(m + n) = \frac{1}{2}(p/q + r/s) = \frac{ps + rq}{2qs}$$

which is rational because $ps + rq$ and $2qs$ are integers.

(II) The average of two irrational numbers can be rational or irrational:

By example,

$\sqrt{2}$, and $-\sqrt{2}$ are irrational and their average is $\dfrac{1}{2}(\sqrt{2} + (-\sqrt{2})) = \dfrac{1}{2}(0) = 0$, which is rational;

$\sqrt{2}$ and $\sqrt{2}$ are irrational and their average is $\dfrac{1}{2}(\sqrt{2} + \sqrt{2}) = \sqrt{2}$, which is irrational.

55. $8x^3 - 4x^2 - 2x + 1$ can be factored by grouping terms:

$(8x^3 - 4x^2) - (2x - 1) = 4x^2(2x - 1) - (2x - 1) = (2x - 1)(4x^2 - 1) = (2x - 1)^2(2x + 1)$. The problem, then, is to solve $(2x - 1)^2(2x + 1) < 0$. By inspection, $x = 1/2$ is not a solution. If $x \neq 1/2$, then $(2x - 1)^2 > 0$ and it follows that $2x + 1 < 0$, $2x < -1$, $x < -1/2$, so $S = (-\infty, -1/2)$.

57. If $a < b$, then $ac < bc$ because c is positive; if $c < d$, then $bc < bd$ because b is positive, so $ac < bd$ (part (a), Theorem 1.1.3.)

EXERCISE SET 1.2

1. **(a)** 7 **(b)** $\sqrt{2}$ **(c)** k^2 **(d)** k^2

3. $|x - 3| = |3 - x| = 3 - x$ if $3 - x \geq 0$, which is true if $x \leq 3$.

5. All real values of x because $x^2 + 9 > 0$.

7. $|3x^2 + 2x| = |x(3x + 2)| = |x||3x + 2|$. If $|x||3x + 2| = x|3x + 2|$, then $|x||3x + 2| - x|3x + 2| = 0$, $(|x| - x)|3x + 2| = 0$, so either $|x| - x = 0$ or $|3x + 2| = 0$. If $|x| - x = 0$, then $|x| = x$, which is true for $x \geq 0$. If $|3x + 2| = 0$, then $x = -2/3$. The statement is true for $x \geq 0$ or $x = -2/3$.

9. $\sqrt{(x+5)^2} = |x+5| = x+5$ if $x+5 \geq 0$, which is true if $x \geq -5$.

13.
(a) $|7-9| = |-2| = 2$
(b) $|3-2| = |1| = 1$
(c) $|6-(-8)| = |14| = 14$
(d) $|-3-\sqrt{2}| = |-(3+\sqrt{2})| = 3+\sqrt{2}$
(e) $|-4-(-11)| = |7| = 7$
(f) $|-5-0| = |-5| = 5$

15.
(a) B is 6 units to the left of A; $b = a - 6 = -3 - 6 = -9$.
(b) B is 9 units to the right of A; $b = a + 9 = -2 + 9 = 7$.
(c) B is 7 units from A; either $b = a + 7 = 5 + 7 = 12$ or $b = a - 7 = 5 - 7 = -2$. Since it is given that $b > 0$, it follows that $b = 12$.

17. $|6x-2| = 7$

Case 1:	Case 2:
$6x-2 = 7$	$6x-2 = -7$
$6x = 9$	$6x = -5$
$x = 3/2$	$x = -5/6$

19. $|6x-7| = |3+2x|$

Case 1:	Case 2:
$6x-7 = 3+2x$	$6x-7 = -(3+2x)$
$4x = 10$	$8x = 4$
$x = 5/2$	$x = 1/2$

21. $|9x| - 11 = x$

Case 1:	Case 2:
$9x - 11 = x$	$-9x - 11 = x$
$8x = 11$	$-10x = 11$
$x = 11/8$	$x = -11/10$

23. $\left|\dfrac{x+5}{2-x}\right| = 6$

Case 1:	Case 2:
$\dfrac{x+5}{2-x} = 6$	$\dfrac{x+5}{2-x} = -6$
$x+5 = 12-6x$	$x+5 = -12+6x$
$7x = 7$	$-5x = -17$
$x = 1$	$x = 17/5$

25.
$$|x+6| < 3$$
$$-3 < x+6 < 3$$
$$-9 < x < -3$$
$$S = (-9, -3)$$

27.
$$|2x-3| \leq 6$$
$$-6 \leq 2x-3 \leq 6$$
$$-3 \leq 2x \leq 9$$
$$-3/2 \leq x \leq 9/2$$
$$S = [-3/2, 9/2]$$

29. $|x+2| > 1$

Case 1:	Case 2:
$x+2 > 1$	$x+2 < -1$
$x > -1$	$x < -3$

$S = (-\infty, -3) \cup (-1, +\infty)$

31. $|5-2x| \geq 4$

Case 1:	Case 2:
$5-2x \geq 4$	$5-2x \leq -4$
$-2x \geq -1$	$-2x \leq -9$
$x \leq 1/2$	$x \geq 9/2$

$S = (-\infty, 1/2] \cup [9/2, +\infty)$

33. $\dfrac{1}{|x-1|} < 2, x \neq 1$

$|x-1| > 1/2$

Case 1: Case 2:

$x-1 > 1/2$ $x-1 < -1/2$

$x > 3/2$ $x < 1/2$

$S = (-\infty, 1/2) \cup (3/2, +\infty)$

35. $\dfrac{3}{|2x-1|} \geq 4, x \neq 1/2$

$\dfrac{|2x-1|}{3} \leq \dfrac{1}{4}$

$|2x-1| \leq 3/4$

$-3/4 \leq 2x-1 \leq 3/4$

$1/4 \leq 2x \leq 7/4$

$1/8 \leq x \leq 7/8$

$S = [1/8, 1/2) \cup (1/2, 7/8]$

37. $|x+3| < |x-8|$

$(x+3)^2 < (x-8)^2$

$x^2 + 6x + 9 < x^2 - 16x + 64$

$22x < 55$

$x < 5/2$

$S = (-\infty, 5/2)$

39. $|4x| \geq |7 - 6x|$

$(4x)^2 \geq (7 - 6x)^2$

$16x^2 \geq 49 - 84x + 36x^2$

$-20x^2 + 84x - 49 \geq 0$

$20x^2 - 84x + 49 \leq 0$

$(2x-7)(10x-7) \leq 0$

$(x - 7/2)(x - 7/10) \leq 0$

$S = [7/10, 7/2]$

41. $\left|\dfrac{x - 1/2}{x + 1/2}\right| < 1, \quad x \neq -1/2$

$|x - 1/2| < |x + 1/2|$

$(x - 1/2)^2 < (x + 1/2)^2$

$x^2 - x + 1/4 < x^2 + x + 1/4$

$-2x < 0$

$x > 0$

$S = (0, +\infty)$

43. $\dfrac{1}{|x-4|} < \dfrac{1}{|x+7|}, \quad x \neq 4 \text{ or } -7$

$|x + 7| < |x - 4|$

$x^2 + 14x + 49 < x^2 - 8x + 16$

$22x < -33$

$x < -3/2;$

but $x = -7$ is excluded

so $S = (-\infty, -7) \cup (-7, -3/2)$

45. $\sqrt{(x^2 - 5x + 6)^2} = x^2 - 5x + 6$ if $x^2 - 5x + 6 \geq 0$ or, equivalently, if $(x - 2)(x - 3) \geq 0$; $x \in (-\infty, 2] \cup [3, +\infty)$.

47. If $u = |x - 3|$ then $u^2 - 4u = 12$, $u^2 - 4u - 12 = 0$, $(u - 6)(u + 2) = 0$, so $u = 6$ or $u = -2$. If $u = 6$ then $|x - 3| = 6$, so $x = 9$ or $x = -3$. If $u = -2$ then $|x - 3| = -2$ which is impossible. The solutions are -3 and 9.

49. If $|3x - 4| < 5$ then $-5 < 3x - 4 < 5$, $-1 < 3x < 9$, $-1/3 < x < 3$, $-2/3 < x - 1/3 < 8/3$, so $|x - 1/3| < 8/3$.

51. $|a - b| = |a + (-b)|$
$\leq |a| + |-b|$ (triangle inequality)
$= |a| + |b|$.

53. From Exercise 52
(i) $|a| - |b| \leq |a - b|$; but $|b| - |a| \leq |b - a| = |a - b|$, so (ii) $|a| - |b| \geq -|a - b|$.
Combining (i) and (ii): $-|a - b| \leq |a| - |b| \leq |a - b|$, so $||a| - |b|| \leq |a - b|$.

55. $-4 < x < 2$, $3 < x + 7 < 9$,

$1/3 > \dfrac{1}{x + 7} > 1/9$, but $\dfrac{1}{x + 7} = \left|\dfrac{1}{x + 7}\right|$ because $\dfrac{1}{x + 7} > 0$, thus $\left|\dfrac{1}{x + 7}\right| < \dfrac{1}{3}$ so $M = \dfrac{1}{3}$.

57. $\left|\dfrac{x + 3}{x - 3}\right| = \dfrac{|x + 3|}{|x - 3|}$; if $-\dfrac{3}{4} \leq x \leq \dfrac{1}{4}$, then

$-\dfrac{15}{4} \leq x - 3 \leq -\dfrac{11}{4}$, $|x - 3| \geq \dfrac{11}{4}$, $\dfrac{1}{|x - 3|} \leq \dfrac{4}{11}$ so (i) $\dfrac{|x + 3|}{|x - 3|} \leq \dfrac{4}{11}|x + 3|$.

But $\dfrac{9}{4} \leq x + 3 \leq \dfrac{13}{4}$, $|x + 3| \leq \dfrac{13}{4}$, $\dfrac{4}{11}|x + 3| \leq \dfrac{4}{11} \cdot \dfrac{13}{4} = \dfrac{13}{11}$, thus from (i) it follows that

$\dfrac{|x + 3|}{|x - 3|} \leq \dfrac{13}{11}$ so $M = \dfrac{13}{11}$.

EXERCISE SET 1.3

1.

3. **(a)** $x = 2$

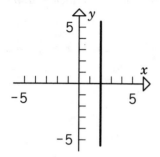

(b) $y = -3$

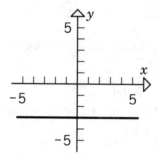

(c) $x \geq 0$

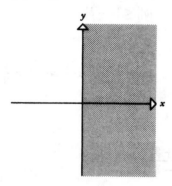

(d) $y = x$

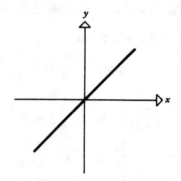

(e) $y \geq x$

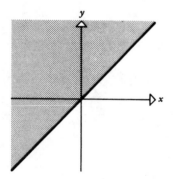

(f) $|x| \geq 1$

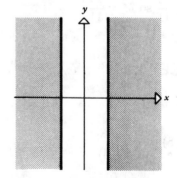

5. **(a)** vertical   **(b)** horizontal   **(c)** vertical

7. **(a)** $0^2 - 2(0) + 4 = 4$, yes   **(b)** $(-3)^2 - 2(-3) + 7 = 22$, no

(c) $(1/2)^2 - 2(1/2) + 19/4 = 1/4 - 1 + 19/4$, yes

(d) $(1 + \sqrt{5-t})^2 - 2(1 + \sqrt{5-t}) + t = 1 + 2\sqrt{5-t} + (5-t) - 2 - 2\sqrt{5-t} + t = 4$, yes

9. **(a)** origin   **(b)** x-axis   **(c)** none   **(d)** y-axis

11. **(a)** $x = y^2$, $x = -y^2$ **(b)** $y = x^2$, $y = -x^2$, $y = 1/x^2$, $y = -1/x^2$

    (c) $y = x^3$, $y = \sqrt[3]{x}$, $y = 1/x$, $y = -1/x$

13. **(a)** y-axis, because $(-x)^4 = 2y^3 + y$ gives $x^4 = 2y^3 + y$.

    (b) origin, because $(-y) = \dfrac{(-x)}{3 + (-x)^2}$ gives $y = \dfrac{x}{3 + x^2}$.

    (c) x-axis, y-axis, and origin because $(-y)^2 = |x| - 5$, $y^2 = |-x| - 5$, and $(-y)^2 = |-x| - 5$ all give $y^2 = |x| - 5$.

15. $y = 6 - x$ 17. $y = 4 - x^2$

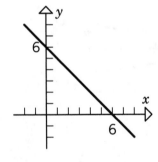
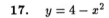

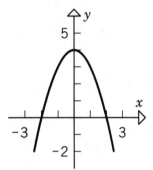

19. $y = \sqrt{x - 4}$ 21. $y = |x - 3|$

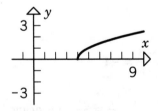
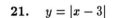

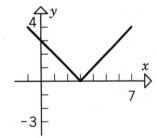

23. $x^2 y = 2$

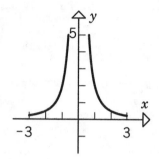

25. $4x^2 + 16y^2 = 16$

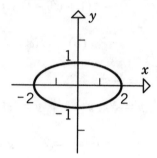

27. $(x - y)(x + y) = 0$
The union of the graphs of
$x - y = 0$ and $x + y = 0$.

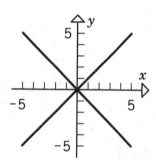

29. $u = 3v^2$

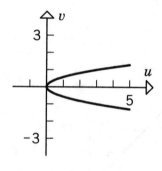

EXERCISE SET 1.4

1. **(a)** $m = \dfrac{4 - 2}{3 - (-1)} = \dfrac{1}{2}$

(b) $m = \dfrac{1 - 3}{7 - 5} = -1$

(c) $m = \dfrac{\sqrt{2} - \sqrt{2}}{-3 - 4} = 0$

(d) $m = \dfrac{12 - (-6)}{-2 - (-2)} = \dfrac{18}{0}$, not defined

3. **(a)** The line through $(1, 1)$ and $(-2, -5)$ has slope $m_1 = \dfrac{-5 - 1}{-2 - 1} = 2$, the line through $(1, 1)$
and $(0, -1)$ has slope $m_2 = \dfrac{-1 - 1}{0 - 1} = 2$. The given points lie on a line because $m_1 = m_2$.

(b)   The line through $(-2,4)$ and $(0,2)$ has slope $m_1 = \dfrac{2-4}{0+2} = -1$, the line through $(-2,4)$ and $(1,5)$ has slope $m_2 = \dfrac{5-4}{1+2} = \dfrac{1}{3}$. The given points do not lie on a line because $m_1 \neq m_2$.

5.

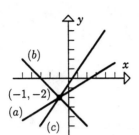

7.   (c), (b), (d), (a)

9.   (a) $m = [0 - (-3)]/(2 - 0) = 3/2$ (b) $m = (-2 - 1)/[3 - (-1)] = -3/4$

11.   slope $\approx 2.4/10 = 0.24$ N/mm $= 240$ N/m.

13.   slope $\approx (354 - 330)/(40 - 0) = 0.6$ m/sec/°C; the speed of sound increases at the rate of 0.6 m/sec per °C.

15.   Use the points $(1, 2)$ and (x, y) to calculate the slope: $(y - 2)/(x - 1) = 3$
      (a) if $x = 5$, then $(y - 2)/(5 - 1) = 3$, $y - 2 = 12$, $y = 14$
      (b) if $y = -2$, then $(-2 - 2)/(x - 1) = 3$, $x - 1 = -4/3$, $x = -1/3$

17.   Using $(3, k)$ and $(-2, 4)$ to calculate the slope, we find $\dfrac{k - 4}{3 - (-2)} = 5$, $k - 4 = 25$, $k = 29$.

19. $(0 - 2)/(x - 1) = -(0 - 5)/(x - 4)$, $-2x + 8 = 5x - 5$, $7x = 13$, $x = 13/7$.

21.   (a) $\tan \dfrac{\pi}{6} = 1/\sqrt{3}$ (b) $\tan 135° = -1$ (c) $\tan 60° = \sqrt{3}$

23.   (a) $153°$ (b) $45°$ (c) $117°$ (d) $89°$

25.   Let m' be the slope of line L'.
      (a) $m' = (5 - 8)/(2 - 1) = -3$, $m' = m$ so L' is parallel to L.
      (b) $m' = (4 - 5)/(3 - 6) = 1/3$, $mm' = -1$ so L' is perpendicular to L.
      (c) $m' = (1 - 0)/(-2 - 1) = -1/3$, $m' \neq m$ and $mm' \neq -1$ so L' is neither parallel nor perpendicular to L.

27. Let $P(x, 0)$ be a point on the x-axis. If AP is perpendicular to BP then $m_{AP} m_{BP} = -1$, so
$\dfrac{0-2}{x-1} \cdot \dfrac{0-3}{x-8} = -1$, thus $x^2 - 9x + 14 = 0$, $(x-7)(x-2) = 0$, thus $x = 2$ or $x = 7$.

29. Show that opposite sides are parallel by showing that they have the same slope:

using $(3, -1)$ and $(6, 4)$, $m_1 = 5/3$; using $(6, 4)$ and $(-3, 2)$, $m_2 = 2/9$;

using $(-3, 2)$ and $(-6, -3)$, $m_3 = 5/3$; using $(-6, -3)$ and $(3, -1)$, $m_4 = 2/9$.

Opposite sides are parallel because $m_1 = m_3$ and $m_2 = m_4$.

31. **(a)** Draw the line through the point of intersection of L_1 and L_2, parallel to the x-axis (see figure). Based on the figure we see that $\phi_2 = \theta + \phi_1$, so $\theta = \phi_2 - \phi_1$.

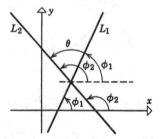

(b) if neither L_1 nor L_2 is vertical, then their slopes m_1 and m_2 both exist and are given by

$$m_1 = \tan \phi_1 \text{ and } m_2 = \tan \phi_2$$

so $\tan \theta = \tan(\phi_2 - \phi_1)$

$$= \frac{\tan \phi_2 - \tan \phi_1}{1 + \tan \phi_2 \tan \phi_1} \text{ (trig identity)}$$

$$= \frac{m_2 - m_1}{1 + m_1 m_2} \text{ if } m_1 m_2 \neq -1$$

(that is, if L_1 and L_2 are not perpendicular).

33. **(a)** $m_1 = \dfrac{1}{3}$, $m_2 = \dfrac{4}{5}$; $\tan \theta = \dfrac{4/5 - 1/3}{1 + (1/3)(4/5)} = \dfrac{7}{19}$

(b) $m_1 = 5$, $m_2 = -0.7$; $\tan \theta = \dfrac{-0.7 - 5}{1 + (5)(-0.7)} = \dfrac{5.7}{2.5} = \dfrac{57}{25}$

(c) $m_1 = -6$, $m_2 = -2$; $\tan \theta = \dfrac{-2 - (-6)}{1 + (-6)(-2)} = \dfrac{4}{13}$

35. **(a)** $20°$ **(b)** $66°$ **(c)** $17°$

37. The slope of the line through A and B is $m_1 = -3/4$ and that through A and C is $m_2 = 1/2$. Let m denote the slope of the line L that bisects the angle A (see figure). Then

$$\theta_1 = \theta_2$$

$$\tan \theta_1 = \tan \theta_2$$

$$\frac{m_1 - m}{1 + m_1 m} = \frac{m - m_2}{1 + m_2 m}$$

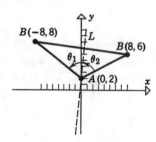

$$\frac{-\frac{3}{4} - m}{1 - \frac{3}{4}m} = \frac{m - \frac{1}{2}}{1 + \frac{1}{2}m}$$

$$\frac{-3 - 4m}{4 - 3m} = \frac{2m - 1}{2 + m}$$

$$(-3 - 4m)(2 + m) = (2m - 1)(4 - 3m)$$

$$-4m^2 - 11m - 6 = -6m^2 + 11m - 4$$

$$2m^2 - 22m - 2 = 0$$

$$m^2 - 11m - 1 = 0$$

and from the quadratic formula $m = \dfrac{11 \pm \sqrt{121 + 4}}{2} = \dfrac{11 \pm \sqrt{125}}{2}$, $m \approx 11.09$ or -0.09, of which only the line with a slope of 11.09 is consistent with the figure.

39. Let L_3 be a line that is perpendicular to L_1, then $m_3 = -1/m_1$. But $m_2 = -1/m_1$, so L_2 is parallel to L_3 because they have the same slope. If L_3 is perpendicular to L_1, then so is L_2.

EXERCISE SET 1.5

1.   **(a)**

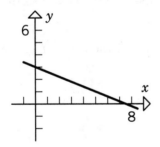

(b)

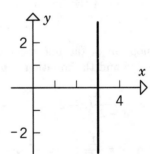

(c)

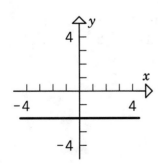

(d)

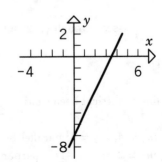

3. **(a)**

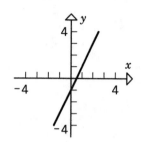

(b)

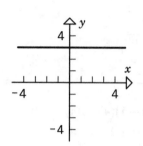

(c)

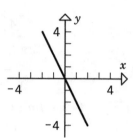

5. **(a)** $m = 3, b = 2$

(b) $m = -\dfrac{1}{4}, b = 3$

(c) $y = -\dfrac{3}{5}x + \dfrac{8}{5}$ so $m = -\dfrac{3}{5}, b = \dfrac{8}{5}$

(d) $m = 0, b = 1$

(e) $y = -\dfrac{b}{a}x + b$ so $m = -\dfrac{b}{a}$, y-intercept b

7. **(a)** $m = \tan\phi = \sqrt{3}, \phi = 60°$

(b) $m = \tan\phi = -2, \phi = 117°$

9. $y = -2x + 4$

11. The slope m of the line must equal the slope of $y = 4x - 2$, thus $m = 4$ so the equation is $y = 4x + 7$.

13. The slope m of the line must be the negative reciprocal of the slope of $y = 5x + 9$, thus $m = -1/5$ and the equation is $y = -x/5 + 6$.

15. $y - 4 = \dfrac{-7 - 4}{1 - 2}(x - 2) = 11(x - 2), y = 11x - 18$.

17. $m = \tan\dfrac{\pi}{6} = \dfrac{1}{\sqrt{3}}$ so $y = \dfrac{1}{\sqrt{3}}x - 3$.

19. The line passes through $(0, 2)$ and $(-4, 0)$, thus $m = \dfrac{0 - 2}{-4 - 0} = \dfrac{1}{2}$ so $y = \dfrac{1}{2}x + 2$.

21. $y = 1$

23. The line is vertical because $\phi = \dfrac{\pi}{2}$, so $x = 5$.

25. **(a)** $m_1 = 4, m_2 = 4$; parallel because $m_1 = m_2$.
   (b) $m_1 = 2, m_2 = -1/2$; perpendicular because $m_1 m_2 = -1$.
   (c) $m_1 = 5/3, m_2 = 5/3$; parallel because $m_1 = m_2$.

(d) If $A \neq 0$ and $B \neq 0$, then $m_1 = -A/B$, $m_2 = B/A$ and the lines are perpendicular because $m_1 m_2 = -1$. If either A or B (but not both) is zero, then the lines are perpendicular because one is horizontal and the other is vertical.

(e) $m_1 = 4$, $m_2 = 1/4$; neither.

27. $y = (-3/k)x + 4/k$, $k \neq 0$

(a) $-3/k = 2$, $k = -3/2$.   (b) $4/k = 5$, $k = 4/5$.

(c) $3(-2) + k(4) = 4$, $k = 5/2$.

(d) The slope of $2x - 5y = 1$ is $2/5$ so $-3/k = 2/5$, $k = -15/2$.

(e) The slope of $4x + 3y = 2$ is $-4/3$ so the slope of the line perpendicular to it is $3/4$; $-3/k = 3/4$, $k = -4$.

29. (a) If we plot T_C along the horizontal axis and T_F along the vertical axis, then any point on the line relating T_C and T_F can be denoted by (T_C, T_F). From the fact that $(0, 32)$ and $(100, 212)$ are on the line, it follows that the slope is $\dfrac{212 - 32}{100 - 0} = \dfrac{180}{100} = \dfrac{9}{5}$ and that the T_F-intercept is 32. An equation of the line is $T_F = \dfrac{9}{5}T_C + 32$.

(b) Solving $T_F = \dfrac{9}{5}T_C + 32$ for T_C we get $T_C = \dfrac{5}{9}T_F - \dfrac{160}{9}$, which identifies the slope as $\dfrac{5}{9}$.

(c) If $T_C = T_F = T$, then $T = \dfrac{9}{5}T + 32$, $T = -40° F = -40°C$.

(d) $T_C = \dfrac{5}{9}(98.6) - \dfrac{160}{9} = 37°C$

31. (a) slope $= (5.9 - 1)/(50 - 0) = 0.098$ atm/m, $p = 0.098h + 1$.

(b) If $p = 2$, then $h = 1/0.098 \approx 10.2$ m.

33. (a) $m = (0.75 - 0.80)/4 = -0.0125$ mm/day, $r = -0.0125t + 0.8$.

(b) If $r = 0$, then $t = 0.8/0.0125 = 64$ days

35. Solve $x = 5t + 2$ for t to get $t = \dfrac{1}{5}x - \dfrac{2}{5}$, so $y = \left(\dfrac{1}{5}x - \dfrac{2}{5}\right) - 3 = \dfrac{1}{5}x - \dfrac{17}{5}$, which is a line.

37. An equation of the line through $(1, 4)$ and $(2, 1)$ is $y = -3x + 7$. It crosses the y-axis at $y = 7$, and the x-axis at $x = 7/3$, so the area of the triangle is $\dfrac{1}{2}(7)(7/3) = 49/6$.

39. (a) Replace y by -1 in the first equation to get $2x - 3 = 5$, $2x = 8$, $x = 4$, so the point is $(4, -1)$.

(b) Multiply the first equation through by 2 and the second by 3 to get the equations $8x + 6y = -4$ and $15x - 6y = 27$; then, by adding these equations, $23x = 23$, $x = 1$, and substitution of this into the first of the original equations gives $4 + 3y = -2$, $3y = -6$, $y = -2$ so the point is $(1, -2)$.

41. Any line parallel to $y = x - 2$ can be written as $y = x + b$ which intersects $y = x^2$ where $x + b = x^2$, $x^2 - x - b = 0$. Solve, using the quadratic formula, to get $x = (1 \pm \sqrt{1 + 4b})/2$, which has real solutions only if $1 + 4b \geq 0$ so $b \geq -1/4$. The line $y = x + b$ intersects $y = x^2$ and is closest to $y = x - 2$ if $b = -1/4$, so $x = 1/2$ and $y = 1/4$. The point $(1/2, 1/4)$ is closest to $y = x - 2$.

43. To prove that the graph of $Ax + By + C = 0$ is a straight line consider two cases, $B = 0$ and $B \neq 0$. If $B \neq 0$, solve for y to get $y = -\dfrac{A}{B}x - \dfrac{C}{B}$ which is the slope-intercept form of a line with $m = -A/B$ and $b = -C/B$. If $B = 0$, then $A \neq 0$ because A and B cannot both be zero. Thus we get $Ax + C = 0$, or $x = -C/A$ which is an equation of a line parallel to the y-axis. In either case the graph of $Ax + By + C = 0$ is a straight line.

Conversely, consider any straight line in the xy-plane. If it is vertical, then it has an equation of the form $x = a$, or $x - a = 0$ which is like $Ax + By + C = 0$ with $A = 1$, $B = 0$, and $C = -a$. If the line is not vertical, then it can be written in slope-intercept form as $y = mx + b$, or $mx - y + b = 0$ which is like $Ax + By + C = 0$ with $A = m$, $B = -1$, and $C = b$.

EXERCISE SET 1.6

1. In the proof of Theorem 1.6.1.

3. (a) $d = \sqrt{(1 - 7)^2 + (9 - 1)^2} = \sqrt{36 + 64} = \sqrt{100} = 10$

(b) $\left(\dfrac{7 + 1}{2}, \dfrac{1 + 9}{2}\right) = (4, 5)$

5. (a) $d = \sqrt{[-7 - (-2)]^2 + [-4 - (-6)]^2} = \sqrt{25 + 4} = \sqrt{29}$

(b) $\left(\dfrac{-2 + (-7)}{2}, \dfrac{-6 + (-4)}{2}\right) = (-9/5, -5)$

7. Let $A(5, -2)$, $B(6, 5)$, and $C(2, 2)$ be the given vertices and a, b, and c the lengths of the sides opposite these vertices; then
$a = \sqrt{(2 - 6)^2 + (2 - 5)^2} = \sqrt{25} = 5$ and $b = \sqrt{(2 - 5)^2 + (2 + 2)^2} = \sqrt{25} = 5$.
Triangle ABC is isosceles because it has two equal sides ($a = b$).

9. $P_1(0, -2)$, $P_2(-4, 8)$, and $P_3(3, 1)$ all lie on a circle whose center is $C(-2, 3)$ if the points P_1, P_2 and P_3 are equidistant from C. Denoting the distances between P_1, P_2, P_3 and C by d_1, d_2 and d_3 we find that $d_1 = \sqrt{(0 + 2)^2 + (-2 - 3)^2} = \sqrt{29}$,
$d_2 = \sqrt{(-4 + 2)^2 + (8 - 3)^2} = \sqrt{29}$, and $d_3 = \sqrt{(3 + 2)^2 + (1 - 3)^2} = \sqrt{29}$.
So P_1, P_2 and P_3 lie on a circle whose center is $C(-2, 3)$ because $d_1 = d_2 = d_3$.

11.   If $(2, k)$ is equidistant from $(3,7)$ and $(9,1)$, then

$\sqrt{(2-3)^2 + (k-7)^2} = \sqrt{(2-9)^2 + (k-1)^2}$, $1 + (k-7)^2 = 49 + (k-1)^2$,

$1 + k^2 - 14k + 49 = 49 + k^2 - 2k + 1$, $-12k = 0$, $k = 0$.

13.   The slope of the line segment joining $(2,8)$ and $(-4,6)$ is $\dfrac{6-8}{-4-2} = \dfrac{1}{3}$ so the slope of the perpendicular bisector is -3. The midpoint of the line segment is $(-1, 7)$ so an equation of the bisector is $y - 7 = -3(x+1)$; $y = -3x + 4$.

15.   Method (see figure): Find an equation of the perpendicular bisector of the line segment joining $A(3,3)$ and $B(7,-3)$. All points on this perpendicular bisector are equidistant from A and B, thus find where it intersects the given line. The midpoint of AB is $(5,0)$, the slope of AB is $-3/2$ thus the slope

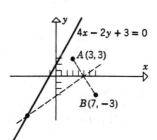

of the perpendicular bisector is $2/3$ so an equation is

$$\begin{aligned} y - 0 &= \frac{2}{3}(x-5) \\ 3y &= 2x - 10 \\ 2x - 3y - 10 &= 0. \end{aligned}$$

The solution of the system

$$\begin{cases} 4x - 2y + 3 &= 0 \\ 2x - 3y - 10 &= 0 \end{cases}$$

gives the point $(-29/8, -23/4)$.

17.   Method (see figure): write an equation of the line that goes through the given point and that is perpendicular to the given line; find the point P where this line intersects the given line; find the distance between P and the given point.

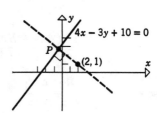

The slope of the given line is $4/3$, so the slope of a line perpendicular to it is $-3/4$.

The line through $(2,1)$ having a slope of $-3/4$ is $y - 1 = -\dfrac{3}{4}(x-2)$ or, after simplification,

$3x + 4y = 10$ which when solved simultaneously with $4x - 3y + 10 = 0$ yields $(-2/5, 14/5)$ as the point of intersection. The distance d between $(-2/5, 14/5)$ and $(2, 1)$ is

$$d = \sqrt{(2 + 2/5)^2 + (1 - 14/5)^2} = 3.$$

19. If $B = 0$, then the line $Ax + C = 0$ is vertical and $x = -C/A$ for each point on the line. The line through (x_0, y_0) and perpendicular to the given line is horizontal and intersects the given line at the point $(-C/A, y_0)$. The distance d between $(-C/A, y_0)$ and (x_0, y_0) is

$$d = \sqrt{(x_0 + C/A)^2 + (y_0 - y_0)^2} = \sqrt{\frac{(Ax_0 + C)^2}{A^2}} = \frac{|Ax_0 + C|}{\sqrt{A^2}}$$

which is the value of $\dfrac{|Ax_0 + By_0 + C|}{\sqrt{A^2 + B^2}}$ for $B = 0$.

If $B \neq 0$, then the slope of the given line is $-A/B$ and the line through (x_0, y_0) and perpendicular to the given line is

$$y - y_0 = \frac{B}{A}(x - x_0), \quad Ay - Ay_0 = Bx - Bx_0, \quad Bx - Ay = Bx_0 - Ay_0.$$

The point of intersection of this line and the given line is obtained by solving

$$Ax + By = -C \text{ and } Bx - Ay = Bx_0 - Ay_0.$$

Multiply the first equation through by A and the second by B and add the results to get

$$(A^2 + B^2)x = B^2 x_0 - ABy_0 - AC \text{ so } x = \frac{B^2 x_0 - ABy_0 - AC}{A^2 + B^2}$$

Similarly, by multiplying by B and $-A$, we get $y = \dfrac{-ABx_0 + A^2 y_0 - BC}{A^2 + B^2}$.

The square of the distance d between (x, y) and (x_0, y_0) is

$$d^2 = \left[x_0 - \frac{B^2 x_0 - ABy_0 - AC}{A^2 + B^2}\right]^2 + \left[y_0 - \frac{-ABx_0 + A^2 y_0 - BC}{A^2 + B^2}\right]^2$$

$$= \frac{(A^2 x_0 + ABy_0 + AC)^2}{(A^2 + B^2)^2} + \frac{(ABx_0 + B^2 y_0 + BC)^2}{(A^2 + B^2)^2}$$

$$= \frac{A^2(Ax_0 + By_0 + C)^2 + B^2(Ax_0 + By_0 + C)^2}{(A^2 + B^2)^2}$$

$$= \frac{(Ax_0 + By_0 + C)^2(A^2 + B^2)}{(A^2 + B^2)^2} = \frac{(Ax_0 + By_0 + C)^2}{A^2 + B^2}$$

so $d = \dfrac{|Ax_0 + By_0 + C|}{\sqrt{A^2 + B^2}}$.

21. $d = \dfrac{|5(8) + 12(4) - 36|}{\sqrt{5^2 + 12^2}} = \dfrac{|52|}{\sqrt{169}} = \dfrac{52}{13} = 4.$

23. **(a)** center $(0,0)$, radius 5   **(b)** center $(1,4)$, radius 4

     (c) center $(-1,-3)$, radius $\sqrt{5}$ **(d)** center $(0,-2)$, radius 1

25. $(x-3)^2 + (y-(-2))^2 = 4^2$, $(x-3)^2 + (y+2)^2 = 16$

27. $r = 8$ because the circle is tangent to the x-axis, so $(x+4)^2 + (y-8)^2 = 64$.

29. $(0,0)$ is on the circle, so $r = \sqrt{(-3-0)^2 + (-4-0)^2} = 5$; $(x+3)^2 + (y+4)^2 = 25$.

31. The center is the midpoint of the line segment joining $(2,0)$ and $(0,2)$ so the center is at $(1,1)$. The radius is $r = \sqrt{(2-1)^2 + (0-1)^2} = \sqrt{2}$, so $(x-1)^2 + (y-1)^2 = 2$.

33. $(x^2 - 2x) + (y^2 - 4y) = 11$, $(x^2 - 2x + 1) + (y^2 - 4y + 4) = 11 + 1 + 4$, $(x-1)^2 + (y-2)^2 = 16$; center $(1,2)$ and radius 4.

35. $2(x^2 + 2x) + 2(y^2 - 2y) = 0$, $2(x^2 + 2x + 1) + 2(y^2 - 2y + 1) = 2 + 2$, $(x+1)^2 + (y-1)^2 = 2$; center $(-1,1)$ and radius $\sqrt{2}$.

37. $(x^2 + 2x) + (y^2 + 2y) = -2$, $(x^2 + 2x + 1) + (y^2 + 2y + 1) = -2 + 1 + 1$, $(x+1)^2 + (y+1)^2 = 0$; the point $(-1,-1)$.

39. $x^2 + y^2 = 1/9$; center $(0,0)$ and radius $1/3$.

41. $x^2 + (y^2 + 10y) = -26$, $x^2 + (y^2 + 10y + 25) = -26 + 25$, $x^2 + (y+5)^2 = -1$; no graph

43. $16\left(x^2 + \dfrac{5}{2}x\right) + 16(y^2 + y) = 7$, $16\left(x^2 + \dfrac{5}{2}x + \dfrac{25}{16}\right) + 16\left(y^2 + y + \dfrac{1}{4}\right) = 7 + 25 + 4$,

$(x + 5/4)^2 + (y + 1/2)^2 = 9/4$; center $(-5/4, -1/2)$ and radius $3/2$.

45. **(a)** $y^2 = 16 - x^2$, so $y = \pm\sqrt{16 - x^2}$. The bottom half is $y = -\sqrt{16 - x^2}$.

     (b) Complete the square in y to get $(y-2)^2 = 3 - 2x - x^2$, so $y - 2 = \pm\sqrt{3 - 2x - x^2}$, or $y = 2 \pm \sqrt{3 - 2x - x^2}$. The top half is $y = 2 + \sqrt{3 - 2x - x^2}$.

47. (a)

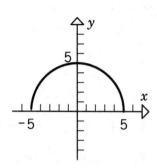

(b) $y = \sqrt{5 + 4x - x^2}$

$\qquad = \sqrt{5 - (x^2 - 4x)}$

$\qquad = \sqrt{5 + 4 - (x^2 - 4x + 4)}$

$\qquad = \sqrt{9 - (x - 2)^2}$

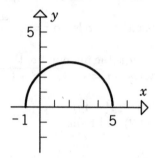

49. The tangent line is perpendicular to the radius at the point. The slope of the radius is 4/3, so the slope of the perpendicular *is* $- 3/4$. An equation of the tangent line is $y - 4 = -\dfrac{3}{4}(x - 3)$, or $y = -\dfrac{3}{4}x + \dfrac{25}{4}$.

51. (a) The center of the circle is at $(0,0)$ and its radius is $\sqrt{20} = 2\sqrt{5}$. The distance between P and the center is $\sqrt{(-1)^2 + (2)^2} = \sqrt{5}$ which is less than $2\sqrt{5}$, so P is inside the circle.

(b) Draw the diameter of the circle that passes through P, then the shorter segment of the diameter is the shortest line that can be drawn from P to the circle, and the longer segment is the longest line that can be drawn from P to the circle (can you prove it?). Thus, the smallest distance is $2\sqrt{5} - \sqrt{5} = \sqrt{5}$, and the largest is $2\sqrt{5} + \sqrt{5} = 3\sqrt{5}$.

53. Let (a, b) be the coordinates of T (or T'). The radius from $(0,0)$ to T (or T') will be perpendicular to L (or L') so, using slopes, $b/a = -(a - 3)/b$, $a^2 + b^2 = 3a$. But (a, b) is on the circle so $a^2 + b^2 = 1$, thus $3a = 1$, $a = 1/3$. Let $a = 1/3$ in $a^2 + b^2 = 1$ to get $b^2 = 8/9$, $b = \pm\sqrt{8}/3$. The coordinates of T and T' are $(1/3, \sqrt{8}/3)$ and $(1/3, -\sqrt{8}/3)$.

55. (a) $[(x - 4)^2 + (y - 1)^2] + [(x - 2)^2 + (y + 5)^2] = 45$

$x^2 - 8x + 16 + y^2 - 2y + 1 + x^2 - 4x + 4 + y^2 + 10y + 25 = 45$

$2x^2 + 2y^2 - 12x + 8y + 1 = 0$, which is a circle.

(b) $2(x^2 - 6x) + 2(y^2 + 4y) = -1$, $2(x^2 - 6x + 9) + 2(y^2 + 4y + 4) = -1 + 18 + 8$,

$(x - 3)^2 + (y + 2)^2 = 25/2$; center $(3, -2)$, radius $5/\sqrt{2}$.

57. $y = x^2 + 2$

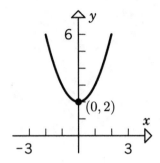

59. $y = x^2 + 2x - 3$

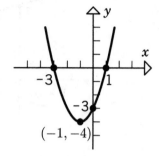

61. $y = -x^2 + 4x + 5$

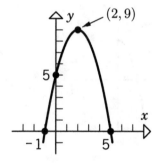

63. $y = (x - 2)^2$

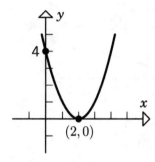

65. $x^2 - 2x + y = 0$

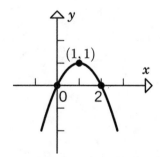

67. $y = 3x^2 - 2x + 1$

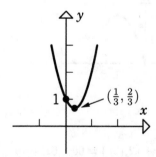

69. $x = -y^2 + 2y + 2$

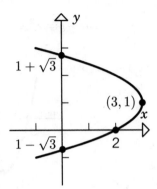

71. (a) $x^2 = 3 - y$, $x = \pm\sqrt{3 - y}$. The right half is $x = \sqrt{3 - y}$.
(b) Complete the square in x to get $(x - 1)^2 = y + 1$, $x = 1 \pm \sqrt{y + 1}$. The left half is
$x = 1 - \sqrt{y + 1}$.

73. (a)

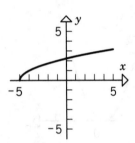

(b)

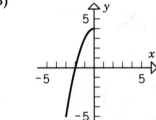

75. (a) $s = 32t - 16t^2$

(b) The ball will be at its
highest point when $t = 1$ sec;
it will rise 16 ft.

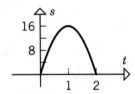

77. (a) $(3)(2x) + (2)(2y) = 600$, $6x + 4y = 600$, $y = 150 - 3x/2$.
(b) $A = xy = x(150 - 3x/2) = 150x - 3x^2/2$.
(c) The graph of A versus x is a parabola with its vertex (high point) at
$x = -b/(2a) = -150/(-3) = 50$, so the maximum value of A is
$A = 150(50) - 3(50)^2/2 = 3{,}750$ ft^2.

SUPPLEMENTARY EXERCISES, CHAPTER 1

1. **(a)** $(-3, 5]$

   (b) $-1 < 0 \leq x^2$ for all x, thus $-1 < x^2 \leq 9$ is equivalent to $x^2 \leq 9$, so $|x| \leq 3$. The interval is $[-3, 3]$.

   (c) $x^2 \geq \dfrac{1}{4}$ is equivalent to $|x| \geq \dfrac{1}{2}$, which in interval notation is $(-\infty, -1/2] \cup [1/2, +\infty)$.

3. **(a)**
$$2x^2 - 5x > 3$$
$$2x^2 - 5x - 3 > 0$$
$$(2x + 1)(x - 3) > 0$$
$$(x + 1/2)(x - 3) > 0$$
$$S = (-\infty, -1/2) \cup (3, +\infty)$$

   (b)
$$x^2 - 5x + 4 \leq 0$$
$$(x - 1)(x - 4) \leq 0$$
$$S = [1, 4]$$

5. **(a)** $\dfrac{|x| - 1}{|x| - 2} \leq 0$

   Case 1: $x \geq 0$,
   $$\frac{x - 1}{x - 2} \leq 0$$

   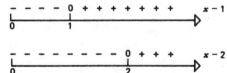

   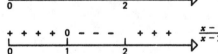

   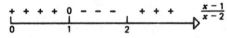

   $S_1 = [1, 2)$

   Case 2: $x < 0$,
   $$\frac{-x - 1}{-x - 2} \leq 0; \quad \frac{x + 1}{x + 2} \leq 0$$

   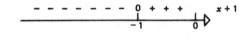

   $S_2 = (-2, -1]$

   $$S = S_1 \cup S_2 = (-2, -1] \cup [1, 2)$$

(b)
$$|x - 1| \le 2|x + 2|$$
$$x^2 - 2x + 1 \le 4(x^2 + 4x + 4)$$
$$-3x^2 - 18x - 15 \le 0$$
$$x^2 + 6x + 5 \ge 0$$
$$(x + 1)(x + 5) \ge 0$$
$$S = (-\infty, -5] \cup [-1, +\infty)$$

```
- - - - - - -  0  + + +      x + 1
                -1
```

```
- - -   0  + + + + + + + +   x + 5
       -5
```

```
+ + +  0  - - - -  0  + + +   (x + 1)(x + 5)
      -5          -1
```

7. **(a)** $a = -2$ and $b = 1$
   (b) $a^2 < b^2$, $a^2 - b^2 < 0$, $(a + b)(a - b) < 0$; if $a < b$ then $a - b < 0$ so $(a + b)(a - b) < 0$ if
   $a + b > 0$

9. $x^2 \le x^2 + y^2$ because $y^2 \ge 0$, thus $\sqrt{x^2} \le \sqrt{x^2 + y^2}$ and so $|x| \le \sqrt{x^2 + y^2}$.

   Similarly, $|y| \le \sqrt{x^2 + y^2}$. The right triangle with vertices $(0,0)$, (x, y), and $(x, 0)$ has legs
   of lengths $|x|$ and $|y|$, and a hypotenuse of length $\sqrt{x^2 + y^2}$. The lengths of the legs cannot
   exceed the length of the hypotenuse.

11. **(a)**
$$xy = x^2$$
$$xy - x^2 = 0$$
$$x(y - x) = 0,$$
so $x = 0$ or $y - x = 0$

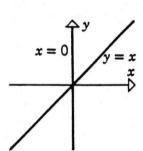

(b)
$$y(x - 1) = x^2 - 1$$
$$y(x - 1) - (x^2 - 1) = 0$$
$$y(x - 1) - (x - 1)(x + 1) = 0$$
$$(x - 1)(y - x - 1) = 0,$$
so $x - 1 = 0$ or $y - x - 1 = 0$

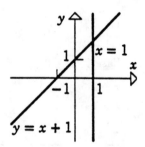

13. **(a)** Complete the square:
$(x^2 - 2x + 1) + (y^2 - 6y + 9) \geq 6 + 1 + 9$
$(x - 1)^2 + (y - 3)^2 \geq 16$,
all points on or outside the circle
with center $(1, 3)$ and radius 4.

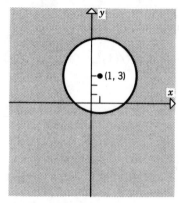

(b) $x + |y - 2| = 1$
If $y \geq 2$, then
$x + y - 2 = 1$
$$y = -x + 3.$$
If $y < 2$, then $x - (y - 2) = 1$
$$y = x + 1.$$

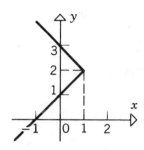

15. The curves intersect at (x, y) where $y = x^2$ and $y = x + 2$, so $x^2 = x + 2$, $x^2 - x - 2 = 0$, $(x + 1)(x - 2) = 0$, $x = -1$ or $x = 2$. The points of intersection are $(-1, 1)$ and $(2, 4)$.

17. If $x \geq 2$, then $y = x - 2$;
if $x < 2$, then $y = -x + 2$.

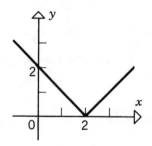

19.

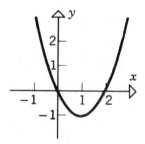

21. The radius is $r = \sqrt{(4 - 1)^2 + (-2 - 2)^2} = 5$, so $(x - 1)^2 + (y - 2)^2 = 25$.

23. Let $C(h, k)$ be the center, then $|h - 6| = 5$ and $|k - 7| = 5$, so $h = 6 \pm 5$ and $k = 7 \pm 5$. There are four circles, $(x - h)^2 + (y - k)^2 = 25$, where $h = 1$ or 11 and $k = 2$ or 12.

25. $(x^2 + 4x + 4) + (y^2 + 2y + 1) = -5 + 4 + 1$, $(x + 2)^2 + (y + 1)^2 = 0$; the point $(-2, -1)$.

27. $(x^2 - 3x + 9/4) + (y^2 + 2y + 1) = -4 + 9/4 + 1$, $(x - 3/2)^2 + (y + 1)^2 = -3/4$; no graph

29. (a) $m = \dfrac{-4-4}{-3-3} = \dfrac{4}{3}$, so $y - 4 = \frac{4}{3}(x-3)$, $y = \frac{4}{3}x$; $d = \sqrt{(-6)^2 + (-8)^2} = 10$;
midpoint: $(0,0)$.

(b) $m = \dfrac{-4-4}{3-3} = \dfrac{-8}{0}$ which is not defined, so the vertical line $x = 3$; $d = |-4-4| = 8$;
midpoint: $(3,0)$.

(c) $m = \dfrac{4-4}{-3-3} = 0$, so $y = 4$; $d = |-3-3| = 6$; midpoint: $(0,4)$.

(d) $m = \dfrac{3-4}{4-3} = -1$, so $y - 4 = -(x - 3)$, $y = -x + 7$; $d = \sqrt{(1)^2 + (-1)^2} = \sqrt{2}$;
midpoint: $(7/2, 7/2)$.

31. Equation of circle is $x^2 + y^2 = 25$, equation of line is $y = -\dfrac{3}{4}x$.

Eliminate y: $x^2 + \left(-\dfrac{3}{4}x\right)^2 = 25$, $x^2 + \dfrac{9}{16}x^2 = 25$, $\dfrac{25}{16}x^2 = 25$, $x^2 = 16$, so $x = \pm 4$.

The points of intersection are $(-4, 3)$ and $(4, -3)$.

33. $m = \dfrac{-3+3}{4-2} = 0$, so $y = -3$.

35. For $x + 2y = 3$, $m = -\dfrac{1}{2}$. A parallel line through the origin is $y - 0 = -\dfrac{1}{2}(x - 0)$, $y = -\dfrac{1}{2}x$.

37. $L : m = -2$, so $y = -2(x - 1)$; $L' : m = \dfrac{1}{2}$ because L' is perpendicular to L, so $y = \dfrac{1}{2}x - 3$. L
meets L' when $-2(x - 1) = \dfrac{1}{2}x - 3$, which gives $x = 2$. But $y = -2$ when $x = 2$, so the point
is $(2, -2)$.

39. $L : y - 1 = \dfrac{2}{5}(x - 3)$, $y = \dfrac{2}{5}x - \dfrac{1}{5}$; $L' : \left(-\dfrac{8}{3}, 0\right)$ and $(0, -4)$ are on the line, so $m = -\dfrac{3}{2}$ and
therefore $y = -\dfrac{3}{2}\left(x + \dfrac{8}{3}\right) = -\dfrac{3}{2}x - 4$. L meets L' when $\dfrac{2}{5}x - \dfrac{1}{5} = -\dfrac{3}{2}x - 4$, which gives
$x = -2$. But $y = -1$ when $x = -2$, so the point is $(-2, -1)$.

41. Label the points as $A(5, 6)$, $B(-4, 3)$, $C(-3, -2)$, and $D(6, 1)$. Then $m_{AB} = 1/3$,
$m_{BC} = -5$, $m_{CD} = -1/3$, and $m_{DA} = -5$, so $ABCD$ is a parallelogram because opposite
sides are parallel ($m_{AB} = m_{CD}$, $m_{BC} = m_{DA}$). It is not a rectangle because sides AB and
BC do not form a right angle ($m_{AB} \neq -1/m_{BC}$).

CHAPTER 2
Functions and Limits

EXERCISE SET 2.1

1. (a) 14 (b) 50 (c) 2

(d) 11 (e) $3a^2 + 6a + 5$ (f) $27t^2 + 2$

3. (a) $2(-4) = -8$ (b) $1/4$ (c) $2(0) = 0$

(d) $2(3) = 6$ (e) $2(2.9) = 5.8$ (f) $1/(t^2 + 5)$

5. $(-\infty, 3) \cup (3, +\infty)$ **7.** $(-\infty, -\sqrt{3}] \cup [\sqrt{3}, +\infty)$

9. $\dfrac{x-1}{x+2} \geq 0$ if $x < -2$ or $x \geq 1$; domain: $(-\infty, -2) \cup [1, +\infty)$

11. $(-\infty, +\infty)$ **13.** $[5, +\infty) \cap (-\infty, 8] = [5, 8]$

15. $x^2 - 2x + 5 = 0$ has no real solutions so $x^2 - 2x + 5$ is always positive or always negative. If $x = 0$, then $x^2 - 2x + 5 = 5 > 0$; domain: $(-\infty, +\infty)$.

17. $(-\infty, 0) \cup (0, +\infty)$ **19.** $[0, +\infty)$

21. all real values of x except those for which $\sin x = 1$; domain: all x except $x = \pi/2 + 2k\pi$, $k = 0, \pm 1, \pm 2, \cdots$

23. domain: $(-\infty, 3]$; range: $[0, +\infty)$ **25.** domain: $[-2, 2]$; range: $[0, 2]$

27. domain: $[0, +\infty)$; range: $[3, +\infty)$ **29.** domain: $(-\infty, +\infty)$; range: $[3, +\infty)$

31. domain: $(-\infty, +\infty)$; range: $(-\infty, +\infty)$ **33.** domain: $(-\infty, +\infty)$; range: $[-3, 3]$

35. domain: $(-\infty, +\infty)$; range: $[1, 3]$

37. If $x < 0$, then $|x| = -x$ so $f(x) = -x + 3x + 1 = 2x + 1$. If $x \geq 0$, then $|x| = x$ so $f(x) = x + 3x + 1 = 4x + 1$;

$$f(x) = \begin{cases} 2x + 1, & x < 0 \\ 4x + 1, & x \geq 0 \end{cases}$$

39. If $x < 0$, then $|x| = -x$ and $|x - 1| = 1 - x$ so $g(x) = -x + 1 - x = 1 - 2x$. If $0 \le x < 1$, then $|x| = x$ and $|x - 1| = 1 - x$ so $g(x) = x + 1 - x = 1$. If $x \ge 1$, then $|x| = x$ and $|x - 1| = x - 1$ so $g(x) = x + x - 1 = 2x - 1$;

$$g(x) = \begin{cases} 1 - 2x, & x < 0 \\ 1, & 0 \le x < 1 \\ 2x - 1, & x \ge 1 \end{cases}.$$

41. $\sqrt{3x - 2} = 6$, $3x - 2 = 36$, $3x = 38$, $x = 38/3$.

43. $x^2 + 5 = 7$, $x^2 = 2$, $x = \pm\sqrt{2}$. **45.** $\cos x = 1$, $x = 2k\pi$, $k = 0, \pm1, \pm2, \cdots$

47. $\sin\sqrt{x} = 1/2$, $\sqrt{x} = \pi/6 + 2k\pi$ or $5\pi/6 + 2k\pi$ for $k = 0, 1, 2, \cdots$ so $x = (1/6 + 2k)^2\pi^2$ or $(5/6 + 2k)^2\pi^2$ for $k = 0, 1, 2, \cdots$

49. $A = \pi r^2$, but $C = 2\pi r$ so $r = \dfrac{C}{2\pi}$; $A = \dfrac{C^2}{4\pi}$.

51. (a) $S = 6x^2$
   (b) $V = x^3$ so $x = V^{1/3}$; substitute into (a) to get $S = 6V^{2/3}$.

53. $V = x(8 - 2x)(15 - 2x) = x(120 - 46x + 4x^2) = 4x^3 - 46x^2 + 120x$.

55. $h = L - L\cos\theta = L(1 - \cos\theta)$.

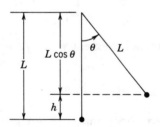

57. (a) $25°$F (b) $2°$F (c) $-15°$F.

59. $91.4 + (91.4 - T)[0.0203(8) - 0.304\sqrt{8} - 0.474] = -10$, solve for T to get $T \approx 5°$F.

61. Multiplication of the numerator and denominator of a fraction by the same number is valid only if the number is not zero, so $(1 - 1/x)/(1 + 1/x) = (x - 1)/(x + 1)$ if $x \ne 0$.

63. $\dfrac{(x + 2)(x^2 - 1)}{(x + 2)(x + 1)} = \dfrac{(x + 2)(x + 1)(x - 1)}{(x + 2)(x + 1)} = x - 1$, $x \ne -1$ or -2

65. $\dfrac{x + 1 + \sqrt{x + 1}}{\sqrt{x + 1}} = \dfrac{\sqrt{x + 1}(\sqrt{x + 1} + 1)}{\sqrt{x + 1}} = \sqrt{x + 1} + 1$, $x \ne -1$

67. $\dfrac{x^3 + 2x^2 - 3x}{(x-1)(x+3)} = \dfrac{x(x^2 + 2x - 3)}{(x-1)(x+3)} = \dfrac{x(x-1)(x+3)}{(x-1)(x+3)} = x,\ x \neq -3$ or 1

EXERCISE SET 2.2

1. (a) $f(t) = t^2 + 1$
    (b) $f(t+2) = (t+2)^2 + 1 = t^2 + 4t + 5$
    (c) $f(x+2) = (x+2)^2 + 1 = x^2 + 4x + 5$
    (d) $f(1/x) = (1/x)^2 + 1 = 1/x^2 + 1$
    (e) $f(x+h) = (x+h)^2 + 1 = x^2 + 2hx + h^2 + 1$
    (f) $f(-x) = (-x)^2 + 1 = x^2 + 1$
    (g) $f(\sqrt{x}) = (\sqrt{x})^2 + 1 = x + 1,\ x \geq 0$
    (h) $f(3x) = (3x)^2 + 1 = 9x^2 + 1$

3. (a) $f(-1) - g(-1) = 4 - 3 = 1$ (b) $f(-1) \cdot g(-1) = (4)(3) = 12$
    (c) $f(2)/g(2) = 5/(-1) = -5$ (d) $f(g(2)) = f(-1) = 4$

5. (a) $x^2 + 2x + 1$ (b) $-x^2 + 2x - 1$ (c) $2x(x^2 + 1)$

    (d) $\dfrac{2x}{x^2 + 1}$ (e) $2(x^2 + 1)$ (f) $4x^2 + 1$

7. (a) $\sqrt{x+1} + x - 2$ (b) $\sqrt{x+1} - x + 2$ (c) $(x-2)\sqrt{x+1}$

    (d) $\dfrac{\sqrt{x+1}}{x-2}$ (e) $\sqrt{x-1}$ (f) $\sqrt{x+1} - 2$

9. (a) $\sqrt{x-2} + \sqrt{x-3}$ (b) $\sqrt{x-2} - \sqrt{x-3}$ (c) $\sqrt{x-2}\sqrt{x-3}$

    (d) $\dfrac{\sqrt{x-2}}{\sqrt{x-3}}$ (e) $\sqrt{\sqrt{x-3} - 2}$ (f) $\sqrt{\sqrt{x-2} - 3}$

11. (a) $\sqrt{1-x^2} + \sin 3x$ (b) $\sqrt{1-x^2} - \sin 3x$ (c) $\sqrt{1-x^2}\sin 3x$
    (d) $\sqrt{1-x^2}/\sin 3x$ (e) $\sqrt{1 - \sin^2 3x} = \sqrt{\cos^2 3x}$ (f) $\sin 3\sqrt{1-x^2}$
                                            $= |\cos 3x|$

13. $(f \circ g)(x) = \dfrac{4}{x+5},\ x \geq 0; (g \circ f)(x) = \dfrac{2}{\sqrt{x^2 + 5}}$

15. (a) $4x - 15$ (b) $4x^2 - 20x + 25$

17. **(a)** $f(g(x)) = 1/(x^2 + 1)$, which is defined for all x.
     (b) $g(x) = x^2 + 2$, for example
     (c) g must be defined for all x and $g(x)$ must never equal zero.

19. Show that $(f \circ (g \circ h))(x) = ((f \circ g) \circ h)(x)$: $(f \circ (g \circ h))(x) = f((g \circ h)(x)) = f(g(h(x)))$, $((f \circ g) \circ h)(x) = (f \circ g)(h(x)) = f(g(h(x)))$. Thus $(f \circ (g \circ h))(x) = ((f \circ g) \circ h)(x)$ so $f \circ (g \circ h) = (f \circ g) \circ h$.

21. $g(x) = \sqrt{x}$, $h(x) = x + 2$ 23. $g(x) = x^7$, $h(x) = x - 5$

25. $g(x) = |x|$, $h(x) = x^2 - 3x + 5$ 27. $g(x) = x^2$, $h(x) = \sin x$

29. $g(x) = \dfrac{3}{5 + x}$, $h(x) = \cos x$ 31. $g(x) = \dfrac{x}{3 + x}$, $h(x) = \tan x$

33. $f(x) = \sqrt{x}$, $g(x) = 3 - x^2$, $h(x) = \sin x$

35. $u = x + 1$ so $x = u - 1$; $f(u) = (u - 1)^2 + 3(u - 1) + 5 = u^2 + u + 3$; $f(x) = x^2 + x + 3$.

37. $f(g(x)) = f(2x - 1) = 0$ only if $2x - 1 = -1$ or $2x - 1 = 2$, so $x = 0$ or $x = 3/2$.

39. $f(g(x)) = \sqrt{g(x) + 5} = 3|x|$, $g(x) + 5 = 9x^2$, $g(x) = 9x^2 - 5$.

41. $f(x) = f(x/2 + x/2) = f(x/2) - f(x/2) = 0$

43. **(a)** monomial, polynomial, rational, explicit algebraic
     (b) explicit algebraic
     (c) rational, explicit algebraic
     (d) polynomial, rational, explicit algebraic

45. **(a)** explicit algebraic
     (b) rational, explicit algebraic
     (c) monomial, polynomial, rational, explicit algebraic
     (d) explicit algebraic $\left(|x| = \sqrt{x^2} \right)$

EXERCISE SET 2.3

1. **(a)** $-4, -3, -2, 2, 3$ **(b)** $0, 4$
    (c) $-4 \le x \le -3$, $-2 \le x \le 2$, $x \ge 3$ **(d)** $x \le -4$, $-3 \le x \le -2$, $2 \le x \le 3$

3.

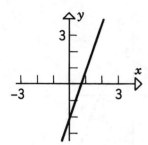

5.

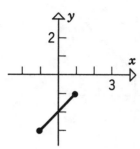

7.

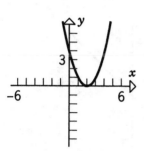

9.

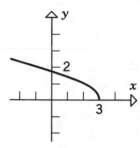

11.

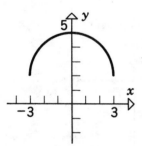

13.

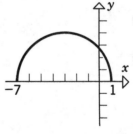

15.

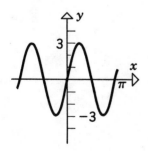

17.

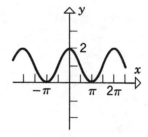

19.

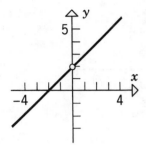

21.

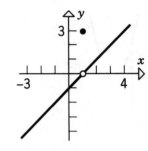

23.

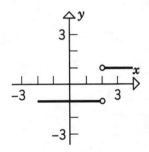

25.

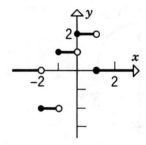

27.

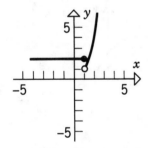

29.

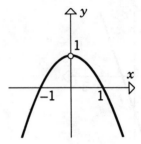

31. $x \leq 2 : f(x) = 2x + (2 - x) = x + 2,$
$x > 2 : f(x) = 2x + (x - 2) = 3x - 2;$

$$f(x) = \begin{cases} x + 2, & x \leq 2 \\ 3x - 2, & x > 2 \end{cases}$$

33. $x < 3 : g(x) = (5 - x) - (3 - x) = 2,$
$3 \leq x < 5 : g(x) = (5 - x) - (x - 3) = 8 - 2x,$
$x \geq 5 : g(x) = (x - 5) - (x - 3) = -2;$

$$g(x) = \begin{cases} 2, & x < 3 \\ 8 - 2x, & 3 \leq x < 5 \\ -2, & x \geq 5 \end{cases}$$

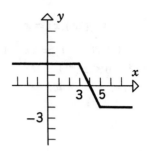

35. **(a)** $(f + g)(x) = \begin{cases} 0, & x < 0 \\ 2x, & x \geq 0 \end{cases}$ **(b)** $(f - g)(x) = \begin{cases} -2x, & x < 0 \\ 0, & x \geq 0 \end{cases}$

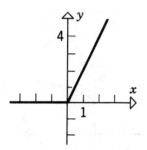 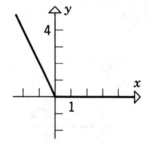

(c) $(f \cdot g)(x) = \begin{cases} -x^2, & x < 0 \\ x^2, & x \geq 0 \end{cases}$ **(d)** $(f/g)(x) = \begin{cases} -1, & x < 0 \\ 1, & x > 0 \end{cases}$

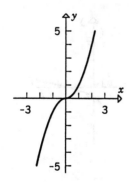 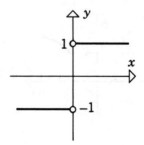

37. If $0 \le x \le 1$, $A = \dfrac{1}{2}(2x)(x) = x^2$;

if $x > 1$, $A = 1 + 2(x - 1) = 2x - 1$;

$$A = \begin{cases} x^2, & 0 \le x \le 1 \\ 2x - 1, & x > 1 \end{cases}$$

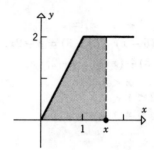

39. $g(x) = \begin{cases} 2, & x < -1 \\ 1 - x, & -1 \le x < 1 \\ \dfrac{1}{2}(x - 1), & x \ge 1 \end{cases}$

41. **(a)**

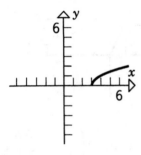

(b)

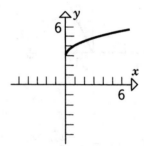

(c)

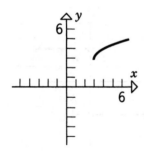

(d)

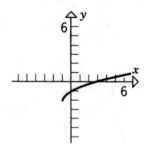

43.

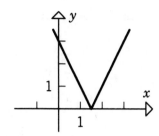

45. (a)

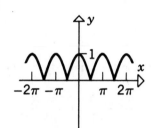

(b)

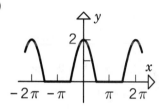

47. **(a)** $f(-x) = (-x)^2 = x^2 = f(x)$, even **(b)** $f(-x) = (-x)^3 = -x^3 = -f(x)$, odd
  (c) $f(-x) = |-x| = |x| = f(x)$, even **(d)** $f(-x) = -x + 1$, neither

  (e) $f(-x) = \dfrac{(-x)^5 - (-x)}{1 + (-x)^2} = \dfrac{-x^5 + x}{1 + x^2} = -\dfrac{x^5 - x}{1 + x^2} = -f(x)$, odd

  (f) $f(-x) = 2 = f(x)$, even

49. **(a)** even **(b)** odd **(c)** odd **(d)** neither

51. **(a)** If f and g are even, then $(f \cdot g)(-x) = f(-x)g(-x) = f(x)g(x) = (f \cdot g)(x)$ so $f \cdot g$ is even.

  (b) If f and g are odd, then
  $(f \cdot g)(-x) = f(-x)g(-x) = [-f(x)][-g(x)] = f(x)g(x) = (f \cdot g)(x)$ so $f \cdot g$ is even.

  (c) If f is even and g is odd, then
  $(f \cdot g)(-x) = f(-x)g(-x) = f(x)[-g(x)] = -f(x)g(x) = -(f \cdot g)(x)$ so $f \cdot g$ is odd.

53. **(a)** If $x_1 = 1.3$, then $x_2 = 1.320006122, x_3 = 1.323822354, \cdots, x_{12} = x_{13} = 1.324717957$

  (b) If $x_1 = 1.3$, then $x_2 = 1.197000000, x_3 = 0.715072373, x_4 = -0.634363117, \cdots$
    The graph of $y = x^3 - 1$ is steeper than that of the line $y = x$ at and near the solution.

55. Let $g(x) = \cos x$. On the calculator used to solve this problem, the successive approximations did not change after 0.739085133 was reached.

57. (a) both; $y = -2x - 4$, $x = -\dfrac{1}{2}y - 2$ (b) y is a function of x; $y = \dfrac{1}{x^{2/3}}$

     (c) neither   (d) both; $y = \dfrac{1}{2x}$, $x = \dfrac{1}{2y}$

59. (a) $y = 1/x^2$ (b)
$$x = \frac{1-y}{1+y}$$
$$x(1+y) = 1-y$$
$$x + xy = 1-y$$
$$xy + y = 1-x$$
$$y(x+1) = 1-x$$
$$y = \frac{1-x}{1+x}$$

     (c) $y^2 + 2xy + x^2 = 0$
$$(y+x)^2 = 0$$
$$y + x = 0$$
$$y = -x$$

61. Treat $y^2 + 4xy + 1 = 0$ as a quadratic equation in y and use the quadratic formula to solve for y :

$$y = \frac{-4x \pm \sqrt{16x^2 - 4}}{2} = -2x \pm \sqrt{4x^2 - 1}.$$

There are two real values of y for each x for which $|x| > 1/2$.

63. (a) both   (b) a function of x

     (c) a function of y (d) neither

EXERCISE SET 2.4

1. (a) -1 (b) 3   (c) does not exist

     (d) 1   (e) -1 (f) 3

3. (a) 1   (b) 1   (c) 1

     (d) 1   (e) $-\infty$ (f) $+\infty$

5. (a) 0   (b) 0   (c) 0

     (d) 3   (e) $+\infty$ (f) $+\infty$

7. (a) $-\infty$ (b) $+\infty$ (c) does not exist

     (d) not defined   (e) 2   (f) 0

9. (a) $-\infty$ (b) $-\infty$ (c) $-\infty$
    (d) 1   (e) 2   (f) 2

11. (a) 0   (b) 0   (c) 0
    (d) 0   (e) does not exist   (f) does not exist

13. all values except -4

EXERCISE SET 2.5

1. 7   3. π 5. 36   7. $\sqrt{109}$ 9. 14   11. 0

13. $\displaystyle\lim_{x\to 4}\frac{x^2-16}{x-4}=\lim_{x\to 4}\frac{(x+4)(x-4)}{x-4}=\lim_{x\to 4}(x+4)=8$

15. $\displaystyle\lim_{x\to 1^+}\frac{x^4-1}{x-1}=\lim_{x\to 1^+}\frac{(x^2-1)(x^2+1)}{x-1}$

    $\displaystyle\qquad\qquad=\lim_{x\to 1^+}\frac{(x-1)(x+1)(x^2+1)}{x-1}=\lim_{x\to 1^+}(x+1)(x^2+1)=4$

17. $\displaystyle\lim_{x\to -1}\frac{x^2+6x+5}{x^2-3x-4}=\lim_{x\to -1}\frac{(x+1)(x+5)}{(x+1)(x-4)}=\lim_{x\to -1}\frac{x+5}{x-4}=-\frac{4}{5}$

19. $\displaystyle\lim_{x\to +\infty}\frac{3x+1}{2x-5}=\lim_{x\to +\infty}\frac{3+1/x}{2-5/x}=\frac{3}{2}$

21. 0

23. $\displaystyle\lim_{x\to -\infty}\frac{x-2}{x^2+2x+1}=\lim_{x\to -\infty}\frac{1/x-2/x^2}{1+2/x+1/x^2}=0$

25. $\displaystyle\lim_{x\to -\infty}\sqrt{\frac{5x^2-2}{x+3}}=\lim_{x\to -\infty}\frac{-\sqrt{5-2/x^2}}{1+3/x}=-\sqrt{5}$

27. $\displaystyle\lim_{y\to -\infty}\frac{2-y}{\sqrt{7+6y^2}}=\lim_{y\to -\infty}\frac{2/y-1}{-\sqrt{7/y^2+6}}=1/\sqrt{6}$

29. $\displaystyle\lim_{x\to -\infty}\frac{\sqrt{3x^4+x}}{x^2-8}=\lim_{x\to -\infty}\frac{\sqrt{3+1/x^3}}{1-8/x^2}=\sqrt{3}$

31. $+\infty$ **33.** does not exist   **35.** $-\infty$

37. $+\infty$ **39.** does not exist

41. $\displaystyle\lim_{x\to4^-}\frac{3-x}{x^2-2x-8}=\lim_{x\to4^-}\frac{3-x}{(x-4)(x+2)}=+\infty$

43. $\displaystyle\lim_{x\to+\infty}\frac{7-6x^5}{x+3}=\lim_{x\to+\infty}\frac{7/x-6x^4}{1+3/x}=-\infty$

45. $\displaystyle\lim_{t\to+\infty}\frac{6-t^3}{7t^3+3}=\lim_{t\to+\infty}\frac{6/t^3-1}{7+3/t^3}=-1/7$

47. $\displaystyle\lim_{x\to0^-}\frac{x}{|x|}=\lim_{x\to0^-}\frac{x}{(-x)}=\lim_{x\to0^-}(-1)=-1$

49. $\displaystyle\lim_{x\to9}\frac{x-9}{\sqrt{x}-3}=\lim_{x\to9}\frac{(\sqrt{x}-3)(\sqrt{x}+3)}{\sqrt{x}-3}=\lim_{x\to9}(\sqrt{x}+3)=6$

51. $+\infty$ **53.** $+\infty$ **55.** $-\infty$

57. If $a\neq0$, then $\displaystyle\lim_{x\to a}\frac{x}{x+a}=\lim_{x\to a}\frac{a}{2a}=\frac{1}{2}$; if $a=0$, then $\displaystyle\lim_{x\to0}\frac{x}{x}=\lim_{x\to0}(1)=1$.

59. (a) $\displaystyle\lim_{x\to3^-}f(x)=\lim_{x\to3^-}(x-1)=2$ (b) $\displaystyle\lim_{x\to3^+}f(x)=\lim_{x\to3^+}(3x-7)=2$

    (c) 2

61. $\displaystyle\lim_{x\to3}h(x)=\lim_{x\to3}(x^2-2x+1)=4$

63. (a) The limit of a difference is the difference of limits if the latter limits exist, in this problem these limits do not exist.

    (b) $\displaystyle\lim_{x\to0^+}\left(\frac{1}{x}-\frac{1}{x^2}\right)=\lim_{x\to0^+}\frac{x-1}{x^2}=-\infty$

65. $\displaystyle\lim_{x\to0}\frac{\sqrt{x+4}-2}{x}=\lim_{x\to0}\frac{(x+4)-4}{x(\sqrt{x+4}+2)}=\lim_{x\to0}\frac{1}{\sqrt{x+4}+2}=\frac{1}{4}$

67. $\displaystyle\lim_{x\to0}\frac{\sqrt{5x+9}-3}{x}=\lim_{x\to0}\frac{5x}{x(\sqrt{5x+9}+3)}=\lim_{x\to0}\frac{5}{\sqrt{5x+9}+3}=\frac{5}{6}$

69. $\displaystyle\lim_{x\to+\infty}(\sqrt{x^2+3}-x)=\lim_{x\to+\infty}\frac{(x^2+3)-x^2}{\sqrt{x^2+3}+x}=\lim_{x\to+\infty}\frac{3}{\sqrt{x^2+3}+x}=0$

71. $\displaystyle\lim_{x\to+\infty}(\sqrt{x^2+5x}-x)=\lim_{x\to+\infty}\frac{5x}{\sqrt{x^2+5x}+x}=\lim_{x\to+\infty}\frac{5}{\sqrt{1+5/x}+1}=\frac{5}{2}$

73. $\displaystyle\lim_{x\to+\infty}(\sqrt{x^2+ax}-x)=\lim_{x\to+\infty}\frac{(x^2+ax)-x^2}{\sqrt{x^2+ax}+x}=\lim_{x\to+\infty}\frac{ax}{\sqrt{x^2+ax}+x}$

$\displaystyle=\lim_{x\to+\infty}\frac{a}{\sqrt{1+a/x}+1}=a/2$

75. $r(x)=p(x)/q(x)$ if $r(x)$ is a rational function, where $p(x)$ and $q(x)$ are polynomials. We know that $\displaystyle\lim_{x\to a}p(x)=p(a)$ and $\displaystyle\lim_{x\to a}q(x)=q(a)$ so if $q(a)\neq 0$ then

$\displaystyle\lim_{x\to a}r(x)=\lim_{x\to a}\frac{p(x)}{q(x)}=\frac{\displaystyle\lim_{x\to a}p(x)}{\displaystyle\lim_{x\to a}q(x)}=\frac{p(a)}{q(a)}=r(a).$ If $q(a)=0$, then $r(a)$ is not defined and

thus cannot equal $\displaystyle\lim_{x\to a}r(x)$, so $\displaystyle\lim_{x\to a}r(x)=r(a)$ only if $r(a)$ is defined.

EXERCISE SET 2.6

1. $|2x-8|<0.1$
if $2|x-4|<0.1$
or if $|x-4|<0.05$,
so $\delta=0.05$

3. $|(7x+5)-(-2)|<0.01$
if $|7x+7|<0.01$
or if $7|x+1|<0.01$
or if $|x+1|<\dfrac{1}{700}$
so $\delta=\dfrac{1}{700}$.

5. Suppose $x\neq 2$, then $\left|\dfrac{x^2-4}{x-2}-4\right|=|(x+2)-4|=|x-2|.$ Thus $\left|\dfrac{x^2-4}{x-2}-4\right|<0.05$ if $0<|x-2|<0.05$, so $\delta=0.05$.

7. $|x^2-16|=|(x+4)(x-4)|=|x+4|\,|x-4|.$ If we restrict δ so that $\delta\le 1$, then

$$|x-4|<1$$
$$3<x<5$$
$$7<x+4<9$$
$$|x+4|<9$$
$$|x+4|\,|x-4|\le 9|x-4|.$$

Thus $|x^2-16|<0.001$ if $9|x-4|<0.001$, or if $|x-4|<1/9000$, so $\delta=1/9000$.

9. $\left|\dfrac{1}{x} - \dfrac{1}{5}\right| = \left|\dfrac{5-x}{5x}\right| = \dfrac{|x-5|}{|5x|}$. If we restrict δ so that $\delta \leq 1$, then

$$|x - 5| < 1$$
$$4 < x < 6$$
$$20 < 5x < 30$$
$$\frac{1}{20} > \frac{1}{5x} > \frac{1}{30}$$
$$\frac{1}{|5x|} < \frac{1}{20}$$
$$\frac{|x-5|}{5|x|} \leq \frac{|x-5|}{20}.$$

     Thus $\left|\dfrac{1}{x} - \dfrac{1}{5}\right| < 0.05$ if $\dfrac{|x-5|}{20} < 0.05$, or if $|x - 5| < 1$, so $\delta = 1$.

11. $|3x - 15| = 3|x - 5| < \epsilon$, if $|x - 5| < \epsilon/3$, so $\delta = \epsilon/3$.

13. $|(2x - 7) + 3| = |2x - 4| = 2|x - 2| < \epsilon$ if $|x - 2| < \epsilon/2$, so $\delta = \epsilon/2$.

15.   Suppose $x \neq 0$, then $\left|\dfrac{x^2 + x}{x} - 1\right| = |(x + 1) - 1| = |x|$. Thus $\left|\dfrac{x^2 + 1}{x} - 1\right| < \epsilon$ if $0 < |x| < \epsilon$, so $\delta = \epsilon$.

17. $|2x^2 - 2| = 2|x + 1||x - 1|$. If we restrict δ so that $\delta \leq 1$, then

$$|x - 1| < 1$$
$$0 < x < 2$$
$$1 < x + 1 < 3$$
$$|x + 1| < 3$$
$$2|x + 1|\,|x - 1| \leq 6|x - 1|.$$

     Thus $|2x^2 - 2| < \epsilon$ if $6|x - 1| < \epsilon$, or if $|x - 1| < \epsilon/6$, so $\delta = \min(\epsilon/6, 1)$.

19. $\left|\dfrac{1}{x} - 3\right| = \left|\dfrac{3}{x}\right|\left|\dfrac{1}{3} - x\right| = \left|\dfrac{3}{x}\right|\left|x - \dfrac{1}{3}\right|$. If we restrict δ so that $\delta \leq \dfrac{1}{4}$, then

$$\left|x - \frac{1}{3}\right| < \frac{1}{4}$$
$$\frac{1}{12} < x < \frac{7}{12}$$
$$12 > \frac{1}{x} > \frac{12}{7}$$

$$36 > \frac{3}{x} > \frac{36}{7}$$

$$\left|\frac{3}{x}\right| < 36$$

$$\left|\frac{3}{x}\right| \left|x - \frac{1}{3}\right| \le 36 \left|x - \frac{1}{3}\right|.$$

Thus $\left|\frac{1}{x} - 3\right| < \epsilon$ if $36\left|x - \frac{1}{3}\right| < \epsilon$, or if $\left|x - \frac{1}{3}\right| < \frac{\epsilon}{36}$, so $\delta = \min(\epsilon/36, 1/4)$.

21. $|\sqrt{x} - 2| = \left|\frac{\sqrt{x} - 2}{1} \frac{\sqrt{x} + 2}{\sqrt{x} + 2}\right| = \frac{|x - 4|}{\sqrt{x} + 2}$. If we restrict δ so that $\delta \le 4$, then

$$|x - 4| < 4$$
$$0 < x < 8$$
$$0 < \sqrt{x} < \sqrt{8}$$
$$2 < \sqrt{x} + 2 < \sqrt{8} + 2$$
$$\frac{1}{2} > \frac{1}{\sqrt{x} + 2} > \frac{1}{\sqrt{8} + 2}$$
$$\frac{1}{\sqrt{x} + 2} < \frac{1}{2}$$
$$\frac{|x - 4|}{\sqrt{x} + 2} \le \frac{1}{2}|x - 4|$$

Thus $|\sqrt{x} - 2| < \epsilon$ if $\frac{1}{2}|x - 4| < \epsilon$, or if $|x - 4| < 2\epsilon$, so $\delta = \min(2\epsilon, 4)$.

23. If $x \ne 1$, then $|f(x) - 3| = |(x + 2) - 3| = |x - 1|$. Thus $|f(x) - 3| < \epsilon$ if $0 < |x - 1| < \epsilon$ so $\delta = \epsilon$.

25. Assume there is a number L such that $\lim_{x \to 0} f(x) = L$. Then there exists a number $\delta > 0$ such that $|f(x) - L| < \frac{1}{8}$ whenever $0 < |x - 0| < \delta$; in particular, $x = \frac{\delta}{2}$ and $x = -\frac{\delta}{2}$ are two such values of x. But $f(\delta/2) = \frac{1}{8}$ and $f(-\delta/2) = -\frac{1}{8}$ so $\left|\frac{1}{8} - L\right| < \frac{1}{8}$ and $\left|-\frac{1}{8} - L\right| < \frac{1}{8}$ or, equivalently, $0 < L < \frac{1}{4}$ and $-\frac{1}{4} < L < 0$, which is impossible.

27. Assume there is a number L such that $\lim_{x \to 1} \frac{1}{x - 1} = L$. Then there exists a number $\delta > 0$ such that $\left|\frac{1}{x - 1} - L\right| < 1$ whenever $0 < |x - 1| < \delta$; in particular, $x = 1 + \frac{\delta}{\delta + 1}$ and

$x = 1 - \dfrac{\delta}{\delta+1}$ are two such values of x. So $\left|\dfrac{\delta+1}{\delta} - L\right| < 1$ and $\left|-\dfrac{\delta+1}{\delta} - L\right| < 1$ or equivalently, $\dfrac{1}{\delta} < L < \dfrac{1}{\delta} + 2$ and $-2 - \dfrac{1}{\delta} < L < -\dfrac{1}{\delta}$, which is impossible.

29. $|x^2 - 9| = |x + 3||x - 3|$. If $\delta \leq 2$, then

$$|x - 3| < 2$$
$$1 < x < 5$$
$$4 < x + 3 < 8$$
$$|x + 3| < 8$$
$$|x + 3|\,|x - 3| \leq 8|x - 3|.$$

Thus $|x^2 - 9| < \epsilon$ if $8|x - 3| < \epsilon$, or if $|x - 3| < \epsilon/8$, so $\delta = \min(\epsilon/8, 2)$

EXERCISE SET 2.7

1. continuous on (d), (e), (f); discontinuous at $x = 2$ on (a), (b), (c)

3. continuous on (b), (d), (f); discontinuous at $x = 1, 3$ on (a), $x = 1$ on (c), $x = 3$ on (e)

5. none **7.** none **9.** $x = \pm 4$

11. $x = \pm 3$ **13.** none **15.** none

17. **(a)** f is continuous everywhere for any k, except perhaps at $x = 1$;

$\displaystyle\lim_{x \to 1^-} f(x) = \lim_{x \to 1^-} (7x - 2) = 5$, $\displaystyle\lim_{x \to 1^+} f(x) = \lim_{x \to 1^+} kx^2 = k$, and $f(1) = 5$ thus

$\displaystyle\lim_{x \to 1} f(x) = f(1)$ if $k = 5$, so f is continuous everywhere if $k = 5$.

(b) $\displaystyle\lim_{x \to 2^-} f(x) = \lim_{x \to 2^-} kx^2 = 4k$, $\displaystyle\lim_{x \to 2^+} f(x) = \lim_{x \to 2^+} (2x + k) = 4 + k$, and $f(2) = 4k$,

so $\displaystyle\lim_{x \to 2} f(x) = f(2)$ if $4k = 4 + k$, $k = 4/3$.

19. **(a)** If $c > 0$, then $\displaystyle\lim_{x \to c} f(x) = \lim_{x \to c} \sqrt{x} = \sqrt{c} = f(c)$; also $\displaystyle\lim_{x \to 0^+} f(x) = \lim_{x \to 0^+} \sqrt{x} = 0 = f(0)$ so $f(x)$ is continuous on $[0, +\infty)$.

(b) $\sqrt{g(x)}$ is continuous by Theorem 2.7.6 because it is the composition of $\sqrt{x}$ with $g(x)$ where $\sqrt{x}$ is continuous on $[0, +\infty)$ and $g(x)$ is continuous and nonnegative.

21. $x = 0, \pm 1, \pm 2, \cdots$

23. **(a)** $x = 0$; not removable because $\displaystyle\lim_{x \to 0} \dfrac{|x|}{x}$ does not exist.

(b) $x = -3$; removable because $\lim\limits_{x \to -3} \dfrac{x^2 + 3x}{x + 3} = -3$.

(c) $x = \pm 2$; removable at $x = 2$ because $\lim\limits_{x \to 2} \dfrac{x - 2}{|x| - 2} = 1$, not removable at $x = -2$ because

$\lim\limits_{x \to -2} \dfrac{x - 2}{|x| - 2}$ does not exist.

25. If f and g are continuous at c, then $\lim\limits_{x \to c} f(x) = f(c)$ and $\lim\limits_{x \to c} g(x) = g(c)$, so

(a) $\lim\limits_{x \to c} (f + g)(x) = \lim\limits_{x \to c} [f(x) + g(x)] = \lim\limits_{x \to c} f(x) + \lim\limits_{x \to c} g(x) = f(c) + g(c) = (f + g)(c)$
therefore $f + g$ is continuous at c.

(b) Similar to part (a) with $+$ replaced by $-$.

(c) $\lim\limits_{x \to c} (f \cdot g)(x) = \lim\limits_{x \to c} [f(x)g(x)] = \left[\lim\limits_{x \to c} f(x) \right] \left[\lim\limits_{x \to c} g(x) \right] = f(c)g(c) = (f \cdot g)(c)$
therefore $f \cdot g$ is continuous at c.

27. (a) Let $f(x) = \begin{cases} 0, & x < 2 \\ 1, & x \geq 2 \end{cases}$ and $g(x) = \begin{cases} 1, & x < 2 \\ 0, & x \geq 2 \end{cases}$; f and g are discontinuous at $x = 2$,

but $f + g$ is continuous at $x = 2$. If $f(x) = \begin{cases} 0, & x < 2 \\ 1, & x \geq 2 \end{cases}$ and $g(x) = \begin{cases} 1, & x < 2 \\ 2, & x \geq 2 \end{cases}$;

then f, g, and $f + g$ are discontinuous at $x = 2$.

(b) Replace $f + g$ by $f \cdot g$ everywhere in part (a).

29. If $f(a)$ and $f(b)$ have opposite signs then 0 is between $f(a)$ and $f(b)$. From Theorem 2.7.9 there is at least one number x in $[a, b]$ such that $f(x) = 0$. But $f(a) \neq 0$ and $f(b) \neq 0$ by assumption, so there is at least one solution of $f(x) = 0$ in the interval (a, b).

31. $f(x) = x^3 + x^2 - 2x - 1$ is continuous on $[-1, 1]$, $f(-1) = 1$ and $f(1) = -1$ have opposite signs so Theorem 2.7.10 applies.

33. From the graph there are two real solutions, one in $(-2, -1.5)$ and the other in $(1, 1.5)$. For the one in $(-2, -1.5)$ we find

x	-2.0	-1.9	-1.8	-1.7	-1.6
y	$-9.$	-6.13	-3.70	-1.65	0.05

so one solution is in $(-1.7, -1.6)$; use -1.65 to approximate it. For the one in $(1, 1.5)$ we find

x	1.0	1.1	1.2	1.3	1.4
y	$3.$	2.44	1.73	0.84	-0.24

so this solution is in $(1.3, 1.4)$; use 1.35 to approximate it.

35.　If $f(x) = x^2 - 5$, then $f(2) = -1$ and $f(3) = 4$ have opposite signs so $\sqrt{5}$ is in $(2, 3)$.

　　(a)　$f(2.2) = -0.16$ and $f(2.3) = 0.29$ so $\sqrt{5}$ is in $(2.2, 2.3)$; use 2.25 to approximate it with an error of at most 0.05.

　　(b)　$f(2.23) = -0.03$ and $f(2.24) = 0.02$ so $\sqrt{5}$ is in $(2.23, 2.24)$; use 2.235 to approximate it with an error of at most 0.005.

37.　$f(x) = \begin{cases} x+1, & 0 \le x \le 1 \\ x-3, & 1 < x \le 2 \end{cases}$, for example.

39.　If $p(x) = a_n x^n + a_{n-1} x^{n-1} + \cdots + a_1 x + a_0$ where $a_n \ne 0$ and n is odd, then either

$$\lim_{x \to +\infty} p(x) = +\infty \text{ and } \lim_{x \to -\infty} p(x) = -\infty, \text{ or } \lim_{x \to +\infty} p(x) = -\infty \text{ and } \lim_{x \to -\infty} p(x) = +\infty,$$

depending on whether $a_n > 0$ or $a_n < 0$, respectively. In either case, there are numbers a and b with $p(x)$ continuous on $[a, b]$ where $p(a)$ and $p(b)$ have opposite signs so that $p(x) = 0$ has at least one real solution in (a, b).

EXERCISE SET 2.8

1.　none

3.　$x = n\pi$; $n = 0, \pm 1, \pm 2, \cdots$

5.　$x = n\pi$; $n = 0, \pm 1, \pm 2, \cdots$

7.　none

9.　discontinuous if $\sin x = 1/2$, so $x = \pi/6 + 2n\pi$ or $x = 5\pi/6 + 2n\pi$; $n = 0, \pm 1, \pm 2, \cdots$

11.　$\sin(g(x))$ is the composition of $\sin x$ with $g(x)$; $\sin x$ is continuous, so by Theorem 2.7.6 $\sin(g(x))$ is continuous at every point where $g(x)$ is continuous.

13.　$\displaystyle \lim_{x \to +\infty} \cos(1/x) = \cos(0) = 1$

15.　$\displaystyle \lim_{x \to +\infty} \sin \left(\frac{\pi x}{2 - 3x} \right) = \sin \left(\lim_{x \to +\infty} \frac{\pi x}{2 - 3x} \right) = \sin(-\pi/3) = -\sqrt{3}/2$

17.　$\displaystyle \lim_{\theta \to 0} \frac{\sin 3\theta}{\theta} = \lim_{\theta \to 0} 3 \frac{\sin 3\theta}{3\theta} = 3 \lim_{\theta \to 0} \frac{\sin 3\theta}{3\theta} = 3(1) = 3.$

19.　$\displaystyle \lim_{x \to 0^-} \frac{\sin x}{|x|} = \lim_{x \to 0^-} \left(-\frac{\sin x}{x} \right) = -1$

21.　$\displaystyle \lim_{x \to 0^+} \frac{\sin x}{5\sqrt{x}} = \frac{1}{5} \lim_{x \to 0^+} \sqrt{x} \left(\frac{\sin x}{x} \right) = 0$

23. $\displaystyle\lim_{x\to 0}\frac{\tan 7x}{\sin 3x}=\lim_{x\to 0}\frac{\dfrac{\sin 7x}{\cos 7x}}{\sin 3x}=\lim_{x\to 0}\frac{1}{\cos 7x}\frac{\sin 7x}{\sin 3x}=\lim_{x\to 0}\frac{1}{\cos 7x}\frac{7\dfrac{\sin 7x}{7x}}{3\dfrac{\sin 3x}{3x}}$

$$=\frac{7}{3}\lim_{x\to 0}\frac{1}{\cos 7x}\frac{\displaystyle\lim_{x\to 0}\frac{\sin 7x}{7x}}{\displaystyle\lim_{x\to 0}\frac{\sin 3x}{3x}}=\frac{7}{3}(1)\frac{(1)}{(1)}=\frac{7}{3}.$$

25. $\displaystyle\lim_{h\to 0}\frac{h}{\tan h}=\lim_{h\to 0}\frac{h}{\dfrac{\sin h}{\cos h}}=\lim_{h\to 0}\frac{h\cos h}{\sin h}=\lim_{h\to 0}\frac{\cos h}{\dfrac{\sin h}{h}}=\frac{\displaystyle\lim_{h\to 0}\cos h}{\displaystyle\lim_{h\to 0}\frac{\sin h}{h}}=\frac{1}{1}=1.$

27. $\displaystyle\lim_{\theta\to 0}\frac{\theta^2}{1-\cos\theta}=\lim_{\theta\to 0}\frac{\theta^2}{1-\cos\theta}\frac{1+\cos\theta}{1+\cos\theta}=\lim_{\theta\to 0}\frac{\theta^2(1+\cos\theta)}{1-\cos^2\theta}$

$$=\lim_{\theta\to 0}\frac{\theta^2(1+\cos\theta)}{\sin^2\theta}=\lim_{\theta\to 0}\frac{1+\cos\theta}{\dfrac{\sin^2\theta}{\theta^2}}=\frac{\displaystyle\lim_{\theta\to 0}(1+\cos\theta)}{\displaystyle\lim_{\theta\to 0}\left(\frac{\sin\theta}{\theta}\right)^2}=\frac{1+1}{1^2}=2.$$

29. $\displaystyle\lim_{\theta\to 0}\frac{\theta}{\cos\theta}=\frac{\displaystyle\lim_{\theta\to 0}\theta}{\displaystyle\lim_{\theta\to 0}\cos\theta}=\frac{0}{1}=0.$

31. Use the identity $1-\cos\theta=2\sin^2\dfrac{\theta}{2}$:

$$\lim_{h\to 0}\frac{1-\cos 5h}{\cos 7h-1}=\lim_{h\to 0}\frac{2\sin^2(5h/2)}{-2\sin^2(7h/2)}=-\lim_{h\to 0}\frac{(5/2)^2\dfrac{\sin^2(5h/2)}{(5h/2)^2}}{(7/2)^2\dfrac{\sin^2(7h/2)}{(7h/2)^2}}$$

$$=-\frac{25}{49}\frac{\displaystyle\lim_{h\to 0}\left[\frac{\sin(5h/2)}{5h/2}\right]^2}{\displaystyle\lim_{h\to 0}\left[\frac{\sin(7h/2)}{7h/2}\right]^2}=-\frac{25}{49}\cdot\frac{1^2}{1^2}=-\frac{25}{49}.$$

33. $\displaystyle\lim_{x\to 0+}\cos(1/x)$ does not exist due to oscillation

35. $\displaystyle\lim_{x\to 0}\frac{2x+\sin x}{x}=\lim_{x\to 0}\left(2+\frac{\sin x}{x}\right)=2+1=3$

37. $\lim\limits_{x \to 0^-} f(x) = \lim\limits_{x \to 0^-} \dfrac{\tan kx}{x} = \lim\limits_{x \to 0^-} \dfrac{k}{\cos kx} \dfrac{\sin kx}{kx} = k; \ \lim\limits_{x \to 0^+} f(x) = \lim\limits_{x \to 0^+} (3x + 2k^2) = 2k^2;$
$f(0) = 2k^2;$ so $\lim\limits_{x \to 0} f(x) = f(0)$ if $2k^2 = k, \ 2k^2 - k = 0, \ k(2k - 1) = 0; \ k = 1/2.$

39. **(a)** If $t = \dfrac{1}{x}$, then $x = \dfrac{1}{t}$ and $x \to +\infty$ as $t \to 0^+$ so $\lim\limits_{x \to +\infty} x \sin \dfrac{1}{x} = \lim\limits_{t \to 0^+} \dfrac{\sin t}{t} = 1.$

(b) If $t = \dfrac{1}{x}$, then $x = \dfrac{1}{t}$ and $x \to -\infty$ as $t \to 0^-$ so $\lim\limits_{x \to -\infty} \left(1 - \cos \dfrac{1}{x} \right) = \lim\limits_{t \to 0^-} \dfrac{1 - \cos t}{t} = 0.$

(c) If $t = \pi - x$, then $x = \pi - t$ and $x \to \pi$ as $t \to 0$
so $\lim\limits_{x \to \pi} \dfrac{\pi - x}{\sin x} = \lim\limits_{t \to 0} \dfrac{t}{\sin(\pi - t)} = \lim\limits_{t \to 0} \dfrac{t}{\sin t} = 1.$

41. Let $t = x - 1$, then $x = t + 1$ and
$\lim\limits_{x \to 1} \dfrac{\sin(\pi x)}{x - 1} = \lim\limits_{t \to 0} \dfrac{\sin(\pi t + \pi)}{t} = \lim\limits_{t \to 0} \dfrac{-\sin \pi t}{t} = -\pi \lim\limits_{t \to 0} \dfrac{\sin \pi t}{\pi t} = -\pi$

43. **(a)** $-1 \le \sin x \le 1$ so $-\dfrac{1}{x} \le \dfrac{\sin x}{x} \le \dfrac{1}{x}$ if $x > 0.$

But $-1/x$ and $1/x \to 0$ as $x \to +\infty$, thus $\lim\limits_{x \to +\infty} \dfrac{\sin x}{x} = 0.$

(b) Replace $\sin x$ by $\cos x$ in part (a).

45. $\lim\limits_{x \to 0} (1 - x^2) = 1$ and $\lim\limits_{x \to 0} \cos x = 1$ so $\lim\limits_{x \to 0} f(x) = 1.$

47. If $x > 0$, then $xL \le xf(x) \le xM$; if $x < 0$, then $xL \ge xf(x) \ge xM$. So $\lim\limits_{x \to 0} xf(x) = 0$
because $\lim\limits_{x \to 0^+} xL = \lim\limits_{x \to 0^+} xM = 0$ and $\lim\limits_{x \to 0^-} xL = \lim\limits_{x \to 0^-} xM = 0.$

49. If $h < 0$, then $-h > 0$ so from (7) $\cos(-h) < \dfrac{\sin(-h)}{-h} < 1, \cos h < \dfrac{-\sin h}{-h} < 1,$
$\cos h < \dfrac{\sin h}{h} < 1.$

51. **(a)** $\sin 10° \approx 0.17365$ **(b)** $\pi/18 \approx 0.17453$

53. **(a)** $\tan 5° \approx 0.08749$ **(b)** $\pi/36 \approx 0.08727$

55. $f(x) = x - \cos x$ is continuous on $[0, \pi/2]$; $f(0) = -1$ and $f(\pi/2) = \pi/2$ have opposite signs so Theorem 2.7.10 applies.

SUPPLEMENTARY EXERCISES CHAPTER 2

1. $\sqrt{4-x^2}$ is real if and only if $4-x^2 \geq 0$, thus $4 \geq x^2$, so the domain is $|x| \leq 2$; $f(-\sqrt{2}) = \sqrt{2}$, $f(0) = 2$, $f(\sqrt{3}) = 1$.

3. $f(x) = \dfrac{(x-1)}{(x+2)(x-1)}$, domain: all x except -2 and 1; $f(0) = 1/2$, $f(1)$ is not defined, $f(2) = 1/4$.

5. domain: all x; $f(0) = -1$, $f(2) = 3$, $f(4) = \sqrt{3}$.

7. (a) $f(x^2) - (f(x))^2 = \dfrac{3-x^2}{x^2} - \left(\dfrac{3-x}{x}\right)^2 = \dfrac{3-x^2}{x^2} - \dfrac{9-6x+x^2}{x^2} = \dfrac{-2x^2+6x-6}{x^2}$

    (b) $f(x+3) - [f(x) + f(3)] = \dfrac{3-(x+3)}{x+3} - \left[\dfrac{3-x}{x} + \dfrac{3-3}{3}\right] = -\dfrac{9}{x(x+3)}$

    (c) $f(1/x) - 1/f(x) = \dfrac{3-1/x}{1/x} - \dfrac{x}{3-x} = 3x - 1 - \dfrac{x}{3-x} = \dfrac{3x^2-9x+3}{x-3}$

    (d) $f(f(x)) = f\left(\dfrac{3-x}{x}\right) = \dfrac{3 - \dfrac{3-x}{x}}{\dfrac{3-x}{x}} = \dfrac{4x-3}{3-x}$

9.

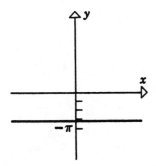

domain: all x
range: $y = -\pi$

11.

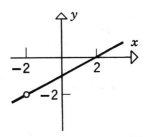

$f(x) = \dfrac{x^2-4}{2x+4} = \dfrac{1}{2}(x-2)$.

$x \neq -2$
domain: all x except -2
range: all y except -2

13.

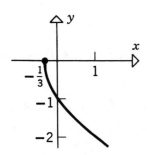

domain: $x \geq -1/3$
range: $y \leq 0$

15.

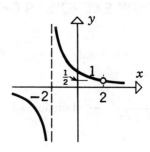

$$f(x) = \frac{2x - 4}{x^2 - 4} = \frac{2}{x + 2}, x \neq 2$$

domain: all x except $-2, 2$
range: all y except $0, 1/2$

17. Some possible answers are:

    (a) $h(x) = x^3$, $g(x) = x^2 + 3$; $h(x) = x^6$, $g(x) = x + 3$
    (b) $h(x) = x^2 + 1$, $g(x) = \sqrt{x}$; $h(x) = x^2$, $g(x) = \sqrt{x + 1}$
    (c) $h(x) = 3x + 2$, $g(x) = \sin x$; $h(x) = 3x$, $g(x) = \sin(x + 2)$

19. **(a)** -1
    (b) does not exist
    (c) 1
    (d) 0
    (e) $-\infty$ (does not exist)
    (f) 0
    (g) 0
    (h) $-\infty$ (does not exist)

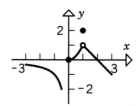

21. $f(x) = \sqrt{2 - x}$ is defined for $x \leq 2$ and $\lim\limits_{x \to a} f(x) = \sqrt{2 - a}$ if $a < 2$, so $\lim\limits_{x \to a} f(x) = 2, 1, 0$ for $a = -2, 1, 2^-$. Because $f(x)$ is not defined for $x > 2$, $\lim\limits_{x \to 2^+} f(x)$ and $\lim\limits_{x \to +\infty} f(x)$ do not exist. Finally, $\lim\limits_{x \to -\infty} f(x) = +\infty$, so this limit does not exist.

23. $f(x) = \dfrac{x^2 - 25}{x - 5} = x + 5$, $x \neq 5$, so $\lim\limits_{x \to a} f(x) = \lim\limits_{x \to a}(x + 5) = a + 5 = 5, 10, 0, 10, 0$ for $a = 0, 5^+, -5^-, 5, -5$. Also, $\lim\limits_{x \to -\infty} f(x) = -\infty$ and $\lim\limits_{x \to +\infty} f(x) = +\infty$, so neither of these limits exist.

25. $\lim\limits_{x\to 0} \dfrac{\tan ax}{\sin bx} = \lim\limits_{x\to 0} \dfrac{\sin ax}{\sin bx}\, \dfrac{1}{\cos ax} = \lim\limits_{x\to 0} \dfrac{a[(\sin ax)/(ax)]}{b[(\sin bx)/(bx)]}\, \dfrac{1}{\cos ax} = \dfrac{a}{b}.$

27. $\lim\limits_{\theta\to 0} \dfrac{\sin 2\theta}{\theta^2} = \lim\limits_{\theta\to 0} \dfrac{\sin 2\theta}{2\theta}\, \dfrac{2}{\theta},$ but $\dfrac{\sin 2\theta}{2\theta} \to 1$ as $\theta \to 0$ and $\left|\dfrac{2}{\theta}\right| \to +\infty$ as $\theta \to 0$ so the limit does not exist.

29. $\lim\limits_{x\to 0+} \dfrac{\sin x}{\sqrt{x}} = \lim\limits_{x\to 0+} \sqrt{x}\left(\dfrac{\sin x}{x}\right) = (0)(1) = 0$

31. $\lim\limits_{x\to 0} \dfrac{3x - \sin(kx)}{x} = \lim\limits_{x\to 0} \left[3 - k\dfrac{\sin(kx)}{kx}\right] = 3 - k$

CHAPTER 3
Differentiation

EXERCISE SET 3.1

1. **(a)** $m_{sec} = \dfrac{f(4) - f(3)}{4 - 3} = \dfrac{\frac{1}{2}(4)^2 - \frac{1}{2}(3)^2}{1} = \dfrac{7}{2}$

(b) $m_{tan} = \lim\limits_{h \to 0} \dfrac{f(3 + h) - f(3)}{h}$

$= \lim\limits_{h \to 0} \dfrac{\frac{1}{2}(3 + h)^2 - \frac{9}{2}}{h}$

$= \lim\limits_{h \to 0} \dfrac{3h + \frac{1}{2}h^2}{h}$

$= \lim\limits_{h \to 0} \left(3 + \dfrac{1}{2}h\right) = 3;$

tangent line: $y - 9/2 = 3(x - 3),$
$$y = 3x - 9/2.$$

(c)

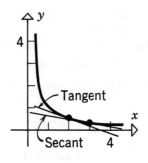

3. **(a)** $m_{sec} = \dfrac{f(3) - f(2)}{3 - 2} = \dfrac{(1/3) - (1/2)}{1} = -\dfrac{1}{6}$

(b) $m_{tan} = \lim\limits_{h \to 0} \dfrac{f(2 + h) - f(2)}{h}$

$= \lim\limits_{h \to 0} \dfrac{\dfrac{1}{2 + h} - \dfrac{1}{2}}{h}$

$= \lim\limits_{h \to 0} \dfrac{2 - (2 + h)}{2h(2 + h)}$

$= \lim\limits_{h \to 0} -\dfrac{1}{2(2 + h)} = -\dfrac{1}{4};$

tangent line: $y - \dfrac{1}{2} = -\dfrac{1}{4}(x - 2),$

$$y = -\dfrac{1}{4}x + 1.$$

(c)

5. (a) $m_{\tan} = \lim\limits_{h \to 0} \dfrac{f(x_0 + h) - f(x_0)}{h} = \lim\limits_{h \to 0} \dfrac{(x_0 + h)^3 - x_0^3}{h}$

$= \lim\limits_{h \to 0} \dfrac{(x_0^3 + 3x_0^2 h + 3x_0 h^2 + h^3) - x_0^3}{h} = \lim\limits_{h \to 0} (3x_0^2 + 3x_0 h + h^2) = 3x_0^2.$

(b) $m_{\tan} = 3(5)^2 = 75$ at $(5, 5^3) = (5, 125)$, so $y - 125 = 75(x - 5)$, or $y = 75x - 250$.

(c) $y - x_0^3 = 3x_0^2(x - x_0)$, $y = 3x_0^2 x - 2x_0^3$.

7. (a) $m_{\tan} = \lim\limits_{h \to 0} \dfrac{f(x_0 + h) - f(x_0)}{h} = \lim\limits_{h \to 0} \dfrac{[(x_0 + h)^2 + (x_0 + h)] - (x_0^2 + x_0)}{h}$

$= \lim\limits_{h \to 0} \dfrac{2x_0 h + h^2 + h}{h} = \lim\limits_{h \to 0} (2x_0 + h + 1) = 2x_0 + 1.$

(b) $m_{\tan} = 2(2) + 1 = 5$ at $(2, 6)$ so $y - 6 = 5(x - 2)$, or $y = 5x - 4$.

(c) $y - (x_0^2 + x_0) = (2x_0 + 1)(x - x_0)$, $y = (2x_0 + 1)x - x_0^2$

9. (a) $m_{\tan} = (50 - 10)/(15 - 5)$
        $= 40/10$
        $= 4$ m/sec

    (b)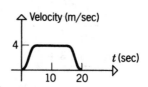

11. From the figure:
    (a) The particle is moving faster at time t_0 because the slope of the tangent to the curve at t_0 is greater than that at t_2.
    (b) The initial velocity is 0 because the slope of a horizontal line is 0.
    (c) The particle is speeding up because the slope increases as t increases from t_0 to t_1.

    (d) The particle is slowing down because the slope decreases as t increases from t_1 to t_2.

13. It is a straight line with slope equal to the velocity.

15. (a) $72°$F at about 4:30 P.M.
    (b) about $(66 - 42)/6 = 4°$F/hr
    (c) decreasing most rapidly at about 9 P.M.; rate of change of temperature is about $-6°$F/hr (slope of estimated tangent line to curve at 9 P.M.)

17. (a) during the first year after birth

(b) about 6 cm/year (slope of estimated tangent line at age 5)

(c) the growth rate is greatest at about age 14; about 10 cm/year

(d)

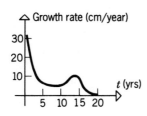

19. (a) $5(40)^3 = 320,000$ ft.

(b) average velocity $= 320,000/40 = 8,000$ ft/sec.

(c) $5t^3 = 135$ when the rocket has gone 135 ft, so $t^3 = 27$, $t = 3$ sec; average velocity $= 135/3 = 45$ ft/sec.

(d) $f'(40) = \lim\limits_{h \to 0} \dfrac{5(40+h)^3 - 5(40)^3}{h} = \lim\limits_{h \to 0} \dfrac{5(4800h + 120h^2 + h^3)}{h}$

$\qquad = \lim\limits_{h \to 0} 5(4800 + 120h + h^2) = 24,000$ ft/sec.

21. (a) average velocity $= \dfrac{6(4)^4 - 6(2)^4}{4-2} = 720$ ft/min.

(b) $f'(2) = \lim\limits_{h \to 0} \dfrac{6(2+h)^4 - 6(2)^4}{h} = \lim\limits_{h \to 0} \dfrac{6(32h + 24h^2 + 8h^3 + h^4)}{h}$

$\qquad = \lim\limits_{h \to 0} 6(32 + 24h + 8h^2 + h^3) = 192$ ft/min.

23. (a) $x_0 = 1$, $x_1 = 4$; $y_0 = 2(1)^2 - 1 = 1$, $y_1 = 2(4)^2 - 1 = 31$, $\dfrac{y_1 - y_0}{x_1 - x_0} = \dfrac{31 - 1}{4 - 1} = 10$.

(b) $y'(1) = \lim\limits_{h \to 0} \dfrac{[2(1+h)^2 - 1] - [2(1)^2 - 1]}{h} = \lim\limits_{h \to 0} (4 + 2h) = 4$.

25. (a) $r_0 = 1$, $r_1 = 2$; $A_0 = \pi(1)^2 = \pi$, $A_1 = \pi(2)^2 = 4\pi$, $\dfrac{A_1 - A_0}{r_1 - r_0} = \dfrac{4\pi - \pi}{2 - 1} = 3\pi$.

(b) $A'(2) = \lim\limits_{h \to 0} \dfrac{\pi(2+h)^2 - \pi(2)^2}{h} = \lim\limits_{h \to 0} \pi(4 + h) = 4\pi$.

27. $m_{sec} = \dfrac{f(x_1) - f(x_0)}{x_1 - x_0} = \dfrac{x_1^2 - x_0^2}{x_1 - x_0} = \dfrac{(x_1 + x_0)(x_1 - x_0)}{x_1 - x_0} = x_1 + x_0$, $x_1 \neq x_0$

m_{sec} approaches $2x_0$ as x_1 approaches x_0, thus $m_{tan} = 2x_0$ so

$|m_{tan} - m_{sec}| = |2x_0 - (x_1 + x_0)| = |x_0 - x_1| = |x_1 - x_0|$.

EXERCISE SET 3.2

1. $f'(x) = \lim\limits_{h \to 0} \dfrac{3(x+h)^2 - 3x^2}{h} = \lim\limits_{h \to 0} \dfrac{3(x^2 + 2xh + h^2) - 3x^2}{h}$

    $= \lim\limits_{h \to 0} \dfrac{6xh + 3h^2}{h} = \lim\limits_{h \to 0}(6x + 3h) = 6x.$

3. $f'(x) = \lim\limits_{h \to 0} \dfrac{(x+h)^3 - x^3}{h} = \lim\limits_{h \to 0} \dfrac{x^3 + 3x^2h + 3xh^2 + h^3 - x^3}{h}$

    $= \lim\limits_{h \to 0} \dfrac{3x^2h + 3xh^2 + h^3}{h} = \lim\limits_{h \to 0}(3x^2 + 3xh + h^2) = 3x^2.$

5. $f'(x) = \lim\limits_{h \to 0} \dfrac{\sqrt{x+h+1} - \sqrt{x+1}}{h}$

    $= \lim\limits_{h \to 0} \dfrac{\sqrt{x+h+1} - \sqrt{x+1}}{h} \dfrac{\sqrt{x+h+1} + \sqrt{x+1}}{\sqrt{x+h+1} + \sqrt{x+1}}$

    $= \lim\limits_{h \to 0} \dfrac{(x+h+1) - (x+1)}{h(\sqrt{x+h+1} + \sqrt{x+1})} = \lim\limits_{h \to 0} \dfrac{h}{h(\sqrt{x+h+1} + \sqrt{x+1})}$

    $= \lim\limits_{h \to 0} \dfrac{1}{\sqrt{x+h+1} + \sqrt{x+1}} = \dfrac{1}{2\sqrt{x+1}}.$

7. $f'(x) = \lim\limits_{h \to 0} \dfrac{\dfrac{1}{x+h} - \dfrac{1}{x}}{h} = \lim\limits_{h \to 0} \dfrac{\dfrac{x - (x+h)}{x(x+h)}}{h}$

    $= \lim\limits_{h \to 0} \dfrac{-h}{hx(x+h)} = \lim\limits_{h \to 0} -\dfrac{1}{x(x+h)} = -\dfrac{1}{x^2}.$

9. $f'(x) = \lim\limits_{h \to 0} \dfrac{[a(x+h)^2 + b] - [ax^2 + b]}{h} = \lim\limits_{h \to 0} \dfrac{ax^2 + 2axh + ah^2 + b - ax^2 - b}{h}$

    $= \lim\limits_{h \to 0} \dfrac{2axh + ah^2}{h} = \lim\limits_{h \to 0}(2ax + ah) = 2ax.$

11. $f'(x) = \lim\limits_{h \to 0} \dfrac{\dfrac{1}{\sqrt{x+h}} - \dfrac{1}{\sqrt{x}}}{h} = \lim\limits_{h \to 0} \dfrac{\sqrt{x} - \sqrt{x+h}}{h\sqrt{x}\sqrt{x+h}}$

    $= \lim\limits_{h \to 0} \dfrac{x - (x+h)}{h\sqrt{x}\sqrt{x+h}(\sqrt{x} + \sqrt{x+h})} = \lim\limits_{h \to 0} \dfrac{-1}{\sqrt{x}\sqrt{x+h}(\sqrt{x} + \sqrt{x+h})} = -\dfrac{1}{2x^{3/2}}.$

13. $f'(3) = 6(3) = 18;\ f(3) = 3(3)^2 = 27$ so $y - 27 = 18(x - 3),\ y = 18x - 27.$

15. $f'(0) = 3(0)^2 = 0$; $f(0) = 0^3 = 0$ so $y - 0 = (0)(x - 0)$, $y = 0$.

17. $f'(8) = \dfrac{1}{2\sqrt{8+1}} = \dfrac{1}{6}$; $f(8) = \sqrt{8+1} = 3$ so $y - 3 = \dfrac{1}{6}(x - 8)$, $y = \dfrac{1}{6}x + \dfrac{5}{3}$.

19. (a) $\dfrac{dy}{dx} = \lim\limits_{h \to 0} \dfrac{[4(x + h)^2 + 2] - [4x^2 + 2]}{h}$

$\quad\quad = \lim\limits_{h \to 0} \dfrac{4x^2 + 8xh + 4h^2 + 2 - 4x^2 - 2}{h} = \lim\limits_{h \to 0}(8x + 4h) = 8x.$

(b) $\dfrac{dy}{dx}\bigg|_{x=1} = 8(1) = 8.$

21. $f'(t) = \lim\limits_{h \to 0} \dfrac{f(t + h) - f(t)}{h} = \lim\limits_{h \to 0} \dfrac{[4(t + h)^2 + (t + h)] - [4t^2 + t]}{h}$

$\quad\quad = \lim\limits_{h \to 0} \dfrac{4t^2 + 8th + 4h^2 + t + h - 4t^2 - t}{h}$

$\quad\quad = \lim\limits_{h \to 0} \dfrac{8th + 4h^2 + h}{h} = \lim\limits_{h \to 0}(8t + 4h + 1) = 8t + 1.$

23. $\dfrac{dA}{d\lambda} = \lim\limits_{h \to 0} \dfrac{[3(\lambda + h)^2 - (\lambda + h)] - [3\lambda^2 - \lambda]}{h} = \lim\limits_{h \to 0} \dfrac{3\lambda^2 + 6\lambda h + 3h^2 - \lambda - h - 3\lambda^2 + \lambda}{h}$

$\quad\quad = \lim\limits_{h \to 0} \dfrac{6\lambda h + 3h^2 - h}{h} = \lim\limits_{h \to 0}(6\lambda + 3h - 1) = 6\lambda - 1.$

25. (a) D   **(b)** F   **(c)** B   **(d)** C   **(e)** A   **(f)** E

27. Estimate the slope of the tangent lines at $t = 0$ and $t = 15$ to get $\dfrac{dc}{dt}\bigg|_{t=0} \approx 0.08$ mol/L per second and $\dfrac{dc}{dt}\bigg|_{t=15} \approx 0.018$ mol/L per second.

29. (a) $F \approx 200$ lb, $dF/d\theta \approx 60$ lb/rad   **(b)** $\mu = (dF/d\theta)/F \approx 60/200 = 0.3$

31.

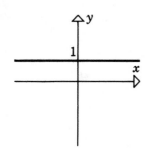

33.

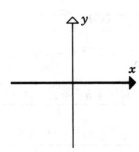

35.

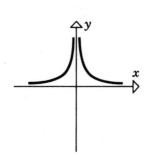

37.

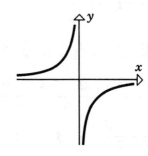

39. $\lim\limits_{x \to 0} f(x) = \lim\limits_{x \to 0} \sqrt[3]{x} = 0 = f(0),$

so f is continuous at $x = 0$.

$$\lim_{h \to 0} \frac{f(0+h) - f(0)}{h} = \lim_{h \to 0} \frac{\sqrt[3]{h} - 0}{h}$$

$$= \lim_{h \to 0} \frac{1}{h^{2/3}} = +\infty,$$

so $f'(0)$ does not exist.

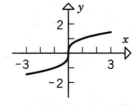

41. $\lim\limits_{x \to 1^-} f(x) = \lim\limits_{x \to 1^+} f(x) = f(1),$ so

f is continuous at $x = 1$.

$$\lim_{h \to 0^-} \frac{f(1+h) - f(1)}{h} = \lim_{h \to 0^-} \frac{[(1+h)^2 + 1] - 2}{h}$$

$$= \lim_{h \to 0^-} (2 + h) = 2;$$

$$\lim_{h \to 0^+} \frac{f(1+h) - f(1)}{h} = \lim_{h \to 0^+} \frac{2(1+h) - 2}{h}$$

$$= \lim_{h \to 0^+} 2 = 2,$$

so $f'(1) = 2$.

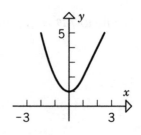

43. f is continuous at $x = 1$ because it is differentiable there, thus $\lim\limits_{h\to 0} f(1 + h) = f(1)$ and so

$f(1) = 0$ because $\lim\limits_{h\to 0} \dfrac{f(1+h)}{h}$ exists; $f'(1) = \lim\limits_{h\to 0} \dfrac{f(1+h) - f(1)}{h} = \lim\limits_{h\to 0} \dfrac{f(1+h)}{h} = 5.$

45. $f'(x) = \lim\limits_{h\to 0} \dfrac{f(x+h) - f(x)}{h}$

$\qquad = \lim\limits_{h\to 0} \dfrac{f(x)f(h) - f(x)}{h}$

$\qquad = \lim\limits_{h\to 0} \dfrac{f(x)[f(h) - 1]}{h} = f(x) \lim\limits_{h\to 0} \dfrac{f(h) - f(0)}{h} = f(x)f'(0) = f(x)$

EXERCISE SET 3.3

1. $28x^6$ **3.** $24x^7 + 2$ **5.** 0

7. $-\dfrac{1}{3}(7x^6 + 2)$ **9.** $3ax^2 + 2bx + c$ **11.** $24x^{-9} + 1/\sqrt{x}$

13. $\quad y = x^{-3} + x^{-7}$ so $\dfrac{dy}{dx} = -3x^{-4} - 7x^{-8}$

15. $\quad \dfrac{dy}{dx} = (3x^2 + 6)\dfrac{d}{dx}\left(2x - \dfrac{1}{4}\right) + \left(2x - \dfrac{1}{4}\right)\dfrac{d}{dx}(3x^2 + 6)$

$\qquad = (3x^2 + 6)(2) + \left(2x - \dfrac{1}{4}\right)(6x) = 18x^2 - \dfrac{3}{2}x + 12$

17. $\quad \dfrac{dy}{dx} = (x^3 + 7x^2 - 8)\dfrac{d}{dx}(2x^{-3} + x^{-4}) + (2x^{-3} + x^{-4})\dfrac{d}{dx}(x^3 + 7x^2 - 8)$

$\qquad = (x^3 + 7x^2 - 8)(-6x^{-4} - 4x^{-5}) + (2x^{-3} + x^{-4})(3x^2 + 14x)$

$\qquad = -15x^{-2} - 14x^{-3} + 48x^{-4} + 32x^{-5}$

19. $12x(3x^2 + 1)$ **21.** $\dfrac{dy}{dx} = -\dfrac{1}{(5x - 3)^2}\dfrac{d}{dx}(5x - 3) = -\dfrac{5}{(5x - 3)^2}$

23. $\quad \dfrac{dy}{dx} = \dfrac{(2x + 1)\dfrac{d}{dx}(3x) - (3x)\dfrac{d}{dx}(2x + 1)}{(2x + 1)^2} = \dfrac{(2x + 1)(3) - (3x)(2)}{(2x + 1)^2} = \dfrac{3}{(2x + 1)^2}$

25. $\quad \dfrac{dy}{dx} = \dfrac{(x + 3)\dfrac{d}{dx}(2x - 1) - (2x - 1)\dfrac{d}{dx}(x + 3)}{(x + 3)^2} = \dfrac{(x + 3)(2) - (2x - 1)(1)}{(x + 3)^2} = \dfrac{7}{(x + 3)^2}$

27. $\dfrac{dy}{dx} = \left(\dfrac{3x+2}{x}\right)\dfrac{d}{dx}\left(x^{-5}+1\right) + \left(x^{-5}+1\right)\dfrac{d}{dx}\left(\dfrac{3x+2}{x}\right)$

$\qquad = \left(\dfrac{3x+2}{x}\right)\left(-5x^{-6}\right) + \left(x^{-5}+1\right)\left[\dfrac{x(3)-(3x+2)(1)}{x^2}\right]$

$\qquad = \left(\dfrac{3x+2}{x}\right)\left(-5x^{-6}\right) + \left(x^{-5}+1\right)\left(-\dfrac{2}{x^2}\right)$

29. **(a)** $g'(x) = \sqrt{x}f'(x) + \dfrac{1}{2\sqrt{x}}f(x)$, $g'(4) = (2)(-5) + \dfrac{1}{4}(3) = -37/4$

$\qquad$ **(b)** $g'(x) = \dfrac{xf'(x)-f(x)}{x^2}$, $g'(4) = \dfrac{(4)(-5)-3}{16} = -23/16$

31. $32t$ $\qquad\qquad\qquad\qquad\qquad$ **33.** $3\pi r^2$

35. $\dfrac{ds}{dt} = \dfrac{(t^3+7)\dfrac{d}{dt}(t) - t\dfrac{d}{dt}(t^3+7)}{(t^3+7)^2} = \dfrac{(t^3+7)(1)-t(3t^2)}{(t^3+7)^2} = \dfrac{7-2t^3}{(t^3+7)^2}$

37. $F = GmMr^{-2}$, $\dfrac{dF}{dr} = -2GmMr^{-3} = -\dfrac{2GmM}{r^3}$

39. **(a)** $dy/dx = 21x^2 - 10x + 1$, $d^2y/dx^2 = 42x - 10$

$\qquad$ **(b)** $dy/dx = 24x - 2$, $d^2y/dx^2 = 24$

$\qquad$ **(c)** $dy/dx = -1/x^2$, $d^2y/dx^2 = 2/x^3$

$\qquad$ **(d)** $y = 35x^5 - 16x^3 - 3x$, $dy/dx = 175x^4 - 48x^2 - 3$, $d^2y/dx^2 = 700x^3 - 96x$

41. **(a)** $y' = -5x^{-6} + 5x^4$, $y'' = 30x^{-7} + 20x^3$, $y''' = -210x^{-8} + 60x^2$

$\qquad$ **(b)** $y = x^{-1}$, $y' = -x^{-2}$, $y'' = 2x^{-3}$, $y''' = -6x^{-4}$

$\qquad$ **(c)** $y' = 3ax^2 + b$, $y'' = 6ax$, $y''' = 6a$

43. **(a)** $f'(x) = 6x$, $f''(x) = 6$, $f'''(x) = 0$, $f'''(2) = 0$

$\qquad$ **(b)** $\dfrac{dy}{dx} = 30x^4 - 8x$, $\dfrac{d^2y}{dx^2} = 120x^3 - 8$, $\dfrac{d^2y}{dx^2}\Big|_{x=1} = 112$

$\qquad$ **(c)** $\dfrac{d}{dx}\left[x^{-3}\right] = -3x^{-4}$, $\dfrac{d^2}{dx^2}\left[x^{-3}\right] = 12x^{-5}$, $\dfrac{d^3}{dx^3}\left[x^{-3}\right] = -60x^{-6}$, $\dfrac{d^4}{dx^4}\left[x^{-3}\right] = 360x^{-7}$,

$\qquad\quad \dfrac{d^4}{dx^4}\left[x^{-3}\right]\Big|_{x=1} = 360$

45. $y' = 3x^2 + 3$, $y'' = 6x$, and $y''' = 6$ so

$\qquad y''' + xy'' - 2y' = 6 + x(6x) - 2(3x^2+3) = 6 + 6x^2 - 6x^2 - 6 = 0$.

47. $F'(x) = xf'(x) + f(x)$, $F''(x) = xf''(x) + f'(x) + f'(x) = xf''(x) + 2f'(x)$

49. The graph has a horizontal tangent at points where $\dfrac{dy}{dx} = 0$, but

$\dfrac{dy}{dx} = x^2 - 3x + 2 = (x-1)(x-2) = 0$ if $x = 1, 2$. The corresponding values of y are 5/6 and 2/3 so the tangent line is horizontal at $(1, 5/6)$ and $(2, 2/3)$.

51. $y - 2 = 5(x + 3)$, $y = 5x + 17$.

53. $m_{\tan} = \dfrac{dy}{dx} = 2ax + b$ so $2a + b = 8$ when $x = 1$. Also $5 = a + b$ because $(1, 5)$ is on the curve. Solve the pair of equations $2a + b = 8$ and $a + b = 5$ to get $a = 3$ and $b = 2$.

55. If the y-intercept is -2, then the point $(0, -2)$ is on the graph so $-2 = a(0)^2 + b(0) + c$ so $c = -2$. If the x-intercept is 1, then the point $(1, 0)$ is on the graph so $0 = a + b - 2$. The slope is $dy/dx = 2ax + b$; at $x = 0$ the slope is b so $b = -1$, thus $a = 3$. The function is $y = 3x^2 - x - 2$.

57. The points $(-1, 1)$ and $(2, 4)$ are on the secant line so its slope is $(4 - 1)/(2 + 1) = 1$. The slope of the tangent line to $y = x^2$ is $y' = 2x$ so $2x = 1$, $x = 1/2$.

59. $y' = -2x$, so at any point (x_0, y_0) on $y = 1 - x^2$ the tangent line is $y - y_0 = -2x_0(x - x_0)$, or $y = -2x_0 x + x_0^2 + 1$. The point $(2, 0)$ is to be on the line, so $0 = -4x_0 + x_0^2 + 1$, $x_0^2 - 4x_0 + 1 = 0$. Use the quadratic formula to get $x_0 = \dfrac{4 \pm \sqrt{16 - 4}}{2} = 2 \pm \sqrt{3}$.

61. $y' = 3ax^2 + b$; the tangent line at $x = x_0$ is $y - y_0 = (3ax_0^2 + b)(x - x_0)$ where $y_0 = ax_0^3 + bx_0$. Solve with $y = ax^3 + bx$ to get

$$(ax^3 + bx) - (ax_0^3 + bx_0) = (3ax_0^2 + b)(x - x_0)$$
$$ax^3 + bx - ax_0^3 - bx_0 = 3ax_0^2 x - 3ax_0^3 + bx - bx_0$$
$$x^3 - 3x_0^2 x + 2x_0^3 = 0$$
$$(x - x_0)(x^2 + 2x_0 x - 2x_0^2) = 0$$
$$(x - x_0)^2(x + 2x_0) = 0, \text{ so } x = -2x_0.$$

63. $y' = -\dfrac{1}{x^2}$; the tangent line at $x = x_0$ is $y - y_0 = -\dfrac{1}{x_0^2}(x - x_0)$, or $y = -\dfrac{x}{x_0^2} + \dfrac{2}{x_0}$. The tangent line crosses the x-axis at $2x_0$, the y-axis at $2/x_0$, so that the area of the triangle is $\dfrac{1}{2}(2/x_0)(2x_0) = 2$.

65. If the graphs of $f(x)$ and $g(x)$ have parallel tangent lines at $x = c$, then $f'(c) = g'(c)$. If $y = f(x) - g(x)$, then $dy/dx = f'(x) - g'(x)$ so at $x = c$, $dy/dx = f'(c) - g'(c) = 0$, hence $y = f(x) - g(x)$ has a horizontal tangent line at $x = c$.

67. **(a)** $2(1 + x^{-1})(x^{-3} + 7) + (2x + 1)(-x^{-2})(x^{-3} + 7) + (2x + 1)(1 + x^{-1})(-3x^{-4})$

(b) $-5x^{-6}(x^2 + 2x)(4 - 3x)(2x^9 + 1) + x^{-5}(2x + 2)(4 - 3x)(2x^9 + 1)$
$$+ x^{-5}(x^2 + 2x)(-3)(2x^9 + 1) + x^{-5}(x^2 + 2x)(4 - 3x)(18x^8)$$

(c) $(x^7 + 2x - 3)^3 = (x^7 + 2x - 3)(x^7 + 2x - 3)(x^7 + 2x - 3)$ so

$$\frac{d}{dx}(x^7 + 2x - 3)^3 = (7x^6 + 2)(x^7 + 2x - 3)(x^7 + 2x - 3)$$
$$+ (x^7 + 2x - 3)(7x^6 + 2)(x^7 + 2x - 3)$$
$$+ (x^7 + 2x - 3)(x^7 + 2x - 3)(7x^6 + 2)$$
$$= 3(7x^6 + 2)(x^7 + 2x - 3)^2$$

(d) $(x^2 + 1)^{50} = (x^2 + 1)(x^2 + 1)\cdots(x^2 + 1)$, where $(x^2 + 1)$ occurs 50 times so

$$\frac{d}{dx}(x^2 + 1)^{50} = [(2x)(x^2 + 1)\cdots(x^2 + 1)] + [(x^2 + 1)(2x)\cdots(x^2 + 1)]$$
$$+ \cdots + [(x^2 + 1)(x^2 + 1)\cdots(2x)]$$
$$= 2x(x^2 + 1)^{49} + 2x(x^2 + 1)^{49} + \cdots + 2x(x^2 + 1)^{49}$$
$$= 100x(x^2 + 1)^{49} \text{ because } 2x(x^2 + 1)^{49} \text{ occurs 50 times.}$$

69. $2(2x^3 - 5x^2 + 7x - 2)(6x^2 - 10x + 7)$

71. f is continuous at 1 because $\lim\limits_{x \to 1^-} f(x) \lim\limits_{x \to 1^+} = f(x) = f(1)$, also $\lim\limits_{x \to 1^-} f'(x) = \lim\limits_{x \to 1^-} 2x = 2$
and $\lim\limits_{x \to 1^+} f'(x) = \lim\limits_{x \to 1^+} \dfrac{1}{2\sqrt{x}} = \dfrac{1}{2}$ so f is not differentiable at 1.

73. If f is differentiable at $x = 1$, then f is continuous there; $\lim\limits_{x \to 1^+} f(x) = \lim\limits_{x \to 1^-} f(x) = f(1) = 3$,
$a + b = 3; \lim_{x \to 1^+} f'(x) = a$ and $\lim\limits_{x \to 1^-} f'(x) = 6$ so $a = 6$ and $b = 3 - 6 = -3$.

75. **(a)** $f(x) = 3x - 2$ if $x \geq 2/3$, $f(x) = -3x + 2$ if $x < 2/3$ so f is differentiable everywhere except perhaps at $2/3$. f is continuous at $2/3$, also $\lim\limits_{x \to 2/3^-} f'(x) = \lim\limits_{x \to 2/3^-} (-3) = -3$ and $\lim\limits_{x \to 2/3^+} f'(x) = \lim\limits_{x \to 2/3^+} (3) = 3$ so f is not differentiable at $x = 2/3$.

(b) $f(x) = x^2 - 4$ if $|x| \geq 2$, $f(x) = -x^2 + 4$ if $|x| < 2$ so f is differentiable everywhere except perhaps at ± 2. f is continuous at -2 and 2, also $\lim\limits_{x \to 2^-} f'(x) = \lim\limits_{x \to 2^-} (-2x) = -4$ and $\lim\limits_{x \to 2^+} f'(x) = \lim\limits_{x \to 2^+} (2x) = 4$ so f is not differentiable at $x = 2$. Similarly, f is not differentiable at $x = -2$.

77. **(a)** $f'(x) = nx^{n-1}$, $f''(x) = n(n-1)x^{n-2}$, $f'''(x) = n(n-1)(n-2)x^{n-3}, \ldots,$
$f^{(n)}(x) = n(n-1)(n-2)\cdots 1$.

(b) From part (a), $f^{(k)}(x) = k(k-1)(k-2)\cdots 1$ so $f^{(k+1)}(x) = 0$ thus $f^{(n)}(x) = 0$ if $n > k$.

(c) From parts (a) and (b), $f^{(n)}(x) = a_n n(n-1)(n-2)\cdots 1$.

79. **(a)** $\dfrac{d^2}{dx^2}[cf(x)] = \dfrac{d}{dx}\left[\dfrac{d}{dx}[cf(x)]\right] = \dfrac{d}{dx}\left[c\dfrac{d}{dx}[f(x)]\right]$

$\qquad\qquad = c\dfrac{d}{dx}\left[\dfrac{d}{dx}[f(x)]\right] = c\dfrac{d^2}{dx^2}[f(x)]$

$\qquad \dfrac{d^2}{dx^2}[f(x) + g(x)] = \dfrac{d}{dx}\left[\dfrac{d}{dx}[f(x) + g(x)]\right]$

$\qquad\qquad = \dfrac{d}{dx}\left[\dfrac{d}{dx}[f(x)] + \dfrac{d}{dx}[g(x)]\right]$

$\qquad\qquad = \dfrac{d^2}{dx^2}[f(x)] + \dfrac{d^2}{dx^2}[g(x)]$

    (b) yes, by repeated application of the procedure illustrated in part (a).

81. **(a)** The argument is based on the assumption that $\lim\limits_{h \to 0} f(x + h) = f(\lim\limits_{h \to 0}(x + h))$ which is not necessarily true.

    (b) If $h \neq 0$, then $1 + h \neq 1$ thus $f(1 + h) = 1 + h$ so $\lim\limits_{h \to 0} f(1 + h) = \lim\limits_{h \to 0}(1 + h) = 1$. But $f(1) = 3$ so $\lim\limits_{h \to 0} f(1 + h) \neq f(1)$.

83. **(a)** If a function is differentiable at a point then it is continuous at that point, thus f' is continuous on (a, b) and consequently so is f.

    (b) f and all its derivatives up to $f^{(n-1)}(x)$ are continuous on (a, b).

EXERCISE SET 3.4

1. $f'(x) = -2\sin x - 3\cos x$

3. $f'(x) = \dfrac{x(\cos x) - \sin x(1)}{x^2} = \dfrac{x\cos x - \sin x}{x^2}$

5. $f'(x) = x^3(\cos x) + (\sin x)(3x^2) - 5(-\sin x) = x^3\cos x + (3x^2 + 5)\sin x$

7. $f'(x) = \sec x \tan x - \sqrt{2}\sec^2 x$

9. $f'(x) = \sec x(\sec^2 x) + (\tan x)(\sec x \tan x) = \sec^3 x + \sec x \tan^2 x$

11. $f'(x) = 1 + 4\csc x \cot x - 2\csc^2 x$

13. $f'(x) = \dfrac{(1 + \csc x)(-\csc^2 x) - \cot x(0 - \csc x \cot x)}{(1 + \csc x)^2}$

$\qquad = \dfrac{\csc x(-\csc x - \csc^2 x + \cot^2 x)}{(1 + \csc x)^2}$

but $1 + \cot^2 x = \csc^2 x$ (identity) thus $\cot^2 x - \csc^2 x = -1$

so $f'(x) = \dfrac{\csc x(-\csc x - 1)}{(1 + \csc x)^2} = -\dfrac{\csc x}{1 + \csc x}$

15. $f(x) = \sin^2 x + \cos^2 x = 1$ (identity) so $f'(x) = 0$

17. $f(x) = \dfrac{\tan x}{1 + x \tan x}$ (because $\sin x \sec x = (\sin x)(1/\cos x) = \tan x$),

$\qquad f'(x) = \dfrac{(1 + x \tan x)(\sec^2 x) - \tan x[x(\sec^2 x) + (\tan x)(1)]}{(1 + x \tan x)^2}$

$\qquad = \dfrac{\sec^2 x - \tan^2 x}{(1 + x \tan x)^2} = \dfrac{1}{(1 + x \tan x)^2}$ (because $\sec^2 x - \tan^2 x = 1$)

19. $dy/dx = -x \sin x + \cos x$,

$\qquad d^2y/dx^2 = -x \cos x - \sin x - \sin x = -x \cos x - 2 \sin x$

21. $dy/dx = x(\cos x) + (\sin x)(1) - 3(-\sin x) = x \cos x + 4 \sin x$,

$\qquad d^2y/dx^2 = x(-\sin x) + (\cos x)(1) + 4 \cos x = -x \sin x + 5 \cos x$

23. $dy/dx = (\sin x)(-\sin x) + (\cos x)(\cos x) = \cos^2 x - \sin^2 x$,

$\qquad d^2y/dx^2 = (\cos x)(-\sin x) + (\cos x)(-\sin x) - [(\sin x)(\cos x) + (\sin x)(\cos x)]$

$\qquad\quad = -4 \sin x \cos x$

25. (a) $f'(x) = -\sin x$; $f'(x) = 0$ when $\sin x = 0$ so $x = n\pi$, $n = 0, \pm 1, \pm 2, \cdots$.

$\quad$ (b) $f'(x) = -\csc^2 x$; $f'(x) = 0$ when $\csc x = 0$, but $\csc x = 0$ has no solutions.

$\quad$ (c) $f'(x) = -\csc x \cot x$; $f'(x) = 0$ when $\cot x = 0$ so $x = \dfrac{\pi}{2} + n\pi$, $n = 0, \pm 1, \pm 2, \cdots$.

27. Let $f(x) = \tan x$, then $f'(x) = \sec^2 x$.

$\quad$ (a) $f(0) = 0$ and $f'(0) = 1$ so $y - 0 = (1)(x - 0)$, $y = x$.

$\quad$ (b) $f\left(\dfrac{\pi}{4}\right) = 1$ and $f'\left(\dfrac{\pi}{4}\right) = 2$ so $y - 1 = 2\left(x - \dfrac{\pi}{4}\right)$, $y = 2x - \dfrac{\pi}{2} + 1$.

$\quad$ (c) $f\left(-\dfrac{\pi}{4}\right) = -1$ and $f'\left(-\dfrac{\pi}{4}\right) = 2$ so $y + 1 = 2\left(x + \dfrac{\pi}{4}\right)$, $y = 2x + \dfrac{\pi}{2} - 1$.

29. $x = 10 \sin \theta$, $dx/d\theta = 10 \cos \theta$; if $\theta = 60°$, then

$\qquad dx/d\theta = 10(1/2) = 5$ ft/rad $= \pi/36$ ft/° ≈ 0.087 ft/°.

31. $D = 50 \tan\theta$, $dD/d\theta = 50 \sec^2\theta$; if $\theta = 30°$, then

$dD/d\theta = 50(\sqrt{2})^2 = 100$ m/rad $= 5\pi/9$ m/° ≈ 1.75 m/°.

33. In each part, f is differentiable throughout its domain because f' can be determined there by use of the various derivative formulas that have been presented in the text; f is not differentiable elsewhere.

(a) all x (b) all x

(c) $x \neq \pi/2 + n\pi$, $n = 0, \pm 1, \pm 2, \cdots$ (d) $x \neq n\pi$, $n = 0, \pm 1, \pm 2, \cdots$

(e) $x \neq \pi/2 + n\pi$, $n = 0, \pm 1, \pm 2, \cdots$ (f) $x \neq n\pi$, $n = 0, \pm 1, \pm 2, \cdots$

(g) $x \neq \pi + 2n\pi$, $n = 0, \pm 1, \pm 2, \cdots$ (h) $x \neq n\pi/2$, $n = 0, \pm 1, \pm 2, \cdots$

(i) all x

35. $f'(x) = -\sin x$, $f''(x) = -\cos x$, $f'''(x) = \sin x$, and $f^{(4)}(x) = \cos x$ with higher order derivatives repeating this pattern, so $f^{(n)}(x) = \sin x$ for $n = 3, 7, 11, \cdots$

37. $\displaystyle\lim_{x \to 0} \frac{\tan(x+y) - \tan y}{x} = \lim_{h \to 0} \frac{\tan(y+h) - \tan y}{h} = \frac{d}{dy}(\tan y) = \sec^2 y$

39. Let t be the radian measure, then $h = \dfrac{180}{\pi} t$ and $\cos h = \cos t$, $\sin h = \sin t$.

(a) $\displaystyle\lim_{h \to 0} \frac{\cos h - 1}{h} = \lim_{t \to 0} \frac{\cos t - 1}{180t/\pi} = \frac{\pi}{180} \lim_{t \to 0} \frac{\cos t - 1}{t} = 0.$

(b) $\displaystyle\lim_{h \to 0} \frac{\sin h}{h} = \lim_{t \to 0} \frac{\sin t}{180t/\pi} = \frac{\pi}{180} \lim_{t \to 0} \frac{\sin t}{t} = \frac{\pi}{180}.$

(c) $\displaystyle\frac{d}{dx}[\sin x] = \sin x \lim_{h \to 0} \frac{\cos h - 1}{h} + \cos x \lim_{h \to 0} \frac{\sin h}{h}$

$\displaystyle = \sin x(0) + \cos x(\pi/180) = \frac{\pi}{180} \cos x.$

EXERCISE SET 3.5

1. $f'(x) = 37(x^3 + 2x)^{36} \dfrac{d}{dx}(x^3 + 2x) = 37(x^3 + 2x)^{36}(3x^2 + 2)$

3. $f'(x) = -2\left(x^3 - \dfrac{7}{x}\right)^{-3} \dfrac{d}{dx}\left(x^3 - \dfrac{7}{x}\right) = -2\left(x^3 - \dfrac{7}{x}\right)^{-3}\left(3x^2 + \dfrac{7}{x^2}\right)$

5. $f(x) = 4(3x^2 - 2x + 1)^{-3}$,

$$f'(x) = -12(3x^2 - 2x + 1)^{-4}\frac{d}{dx}(3x^2 - 2x + 1)$$

$$= -12(3x^2 - 2x + 1)^{-4}(6x - 2) = \frac{24(1 - 3x)}{(3x^2 - 2x + 1)^4}$$

7. $f'(x) = \dfrac{1}{2\sqrt{4 + 3\sqrt{x}}}\dfrac{d}{dx}(4 + 3\sqrt{x}) = \dfrac{3}{4\sqrt{x}\sqrt{4 + 3\sqrt{x}}}$

9. $f'(x) = \cos(x^3)\dfrac{d}{dx}(x^3) = 3x^2\cos(x^3)$

11. $f'(x) = \sec^2(4x^2)\dfrac{d}{dx}(4x^2) = 8x\sec^2(4x^2)$

13. $f'(x) = 20\cos^4 x\dfrac{d}{dx}(\cos x) = 20\cos^4 x(-\sin x) = -20\cos^4 x\sin x$

15. $f'(x) = \cos(1/x^2)\dfrac{d}{dx}(1/x^2) = -\dfrac{2}{x^3}\cos(1/x^2)$

17. $f'(x) = 4\sec(x^7)\dfrac{d}{dx}[\sec(x^7)] = 4\sec(x^7)\sec(x^7)\tan(x^7)\dfrac{d}{dx}(x^7) = 28x^6\sec^2(x^7)\tan(x^7)$

19. $f'(x) = \dfrac{1}{2\sqrt{\cos(5x)}}\dfrac{d}{dx}[\cos(5x)] = -\dfrac{5\sin(5x)}{2\sqrt{\cos(5x)}}$

21. $f'(x) = -3\left[x + \csc(x^3 + 3)\right]^{-4}\dfrac{d}{dx}\left[x + \csc(x^3 + 3)\right]$

$$= -3\left[x + \csc(x^3 + 3)\right]^{-4}\left[1 - \csc(x^3 + 3)\cot(x^3 + 3)\dfrac{d}{dx}(x^3 + 3)\right]$$

$$= -3\left[x + \csc(x^3 + 3)\right]^{-4}\left[1 - 3x^2\csc(x^3 + 3)\cot(x^3 + 3)\right]$$

23. $f'(x) = x^2\cdot\dfrac{-2x}{2\sqrt{5 - x^2}} + 2x\sqrt{5 - x^2} = \dfrac{x(10 - 3x^2)}{\sqrt{5 - x^2}}$

25. $f'(x) = x^3(2\sin 5x)\dfrac{d}{dx}(\sin 5x) + 3x^2\sin^2 5x = 10x^3\sin 5x\cos 5x + 3x^2\sin^2 5x$

27. $f'(x) = x^5 \sec\left(\dfrac{1}{x}\right) \tan\left(\dfrac{1}{x}\right) \dfrac{d}{dx}\left(\dfrac{1}{x}\right) + \sec\left(\dfrac{1}{x}\right)(5x^4)$

$\quad = x^5 \sec\left(\dfrac{1}{x}\right) \tan\left(\dfrac{1}{x}\right)\left(-\dfrac{1}{x^2}\right) + 5x^4 \sec\left(\dfrac{1}{x}\right)$

$\quad = -x^3 \sec\left(\dfrac{1}{x}\right) \tan\left(\dfrac{1}{x}\right) + 5x^4 \sec\left(\dfrac{1}{x}\right)$

29. $f'(x) = -\sin(\cos x)\dfrac{d}{dx}(\cos x) = -\sin(\cos x)(-\sin x) = \sin(\cos x)\sin x$

31. $f'(x) = 3\cos^2(\sin 2x)\dfrac{d}{dx}[\cos(\sin 2x)]$

$\quad = 3\cos^2(\sin 2x)[-\sin(\sin 2x)]\dfrac{d}{dx}(\sin 2x)$

$\quad = -6\cos^2(\sin 2x)\sin(\sin 2x)\cos 2x$

33. $f'(x) = (5x+8)^{13}12(x^3+7x)^{11}\dfrac{d}{dx}(x^3+7x) + (x^3+7x)^{12}13(5x+8)^{12}\dfrac{d}{dx}(5x+8)$

$\quad = 12(5x+8)^{13}(x^3+7x)^{11}(3x^2+7) + 65(x^3+7x)^{12}(5x+8)^{12}$

35. $f'(x) = 3\left[\dfrac{x-5}{2x+1}\right]^2 \dfrac{d}{dx}\left[\dfrac{x-5}{2x+1}\right] = 3\left[\dfrac{x-5}{2x+1}\right]^2 \cdot \dfrac{11}{(2x+1)^2} = \dfrac{33(x-5)^2}{(2x+1)^4}$

37. $f'(x) = \dfrac{(4x^2-1)^8(3)(2x+3)^2)(2) - (2x+3)^3(8)(4x^2-1)^7(8x)}{(4x^2-1)^{16}}$

$\quad = \dfrac{2(2x+3)^2(4x^2-1)^7[3(4x^2-1) - 32x(2x+3)]}{(4x^2-1)^{16}}$

$\quad = -\dfrac{2(2x+3)^2(52x^2+96x+3)}{(4x^2-1)^9}$

39. $f'(x) = 5\left[x\sin 2x \tan^4(x^7)\right]^4 \dfrac{d}{dx}\left[x\sin 2x \tan^4(x^7)\right]$

$\quad = 5\left[x\sin 2x \tan^4(x^7)\right]^4 \left[x\cos 2x\dfrac{d}{dx}(2x) + \sin 2x + 4\tan^3(x^7)\dfrac{d}{dx}\tan(x^7)\right]$

$\quad = 5\left[x\sin 2x \tan^4(x^7)\right]^4 \left[2x\cos 2x + \sin 2x + 28x^6 \tan^3(x^7)\sec^2(x^7)\right]$

41. $\dfrac{dy}{dx} = x(-\sin(5x))\dfrac{d}{dx}(5x) + \cos(5x) - 2\sin x\dfrac{d}{dx}(\sin x)$

$\quad = -5x\sin(5x) + \cos(5x) - 2\sin x\cos x = -5x\sin(5x) + \cos(5x) - \sin(2x),$

$$\frac{d^2y}{dx^2} = -5x\cos(5x)\frac{d}{dx}(5x) - 5\sin(5x) - \sin(5x)\frac{d}{dx}(5x) - \cos(2x)\frac{d}{dx}(2x)$$

$$= -25x\cos(5x) - 10\sin(5x) - 2\cos(2x)$$

43. $\dfrac{dy}{dx} = -3x\sin 3x + \cos 3x$; if $x = \pi$ then $y = -\pi$ and $\dfrac{dy}{dx} = -1$ so $y + \pi = -(x - \pi)$, $y = -x$.

45. $\dfrac{dy}{dx} = -3\sec^3(\pi/2 - x)\tan(\pi/2 - x)$; if $x = -\pi/2$ then $y = -1$ and $\dfrac{dy}{dx} = 0$

so $y + 1 = (0)(x + \pi/2)$, $y = -1$.

47. $y = \cot^3(\pi - \theta) = -\cot^3\theta$ so $dy/dx = 3\cot^2\theta\csc^2\theta$.

49. $\dfrac{d}{d\omega}[a\cos^2\pi\omega + b\sin^2\pi\omega] = -2\pi a\cos\pi\omega\sin\pi\omega + 2\pi b\sin\pi\omega\cos\pi\omega$

$$= \pi(b - a)(2\sin\pi\omega\cos\pi\omega) = \pi(b - a)\sin 2\pi\omega$$

51. (a) $dy/dt = -A\omega\sin\omega t$, $d^2y/dt^2 = -A\omega^2\cos\omega t = -\omega^2 y$

(b) One complete oscillation occurs when ωt increases over an interval of length 2π, or if t increases over an interval of length $2\pi/\omega$.

(c) $f = 1/T$

(d) amplitude $= 0.6$ cm, $T = 2\pi/15$ sec/oscillation, $f = 15/(2\pi)$ oscillations/sec.

53. (a) $p \approx 10$ lb/in^2, $dp/dh \approx -2$ lb/in^2 per mi

(b) $\dfrac{dp}{dt} = \dfrac{dp}{dh}\dfrac{dh}{dt} \approx (-2)(0.3) = -0.6$ lb/in^2 per sec

55. $3x^2y^2\dfrac{dy}{dx} + 2xy^3$

57. $\left(x\dfrac{dy}{dx} + y\right)\cos(xy)$

59. $2x\dfrac{dx}{dt} + 2y\dfrac{dy}{dt}$

61. $\dfrac{x^2}{2\sqrt{y}}\dfrac{dy}{dt} + 2x\sqrt{y}\dfrac{dx}{dt}$

63. With $u = \sin x$, $\dfrac{d}{dx}(|\sin x|) = \dfrac{d}{dx}(|u|) = \dfrac{d}{du}(|u|)\dfrac{du}{dx} = \dfrac{d}{du}(|u|)\cos x = \begin{cases} \cos x, & u > 0 \\ -\cos x, & u < 0 \end{cases}$

$$= \begin{cases} \cos x, & \sin x > 0 \\ -\cos x, & \sin x < 0 \end{cases} = \begin{cases} \cos x, & 0 < x < \pi \\ -\cos x, & -\pi < x < 0 \end{cases}.$$

65. (a) For $x \neq 0$, $f'(x) = x\left(\cos\dfrac{1}{x}\right)\left(-\dfrac{1}{x^2}\right) + \sin\dfrac{1}{x} = -\dfrac{1}{x}\cos\dfrac{1}{x} + \sin\dfrac{1}{x}$

(b) $\lim\limits_{x\to 0} x\sin\dfrac{1}{x} = 0 = f(0)$

(c) $\displaystyle\lim_{h\to 0}\frac{f(0+h)-f(0)}{h}=\lim_{h\to 0}\frac{h\sin\dfrac{1}{h}}{h}=\lim_{h\to 0}\sin\frac{1}{h}$, which does not exist

67. **(a)** $g'(x)=3[f(x)]^2 f'(x)$,

$\qquad g'(2)=3[f(2)]^2 f'(2)=3(1)^2(7)=21$

    (b) $h'(x)=f'(x^3)(3x^2)$,

$\qquad h'(2)=f'(8)(12)=(-3)(12)=-36$

69. $(f\circ g)'(x)=f'(g(x))g'(x)$ so $(f\circ g)'(0)=f'(g(0))g'(0)=f'(0)(3)=(2)(3)=6$.

71. $F'(x)=f'(g(x))g'(x)=f'(\sqrt{3x-1})\dfrac{3}{2\sqrt{3x-1}}=\dfrac{\sqrt{3x-1}}{(3x-1)+1}\dfrac{3}{2\sqrt{3x-1}}=\dfrac{1}{2x}$

73. $\dfrac{d}{dx}[f(3x)]=f'(3x)\dfrac{d}{dx}(3x)=3f'(3x)=6x$, so $f'(3x)=2x$. Let $u=3x$ to get $f'(u)=\dfrac{2}{3}u$;

$\qquad \dfrac{d}{dx}[f(x)]=f'(x)=\dfrac{2}{3}x$.

75. $\dfrac{d}{dx}[f(g(h(x)))]=\dfrac{d}{dx}[f(g(u))],\quad u=h(x)$

$\qquad\qquad =\dfrac{d}{du}[f(g(u))]\dfrac{du}{dx}=f'(g(u))g'(u)\dfrac{du}{dx}=f'(g(h(x)))g'(h(x))h'(x)$

EXERCISE SET 3.6

1. $y=(2x-5)^{1/3};\ dy/dx=\dfrac{2}{3}(2x-5)^{-2/3}$

3. $dy/dx=\dfrac{3}{2}\left[\dfrac{x-1}{x+2}\right]^{1/2}\dfrac{d}{dx}\left[\dfrac{x-1}{x+2}\right]=\dfrac{9}{2(x+2)^2}\left[\dfrac{x-1}{x+2}\right]^{1/2}$

5. $dy/dx=x^3\left(-\dfrac{2}{3}\right)(5x^2+1)^{-5/3}(10x)+3x^2(5x^2+1)^{-2/3}=\dfrac{1}{3}x^2(5x^2+1)^{-5/3}(25x^2+9)$

7. $dy/dx=\dfrac{5}{2}[\sin(3/x)]^{3/2}[\cos(3/x)](-3/x^2)=-\dfrac{15[\sin(3/x)]^{3/2}\cos(3/x)}{2x^2}$

9. $dy/dx=\sec^2\left[(2x-1)^{-1/3}\right]\left(-\dfrac{1}{3}\right)(2x-1)^{-4/3}(2)=-\dfrac{2}{3}(2x-1)^{-4/3}\sec^2\left[(2x-1)^{-1/3}\right]$

11. $y' = rx^{r-1}, y'' = r(r-1)x^{r-2}$ so $3x^2\left[r(r-1)x^{r-2}\right] + 4x\left(rx^{r-1}\right) - 2x^r = 0$,

$3r(r-1)x^r + 4rx^r - 2x^r = 0, (3r^2 + r - 2)x^r = 0$,

$3r^2 + r - 2 = 0, (3r-2)(r+1) = 0; r = -1, 2/3$.

13. $2x + 2y\dfrac{dy}{dx} = 0$ so $\dfrac{dy}{dx} = -\dfrac{x}{y}$

15. $x^2\dfrac{dy}{dx} + 2xy + 3x(3y^2)\dfrac{dy}{dx} + 3y^3 - 1 = 0$

$(x^2 + 9xy^2)\dfrac{dy}{dx} = 1 - 2xy - 3y^3$ so $\dfrac{dy}{dx} = \dfrac{1 - 2xy - 3y^3}{x^2 + 9xy^2}$

17. $-\dfrac{1}{y^2}\dfrac{dy}{dx} - \dfrac{1}{x^2} = 0$ so $\dfrac{dy}{dx} = -\dfrac{y^2}{x^2}$ **19.** $\dfrac{1}{2\sqrt{x}} + \dfrac{1}{2\sqrt{y}}\dfrac{dy}{dx} = 0$ so $\dfrac{dy}{dx} = -\dfrac{\sqrt{y}}{\sqrt{x}}$

21. $35\left(x^2 + 3y^2\right)^{34}\left(2x + 6y\dfrac{dy}{dx}\right) = 1$,

$70x(x^2 + 3y^2)^{34} + 210y(x^2 + 3y^2)^{34}\dfrac{dy}{dx} = 1$ so $\dfrac{dy}{dx} = \dfrac{1 - 70x(x^2 + 3y^2)^{34}}{210y(x^2 + 3y^2)^{34}}$

23. $3x\dfrac{dy}{dx} + 3y = \dfrac{3}{2}\left(x^3 + y^2\right)^{1/2}\left(3x^2 + 2y\dfrac{dy}{dx}\right)$,

$\left[3x - 3y(x^3 + y^2)^{1/2}\right]\dfrac{dy}{dx} = \dfrac{9}{2}x^2(x^3 + y^2)^{1/2} - 3y$ so $\dfrac{dy}{dx} = \dfrac{(3/2)x^2(x^3 + y^2)^{1/2} - y}{x - y(x^3 + y^2)^{1/2}}$

25. $\cos(x^2y^2)\left[x^2(2y)\dfrac{dy}{dx} + 2xy^2\right] = 1$, $\dfrac{dy}{dx} = \dfrac{1 - 2xy^2\cos(x^2y^2)}{2x^2y\cos(x^2y^2)}$

27. $3\tan^2(xy^2 + y)\sec^2(xy^2 + y)\left(2xy\dfrac{dy}{dx} + y^2 + \dfrac{dy}{dx}\right) = 1$

so $\dfrac{dy}{dx} = \dfrac{1 - 3y^2\tan^2(xy^2 + y)\sec^2(xy^2 + y)}{3(2xy + 1)\tan^2(xy^2 + y)\sec^2(xy^2 + y)}$

29. $\dfrac{1}{2}\left[1 + \sin^3(xy^2)\right]^{-1/2}\left[3\sin^2(xy^2)\right]\left[\cos(xy^2)\right]\left(2xy\dfrac{dy}{dx} + y^2\right) = \dfrac{dy}{dx}$,

multiply through by $2\sqrt{1 + \sin^3(xy^2)}$ and solve for $\dfrac{dy}{dx}$

to get $\dfrac{dy}{dx} = \dfrac{3y^2\sin^2(xy^2)\cos(xy^2)}{2\sqrt{1 + \sin^3(xy^2)} - 6xy\sin^2(xy^2)\cos(xy^2)}$

31. $\dfrac{dy}{dx} = -\dfrac{3x^2y + y^3}{x^3 + 3y^2x}$; $\dfrac{dy}{dx}\Big|_{(1,2)} = -\dfrac{14}{13}$ **33.** $\dfrac{dy}{dx} = \dfrac{2y^{1/3}}{2x^{1/3} + 3x^{1/3}y^{1/3}}$; $\dfrac{dy}{dx}\Big|_{(1,-1)} = 2$

35. If $xy = 8$, then $y = 8/x$ so $dy/dx = -8/x^2$. By implicit differentiation we get $dy/dx = -y/x$. In both cases, $dy/dx\big|_{(2,4)} = -2$.

37. If $x^2 + y^2 = 1$, then $y = -\sqrt{1 - x^2}$ goes through the point $(1/\sqrt{2},\, -1/\sqrt{2})$ so $dy/dx = x/\sqrt{1 - x^2}$. By implicit differentiation we get $dy/dx = -x/y$. In both cases, $dy/dx\big|_{(1/\sqrt{2},-1/\sqrt{2})} = 1$.

39. Apply the quadratic formula to $y^2 - 3xy + 2x^2 - 4 = 0$ to get
$y = (3x \pm \sqrt{9x^2 - 4(2x^2 - 4)})/2 = (3x \pm \sqrt{x^2 + 16})/2$, of which only $y = (3x - \sqrt{x^2 + 16})/2$ contains the point $(3,2)$ so $dy/dx = [3 - x(x^2 + 16)^{-1/2}]/2$. By implicit differentiation we get $dy/dx = (3y - 4x)/(2y - 3x)$. In both cases, $dy/dx\big|_{(3,2)} = 6/5$.

41. $\dfrac{dy}{dx} = -\dfrac{x^2}{y^2}$, $\dfrac{d^2y}{dx^2} = -\dfrac{y^2(2x) - x^2(2y\,dy/dx)}{y^4} = -\dfrac{2xy^2 - 2x^2y(-x^2/y^2)}{y^4} = -\dfrac{2x(y^3 + x^3)}{y^5}$,

but $x^3 + y^3 = 1$ so $\dfrac{d^2y}{dx^2} = -\dfrac{2x}{y^5}$.

43. $\dfrac{dy}{dx} = \dfrac{y}{y - x}$,

$\dfrac{d^2y}{dx^2} = \dfrac{(y - x)(dy/dx) - y(dy/dx - 1)}{(y - x)^2} = \dfrac{(y - x)\left(\dfrac{y}{y - x}\right) - y\left(\dfrac{y}{y - x} - 1\right)}{(y - x)^2}$

$= \dfrac{y^2 - 2xy}{(y - x)^3}$ but $y^2 - 2xy = -3$, so $\dfrac{d^2y}{dx^2} = -\dfrac{3}{(y - x)^3}$

45. $\dfrac{dy}{dx} = \dfrac{\cos y}{1 + x \sin y}$,

$\dfrac{d^2y}{dx^2} = \dfrac{(1 + x \sin y)(-\sin y)(dy/dx) - (\cos y)[(x \cos y)(dy/dx) + \sin y]}{(1 + x \sin y)^2}$

$= -\dfrac{2 \sin y \cos y + (x \cos y)(2 \sin^2 y + \cos^2 y)}{(1 + x \sin y)^3}$,

but $x \cos y = y$, $2 \sin y \cos y = \sin 2y$, and $\sin^2 y + \cos^2 y = 1$ so

$$\dfrac{d^2y}{dx^2} = -\dfrac{\sin 2y + y(\sin^2 y + 1)}{(1 + x \sin y)^3}.$$

47. $4a^3 \dfrac{da}{dt} - 4t^3 = 6\left(a^2 + 2at\dfrac{da}{dt}\right)$, solve for $\dfrac{da}{dt}$ to get $\dfrac{da}{dt} = \dfrac{2t^3 + 3a^2}{2a^3 - 6at}$

49. $2a^2\omega \dfrac{d\omega}{d\lambda} + 2b^2\lambda = 0$ so $\dfrac{d\omega}{d\lambda} = -\dfrac{b^2\lambda}{a^2\omega}$

51. By the chain rule, $\dfrac{dy}{dx} = \dfrac{dy}{dt}\dfrac{dt}{dx}$. Use implicit differentiation on $2y^3t + t^3y = 1$ to get

    $\dfrac{dy}{dt} = -\dfrac{2y^3 + 3t^2y}{6ty^2 + t^3}$, but $\dfrac{dt}{dx} = \dfrac{1}{\cos t}$ so $\dfrac{dy}{dx} = -\dfrac{2y^3 + 3t^2y}{(6ty^2 + t^3)\cos t}$.

53. $2xy\dfrac{dy}{dt} = y^2\dfrac{dx}{dt} = 3(\cos 3x)\dfrac{dx}{dt}, \dfrac{dy}{dt} = \dfrac{3\cos 3x - y^2}{2xy}\dfrac{dx}{dt}$

55. The point $(1,1)$ is on the graph, so $1 + a = b$. The slope of the tangent line at $(1,1)$ is $-4/3$; use implicit differentiation to get $\dfrac{dy}{dx} = -\dfrac{2xy}{x^2 + 2ay}$ so at $(1,1)$, $-\dfrac{2}{1 + 2a} = -\dfrac{4}{3}$, $1 + 2a = 3/2$, $a = 1/4$ and hence $b = 1 + 1/4 = 5/4$.

57. Let $P(x_0, y_0)$ be a point where a line through the orgin is tangent to the curve $x^2 - 4x + y^2 + 3 = 0$. Implicit differentiation applied to the equation of the curve gives $dy/dx = (2 - x)/y$. At P the slope of the curve must equal the slope of the line so $(2 - x_0)/y_0 = y_0/x_0$, or $y_0^2 = 2x_0 - x_0^2$. But $x_0^2 - 4x_0 + y_0^2 + 3 = 0$ because (x_0, y_0) is on the curve, and elimination of y_0^2 in the latter two equations gives $x_0^2 - 4x_0 + (2x_0 - x_0^2) + 3 = 0$, $x_0 = 3/2$ which when substituted into $y_0^2 = 2x_0 - x_0^2$ yields $y_0^2 = 3/4$, so $y_0 = \pm\sqrt{3}/2$. The slopes of the lines are $(\pm\sqrt{3}/2)/(3/2) = \pm\sqrt{3}/3$ and their equations are $y = (\sqrt{3}/3)x$ and $y = -(\sqrt{3}/3)x$.

59. **(a)** The equation does not define y as an implicit function of x at the points where the curve crosses the x-axis. If $y = 0$ then $8x^4 = 100x^2$, $8x^2(x^2 - 25/2) = 0$ so $x = 0, \pm5/\sqrt{2}$. The points are $(0,0)$, $(-5/\sqrt{2}, 0)$, and $(5/\sqrt{2}, 0)$.

    (b) $16(x^2 + y^2)\left(2x + 2y\dfrac{dy}{dx}\right) = 100\left(2x - 2y\dfrac{dy}{dx}\right)$,

    $\dfrac{dy}{dx} = \dfrac{x[25 - 4(x^2 + y^2)]}{y[25 + 4(x^2 + y^2)]}$; at $(3,1)$ $\dfrac{dy}{dx} = -9/13$ so the equation of the tangent line is

    $y - 1 = (-9/13)(x - 3)$, $9x + 13y = 40$.

EXERCISE SET 3.7

1. (a) $\Delta y = (x + \Delta x)^2 - x^2$
    $= (2 + 1)^2 - 2^2$
    $= 9 - 4 = 5$

    (b) $dy = 2x\,dx$
    $= 2(2)(1) = 4$

    (c)

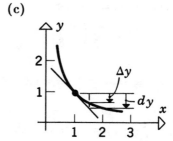

3. (a) $\Delta y = \dfrac{1}{x + \Delta x} - \dfrac{1}{x}$
    $= \dfrac{1}{1 + 0.5} - \dfrac{1}{1}$
    $= -\dfrac{1}{3}$

    (b) $dy = -\dfrac{1}{x^2}dx$
    $= -\dfrac{1}{1^2}(0.5) = -0.5$

    (c)

5. $dy = 3x^2 dx;$

    $\Delta y = (x + \Delta x)^3 - x^3 = x^3 3x^2\Delta x + 3x(\Delta x)^2 + (\Delta x)^3 - x^3 = 3x^2\Delta x + 3x(\Delta x)^2 + (\Delta x)^3$

7. $dy = (2x - 2)dx;$

    $\Delta y = [(x + \Delta x)^2 - 2(x + \Delta x) + 1] - [x^2 - 2x + 1]$
    $= x^2 + 2x\,\Delta x + (\Delta x)^2 - 2x - 2\Delta x + 1 - x^2 + 2x - 1 = 2x\,\Delta x + (\Delta x)^2 - 2\Delta x$

9. $dy = (12x^2 - 14x + 2)dx$

11. $dy = x\,d(\cos x) + \cos x\,dx = x(-\sin x)dx + \cos x\,dx = (-x\sin x + \cos x)dx$

13. $\displaystyle\lim_{\Delta x \to 0}\dfrac{(x + \Delta x)^2 - x^2}{\Delta x} = \dfrac{d}{dx}(x^2) = 2x$

15. $\displaystyle\lim_{\Delta x \to 0}\dfrac{\sin(\pi + \Delta x) - \sin \pi}{\Delta x} = \dfrac{d}{dx}(\sin x)\bigg|_{x=\pi} = \cos \pi = -1$

17. $f(x) = x^4$, $f'(x) = 4x^3$, $x_0 = 3$, $\Delta x = 0.02$; $(3.02)^4 \approx 3^4 + (108)(0.02) = 81 + 2.16 = 83.16$.

19. $f(x) = \sqrt{x}$, $f'(x) = \dfrac{1}{2\sqrt{x}}$, $x_0 = 64$, $\Delta x = 1$; $\sqrt{65} \approx \sqrt{64} + \dfrac{1}{16}(1) = 8 + \dfrac{1}{16} = 8.0625$.

21. $f(x) = \sqrt{x}$, $f'(x) = \dfrac{1}{2\sqrt{x}}$, $x_0 = 81$, $\Delta x = -0.1$; $\sqrt{80.9} \approx \sqrt{81} + \dfrac{1}{18}(-0.1) \approx 8.9944$.

23. $f(x) = \sqrt[3]{x}$, $f'(x) = \dfrac{1}{3}x^{-2/3}$, $x_0 = 8$, $\Delta x = 0.06$; $\sqrt[3]{8.06} \approx \sqrt[3]{8} + \dfrac{1}{12}(0.06) = 2.005$.

25. $f(x) = \cos x$, $f'(x) = -\sin x$, $x_0 = \pi/6$, $\Delta x = \pi/180$;

$\cos 31° \approx \cos 30° + \left(-\dfrac{1}{2}\right)\left(\dfrac{\pi}{180}\right) = \dfrac{\sqrt{3}}{2} - \dfrac{\pi}{360} \approx 0.8573$.

27. $f(x) = \sin x$, $f'(x) = \cos x$, $x_0 = \pi/4$, $\Delta x = -\pi/180$;

$\sin 44° \approx \sin 45° + \dfrac{1}{\sqrt{2}}\left(-\dfrac{\pi}{180}\right) = \dfrac{1}{\sqrt{2}} - \dfrac{\pi}{180\sqrt{2}} \approx 0.6947$.

29. $dy = \dfrac{3}{2\sqrt{3x-2}}dx$, $x = 2$, $dx = 0.03$; $\Delta y \approx dy = \dfrac{3}{4}(0.03) = 0.0225$.

31. $dy = \dfrac{1-x^2}{(x^2+1)^2}dx$, $x = 2$, $dx = -0.04$; $\Delta y \approx dy = \left(-\dfrac{3}{25}\right)(-0.04) = 0.0048$.

33. **(a)** $A = x^2$ where x is the length of a side; $dA = 2x\,dx = 2(10)(\pm 0.1) = \pm 2\text{ ft}^2$.

(b) relative error in $x \approx \dfrac{dx}{x} = \dfrac{\pm 0.1}{10} = \pm 0.01$ so percentage error in $x \approx \pm 1\%$; relative error

in $A \approx \dfrac{dA}{A} = \dfrac{2x\,dx}{x^2} = 2\dfrac{dx}{x} = 2(\pm 0.01) = \pm 0.02$ so percentage error in $A \approx \pm 2\%$.

35. **(a)** $x = 10\sin\theta$, $y = 10\cos\theta$ (see figure),

$dx = 10\cos\theta\,d\theta$

$= 10\left(\cos\dfrac{\pi}{6}\right)\left(\pm\dfrac{\pi}{180}\right)$

$= 10\left(\dfrac{\sqrt{3}}{2}\right)\left(\pm\dfrac{\pi}{180}\right) \approx \pm 0.151''$,

$dy = -10(\sin\theta)d\theta = -10\left(\sin\dfrac{\pi}{6}\right)\left(\pm\dfrac{\pi}{180}\right) = -10\left(\dfrac{1}{2}\right)\left(\pm\dfrac{\pi}{180}\right) \approx \pm 0.087''$.

(b) relative error in $x \approx \dfrac{dx}{x} = (\cot\theta)d\theta = \left(\cot\dfrac{\pi}{6}\right)\left(\pm\dfrac{\pi}{180}\right) = \sqrt{3}\left(\pm\dfrac{\pi}{180}\right) \approx \pm 0.030$

so percentage error in $x \approx \pm 3.0\%$;

relative error in $y \approx \dfrac{dy}{y} = -\tan\theta \, d\theta = -\left(\tan\dfrac{\pi}{6}\right)\left(\pm\dfrac{\pi}{180}\right) = -\dfrac{1}{\sqrt{3}}\left(\pm\dfrac{\pi}{180}\right) \approx \pm 0.010$

so percentage error in $y \approx \pm 1.0\%$.

37. $\dfrac{dR}{R} = \dfrac{(-2k/r^3)dr}{(k/r^2)} = -2\dfrac{dr}{r}$, but $\dfrac{dr}{r} \approx \pm 0.05$ so $\dfrac{dR}{R} \approx -2(\pm 0.05) = \pm 0.10$; percentage error in $R \approx \pm 10\%$.

39. $V = x^3$ where x is the length of a side; $\dfrac{dV}{V} = \dfrac{3x^2 dx}{x^3} = 3\dfrac{dx}{x}$, but $\dfrac{dx}{x} \approx \pm 0.02$

so $\dfrac{dV}{V} \approx 3(\pm 0.02) = \pm 0.06$; percentage error in $V \approx \pm 6\%$.

41. $A = \dfrac{1}{4}\pi D^2$ where D is the diameter of the circle; $\dfrac{dA}{A} = \dfrac{(\pi D/2)dD}{\pi D^2/4} = 2\dfrac{dD}{D}$, but $\dfrac{dA}{A} \approx \pm 0.01$

so $2\dfrac{dD}{D} \approx \pm 0.01$, $\dfrac{dD}{D} \approx \pm 0.005$; maximum permissible percentage error in $D \approx \pm 0.5\%$.

43. $V = $ volume of cylindrical rod $= \pi r^2 h = \pi r^2(15) = 15\pi r^2$; approximate ΔV by dV if $r = 2.5$ and $dr = \Delta r = 0.001$. $dV = 30\pi r \, dr = 30\pi(2.5)(0.001) \approx 0.236$ cm^3.

45. (a) $\alpha = \Delta L/(L\Delta T) = 0.006/(40 \times 10) = 1.5 \times 10^{-5}/°C$

(b) $\Delta L = 2.3 \times 10^{-5}(180)(25) \approx 0.1$ cm, so the pole is about 180.1 cm long.

47. $f(x) = \dfrac{1}{1+x}$, $f'(x) = -\dfrac{1}{(1+x)^2}$, $x_0 = 0$, $\Delta x = x$;

$f(x_0 + \Delta x) \approx f(x_0) + f'(x_0)\Delta x$ so $f(x) \approx 1 - x$.

$x = 0.1 : 1/(1+x) = 0.9091, \ 1 - x = 0.9$

$x = -0.02 : 1/(1+x) = 1.0204, \ 1 - x = 1.02$

49. $y = x^k$, $dy = kx^{k-1}dx$; $\dfrac{dy}{y} = \dfrac{kx^{k-1}dx}{x^k} = k\dfrac{dx}{x}$

SUPPLEMENTARY EXERCISES CHAPTER 3

1. $f'(x) = \lim\limits_{h\to 0}\dfrac{k(x+h) - kx}{h} = \lim\limits_{h\to 0} k = k$

3. $f'(x) = \lim\limits_{h\to 0}\dfrac{\sqrt{9 - 4(x+h)} - \sqrt{9 - 4x}}{h} = \lim\limits_{h\to 0}\dfrac{[9 - 4(x+h)] - [9 - 4x]}{h(\sqrt{9 - 4(x+h)} + \sqrt{9 - 4x})}$

$= \lim\limits_{h\to 0}\dfrac{-4}{\sqrt{9 - 4(x+h)} + \sqrt{9 - 4x}} = -\dfrac{2}{\sqrt{9 - 4x}}$

5. $\dfrac{d}{dx}\left(|x|^3\right)\bigg|_{x=0} = \lim\limits_{h\to 0}\dfrac{|0+h|^3 - |0|^3}{h} = \lim\limits_{h\to 0}\dfrac{|h|^3}{h} = \lim\limits_{h\to 0} h|h| = 0$

7. $y - (-1) = 5(x - 3)$, $y = 5x - 16$.

9. **(a)** $2f(x)f'(x) - 3g'(x^2)(2x)\big|_{x=1} = 12$ **(b)** $f(x)g'(x) + f'(x)g(x)\big|_{x=1} = -7$

    (c) $\dfrac{g(x)f'(x) - f(x)g'(x)}{g^2(x)}\bigg|_{x=-2} = 9$ **(d)** $\dfrac{f(x)g'(x) - g(x)f'(x)}{f^2(x)}\bigg|_{x=-2} = -\dfrac{9}{4}$

    (e) $f'(g(x))g'(x)\big|_{x=1} = f'(g(1))g'(1) = f'(-2)(-1) = 5$

    (f) $f'(g(x))g'(x)\big|_{x=-2} = f'(g(-2))g'(-2) = f'(1)(7) = 21$

    (g) $g'(f(x))f'(x)\big|_{x=-2} = g'(f(-2))f'(-2) = g'(-2)(-5) = -35$

    (h) $g'(g(x))g'(x)\big|_{x=-2} = g'(g(-2))g'(-2) = g'(1)(7) = -7$

    (i) $f'(g(4-6x)g'(4-6x)(-6)\big|_{x=1} = f'(g(-2))g'(-2)(-6) = f'(1)(7)(-6) = -126$

    (j) $3g^2(x)g'(x)\big|_{x=1} = 3(-2)^2(-1) = -12$

    (k) $\dfrac{1}{2}[f(x)]^{-1/2}f'(x)\bigg|_{x=1} = \dfrac{1}{2}(1)^{-1/2}(3) = \dfrac{3}{2}$

    (l) $f'(-x/2)(-1/2)\big|_{x=-2} = -\dfrac{3}{2}$

11. $f'(x) = \dfrac{(x^2 + 2x)4(x-3)^3 - (x-3)^4(2x+2)}{(x^2+2x)^2}$

    $= \dfrac{(x-3)^3[4(x^2+2x) - (x-3)(2x+2)]}{(x^2+2x)^2}$

    $= \dfrac{(x-3)^3(2x^2+12x+6)}{(x^2+2x)^2} = \dfrac{2(x-3)^3(x^2+6x+3)}{(x^2+2x)^2}$

so $f'(x) = 0$ if $x - 3 = 0$ or if $x^2 + 6x + 3 = 0$; the solution of $x - 3 = 0$ is $x = 3$, and the solution of $x^2 + 6x + 3 = 0$ is $x = -3 \pm \sqrt{6}$.

13. $f'(x) = 3\left[\dfrac{3x+1}{x^2}\right]^2 \dfrac{x^2(3) - (3x+1)(2x)}{x^4} = -\dfrac{3(3x+2)(3x+1)^2}{x^7}$ so $f'(x) = 0$ if $x = -2/3, -1/3$.

15. $f'(x) = x^{1/2}(1/3)(x^2 + x + 1)^{-2/3}(2x+1) + (x^2 + x + 1)^{1/3}(1/2)x^{-1/2}$

    $= \dfrac{1}{6}x^{-1/2}(x^2 + x + 1)^{-2/3}[2x(2x+1) + 3(x^2 + x + 1)]$

    $= \dfrac{7x^2 + 5x + 3}{6x^{1/2}(x^2 + x + 1)^{2/3}}$

but $7x^2 + 5x + 3 = 0$ has no real solutions so there are no values of x for which $f'(x) = 0$.

17. $\dfrac{d}{dx}(\sqrt{2}x^{-2} - \dfrac{2}{5}x^{-1}) = -2\sqrt{2}x^{-3} + \dfrac{2}{5}x^{-2}$

19. $z = (2\sin r\cos r)^2 = \sin^2 2r$ so $\dfrac{dz}{dr} = 2(\sin 2r)(\cos 2r)(2) = 2\sin 4r$

    and $\dfrac{dz}{dr}\bigg|_{r=\pi/6} = 2\sin(2\pi/3) = \sqrt{3}$

21. $u = \left[\dfrac{x-1}{x}\right]^2 = (1 - x^{-1})^2$ so $\dfrac{du}{dx} = 2(1 - x^{-1})(x^{-2}) = 2(x-1)/x^3$.

23. $\dfrac{d}{dx}(\sec^2 x - \tan^2 x) = \dfrac{d}{dx}(1) = 0$

25. $F(x) = \dfrac{2x + 4x^3}{1 + 2x^2} = \dfrac{2x(1 + 2x^2)}{1 + 2x^2} = 2x$ so $F'(x) = 2$.

27. $y' = 1 + x^{-2}$, and the slope of $2x - y = 5$ is 2 so we want $1 + x^{-2} = 2$ which gives $x^2 = 1$, $x = \pm 1$.

29. $y' = 2(x + 2)$ so at $(x_0, f(x_0))$ the tangent line is $y - f(x_0) = 2(x_0 + 2)(x - x_0)$, or $y - (x_0 + 2)^2 = 2(x_0 + 2)(x - x_0)$. But if the line passes through the orgin then $x = 0$, $y = 0$ must satisfy the latter equation thus $-(x_0 + 2)^2 = -2x_0(x_0 + 2)$ which leads to $(x_0 + 2)(x_0 - 2) = 0$ so $x_0 = -2, 2$.

31. $y' = 3 - \sec^2 x$, and the slope of $y - x = 2$ is 1 so we want $3 - \sec^2 x = 1$ which gives $\sec^2 x = 2, \sec x = \pm\sqrt{2}, x = \pi/4 + k\pi/2$ where $k = 0, \pm 1, \pm 2, \cdots$.

33. $\Delta x = 0 - (-\pi/4) = \pi/4, \Delta y = y|_{x=0} - y|_{x=-\pi/4} = 0 - (-1) = 1,$

    $dy = \sec^2(-\pi/4)(\pi/4) = \pi/2.$

35. (a) Consider $y = f(x) = \sqrt[3]{x}$ with $x = -8$ and $dx = -0.25 = -1/4$, then

    $f(-8.25) \approx f(-8) + dy, \sqrt[3]{-8.25} \approx \sqrt[3]{-8} + \dfrac{1}{3}(-8)^{-2/3}(-1/4) = -2 - 1/48 = -97/48.$

    (b) Consider $y = f(x) = \cot x$ (x in radians) with $x = 45° = \pi/4$ radians and $dx = 1° = \pi/180$ radians, then $f(\pi/4 + 180/\pi) \approx f(\pi/4) + dy,$

    $\cot 46° \approx \cot 45° + (-\csc^2 45°)(\pi/180) = 1 - \pi/90.$

37. $h = 12\sin\theta$ thus $dh = 12\cos\theta\, d\theta$ so, with $\theta = 60° = \pi/3$ radians and $d\theta = -1° = -\pi/180$ radians, $dh = 12\cos(\pi/3)(-\pi/180) = -\pi/30$ ft.

39. (a) $dW/dt|_{t=5} = 200(t - 15)|_{t=5} = -2000$ so water is running out at the rate of 2000 gal/min.

(b) average rate of change of $W = (W|_{t=5} - W|_{t=0})/5 = (10,000 - 22,500)/5 = -2500$ so water flows out at an average rate of 2500 gal/min during the first 5 minutes.

41. $2xy\dfrac{dy}{dx} + y^2 = \cos(x + 2y)\left(1 + 2\dfrac{dy}{dx}\right)$ so $\dfrac{dy}{dx} = \dfrac{\cos(x + 2y) - y^2}{2xy - 2\cos(x + 2y)}$.

$\dfrac{dy}{dx}\bigg|_{(0,0)} = -\dfrac{1}{2}$, the tangent line is $y = -x/2$.

43. $x = y = 1$ satisfies both equations so they intersect at the point $(1,1)$. For $2x^2 + 3y^2 = 5$, $\dfrac{dy}{dx} = -\dfrac{2x}{3y}$ so $\dfrac{dy}{dx}\bigg|_{(1,1)} = -\dfrac{2}{3}$. For $y^2 = x^3$, $\dfrac{dy}{dx} = \dfrac{3x^2}{2y}$ so $\dfrac{dy}{dx}\bigg|_{(1,1)} = \dfrac{3}{2}$. The tangent lines are perpendicular at $(1,1)$ because the slope of one curve is the negative reciprocal of the slope of the other curve.

45. $y = \cos x - 3\sin x$, $y' = -\sin x - 3\cos x$, $y'' = -\cos x + 3\sin x$, $y''' = \sin x + 3\cos x$ so $y''' + y'' + y' + y = (-3 - 1 + 3 + 1)\sin x + (1 - 3 - 1 + 3)\cos x = 0$.

CHAPTER 4
Applications of Differentiation

EXERCISE SET 4.1

1.   **(a)** $A = x^2$, so $\dfrac{dA}{dt} = 2x\dfrac{dx}{dt}$

     (b)   Find $\dfrac{dA}{dt}\Big|_{x=3}$ given that $\dfrac{dx}{dt}\Big|_{x=3} = 2.$

          From part (a), $\dfrac{dA}{dt}\Big|_{x=3} = 2(3)(2) = 12 \text{ ft}^2/\text{min}.$

3.   **(a)** $V = \pi r^2 h$, so $\dfrac{dV}{dt} = \pi\left(r^2\dfrac{dh}{dt} + 2rh\dfrac{dr}{dt}\right).$

     (b)   Find $\dfrac{dV}{dt}\Big|_{\substack{h=6,\\r=10}}$ given that $\dfrac{dh}{dt}\Big|_{\substack{h=6,\\r=10}} = 1$ and $\dfrac{dr}{dt}\Big|_{\substack{h=6,\\r=10}} = -1.$

          From part (a), $\dfrac{dV}{dt}\Big|_{\substack{h=6,\\r=10}} = \pi[10^2(1) + 2(10)(6)(-1)] = -20\pi \text{ in}^3/\text{sec};$

          the volume is decreasing.

5.   **(a)** $\tan\theta = \dfrac{y}{x}$, so $\sec^2\theta\dfrac{d\theta}{dt} = \dfrac{x\dfrac{dy}{dt} - y\dfrac{dx}{dt}}{x^2}$, $\dfrac{d\theta}{dt} = \dfrac{\cos^2\theta}{x^2}\left(x\dfrac{dy}{dt} - y\dfrac{dx}{dt}\right).$

     (b)   Find $\dfrac{d\theta}{dt}\Big|_{\substack{x=2,\\y=2}}$ given that $\dfrac{dx}{dt}\Big|_{\substack{x=2,\\y=2}} = 1$ and $\dfrac{dy}{dt}\Big|_{\substack{x=2,\\y=2}} = -\dfrac{1}{4}.$

          When $x = 2$ and $y = 2$, $\tan\theta = 2/2 = 1$ so $\theta = \dfrac{\pi}{4}$ and $\cos\theta = \cos\dfrac{\pi}{4} = \dfrac{1}{\sqrt{2}}.$ Thus

          from part (a), $\dfrac{d\theta}{dt}\Big|_{\substack{x=2,\\y=2}} = \dfrac{(1/\sqrt{2})^2}{2^2}\left[2\left(-\dfrac{1}{4}\right) - 2(1)\right] = -\dfrac{5}{16}$ radians/sec; θ is decreasing.

7.   Let A be the area swept out, and θ the angle through which the minute hand has rotated.
    Find $\dfrac{dA}{dt}$ given that $\dfrac{d\theta}{dt} = \dfrac{\pi}{30}$ radians/min; $A = \dfrac{1}{2}r^2\theta = 8\theta$, so $\dfrac{dA}{dt} = 8\dfrac{d\theta}{dt} = \dfrac{4\pi}{15}$ in^2/min.

9.   Find $\dfrac{dr}{dt}\Big|_{A=9}$ given that $\dfrac{dA}{dt} = 6.$ From $A = \pi r^2$ we get $\dfrac{dA}{dt} = 2\pi r\dfrac{dr}{dt}$ so $\dfrac{dr}{dt} = \dfrac{1}{2\pi r}\dfrac{dA}{dt}.$ If
    $A = 9$ then $\pi r^2 = 9$, $r = 3/\sqrt{\pi}$ so $\dfrac{dr}{dt}\Big|_{A=9} = \dfrac{1}{2\pi(3/\sqrt{\pi})}(6) = 1/\sqrt{\pi}$ mph.

11. Find $\dfrac{dV}{dt}\bigg|_{r=9}$ given that $\dfrac{dr}{dt} = -15$. From $V = \dfrac{4}{3}\pi r^3$ we get $\dfrac{dV}{dt} = 4\pi r^2 \dfrac{dr}{dt}$ so

$\dfrac{dV}{dt}\bigg|_{r=9} = 4\pi(9)^2(-15) = -4860\pi$. Air must be removed at the rate of 4860π cm^3/min.

13. Find $\dfrac{dx}{dt}\bigg|_{y=5}$ given that $\dfrac{dy}{dt} = -2$. From $x^2 + y^2 =$

13^2 we get $2x\dfrac{dx}{dt} + 2y\dfrac{dy}{dt} = 0$ so $\dfrac{dx}{dt} = -\dfrac{y}{x}\dfrac{dy}{dt}$. Use
$x^2 + y^2 = 169$ to find that $x = 12$ when $y = 5$ so
$\dfrac{dx}{dt}\bigg|_{y=5} = -\dfrac{5}{12}(-2) = \dfrac{5}{6}$ ft/sec.

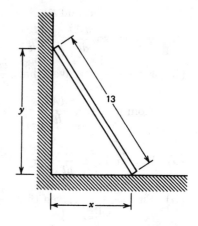

15. Let x be the length of each edge, S the surface area, and V the volume. Find $\dfrac{dS}{dt}\bigg|_{x=5}$ given

that $\dfrac{dV}{dt}\bigg|_{x=5} = 2$. $S = 6x^2$ and $V = x^3$, so $x = V^{1/3}$, $S = 6V^{2/3}$,

$\dfrac{dS}{dt} = 4V^{-1/3}\dfrac{dV}{dt} = \dfrac{4}{x}\dfrac{dV}{dt} = \dfrac{4}{5}(2) = 8/5$ in^2/min.

17. With ϕ and x as shown in the figure,

find $\dfrac{d\phi}{dt}\bigg|_{x=3000}$ given that

$\dfrac{dx}{dt}\bigg|_{x=3000} = 500$. But $\tan\phi = \dfrac{x}{3000}$

thus $\sec^2\phi\dfrac{d\phi}{dt} = \dfrac{1}{3000}\dfrac{dx}{dt}$

$\dfrac{d\phi}{dt} = \dfrac{\cos^2\phi}{3000}\dfrac{dx}{dt}$,

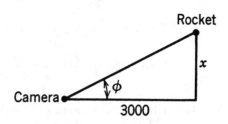

$\phi = \pi/4$ when $x = 3000$ so $\cos\phi = 1/\sqrt{2}$ and $\dfrac{d\phi}{dt}\bigg|_{x=3000} = \dfrac{1/2}{3000}(500) = 1/12$ radian/sec.

19. **(a)** $\theta = 0$ at perigee, so $r = 4995/1.12 \approx 4460$; the altitude is $4460 - 3960 = 500$ miles. $\theta = \pi$
at apogee, so $r = 4995/0.88 \approx 5676$; the altitude is $5676 - 3960 = 1716$ miles.

(b) If $\theta = 120°$, then $r = 4995/0.94 \approx 5314$; the altitude is $5314 - 3960 = 1354$ miles. The rate of change of the altitude is given by

$$\frac{dr}{dt} = \frac{4995(0.12\sin\theta)}{(1+0.12\cos\theta)^2}\frac{d\theta}{dt}.$$

Use $\theta = 120°$ and $d\theta/dt = 2.7°/\min = (2.7)(\pi/180)$ rad/min to get $dr/dt \approx 27.7$ mi/min.

21. Find $\dfrac{dh}{dt}\Big|_{h=16}$ given that $\dfrac{dV}{dt} = 20$. The volume of water in the tank at a depth h is $V = \dfrac{1}{3}\pi r^2 h$. Use similar triangles (see figure) to get $\dfrac{r}{h} = \dfrac{10}{24}$

so $r = \dfrac{5}{12}h$ thus $V = \dfrac{1}{3}\pi\left(\dfrac{5}{12}h\right)^2 h = \dfrac{25}{432}\pi h^3$,

$\dfrac{dV}{dt} = \dfrac{25}{144}\pi h^2 \dfrac{dh}{dt}; \dfrac{dh}{dt} = \dfrac{144}{25\pi h^2}\dfrac{dV}{dt}$,

$\dfrac{dh}{dt}\Big|_{h=16} = \dfrac{144}{25\pi(16)^2}(20) = \dfrac{9}{20\pi}$ ft/min.

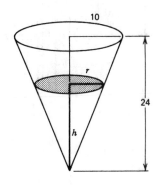

23. Find $\dfrac{dV}{dt}\Big|_{h=10}$ given that $\dfrac{dh}{dt} = 5$. $V = \dfrac{1}{3}\pi r^2 h$,

but $r = \dfrac{1}{2}h$ so $V = \dfrac{1}{3}\pi\left(\dfrac{h}{2}\right)^2 h = \dfrac{1}{12}\pi h^3$,

$\dfrac{dV}{dt} = \dfrac{1}{4}\pi h^2 \dfrac{dh}{dt}, \dfrac{dV}{dt}\Big|_{h=10}$

$= \dfrac{1}{4}\pi(10)^2(5) = 125\pi$ ft^3/min.

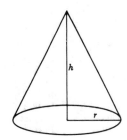

25. With s and h as shown in the figure, we want to find $\dfrac{dh}{dt}$ given that

$\dfrac{ds}{dt} = 500$. From the figure, $h = s\sin 30° = \dfrac{1}{2}s$

so $\dfrac{dh}{dt} = \dfrac{1}{2}\dfrac{ds}{dt} = \dfrac{1}{2}(500) = 250$ mph.

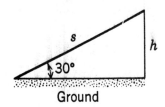

27. Find $\dfrac{dy}{dt}$ given that $\dfrac{dx}{dt}\Big|_{y=125} = -12$.

From $x^2 + 10^2 = y^2$ we get

$2x\dfrac{dx}{dt} = 2y\dfrac{dy}{dt}$ so $\dfrac{dy}{dt} = \dfrac{x}{y}\dfrac{dx}{dt}.$

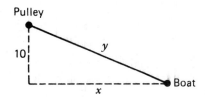

Use $x^2 + 100 = y^2$ to find that

$x = \sqrt{15,525} = 15\sqrt{69}$ when $y = 125$ so $\dfrac{dy}{dt} = \dfrac{15\sqrt{69}}{125}(-12) = -\dfrac{36\sqrt{69}}{25}$. The rope must be pulled at the rate of $\dfrac{36\sqrt{69}}{25}$ ft/min.

29. Find $\dfrac{dx}{dt}\bigg|_{\theta=\pi/4}$ given that $\dfrac{d\theta}{dt} = \dfrac{2\pi}{10} = \dfrac{\pi}{5}$

radians/sec. $x = 4\tan\theta$ (see figure)

so $\dfrac{dx}{dt} = 4\sec^2\theta\dfrac{d\theta}{dt}$,

$\dfrac{dx}{dt}\bigg|_{\theta=\pi/4} = 4\left(\sec^2\dfrac{\pi}{4}\right)\left(\dfrac{\pi}{5}\right) = 8\pi/5$ kilometers/sec.

31. We wish to find $\dfrac{dz}{dt}\bigg|_{\substack{x=2,\\y=4}}$ given that

$\dfrac{dx}{dt} = -600$ and $\dfrac{dy}{dt}\bigg|_{\substack{x=2,\\y=4}} = -1200$

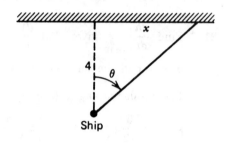

(see figure). From the law of cosines,

$z^2 = x^2 + y^2 - 2xy\cos 120°$
$\quad = x^2 + y^2 - 2xy(-1/2)$
$\quad = x^2 + y^2 + xy$,

so $2z\dfrac{dz}{dt} = 2x\dfrac{dx}{dt} + 2y\dfrac{dy}{dt} + x\dfrac{dy}{dt} + y\dfrac{dx}{dt}$,

$\dfrac{dz}{dt} = \dfrac{1}{2z}\left[(2x+y)\dfrac{dx}{dt} + (2y+x)\dfrac{dy}{dt}\right]$. When $x = 2$ and $y = 4$,

$z^2 = 2^2 + 4^2 + (2)(4) = 28$, so $z = \sqrt{28} = 2\sqrt{7}$, thus

$\dfrac{dz}{dt}\bigg|_{\substack{x=2,\\y=4}} = \dfrac{1}{2(2\sqrt{7})}[(2(2)+4)(-600) + (2(4)+2)(-1200)] = -\dfrac{4200}{\sqrt{7}} = -600\sqrt{7}$ mph;

the distance between missile and aircraft is decreasing at the rate of $600\sqrt{7}$ mph.

33. (a) We want $\dfrac{dy}{dt}\bigg|_{\substack{x=1,\\y=2}}$ given that $\dfrac{dx}{dt}\bigg|_{\substack{x=1,\\y=2}} = 6$. For convenience, first rewrite the equation as

$xy^3 = \dfrac{8}{5} + \dfrac{8}{5}y^2$ then $3xy^2\dfrac{dy}{dt} + y^3\dfrac{dx}{dt} = \dfrac{16}{5}y\dfrac{dy}{dt}$, $\dfrac{dy}{dt} = \dfrac{y^3}{\frac{16}{5}y - 3xy^2}\dfrac{dx}{dt}$ so

$\dfrac{dy}{dt}\bigg|_{\substack{x=1,\\y=2}} = \dfrac{2^3}{\frac{16}{5}(2) - 3(1)2^2}(6) = -60/7$ units/sec.

(b) falling, because $\dfrac{dy}{dt} < 0$.

35. The coordinates of P are $(x, 2x)$, so the distance between P and the point $(3, 0)$ is

$$D = \sqrt{(x-3)^2 + (2x - 0)^2} = \sqrt{5x^2 - 6x + 9}. \text{ Find } \dfrac{dD}{dt}\bigg|_{x=3} \text{ given that } \dfrac{dx}{dt}\bigg|_{x=3} = -2.$$

$$\dfrac{dD}{dt} = \dfrac{5x-3}{\sqrt{5x^2 - 6x + 9}} \dfrac{dx}{dt}, \text{ so } \dfrac{dD}{dt}\bigg|_{x=3} = \dfrac{12}{\sqrt{36}}(-2) = -4 \text{ units/sec.}$$

37. $dy/dt = 2x\, dx/dt$, but $dy/dt = 3\, dx/dt$ so $3\, dx/dt = 2x\, dx/dt$, $(3 - 2x)dx/dt = 0$, $3 - 2x = 0$, $x = 3/2$.

39. Find $\dfrac{dS}{dt}\bigg|_{s=10}$ given that $\dfrac{ds}{dt}\bigg|_{s=10} = -2$. From $\dfrac{1}{s} + \dfrac{1}{S} = \dfrac{1}{6}$ we get $-\dfrac{1}{s^2}\dfrac{ds}{dt} - \dfrac{1}{S^2}\dfrac{dS}{dt} = 0$, so

$\dfrac{dS}{dt} = -\dfrac{S^2}{s^2}\dfrac{ds}{dt}$. If $s = 10$, then $\dfrac{1}{10} + \dfrac{1}{S} = \dfrac{1}{6}$ which gives $S = 15$. $\dfrac{dS}{dt}\bigg|_{s=10} = -\dfrac{225}{100}(-2) = 4.5$

cm/sec. The image is moving away from the lens.

41. Let r be the radius, V the volume, and A the surface area of a sphere. Show that $\dfrac{dr}{dt}$ is a constant given that $\dfrac{dV}{dt} = -kA$, where k is a positive constant. Because $V = \dfrac{4}{3}\pi r^3$,

$$\dfrac{dV}{dt} = 4\pi r^2 \dfrac{dr}{dt} \qquad\qquad (1)$$

But it is given that $\dfrac{dV}{dt} = -kA$ or, because $A = 4\pi r^2$, $\dfrac{dV}{dt} = -4r^2 k$ which when substituted into equation (1) gives $-4\pi r^2 k = 4\pi r^2 \dfrac{dr}{dt}$, $\dfrac{dr}{dt} = -k$.

43. Extend sides of cup to complete the cone and let V_0 be the volume of the portion added, then (see figure)

$V = \dfrac{1}{3}\pi r^2 h - V_0$ where

$\dfrac{r}{h} = \dfrac{4}{12} = \dfrac{1}{3}$ so $r = \dfrac{1}{3}h$ and

$V = \dfrac{1}{3}\pi \left(\dfrac{h}{3}\right)^2 h - V_0 = \dfrac{1}{27}\pi h^3 - V_0$,

$\dfrac{dV}{dt} = \dfrac{1}{9}\pi h^2 \dfrac{dh}{dt}, \dfrac{dh}{dt} = \dfrac{9}{\pi h^2}\dfrac{dV}{dt}$,

$\dfrac{dh}{dt}\bigg|_{h=9} = \dfrac{9}{\pi (9)^2}(2) = \dfrac{2}{9\pi} \text{ cm/sec.}$

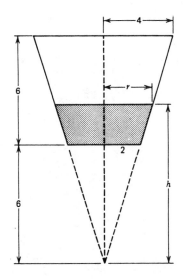

EXERCISE SET 4.2

1. **(a)** (d, f) **(b)** $(a, d), (f, g)$ **(c)** $(a, b), (c, e)$ **(d)** $(b, c), (e, g)$

3.   A: $dy/dx < 0$, $d^2y/dx^2 > 0$
      B: $dy/dx > 0$, $d^2y/dx^2 < 0$
      C: $dy/dx < 0$, $d^2y/dx^2 < 0$

5. $f'(x) = 2x - 5$
      $f''(x) = 2$

      (a) $[5/2, +\infty)$
      (b) $(-\infty, 5/2]$
      (c) $(-\infty, +\infty)$
      (d) none
      (e) none

7. $f'(x) = 3(x + 2)^2$
      $f''(x) = 6(x + 2)$

      (a) $(-\infty, +\infty)$
      (b) none
      (c) $(-2, +\infty)$
      (d) $(-\infty, -2)$
      (e) -2

9. $f'(x) = 9(x^2 - 4/9)$
      $f''(x) = 18x$

      (a) $(-\infty, -2/3], [2/3, +\infty)$
      (b) $[-2/3, 2/3]$
      (c) $(0, +\infty)$
      (d) $(-\infty, 0)$
      (e) 0

11. $f'(x) = 12x^2(x - 1)$
      $f''(x) = 36x(x - 2/3)$

      (a) $[1, +\infty)$
      (b) $(-\infty, 1]$
      (c) $(-\infty, 0), (2/3, +\infty)$
      (d) $(0, 2/3)$
      (e) $0, 2/3$

13. $f'(x) = -\sin x$
      $f''(x) = -\cos x$

      (a) $[\pi, 2\pi)$
      (b) $(0, \pi]$
      (c) $(\pi/2, 3\pi/2)$
      (d) $(0, \pi/2), (3\pi/2, 2\pi)$
      (e) $\pi/2, 3\pi/2$

15. $f'(x) = \sec^2 x$
      $f''(x) = 2\sec^2 x \tan x$

      (a) $(-\pi/2, \pi/2)$
      (b) none
      (c) $(0, \pi/2)$
      (d) $(-\pi/2, 0)$
      (e) 0

17. $f'(x) = \dfrac{1}{3}(x+2)^{-2/3}$

$f''(x) = -\dfrac{2}{9}(x+2)^{-5/3}$

(a) $(-\infty, +\infty)$
(b) none
(c) $(-\infty, -2)$
(d) $(-2, +\infty)$
(e) -2

19. $f'(x) = \dfrac{4(x+1)}{3x^{2/3}}$

$f''(x) = \dfrac{4(x-2)}{9x^{5/3}}$

(a) $[-1, +\infty)$
(b) $(-\infty, -1]$
(c) $(-\infty, 0), (2, +\infty)$
(d) $(0, 2)$
(e) $0, 2$

21. $f'(x) = 1 + \cos x \geq 0$; $f'(x) = 0$ when $\cos x = -1$, $x = \pm\pi, \pm3\pi, \pm5\pi, \cdots$ so f is increasing on the intervals $\cdots [-3\pi, -\pi], [-\pi, \pi], [\pi, 3\pi], \cdots$ and putting all of these intervals together gives $(-\infty, +\infty)$.

23. $f'(x) = 1 - 1/x^2$ so f is increasing on $[1, +\infty)$ thus if $x > 1$, then $f(x) > f(1) = 2$, $x + 1/x > 2$.

25. $f'(x) = 1/3 - 1/[3(1+x)^{2/3}]$ so f is increasing on $[0, +\infty)$ thus if $x > 0$, then $f(x) > f(0) = 0$, $1 + x/3 - \sqrt[3]{1+x} > 0$, $\sqrt[3]{1+x} < 1 + x/3$.

27. (a)

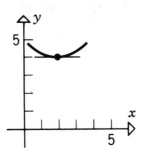

(b)

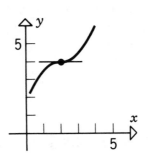

(c)

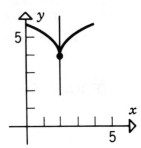

29. (a)

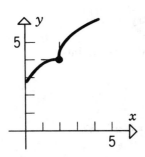

(b)

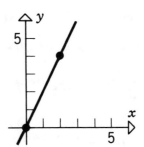

31. $f'(x) = 4(x-a)^3$, $f''(x) = 12(x-a)^2$; no inflection points.

33. $f(x_1) - f(x_2) = x_1^2 - x_2^2 = (x_1 + x_2)(x_1 - x_2) < 0$ if $x_1 < x_2$ for x_1, x_2 in $[0, +\infty)$, so $f(x_1) < f(x_2)$ and thus increasing.

35. $f(x_1) - f(x_2) = \sqrt{x_1} - \sqrt{x_2} = \dfrac{x_1 - x_2}{\sqrt{x_1} + \sqrt{x_2}} < 0$ if $x_1 < x_2$ for x_1, x_2 in $[0, +\infty)$, so

   $f(x_1) < f(x_2)$ and thus increasing.

37. If $x_1 < x_2$ where x_1 and x_2 are in I, then $f(x_1) < f(x_2)$ and $g(x_1) < g(x_2)$, so $f(x_1) + g(x_1) < f(x_2) + g(x_2)$, $(f + g)(x_1) < (f + g)(x_2)$. Thus $f + g$ is increasing on I.

39. For example, $f(x) = x$ and $g(x) = 2x$ on $(-\infty, +\infty)$.

41. $f''(x) = 6ax + 2b = 6a(x + \dfrac{b}{3a})$, $f''(x) = 0$ when $x = -\dfrac{b}{3a}$. f changes its direction of concavity

   at $x = -\dfrac{b}{3a}$ so $-\dfrac{b}{3a}$ is an inflection point.

43. **(a)** Let x_1 and x_2 be any points in (a, b) where $x_1 < x_2$. If x_1 and x_2 are both in $(a, c]$ or both in $[c, b)$, then $f(x_1) < f(x_2)$ because f is increasing on these intervals. If x_1 is in $(a, c]$ and x_2 is in (c, b), then $f(x_1) \leq f(c) < f(x_2)$ so $f(x_1) < f(x_2)$. Thus in all cases, $f(x_1) < f(x_2)$.

   (b) Similar to the proof in part (a).

EXERCISE SET 4.3

1. $f'(x) = 2x - 5$, $f'(x) = 0$ when $x = 5/2$ (stationary point).

3. $f'(x) = 3x^2 + 6x - 9 = 3(x + 3)(x - 1)$, $f'(x) = 0$ when $x = -3, 1$ (stationary points).

5. $f'(x) = 4x(x^2 - 3)$, $f'(x) = 0$ when $x = 0, \pm\sqrt{3}$ (stationary points).

7. $f'(x) = (2 - x^2)/(x^2 + 2)^2$, $f'(x) = 0$ when $x = \pm\sqrt{2}$ (stationary points).

9. $f'(x) = \dfrac{2}{3}x^{-1/3} = 2/(3x^{1/3})$, $f'(x)$ does not exist when $x = 0$.

11. $f'(x) = -3\sin 3x$, $f'(x) = 0$ when $\sin 3x = 0$, $3x = n\pi$, $n = 0, \pm 1, \pm 2, \cdots$
    $x = n\pi/3$, $n = 0, \pm 1, \pm 2, \cdots$ (stationary points).

13. $f'(x) = 4\sin 2x \cos 2x = 2\sin 4x$,
    $f'(x) = 0$ when $4x = n\pi$, $x = n\pi/4$, $n = 1, 2, 3, \cdots, 7$ (stationary points)

15. $f'(x) = \dfrac{4(x+1)}{3x^{2/3}}$, $f'(x) = 0$ when $x = -1$ (stationary point), $f'(x)$ does not exist when $x = 0$.

17. (a) $x = 2$ because $f'(x)$ changes sign from $-$ to $+$ there.
   (b) $x = 0$ because $f'(x)$ changes sign from $+$ to $-$ there.
   (c) $x = 1, 3$ because $f''(x)$ (the slope of the graph of $f'(x)$) changes sign at these points.

19. critical points $x = 0, \pm\sqrt{5}$; $f'(x)$:

$$\begin{array}{ccccc} - & 0 & +0- & 0 & + \\ \hline & -\sqrt{5} & 0 & \sqrt{5} & \end{array}$$

$x = 0$: relative maximum; $x = \pm\sqrt{5}$: relative minimum.

21. critical points: $\pm 3/2, -1$; $f'(x)$:

$$\begin{array}{ccccc} + & 0 & - & ? & + & 0 & - \\ \hline & -3/2 & & -1 & & 3/2 & \end{array}$$

$x = \pm 3/2$: relative maximum; $x = -1$: relative minimum.

23. $f'(x) = -2(x+2)$; critical point $x = -2$

   (a) $f'(x)$:

$$\begin{array}{ccc} +++ & 0 & --- \\ \hline & -2 & \end{array}$$

   (b) $f''(x) = -2$; $f''(-2) < 0$, $f(-2) = 5$; relative max of 5 at $x = -2$

25. $f'(x) = 2\sin x \cos x = \sin 2x$; critical points $x = \pi/2, \pi, 3\pi/2$

   (a) $f'(x)$:

$$\begin{array}{ccccccc} +++ & 0 & --- & 0 & +++ & 0 & --- \\ \hline & \pi/2 & & \pi & & 3\pi/2 & \end{array}$$

   (b) $f''(x) = 2\cos 2x$; $f''(\pi/2) < 0$, $f''(\pi) > 0$, $f''(3\pi/2) < 0$, $f(\pi/2) = f(3\pi/2) = 1$, $f(\pi) = 0$; relative min of 0 at $x = \pi$, relative max of 1 at $x = \pi/2, 3\pi/2$

27. $f'(x) = 3x^2 + 5$; no relative extrema because there are no critical points.

29. $f'(x) = (x-1)(3x-1)$; critical points $x = 1, 1/3$
   $f''(x) = 6x - 4$; $f''(1) > 0$, $f''(1/3) < 0$
   relative min of 0 at $x = 1$, relative max of 4/27 at $x = 1/3$

31. $f'(x) = 4x(1 - x^2)$; critical points $x = 0, 1, -1$
$f''(x) = 4 - 12x^2$; $f''(0) > 0$, $f''(1) < 0$, $f''(-1) < 0$
relative min of 0 at $x = 0$, relative max of 1 at $x = 1, -1$

33. $f'(x) = \frac{4}{5}x^{-1/5}$; critical point $x = 0$; relative min of 0 at $x = 0$ (first derivative test)

35. $f'(x) = 2x/(x^2 + 1)^2$; critical point $x = 0$; relative min of 0 at $x = 0$

37. $f'(x) = 2x$ if $|x| > 2$, $f'(x) = -2x$ if $|x| < 2$,
$f'(x)$ does not exist when $x = \pm 2$; critical points $x = 0, 2, -2$
relative min of 0 at $x = 2, -2$, relative max of 4 at $x = 0$

39. $f'(x) = -\sin 2x$; critical points $x = 0, \pm\pi/2, \pm\pi, \pm3\pi/2, \cdots$
relative min of 0 at $x = \pm\pi/2, \pm3\pi/2, \cdots$; relative max of 1 at $x = 0, \pm\pi, \pm2\pi, \cdots$

41. $f'(x) = 2x \sec^2(x^2 + 1)$; critical point $x = 0$; relative min of $\tan 1$ at $x = 0$

43. $f'(x) = 2\cos 2x$ if $\sin 2x > 0$, $f'(x) = -2\cos 2x$ if $\sin 2x < 0$,
$f'(x)$ does not exist when $x = \pi/2, \pi, 3\pi/2$;
critical points $x = \pi/4, 3\pi/4, 5\pi/4, 7\pi/4, \pi/2, \pi, 3\pi/2$
relative min of 0 at $x = \pi/2, \pi, 3\pi/2$; relative max of 1 at $x = \pi/4, 3\pi/4, 5\pi/4, 7\pi/4$

45. Let $f(x) = x^2 + \frac{k}{x}$, then $f'(x) = 2x - \frac{k}{x^2} = \frac{2x^3 - k}{x^2}$. f has a relative extremum when $2x^3 - k = 0$, so $k = 2x^3 = 2(3)^3 = 54$.

47. $f(x) = -x^4$ has a relative maximum at $x = 0$, $f(x) = x^4$ has a relative minimum at $x = 0$, $f(x) = x^3$ has neither at $x = 0$; $f'(0) = 0$ for all three functions.

49. **(a)**   **(b)**

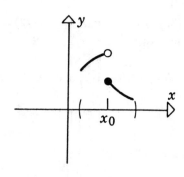

$f(x_0)$ is not an extreme value

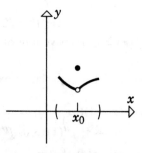

$f(x_0)$ is a relative maximum.

(c)

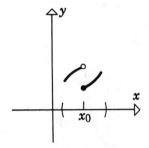

$f(x_0)$ is a relative
minimum.

EXERCISE SET 4.4

1. $y = x^2 - 2x - 3$
    $y' = 2(x - 1)$
    $y'' = 2$

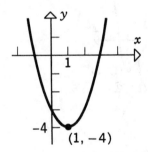

3. $y = x^3 - 3x + 1$
    $y' = 3(x^2 - 1)$
    $y'' = 6x$

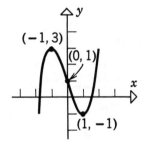

5. $y = x^3 + 3x^2 + 5$
    $y' = 3x(x + 2)$
    $y'' = 6(x + 1)$

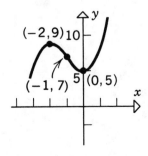

7. $y = 2x^3 - 3x^2 + 12x + 9$
    $y' = 6(x^2 - x + 2)$
    $y'' = 12(x - 1/2)$

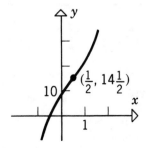

9. $y = (x - 1)^4$
     $y' = 4(x - 1)^3$
     $y'' = 12(x - 1)^2$

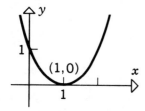

11. $y = x^4 + 2x^3 - 1$
      $y' = 4x^2(x + 3/2)$
      $y'' = 12x(x + 1)$

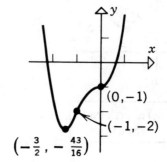

13. $y = x^4 - 3x^3 + 3x^2 + 1$
      $y' = x(4x^2 - 9x + 6)$
      $y'' = 12(x - 1/2)(x - 1)$

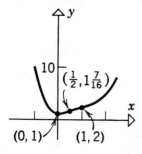

15. $y = x^3(3x^2 - 5)$
      $y' = 15x^2(x^2 - 1)$
      $y'' = 30x(2x^2 - 1)$

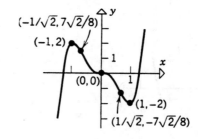

17. $y = x(x - 1)^3$
      $y' = (4x - 1)(x - 1)^2$
      $y'' = 6(2x - 1)(x - 1)$

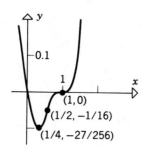

19.   vertical: $x = 2$; horizontal: $y = 3$

21. vertical: $x = -\sqrt{5}$, $x = \sqrt{5}$; horizontal: none

23. vertical: $x = -1$, $x = 3$; horizontal: $y = 1$

25. $x^2/(x^2 + 2x + 5) = 1$, $2x + 5 = 0$, $x = -5/2$.

27. $(x^2 + 1)/(2x^2 - 6x) = 1/2$, $-3x = 1$, $x = -1/3$.

29. $y = 2x/(x-3)$
$y' = -6(x-3)^2$
$y'' = 12/(x-3)^3$

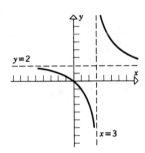

31. $y = \dfrac{x^2}{x^2 - 1}$

$y' = -\dfrac{2x}{(x^2 - 1)^2}$

$y'' = \dfrac{2(3x^2 + 1)}{(x^2 - 1)^3}$

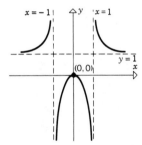

33. $y = \dfrac{x}{x^2 + 1}$

$y' = \dfrac{1 - x^2}{(x^2 + 1)^2}$

$y'' = \dfrac{2x(x^2 - 3)}{(x^2 + 1)^3}$

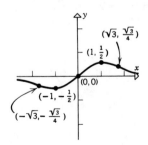

35. $y = (x-1)/(x-2)$
$y' = -1/(x-2)^2$
$y'' = 2/(x-2)^3$

37. $y = x^2 - \dfrac{1}{x} = \dfrac{x^3 - 1}{x}$

$y' = \dfrac{2x^3 + 1}{x^2}$,

$y' = 0$ when $x = -\sqrt[3]{\dfrac{1}{2}} \approx -0.8$

$y'' = \dfrac{2(x^3 - 1)}{x^3}$

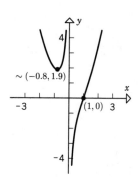

39. $y = \dfrac{1 - x}{x^2}$

$y' = \dfrac{x - 2}{x^3}$

$y'' = \dfrac{2(3 - x)}{x^4}$

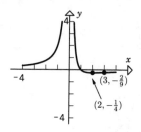

41. $y = \dfrac{x - 1}{x^2 - 4}$

$y' = -\dfrac{x^2 - 2x + 4}{(x^2 - 4)^2}$

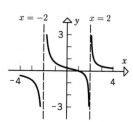

43. $y = \dfrac{(x - 1)^2}{x^2}$

$y' = \dfrac{2(x - 1)}{x^3}$

$y'' = \dfrac{2(3 - 2x)}{x^4}$

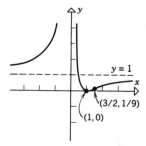

45. $y = 3 - \dfrac{4}{x} - \dfrac{4}{x^2}$

$y' = \dfrac{4(x+2)}{x^3}$

$y'' = -\dfrac{8(x+3)}{x^4}$

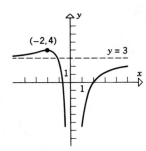

47. $y = \dfrac{x^3 - 1}{x^3 + 1}$

$y' = \dfrac{6x^2}{(x^3+1)^2}$

$y'' = \dfrac{12x(1 - 2x^3)}{(x^3+1)^3}$

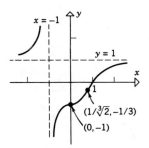

49. $y = \dfrac{x^2 - 2}{x} = x - \dfrac{2}{x}$

so $y = x$ is an oblique asymptote

$y' = \dfrac{x^2 + 2}{x^2}$

$y'' = -\dfrac{4}{x^3}$

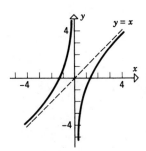

51. $y = \dfrac{(x-2)^3}{x^2} = x - 6 + \dfrac{12x - 8}{x^2}$

so $y = x - 6$ is an oblique asymptote

$y' = \dfrac{(x-2)^2(x+4)}{x^3}$

$y'' = \dfrac{24(x-2)}{x^4}$

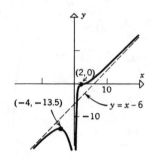

53. $y = x + 1 - \dfrac{1}{x} - \dfrac{1}{x^2} = \dfrac{(x-1)(x+1)^2}{x^2}$

$y = x + 1$ is an oblique asymptote

$y' = \dfrac{(x+1)(x^2 - x + 2)}{x^3}$

$y'' = -\dfrac{2(x+3)}{x^4}$

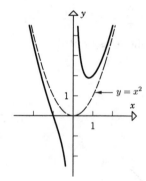

55. $\displaystyle\lim_{x \to \pm\infty} [f(x) - x^2] = \lim_{x \to \pm\infty} (1/x) = 0$

$y = x^2 + \dfrac{1}{x} = \dfrac{x^3 + 1}{x}$

$y' = 2x - \dfrac{1}{x^2} = \dfrac{2x^3 - 1}{x^2}$

$y'' = 2 + \dfrac{2}{x^3} = \dfrac{2(x^3 + 1)}{x^3}$

$y' = 0$ when $x = 1/\sqrt[3]{2} \approx 0.8$,
$\qquad\qquad y = 3\sqrt[3]{2}/2 \approx 1.9$

$y'' = 0$ when $x = -1, y = 0$

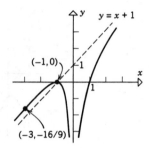

57. Let y be the length of the other side of the rectangle, then $L = 2x + 2y$ and $xy = 400$ so $y = 400/x$ and hence $L = 2x + 800/x$.

$L = 2x$ is an oblique asymptote

(see Exercise 48)

$$L = 2x + \frac{800}{x} = \frac{2(x^2 + 400)}{x}$$

$$L' = 2 - \frac{800}{x^2} = \frac{2(x^2 - 400)}{x^2}$$

$$L'' = \frac{1600}{x^3}$$

$L' = 0$ when $x = 20, L = 80$

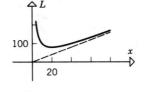

59. $y' = 0.1x^4(6x - 5)$

critical points: $x = 0$, $x = 5/6$

relative minimum at $x = 5/6$,

$y \approx -6.7 \times 10^{-3}$

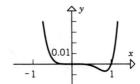

EXERCISE SET 4.5

1. $y = (x - 2)^{1/3}$

$y' = \frac{1}{3}(x - 2)^{-2/3}$

$y'' = -\frac{2}{9}(x - 2)^{-5/3}$

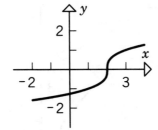

3. $y = x^{1/5}$

$y' = \dfrac{1}{5}x^{-4/5}$

$y'' = -\dfrac{4}{25}x^{-9/5}$

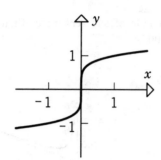

5. $y = x^{4/3}$

$y' = \dfrac{4}{3}x^{1/3}$

$y'' = \dfrac{4}{9}x^{-2/3}$

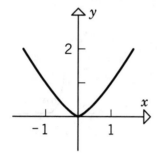

7. $y = 1 - x^{2/3}$

$y' = -\dfrac{2}{3}x^{-1/3}$

$y'' = \dfrac{2}{9}x^{-4/3}$

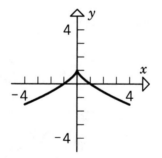

9. $y = \sqrt{x^2 - 1}$

$y' = \dfrac{x}{\sqrt{x^2 - 1}}$

$y'' = -\dfrac{1}{(x^2 - 1)^{3/2}}$

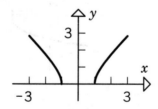

11. $y = 2x + 3x^{2/3}$

      $y' = 2 + 2x^{-1/3}$,

      $y'' = -\dfrac{2}{3}x^{-4/3}$

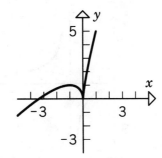

13. $y = x(3 - x)^{1/2}$

      $y' = \dfrac{3(2 - x)}{2\sqrt{3 - x}}$

      $y'' = \dfrac{3(x - 4)}{4(3 - x)^{3/2}}$

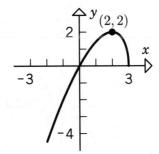

15. $y = \dfrac{8(\sqrt{x} - 1)}{x}$

      $y' = \dfrac{4(2 - \sqrt{x})}{x^2}$

      $y'' = \dfrac{2(3\sqrt{x} - 8)}{x^3}$

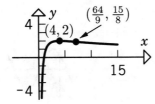

17. $y = \dfrac{\sqrt{x}}{x-3}$

$y' = -\dfrac{x+3}{2\sqrt{x}(x-3)^2}$

$y'' = \dfrac{3(x^2+6x-3)}{4x^{3/2}(x-3)^3}$

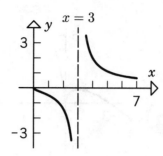

19. $y = x - \cos x$
$y' = 1 + \sin x,\ \ y' = 0$
   when $x = -\pi/2 + 2n\pi$

$y'' = \cos x,\ \ y'' = 0$
   when $x = \pi/2 + n\pi$

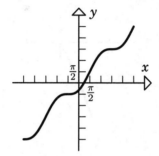

21. $y = \sin x + \cos x$
$y' = \cos x - \sin x,\ \ y' = 0$
   when $x = \pi/4 + n\pi$

$y'' = -\sin x - \cos x,\ \ y'' = 0$
   when $x = 3\pi/4 + n\pi$

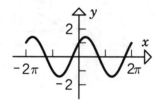

23. $y = \sin^2 x,\; 0 \le x \le 2\pi$
      $y' = 2\sin x \cos x = \sin 2x$
      $y'' = 2\cos 2x$

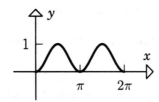

EXERCISE SET 4.6

1. $f'(x) = 8x - 4$, $f'(x) = 0$ when $x = 1/2$; $f(0) = 1$, $f(1/2) = 0$, $f(1) = 1$ so the maximum value is 1 at $x = 0, 1$ and the minimum value is 0 at $x = 1/2$.

3. $f'(x) = 3(x-1)^2$, $f'(x) = 0$ when $x = 1$; $f(0) = -1$, $f(1) = 0$, $f(4) = 27$ so the maximum value is 27 at $x = 4$ and the minimum value is -1 at $x = 0$.

5. $f'(x) = 3/(4x^2 + 1)^{3/2}$, thus there are no critical points; $f(-1) = -3/\sqrt{5}$, $f(1) = 3/\sqrt{5}$ so the maximum value is $3/\sqrt{5}$ at $x = 1$ and the minimum value is $-3/\sqrt{5}$ at $x = -1$.

7. $f'(x) = \dfrac{5(8-x)}{3x^{1/3}}$, $f'(x) = 0$ when $x = 8$ and $f'(x)$ does not exist when $x = 0$; $f(-1) = 21$, $f(0) = 0$, $f(8) = 48$, $f(20) = 0$ so the maximum value is 48 at $x = 8$ and the minimum value is 0 at $x = 0, 20$.

9. $f'(x) = 1 - \sec^2 x$, $f'(x) = 0$ for x in $(-\pi/4, \pi/4)$ when $x = 0$; $f(-\pi/4) = 1 - \pi/4$, $f(0) = 0$, $f(\pi/4) = \pi/4 - 1$ so the maximum value is $1 - \pi/4$ at $x = -\pi/4$ and the minimum value is $\pi/4 - 1$ at $x = \pi/4$.

11. $f'(x) = 2\sec x \tan x - \sec^2 x = (2\sin x - 1)/\cos^2 x$, $f'(x) = 0$ for x in $(0, \pi/4)$ when $x = \pi/6$; $f(0) = 2$, $f(\pi/6) = \sqrt{3}$, $f(\pi/4) = 2\sqrt{2} - 1$ so the maximum value is 2 at $x = 0$ and the minimum value is $\sqrt{3}$ at $x = \pi/6$.

13. $f(x) = 1 + |9 - x^2| = \begin{cases} 10 - x^2, & |x| \le 3 \\ -8 + x^2, & |x| > 3 \end{cases}$, $f'(x) = \begin{cases} -2x, & |x| < 3 \\ 2x, & |x| > 3 \end{cases}$ thus $f'(x) = 0$ when $x = 0$, $f'(x)$ does not exist for x in $(-5, 1)$ when $x = -3$ because $\lim\limits_{x \to -3^-} f'(x) \neq \lim\limits_{x \to -3^+} f'(x)$ (see Theorem preceding Exercise 71, Section 3.3); $f(-5) = 17$, $f(-3) = 1$, $f(0) = 10$, $f(1) = 9$ so the maximum value is 17 at $x = -5$ and the minimum value is 1 at $x = -3$.

15. $f'(x) = 2x - 3$; critical point $x = 3/2$. Minimum value $f(3/2) = -13/4$, no maximum.

17. $f'(x) = 12x^2(1-x)$; critical points $x = 0, 1$. Maximum value $f(1) = 1$, no minimum because $\lim_{x \to +\infty} f(x) = -\infty$.

19. $(x^2 - 1)^2$ can never be less than zero because it is the square of $x^2 - 1$; the minimum value is 0 for $x = \pm 1$, no maximum because $\lim_{x \to +\infty} f(x) = +\infty$.

21. No maximum or minimum because $\lim_{x \to +\infty} f(x) = +\infty$ and $\lim_{x \to -\infty} f(x) = -\infty$.

23. $f'(x) = -1/x^2$; no maximum or minimum because there are no critical points in $(0, +\infty)$.

25. $f'(x) = x(x+2)/(x+1)^2$; critical point $x = -2$ in $(-5, -1)$. Maximum value $f(-2) = -4$, no minimum.

27. $\sin 2x$ has a period of π, and $\sin 4x$ a period of $\pi/2$ so $f(x)$ is periodic with period π. Consider the interval $[0, \pi]$. $f'(x) = 4\cos 2x + 4\cos 4x$, $f'(x) = 0$ when $\cos 2x + \cos 4x = 0$, but $\cos 4x = 2\cos^2 2x - 1$ (trig identity) so

$$2\cos^2 2x + \cos 2x - 1 = 0$$
$$(2\cos 2x - 1)(\cos 2x + 1) = 0$$
$$\cos 2x = 1/2 \quad \text{or} \quad \cos 2x = -1.$$

From $\cos 2x = 1/2$, $2x = \pi/3$ or $5\pi/3$ so $x = \pi/6$ or $5\pi/6$. From $\cos 2x = -1$, $2x = \pi$ so $x = \pi/2$. $f(0) = 0$, $f(\pi/6) = 3\sqrt{3}/2$, $f(\pi/2) = 0$, $f(5\pi/6) = -3\sqrt{3}/2$, $f(\pi) = 0$. The maximum value is $3\sqrt{3}/2$ at $x = \pi/6 + n\pi$ and the minimum value is $-3\sqrt{3}/2$ at $x = 5\pi/6 + n\pi$, $n = 0, \pm 1, \pm 2, \cdots$.

29. $f'(x) = -[\cos(\cos x)]\sin x$; $f'(x) = 0$ if $\sin x = 0$ or if $\cos(\cos x) = 0$. If $\sin x = 0$, then $x = \pi$ is the critical point in $(0, 2\pi)$; $\cos(\cos x) = 0$ has no solutions because $-1 \leq \cos x \leq 1$. Thus $f(0) = \sin(1)$, $f(\pi) = \sin(-1) = -\sin(1)$, and $f(2\pi) = \sin(1)$ so the maximum value is $\sin(1) \approx 0.84147$ and the minimum value is $-\sin(1) \approx -0.84147$.

31. $f'(x) = \begin{cases} 4, & x < 1 \\ 2x - 5, & x > 1 \end{cases}$ so $f'(x) = 0$ when $x = 5/2$, and $f'(x)$ does not exist when $x = 1$ because $\lim_{x \to 1^-} f'(x) \neq \lim_{x \to 1^+} f'(x)$ (see Theorem preceding Exercise 71, Section 3.3); $f(1/2) = 0$, $f(1) = 2$, $f(5/2) = -1/4$, $f(7/2) = 3/4$ so the maximum value is 2 and the minimum value is $-1/4$.

33. $f'(x) = p(x - a)^{p-1}$; critical point $x = a$

(a) if p is even then $p-1$ is odd and $f'(x) < 0$ when $x < a$, $f'(x) > 0$ when $x > a$ so $f(a) = 0$ is a relative minimum.

(b) if p is odd then $p-1$ is even and $f'(x)$ does not change sign at $x=a$ so f does not have relative extrema.

35. (a) $f'(x) = -\dfrac{64\cos x}{\sin^2 x} + \dfrac{27\sin x}{\cos^2 x} = \dfrac{-64\cos^3 x + 27\sin^3 x}{\sin^2 x \cos^2 x}$, $f'(x) = 0$ when

$27\sin^3 x = 64\cos^3 x$, $\tan^3 x = 64/27$, $\tan x = 4/3$ so the critical point is $x = x_0$ where $\tan x_0 = 4/3$ and $0 < x_0 < \pi/2$. To test x_0 first rewrite $f'(x)$ as

$$f'(x) = \frac{27\cos^3 x(\tan^3 x - 64/27)}{\sin^2 x \cos^2 x} = \frac{27\cos x(\tan^3 x - 64/27)}{\sin^2 x};$$

if $x < x_0$ then $\tan x < 4/3$ and $f'(x) < 0$, if $x > x_0$ then $\tan x > 4/3$ and $f'(x) > 0$ so $f(x_0)$ is the minimum value. f has no maximum because $\lim\limits_{x \to 0^+} f(x) = +\infty$.

(b) If $\tan x_0 = 4/3$ then (see figure)

$\sin x_0 = 4/5$ and $\cos x_0 = 3/5$

so $f(x_0) = 64/\sin x_0 + 27/\cos x_0$

$= 64/(4/5) + 27/(3/5)$

$= 80 + 45 = 125$

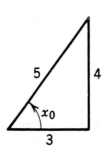

37. $f(\theta) = \sin^2 \theta \cos \theta$, $f'(\theta) = \sin\theta(2\cos^2\theta - \sin^2\theta) = \sin\theta(3\cos^2\theta - 1)$; $f'(\theta) = 0$ for θ in $(0, \pi/2)$ if $\cos\theta = 1/\sqrt{3}$. $f(0) = 0$ and $f(\pi/2) = 0$, so the maximum value occurs when $\cos\theta = 1/\sqrt{3}$ where $f(\theta) = \sin^2 \theta \cos\theta = (1 - \cos^2\theta)\cos\theta = (1 - 1/3)(1/\sqrt{3}) = 2/(3\sqrt{3})$.

39. $(0, 9)$ is on the graph so $f(0) = a_0 + a_1(0) + a_2(0)^2 = 9$, $a_0 = 9$ thus $f(x) = 9 + a_1 x + a_2 x^2$. $(2, 1)$ is on the graph so $f(2) = 9 + 2a_1 + 4a_2 = 1$,

$$a_1 + 2a_2 = -4 \tag{i}.$$

$f'(x) = a_1 + 2a_2 x$, but $f'(2) = 0$ because it is given that $f(2)$ is to be an extreme value, so

$$f'(2) = a_1 + 4a_2 = 0 \tag{ii}.$$

Solve (i) and (ii) to get $a_1 = -8$ and $a_2 = 2$, thus $f(x) = 9 - 8x + 2x^2$. As a check, we find that $f''(x) = 4 > 0$ so $f(2)$ is a minimum.

41. Let $f(x) = 1 - x^2/2 - \cos x$, then $f'(x) = -x + \sin x$ so $f'(x) = 0$ when $\sin x = x$ which has no solution for $0 < x < 2\pi$ thus the maximum and minimum values must occur at the endpoints of $[0, 2\pi]$. $f(0) = 0$, $f(2\pi) = -2\pi^2$, so 0 is the maximum value, thus $1 - x^2/2 - \cos x \le 0$, $1 - x^2/2 \le \cos x$ for all x in $[0, 2\pi]$.

43. By the quadratic formula, the roots are $x_1 = \dfrac{-b - \sqrt{b^2 - 4ac}}{2a}$ and $x_2 = \dfrac{-b + \sqrt{b^2 - 4ac}}{2a}$. The midpoint is $\dfrac{1}{2}(x_1 + x_2) = -\dfrac{b}{2a}$. But $f'(x) = 2ax + b$ so $f'\left(-\dfrac{b}{2a}\right) = 2a\left(-\dfrac{b}{2a}\right) + b = 0$.

45. $f'(x) = 2ax + b$; critical point is $x = -\dfrac{b}{2a}$

$f''(x) = 2a > 0$ so $f\left(-\dfrac{b}{2a}\right)$ is the minimum value of f, but

$f\left(-\dfrac{b}{2a}\right) = a\left(-\dfrac{b}{2a}\right)^2 + b\left(-\dfrac{b}{2a}\right) + c = \dfrac{-b^2 + 4ac}{4a}$ thus $f(x) \geq 0$ if and only if

$f\left(-\dfrac{b}{2a}\right) \geq 0,\ \dfrac{-b^2 + 4ac}{4a} \geq 0,\ -b^2 + 4ac \geq 0,\ b^2 - 4ac \leq 0$

47. The slope of the line is -1, and the slope of the tangent to $y = -x^2$ is $-2x$ so $-2x = -1$, $x = 1/2$. The line lies above the curve so the vertical distance is given by $F(x) = 2 - x + x^2$; $F(-1) = 4$, $F(1/2) = 7/4$, $F(3/2) = 11/4$. The point $(1/2, -1/4)$ is closest, the point $(-1, -1)$ farthest.

EXERCISE SET 4.7

1. Let $x =$ one number, $y =$ the other number, and $P = xy$ where $x + y = 10$. Thus $y = 10 - x$ so $P = x(10 - x) = 10x - x^2$ for x in $[0, 10]$. $dP/dx = 10 - 2x$, $dP/dx = 0$ when $x = 5$. If $x = 0, 5, 10$ then $P = 0, 25, 0$ so P is maximum when $x = 5$ and, from $y = 10 - x$, when $y = 5$.

3. If $y = x + 1/x$ for $1/2 \leq x \leq 3/2$ then $dy/dx = 1 - 1/x^2 = (x^2 - 1)/x^2$, $dy/dx = 0$ when $x = 1$. If $x = 1/2, 1, 3/2$ then $y = 5/2, 2, 13/6$ so

    (a) y is as small as possible when $x = 1$.   **(b)** y is as large as possible when $x = 1/2$.

5. Let x and y be the dimensions shown in the figure and A the area, then

$A = xy$ subject to the cost condition

$3(2x) + 2(2y) = 6000$, or $y = 1500 - 3x/2$.

Thus $A = x(1500 - 3x/2) = 1500x - 3x^2/2$

for x in $[0, 1000]$. $dA/dx = 1500 - 3x$, $dA/dx = 0$ when $x = 500$. If $x = 0$ or 1000 then $A = 0$, if $x = 500$ then $A = 375,000$ so the area is greatest when $x = 500$ ft and (from $y = 1500 - 3x/2$) when $y = 750$ ft.

7. $A = xy$ where $x^2 + y^2 = 20^2 = 400$ so
   $y = \sqrt{400 - x^2}$ and $A = x\sqrt{400 - x^2}$ for
   $0 \le x \le 20$; $dA/dx = 2(200 - x^2)/\sqrt{400 - x^2}$,
   $dA/dx = 0$ when $x = \sqrt{200} = 10\sqrt{2}$. If
   $x = 0, 10\sqrt{2}, 20$ then $A = 0, 200, 0$ so
   the area is maximum when $x = 10\sqrt{2}$ and
   $y = \sqrt{400 - 200} = 10\sqrt{2}$.

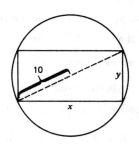

9. Let x, y, and z be as shown in the figure
   and A the area of the rectangle, then
   $A = xy$ and, by similar triangles, $z/10 = y/6$,
   $z = 5y/3$; also $x/10 = (8 - z)/8 = (8 - 5y/3)/8$
   thus $y = 24/5 - 12x/25$ so
   $A = x(24/5 - 12x/25) = 24x/5 - 12x^2/25$
   for x in $[0, 10]$. $dA/dx = 24/5 - 24x/25$,
   $dA/dx = 0$ when $x = 5$. If $x = 0, 5, 10$ then
   $A = 0, 12, 0$ so the area is greatest when $x = 5$ in. and $y = 12/5$ in.

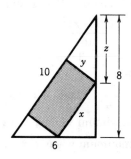

11. $V = x(12 - 2x)^2$ for $0 \le x \le 6$;
    $dV/dx = 12(x - 2)(x - 6)$, $dV/dx = 0$
    when $x = 2$ for $0 < x < 6$. If $x = 0, 2, 6$
    then $V = 0, 128, 0$ so the volume
    is largest when $x = 2$ in.

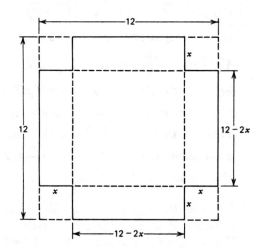

13. Let x be the length of each side of a square, then $V = x(3 - 2x)(8 - 2x) = 4x^3 - 22x^2 + 24x$
    for $0 \le x \le 3/2$; $dV/dx = 12x^2 - 44x + 24 = 4(3x - 2)(x - 3)$, $dV/dx = 0$ when $x = 2/3$ for
    $0 < x < 3/2$. If $x = 0, 2/3, 3/2$ then $V = 0, 200/27, 0$ so the maximum volume is $200/27$ ft^3.

15. With x, y, r, and s as shown in the figure, the sum of the enclosed areas is $A = \pi r^2 + s^2$ where $r = \dfrac{x}{2\pi}$ and $s = \dfrac{y}{4}$ because x is the circumference of the circle and y is the perimeter of the square, thus $A = \dfrac{x^2}{4\pi} + \dfrac{y^2}{16}$. But $x + y = 12$, so $y = 12 - x$ and

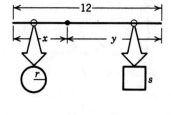

$$A = \frac{x^2}{4\pi} + \frac{(12-x)^2}{16}$$

$$= \frac{\pi+4}{16\pi}x^2 - \frac{3}{2}x + 9 \quad \text{for } 0 \le x \le 12. \quad \frac{dA}{dx} = \frac{\pi+4}{8\pi}x - \frac{3}{2}, \quad \frac{dA}{dx} = 0 \text{ when } x = \frac{12\pi}{\pi+4}. \text{ If}$$

$x = 0, \dfrac{12\pi}{\pi+4}, 12$ then $A = 9, \dfrac{36}{\pi+4}, \dfrac{36}{\pi}$ so the sum of the enclosed areas is

(a) a maximum when $x = 12$ in. (when all of the wire is used for the circle)

(b) a minimum when $x = 12\pi/(\pi+4)$ in.

17. The altitude of the triangle (see figure) is $\sqrt{y^2 - x^2/4}$ so $A = \dfrac{1}{2}x\sqrt{y^2 - x^2/4}$ where $x + 2y = 12$ thus $y = 6 - x/2$ and

$$A = \frac{1}{2}x\sqrt{(6-x/2)^2 - x^2/4} = \frac{1}{2}x\sqrt{36-6x}$$

for $0 \le x \le 6$; $dA/dx = 9(4-x)/(2\sqrt{36-6x})$, $dA/dx = 0$ when $x = 4$ for $0 < x < 6$. If $x = 0, 4, 6$ then $A = 0, 4\sqrt{3}, 0$ so the area is maximum when $x = 4$ and $y = 6 - 4/2 = 4$, which is an equilateral triangle.

19. Let x be the length of each side of the squares and y the height of the frame, then the volume is $V = x^2 y$. The total length of the wire is L thus $8x + 4y = L$, $y = (L - 8x)/4$ so $V = x^2(L - 8x)/4 = (Lx^2 - 8x^3)/4$ for $0 \le x \le L/8$. $dV/dx = (2Lx - 24x^2)/4$, $dV/dx = 0$ for $0 < x < L/8$ when $x = L/12$. If $x = 0, L/12, L/8$ then $V = 0, L^3/1728, 0$ so the volume is greatest when $x = L/12$ and $y = L/12$.

21. (a) The daily profit is
$$P = (\text{revenue}) - (\text{production cost}) = 100x - (100{,}000 + 50x + 0.0025x^2)$$
$$= -100{,}000 + 50x - 0.0025x^2$$

for $0 \le x \le 7000$, so $dP/dx = 50 - 0.005x$ and $dP/dx = 0$ when $x = 10,000$. Because 10,000 is not in the interval $[0, 7000]$, the maximum profit must occur at an endpoint. When $x = 0$, $P = -100,000$; when $x = 7000$, $P = 127,500$ so 7000 units should be manufactured and sold daily.

(b) Yes, because $dP/dx > 0$ when $x = 7000$ so profit is increasing at this production level.

23. Let h and r be the dimensions shown in the figure, then the volume is $V = \dfrac{1}{3}\pi r^2 h$.

But $r^2 + h^2 = L^2$ thus $r^2 = L^2 - h^2$ so

$V = \dfrac{1}{3}\pi(L^2 - h^2)h = \dfrac{1}{3}\pi(L^2 h - h^3)$

for $0 \le h \le L$. $\dfrac{dV}{dh} = \dfrac{1}{3}\pi(L^2 - 3h^2)$.

$\dfrac{dV}{dh} = 0$ when $h = L/\sqrt{3}$. If $h = 0, L/\sqrt{3}, 0$

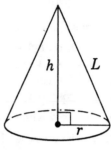

then $V = 0, \dfrac{2\pi}{9\sqrt{3}}L^3, 0$ so the volume is as large as possible when $h = L/\sqrt{3}$ and $r = \sqrt{2/3}\,L$.

25. Let r and h be the dimensions shown in the figure, then the surface area is $S = 2\pi rh + 2\pi r^2$.

But $r^2 + \left(\dfrac{h}{2}\right)^2 = R^2$ thus $h = 2\sqrt{R^2 - r^2}$ so

$S = 4\pi r\sqrt{R^2 - r^2} + 2\pi r^2$ for $0 \le r \le R$,

$\dfrac{dS}{dr} = \dfrac{4\pi(R^2 - 2r^2)}{\sqrt{R^2 - r^2}} + 4\pi r$. $\dfrac{dS}{dr} = 0$ when

$$\dfrac{R^2 - 2r^2}{\sqrt{R^2 - r^2}} = -r \qquad \text{(i)}$$

$$R^2 - 2r^2 = -r\sqrt{R^2 - r^2}$$

$R^4 - 4R^2 r^2 + 4r^4 = r^2(R^2 - r^2)$

$5r^2 - 5R^2 r^2 + R^4 = 0$

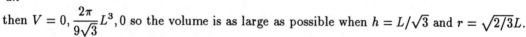

and using the quadratic formula $r^2 = \dfrac{5R^2 \pm \sqrt{25R^4 - 20R^4}}{10} = \dfrac{5 \pm \sqrt{5}}{10}R^2$, $r = \sqrt{\dfrac{5 \pm \sqrt{5}}{10}}R$, of

which only $r = \sqrt{\dfrac{5 + \sqrt{5}}{10}}R$ satisfies (i). If $r = 0, \sqrt{\dfrac{5 + \sqrt{5}}{10}}R, 0$ then $S = 0, (5 + \sqrt{5})\pi R^2, 2\pi R^2$

so the surface area is greatest when $r = \sqrt{\dfrac{5 + \sqrt{5}}{10}}R$ and, from $h = 2\sqrt{R^2 - r^2}$,

$h = 2\sqrt{\dfrac{5 - \sqrt{5}}{10}}R.$

27. Let b and h be the dimensions shown in
the figure, then the cross-sectional
area is $A = \dfrac{1}{2}h(5 + b)$. But $h = 5\sin\theta$

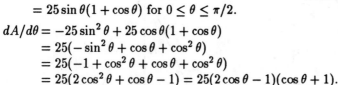

and $b = 5 + 2(5\cos\theta) = 5 + 10\cos\theta$

so $A = \dfrac{5}{2}\sin\theta(10 + 10\cos\theta)$

$\qquad = 25\sin\theta(1 + \cos\theta)$ for $0 \le \theta \le \pi/2$.

$dA/d\theta = -25\sin^2\theta + 25\cos\theta(1 + \cos\theta)$
$\qquad = 25(-\sin^2\theta + \cos\theta + \cos^2\theta)$
$\qquad = 25(-1 + \cos^2\theta + \cos\theta + \cos^2\theta)$
$\qquad = 25(2\cos^2\theta + \cos\theta - 1) = 25(2\cos\theta - 1)(\cos\theta + 1)$.

$dA/d\theta = 0$ for $0 < \theta < \pi/2$ when $\cos\theta = 1/2$, $\theta = \pi/3$. If $\theta = 0, \pi/3, \pi/2$ then
$A = 0, 75\sqrt{3}/4, 25$ so the cross-sectional area is greatest when $\theta = \pi/3$.

29. Let r and h be the radius and height of the cone
(see figure). The slant height of any such cone
will be R, the radius of the circular sheet.
Refer to the solution of Exercise 23 to find
that the largest volume is $\dfrac{2\pi}{9\sqrt{3}}R^3$.

31. (a) $C'(x) = 4 + 0.2x$, $C'(100) = 24$ (b) $C'(100) = 24$
 (c) $C(101) - C(100) = 24.1$
 (d) $R(x) = 10x$, $R'(x) = 10$;
 $\qquad P(x) = R(x) - C(x)$, $P'(x) = R'(x) - C'(x) = 10 - (4 + 0.2x) = 6 - 0.2x$

33. Let $P(x, y)$ be a point on the curve $x^2 + y^2 = 1$. The distance between $P(x, y)$ and $P_0(2, 0)$ is
$D = \sqrt{(x - 2)^2 + y^2}$, but $y^2 = 1 - x^2$ so $D = \sqrt{(x - 2)^2 + 1 - x^2} = \sqrt{5 - 4x}$ for $-1 \le x \le 1$,
$\dfrac{dD}{dx} = -\dfrac{2}{\sqrt{5 - 4x}}$ which has no critical points for $-1 < x < 1$. If $x = -1, 1$ then $D = 3, 1$ so
the closest point occurs when $x = 1$ and $y = 0$.

35. The area of the window is $A = 2rh + \pi r^2/2$,

the perimeter is $p = 2r + 2h + \pi r$ thus

$h = \dfrac{1}{2}[p - (2 + \pi)r]$ so

$A = r[p - (2 + \pi)r] + \pi r^2/2$
$\quad = pr - (2 + \pi/2)r^2$ for $0 \le r \le p/(2 + \pi)$,

$dA/dr = p - (4 + \pi)r$, $dA/dr = 0$ when

$r = p/(4 + \pi)$. $d^2A/dr^2 < 0$, so A is

maximum when $r = p/(4 + \pi)$.

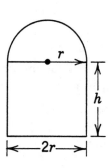

37. **(a)** Let $x =$ diameter of the sphere, $y =$ length of an edge of the cube. The combined

volume is $V = \dfrac{1}{6}\pi x^3 + y^3$ and the surface area is $S = \pi x^2 + 6y^2 =$ constant. Thus

$y = \dfrac{(S - \pi x^2)^{1/2}}{6^{1/2}}$ and $V = \dfrac{\pi}{6}x^3 + \dfrac{(S - \pi x^2)^{3/2}}{6^{3/2}}$ for $0 \le x \le \sqrt{\dfrac{S}{\pi}}$;

$\dfrac{dV}{dx} = \dfrac{\pi}{2}x^2 - \dfrac{3\pi}{6^{3/2}}x(S - \pi x^2)^{1/2} = \dfrac{\pi}{2\sqrt{6}}x(\sqrt{6}x - \sqrt{S - \pi x^2})$. $\dfrac{dV}{dx} = 0$ when $x = 0$, or

when $\sqrt{6}x = \sqrt{S - \pi x^2}$, $6x^2 = S - \pi x^2$, $x^2 = \dfrac{S}{6 + \pi}$, $x = \sqrt{\dfrac{S}{6 + \pi}}$. If $x = 0$, $\sqrt{\dfrac{S}{6 + \pi}}$,

$\sqrt{\dfrac{S}{\pi}}$, then $V = \dfrac{S^{3/2}}{6^{3/2}}, \dfrac{S^{3/2}}{6\sqrt{6 + \pi}}, \dfrac{S^{3/2}}{6\sqrt{\pi}}$ so that V is smallest when $x = \sqrt{\dfrac{S}{6 + \pi}}$, and hence

when $y = \sqrt{\dfrac{S}{6 + \pi}}$, thus $x = y$.

(b) From part (a), the sum of the volumes is greatest when there is no cube.

39. **(a)** Let $y = x^2 - 3x + 2$ for $1 \le x \le 5/2$, then $dy/dx = 2x - 3$, $dy/dx = 0$ when $x = 3/2$.
If $x = 1, 3/2, 5/2$ then $y = 0, -1/4, 3/4$ so $-1/4 \le x^2 - 3x + 2 \le 3/4$ for $1 \le x \le 5/2$.
Thus $|x^2 - 3x + 2| \le 3/4$ so $M = 3/4$.

(b) There are no critical values for $3/2 < x < 7/4$. If $x = 3/2, 7/4$ then $y = -1/4, -3/16$ so
$-1/4 \le x^2 - 3x + 2 \le -3/16$. Thus $|x^2 - 3x + 2| \ge 3/16$ so $m = 3/16$.

41. Let $v =$ speed of light in the medium. The total time required for the light to travel from A
to P to B is

$t = $ (total distance from A to P to B)$/v = \dfrac{1}{v}(\sqrt{(c - x)^2 + a^2} + \sqrt{x^2 + b^2})$,

$\dfrac{dt}{dx} = \dfrac{1}{v}\left[-\dfrac{c - x}{\sqrt{(c - x)^2 + a^2}} + \dfrac{x}{\sqrt{x^2 + b^2}}\right]$

and $\dfrac{dt}{dx} = 0$ when $\dfrac{x}{\sqrt{x^2 + b^2}} = \dfrac{c - x}{\sqrt{(c - x)^2 + a^2}}$. But $x/\sqrt{x^2 + b^2} = \sin\theta_2$ and

$(c - x)/\sqrt{(c - x)^2 + a^2} = \sin\theta_1$ thus $dt/dx = 0$ when $\sin\theta_2 = \sin\theta_1$ so $\theta_2 = \theta_1$.

43. **(a)** The rate at which the farmer walks is analogous to the speed of light in Fermat's principle.

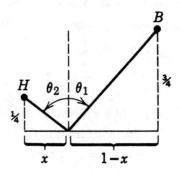

(b) the best path occurs when $\theta_1 = \theta_2$ (see figure).

(c) by similar triangles,
$$x/(1/4) = (1-x)/(3/4)$$
$$3x = 1-x$$
$$4x = 1$$
$$x = 1/4.$$

EXERCISE SET 4.8

1. Let x and y be two numbers, then their product is $P = xy$. But $x + y = 20$ thus $y = 20 - x$ so
$$P = x(20 - x) = 20x - x^2 \text{ for } -\infty < x < +\infty,$$
$$dP/dx = 20 - 2x, \ dP/dx = 0 \text{ when } x = 10, \ d^2P/dx^2 = -2 \text{ so}$$

(a) P is a maximum when $x = 10$ and $y = 10$ and

(b) P has no minimum.

3. Let $x =$ length of each side that uses the \$1 per foot fencing,
$y =$ length of each side that uses the \$2 per foot fencing.

The cost is $C = (1)(2x) + (2)(2y) = 2x + 4y$, but $A = xy = 3200$ thus $y = 3200/x$ so
$$C = 2x + 12800/x \text{ for } x > 0,$$
$$dC/dx = 2 - 12800/x^2, \ dC/dx = 0 \text{ when } x = 80, \ d^2C/dx^2 > 0 \text{ so}$$

C is least when $x = 80$, $y = 40$.

5. Let $x =$ length of each edge of base, $y =$ height, $k =$ \$/cm² for the sides. The cost is
$$C = (2k)(2x^2) + (k)(4xy) = 4k(x^2 + xy), \text{ but } V = x^2y = 2000 \text{ thus } y = 2000/x^2 \text{ so}$$
$$C = 4k(x^2 + 2000/x) \text{ for } x > 0 \ dC/dx = 4k(2x - 2000/x^2), \ dC/dx = 0 \text{ when}$$
$x = \sqrt[3]{1000} = 10, \ d^2C/dx^2 > 0 \text{ so } C \text{ is least when } x = 10, \ y = 20.$

7. Let $x =$ height and width, $y =$ length. The surface area is $S = 2x^2 + 3xy$ where $x^2y = V$, so
$y = V/x^2$ and $S = 2x^2 + 3V/x$ for $x > 0$; $dS/dx = 4x - 3V/x^2$, $dS/dx = 0$ when $x = \sqrt[3]{3V/4}$,
$d^2S/dx^2 > 0$ so S is minimum when $x = \sqrt[3]{\dfrac{3V}{4}}$, $y = \dfrac{4}{3}\sqrt[3]{\dfrac{3V}{4}}$.

9. The surface area is $S = \pi r^2 + 2\pi rh$

where $V = \pi r^2 h = 500$ so $h = 500/(\pi r^2)$

and $S = \pi r^2 + 1000/r$ for $r > 0$;

$dS/dr = 2\pi r - 1000/r^2 = (2\pi r^3 - 1000)/r^2$,

$dS/dr = 0$ when $r = \sqrt[3]{500/\pi}$, $d^2S/dr^2 > 0$

for $r > 0$ so S is minimum when

$r = \sqrt[3]{500/\pi}$ and

$h = \dfrac{500}{\pi r^2} = \dfrac{500}{\pi r^3} \, r = \dfrac{500}{\pi(500/\pi)} \sqrt[3]{500/\pi}$

$\qquad = \sqrt[3]{500/\pi}.$

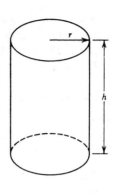

11. The area of the paper is

$A = \pi r L = \pi r \sqrt{r^2 + h^2}$, but

$V = \dfrac{1}{3}\pi r^2 h = 10$ thus $h = 30/(\pi r^2)$

so $A = \pi r \sqrt{r^2 + 900/(\pi^2 r^4)}$.

To simplify the computations let $S = A^2$,

$S = \pi^2 r^2 \left(r^2 + \dfrac{900}{\pi^2 r^4} \right) = \pi^2 r^4 + \dfrac{900}{r^2}$ for $r > 0$,

$\dfrac{dS}{dr} = 4\pi^2 r^3 - \dfrac{1800}{r^3} = \dfrac{4(\pi^2 r^6 - 450)}{r^3}$, $dS/dr = 0$ when $r = \sqrt[6]{450/\pi^2}$,

$d^2S/dr^2 > 0$, so S and hence A is least when $r = \sqrt[6]{450/\pi^2}$, $h = \dfrac{30}{\pi}\sqrt[3]{\pi^2/450}$.

13. Let (x, y) be a point on the curve, then the square of the distance between (x, y) and $(0, 2)$ is
$S = x^2 + (y - 2)^2$ where $x^2 - y^2 = 1$, $x^2 = y^2 + 1$ so
$S = (y^2 + 1) + (y - 2)^2 = 2y^2 - 4y + 5$ for any y, $dS/dy = 4y - 4$, $dS/dy = 0$ when $y = 1$,
$d^2S/dy^2 > 0$ so S is least when $y = 1$ and $x = \pm\sqrt{2}$.

15. The area of the triangle is $A = \dfrac{1}{2}ab$. Equate slopes to get $\dfrac{b - 3}{0 - 1} = \dfrac{0 - 3}{a - 1}$, $b = \dfrac{3a}{a - 1}$ so

$A = \dfrac{3}{2}\dfrac{a^2}{a - 1}$ for $a > 1$, $\dfrac{dA}{da} = \dfrac{3a(a - 2)}{(a - 1)^2}$, $\dfrac{dA}{da} = 0$ for $a > 1$ when $a = 2$.

(a) there is no maximum for A because $\displaystyle\lim_{a \to 1+} A = +\infty$.

(b) by the first derivative test A is minimum when $a = 2$ so the slope is $m = \dfrac{0 - 3}{2 - 1} = -3$.

17. The distance between the particles is $D = \sqrt{(1 - t - t)^2 + (t - 2t)^2} = \sqrt{5t^2 - 4t + 1}$ for $t \geq 0$. For convenience, we minimize D^2 instead, so $D^2 = 5t^2 - 4t + 1$, $dD^2/dt = 10t - 4$, which is 0 when $t = 2/5$. $d^2 D^2/dt^2 > 0$ so D^2 and hence D is minimum when $t = 2/5$. The minimum distance is $D = 1/\sqrt{5}$.

19. If $P(x_0, y_0)$ is on the curve $y = 1/x^2$, then $y_0 = 1/x_0^2$. At P the slope of the tangent line is $-2/x_0^3$ so its equation is $y - \dfrac{1}{x_0^2} = -\dfrac{2}{x_0^3}(x - x_0)$, or $y = -\dfrac{2}{x_0^3}x + \dfrac{3}{x_0^2}$. The tangent line crosses the y-axis at $\dfrac{3}{x_0^2}$, and the x-axis at $\dfrac{3}{2}x_0$. The length of the segment then is

$$L = \sqrt{\frac{9}{x_0^4} + \frac{9}{4}x_0^2}$$ for $x_0 > 0$. For convenience, we minimize L^2 instead, so $L^2 = \dfrac{9}{x_0^4} + \dfrac{9}{4}x_0^2$,

$\dfrac{dL^2}{dx_0} = -\dfrac{36}{x_0^5} + \dfrac{9}{2}x_0 = \dfrac{9(x_0^6 - 8)}{2x_0^5}$, which is 0 when $x_0^6 = 8$, $x_0 = \sqrt{2}$. $\dfrac{d^2 L^2}{dx_0^2} > 0$ so L^2 and hence L is minimum when $x_0 = \sqrt{2}$, $y_0 = 1/2$.

21. At each point (x, y) on the curve the slope of the tangent line is $m = \dfrac{dy}{dx} = -\dfrac{2x}{(1 + x^2)^2}$ for any x, $\dfrac{dm}{dx} = \dfrac{2(3x^2 - 1)}{(1 + x^2)^3}$, $\dfrac{dm}{dx} = 0$ when $x = \pm 1/\sqrt{3}$, by the first derivative test the only relative maximum occurs at $x = -1/\sqrt{3}$, which is the absolute maximum because $\lim\limits_{x \to \pm\infty} m = 0$. The tangent line has greatest slope at the point $(-1/\sqrt{3}, 3/4)$.

23. The volume of the cone is $V = \dfrac{1}{3}\pi r^2 h$.

By similar triangles (see figure)

$\dfrac{r}{h} = \dfrac{R}{\sqrt{h^2 - 2Rh}}$, $r = \dfrac{Rh}{\sqrt{h^2 - 2Rh}}$ so

$V = \dfrac{1}{3}\pi R^2 \dfrac{h^3}{h^2 - 2Rh} = \dfrac{1}{3}\pi R^2 \dfrac{h^2}{h - 2R}$

for $h > 2R$, $\dfrac{dV}{dh} = \dfrac{1}{3}\pi R^2 \dfrac{h(h - 4R)}{(h - 2R)^2}$,

$\dfrac{dV}{dh} = 0$ for $h > 2R$ when $h = 4R$,

by the first derivative test V is

minimum when $h = 4R$. If $h = 4R$ then $r = \sqrt{2}R$.

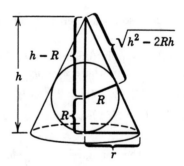

25. With x and y as shown in the figure, the maximum length of pipe will be the smallest value of $L = x + y$. By similar triangles

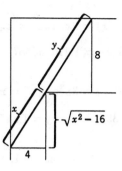

$$\frac{y}{8} = \frac{x}{\sqrt{x^2 - 16}}, \quad y = \frac{8x}{\sqrt{x^2 - 16}} \text{ so}$$

$$L = x + \frac{8x}{\sqrt{x^2 - 16}} \text{ for } x > 4,$$

$$\frac{dL}{dx} = 1 - \frac{128}{(x^2 - 16)^{3/2}}, \quad \frac{dL}{dx} = 0 \text{ when}$$

$$(x^2 - 16)^{3/2} = 128$$
$$x^2 - 16 = 128^{2/3} = 16(2^{2/3})$$
$$x^2 = 16(1 + 2^{2/3})$$
$$x = 4(1 + 2^{2/3})^{1/2},$$

$d^2 L/dx^2 = 384x/(x^2 - 16)^{5/2} > 0$ if $x > 4$ so L is smallest when $x = 4(1 + 2^{2/3})^{1/2}$. For this value of x, $L = 4(1 + 2^{2/3})^{3/2}$.

27. Let $x = $ distance from the weaker light source, $I = $ the intensity at that point, and k the constant of proportionality. Then

$$I = \frac{kS}{x^2} + \frac{8kS}{(90 - x)^2} \text{ for } 0 < x < 90; \quad \frac{dI}{dx} = -\frac{2kS}{x^3} + \frac{16kS}{(90 - x)^3} = \frac{2kS[8x^3 - (90 - x)^3]}{x^3(90 - x)^3},$$

which is 0 when $8x^3 = (90 - x)^3$, $2x = 90 - x$, $x = 30$. $\dfrac{dI}{dx} < 0$ if $x < 30$, and $\dfrac{dI}{dx} > 0$ if $x > 30$, so the intensity is minimum at a distance of 30 cm from the weaker source.

29. Minimize $S = L^2 = (x - x_1)^2 + (y - y_1)^2$ where $ax + by + c = 0$. If $b \neq 0$ then $y = -\dfrac{a}{b}x - \dfrac{c}{b}$ so

$$S = (x - x_1)^2 + \left(-\frac{a}{b}x - \frac{c}{b} - y_1\right)^2, \quad dS/dx = 2(x - x_1)^2 + 2\left(-\frac{a}{b}x - \frac{c}{b} - y_1\right)^2\left(-\frac{a}{b}\right),$$

$dS/dx = 0$ when $x = \dfrac{b^2 x_1 - aby_1 - ac}{a^2 + b^2}$, $d^2 S/dx^2 = 2(1 + a^2/b^2) > 0$ so S and hence L is minimum. Substitution of this value into the formula for S and simplification eventually gives $S = \dfrac{(ax_1 + by_1 + c)^2}{a^2 + b^2}$ so $L = \dfrac{|ax_1 + by_1 + c|}{\sqrt{a^2 + b^2}}$. The special case for $b = 0$ is treated in a similar way.

EXERCISE SET 4.9

1. $f(x) = x^2 - 2$, $f'(x) = 2x$, $x_{n+1} = x_n - \dfrac{x_n^2 - 2}{2x_n}$

    $x_1 = 1$, $x_2 = 1.5$, $x_3 = 1.416666667, \cdots$, $x_5 = x_6 = 1.414213562$

3. $f(x) = x^3 - 6$, $f'(x) = 3x^2$, $x_{n+1} = x_n - \dfrac{x_n^3 - 6}{3x_n^2}$

    $x_1 = 2$, $x_2 = 1.833333333$, $x_3 = 1.817263545, \cdots$, $x_5 = x_6 = 1.817120593$

5. $f(x) = x^3 - x + 3$, $f'(x) = 3x^2 - 1$, $x_{n+1} = x_n - \dfrac{x_n^3 - x_n + 3}{3x_n^2 - 1}$

    $x_1 = -2$, $x_2 = -1.727272727$, $x_3 = -1.673691174, \cdots$, $x_5 = x_6 = -1.671699882$

7. $f(x) = x^5 + x^4 - 5$, $f'(x) = 5x^4 + 4x^3$, $x_{n+1} = x_n - \dfrac{x_n^5 + x_n^4 - 5}{5x_n^4 + 4x_n^3}$

    $x_1 = 1$, $x_2 = 1.333333333$, $x_3 = 1.239420573, \cdots$, $x_6 = x_7 = 1.224439550$

9. $f(x) = 2x^2 + 4x - 3$, $f'(x) = 4x + 4$, $x_{n+1} = x_n - \dfrac{2x_n^2 + 4x_n - 3}{4x_n + 4}$

    $x_1 = 1$, $x_2 = 0.625$, $x_3 = 0.581730769, \cdots$, $x_5 = x_6 = 0.581138830$

11. $f(x) = x^4 + x - 3$, $f'(x) = 4x^3 + 1$, $x_{n+1} = x_n - \dfrac{x_n^4 + x_n - 3}{4x_n^3 + 1}$

    $x_1 = -2$, $x_2 = -1.645161290$, $x_3 = -1.485723955, \cdots$, $x_6 = x_7 = -1.452626879$

13. $f(x) = 2\sin x - x$, $f'(x) = 2\cos x - 1$, $x_{n+1} = x_n - \dfrac{2\sin x_n - x_n}{2\cos x_n - 1}$

    $x_1 = 2$, $x_2 = 1.900995594$, $x_3 = 1.895511645$, $x_4 = x_5 = 1.895494267$

15. $f(x) = x - \tan x$, $f'(x) = 1 - \sec^2 x = -\tan^2 x$, $x_{n+1} = x_n + \dfrac{x_n - \tan x_n}{\tan^2 x_n}$

    $x_1 = 4.5$, $x_2 = 4.493613903$, $x_3 = 4.493409655$, $x_4 = x_5 = 4.493409458$

17. (a) $f(x) = x^2 - a$, $f'(x) = 2x$, $x_{n+1} = \dfrac{1}{2}\left(x_n + \dfrac{a}{x_n}\right)$

    (b) $a = 10$; $x_1 = 3$, $x_2 = 3.166666667$, $x_3 = 3.162280702$, $x_4 = x_5 = 3.162277660$

19. At the point of intersection, $x^3 = 0.5x - 1$, $x^3 - 0.5x + 1 = 0$. Let $f(x) = x^3 - 0.5x + 1$. By graphing $y = x^3$ and $y = 0.5x - 1$ it is evident that there is only one point of intersection and it occurs in the interval $[-2, -1]$; note that $f(-2) < 0$ and $f(-1) > 0$. $f'(x) = 3x^2 - 0.5$ so

$$x_{n+1} = x_n - \frac{x_n^3 - 0.5x + 1}{3x_n^2 - 0.5}; \ x_1 = -1, \ x_2 = -1.2, \ x_3 = -1.166492147, \cdots,$$

$$x_5 = x_6 = -1.165373043$$

21. The graphs of $y = x^2$ and $y = \sqrt{2x+1}$ intersect at points near $x = -0.5$ and $x = 1$; $x^2 = \sqrt{2x+1}$, $x^4 - 2x - 1 = 0$. Let $f(x) = x^4 - 2x - 1$, then $f'(x) = 4x^3 - 2$ so

$$x_{n+1} = x_n - \frac{x_n^4 - 2x_n - 1}{4x_n^3 - 2}.$$

If $x_1 = -0.5$, then $x_2 = -0.475$, $x_3 = -0.474626695$, $x_4 = x_5 = -0.474626618$;
if $x_1 = 1$, then $x_2 = 2$, $x_3 = 1.633333333, \cdots, x_8 = x_9 = 1.395336994$.

23. If $x = 1$, then $y^4 + y = 1$, $y^4 + y - 1 = 0$. Graph $z = y^4$ and $z = 1 - y$ to see that they intersect near $y = -1$ and $y = 1$. Let $f(y) = y^4 + y - 1$, then $f'(y) = 4y^3 + 1$ so $y_{n+1} = y_n - \dfrac{y_n^4 + y_n - 1}{4y_n^3 + 1}$.

If $y_1 = -1$, then $y_2 = -1.333333333$, $y_3 = -1.235807860, \cdots, y_6 = y_7 = -1.220744085$;
if $y_1 = 1$, then $y_2 = 0.8$, $y_3 = 0.731233596, \cdots, y_6 = y_7 = 0.724491959$.

25. $f'(x) = x^3 + 2x + 5$; solve $f'(x) = 0$ to find the critical points. Graph $y = x^3$ and $y = -2x - 5$ to see that they intersect at a point near $x = -1$; $f''(x) = 3x^2 + 2$ so $x_{n+1} = x_n - \dfrac{x_n^3 + 2x_n + 5}{3x_n^2 + 2}$.

$x_1 = -1$, $x_2 = -1.4$, $x_3 = -1.330964467, \cdots, x_5 = x_6 = -1.328268856$ so the minimum value of $f(x)$ occurs at $x \approx -1.328268856$ because $f''(x) > 0$; its value is approximately -4.098859132.

27. Let $f(x)$ be the square of the distance between $(1,0)$ and any point (x, x^2) on the parabola, then $f(x) = (x-1)^2 + (x^2 - 0)^2 = x^4 + x^2 - 2x + 1$ and $f'(x) = 4x^3 + 2x - 2$. Solve $f'(x) = 0$ to find the critical points; $f''(x) = 12x^2 + 2$ so $x_{n+1} = x_n - \dfrac{4x_n^3 + 2x_n - 2}{12x_n^2 + 2} = x_n - \dfrac{2x_n^3 + x_n - 1}{6x_n^2 + 1}$.

$x_1 = 1$, $x_2 = 0.714285714$, $x_3 = 0.605168701, \cdots, x_6 = x_7 = 0.589754512$; the coordinates are approximately $(0.589754512, 0.347810385)$.

29. Let s be the arc length, and L the length of the chord, then $s = 1.5L$. But $s = r\theta$ and $L = 2r\sin(\theta/2)$ so $r\theta = 3r\sin(\theta/2)$, $\theta - 3\sin(\theta/2) = 0$. Let $f(\theta) = \theta - 3\sin(\theta/2)$, then $f'(\theta) = 1 - 1.5\cos(\theta/2)$ so $\theta_{n+1} = \theta_n - \dfrac{\theta_n - 3\sin(\theta_n/2)}{1 - 1.5\cos(\theta_n/2)}$.

$\theta_1 = 3$, $\theta_2 = 2.991592920$, $\theta_3 = 2.991563137$, $\theta_4 = \theta_5 = 2.991563136$ rad so $\theta \approx 171°$.

EXERCISE SET 4.10

1. $f(2) = f(4) = 0$, $f'(x) = 2x - 6$, $2c - 6 = 0$, $c = 3$

3. $f(\pi/2) = f(3\pi/2) = 0$, $f'(x) = -\sin x$, $-\sin c = 0$, $c = \pi$

5. $f(0) = f(4) = 0$, $f'(x) = \dfrac{1}{2} - \dfrac{1}{2\sqrt{x}}$, $\dfrac{1}{2} - \dfrac{1}{2\sqrt{c}} = 0$, $c = 1$

7. $f(-4) = 12$, $f(6) = 42$, $f'(x) = 2x + 1$, $2c + 1 = \dfrac{42 - 12}{6 - (-4)} = 3$, $c = 1$

9. $f(0) = 1$, $f(3) = 2$, $f'(x) = \dfrac{1}{2\sqrt{x+1}}$, $\dfrac{1}{2\sqrt{c+1}} = \dfrac{2-1}{3-0} = \dfrac{1}{3}$
     $\sqrt{c+1} = 3/2$, $c + 1 = 9/4$, $c = 5/4$

11. $f(-5) = 0$, $f(3) = 4$, $f'(x) = -\dfrac{x}{\sqrt{25 - x^2}}$, $-\dfrac{c}{\sqrt{25 - c^2}} = \dfrac{4-0}{3-(-5)} = \dfrac{1}{2}$, $-2c = \sqrt{25 - c^2}$,
     $4c^2 = 25 - c^2$, $c^2 = 5$, $c = -\sqrt{5}$
     (we reject $c = \sqrt{5}$ because it does not satisfy the equation $-2c = \sqrt{25 - c^2}$)

13. **(a)** $f'(x) = \sec^2 x$, $\sec^2 c = 0$ has no solution   **(b)** $\tan x$ is not continuous on $[0, \pi]$

15. Let $f(x) = \sin x$ and $x \neq y$. By the Mean-Value Theorem there is a number c between x and
     y such that
     $$\frac{\sin x - \sin y}{x - y} = \cos c, \quad \frac{|\sin x - \sin y|}{|x - y|} = |\cos c| \leq 1$$
     so $|\sin x - \sin y| \leq |x - y|$, which also holds when $x = y$.

17. Let $f(x) = \sqrt{x}$. By the Mean-Value Theorem there is a number c between x and y such that
     $$\frac{\sqrt{y} - \sqrt{x}}{y - x} = \frac{1}{2\sqrt{c}} < \frac{1}{2\sqrt{x}} \text{ for } c \text{ in } (x, y), \text{ thus } \sqrt{y} - \sqrt{x} < \frac{y - x}{2\sqrt{x}};$$
     multiply through and rearrange to get $\sqrt{xy} < \dfrac{1}{2}(x + y)$.

19. $f'(x) = 2a_2 x + a_1$,
     $$2a_2 c + a_1 = \frac{(a_2 b^2 + a_1 b + a_0) - (a_2 a^2 + a_1 a + a_0)}{b - a}$$
     $$= \frac{a_2(b^2 - a^2) + a_1(b - a)}{b - a} = a_2(b + a) + a_1, \text{ so } c = \frac{1}{2}(b + a).$$

21. $f(0) = f(1) = 0$, $f'(x) = 3ax^2 + 2bx - (a + b)$, so there is at least one number c in $(0, 1)$ where $f'(c) = 0$.

23. Let $f(x) = x^3 + 4x - 1$. Assume that $f(x) = 0$ has at least two distinct real solutions r_1 and r_2. Then $f(r_1) = f(r_2) = 0$ and so by Rolle's Theorem there is at least one number c between r_1 and r_2 where $f'(c) = 0$. But $f'(x) = 3x^2 + 4$ is never zero, so $f(x) = 0$ must have fewer than two distinct real solutions.

25. Assume that $f(x) = 0$ has at least four distinct real solutions $r_1 < r_2 < r_3 < r_4$, then by Rolle's Theorem there is at least one number in each of the intervals (r_1, r_2), (r_2, r_3) and (r_3, r_4) so that $f'(x) = 0$ at least three times. Apply Rolle's Theorem to $f'(x)$ to show that $f''(x) = 0$ at least twice; again to $f''(x)$ to show that $f'''(x) = 0$ at least once. But $f'''(x) = 60x^2 + 24ax + 6b$, and $f'''(x) = 0$ if $10x^2 + 4ax + b = 0$. Use the quadratic formula to get $x = \dfrac{-4a \pm \sqrt{16a^2 - 40b}}{20}$, which has no real solutions if $16a^2 - 40b < 0$, $16a^2 < 40b$, $2a^2 < 5b$.

27. $\dfrac{d}{dx}[f^2(x) + g^2(x)] = 2f(x)f'(x) + 2g(x)g'(x) = 2f(x)g(x) + 2g(x)[-f(x)] = 0$, so $f^2(x) + g^2(x)$ is constant.

29. $f'(x) = 3(x - 1)^2$, $g'(x) = (x^2 + 3) + 2x(x - 3) = 3x^2 - 6x + 3 = 3(x^2 - 2x + 1) = 3(x - 1)^2$, so $f'(x) = g'(x)$ and hence $f(x) - g(x) = k$. Expand $f(x)$ and $g(x)$ to get $f(x) - g(x) = (x^3 - 3x^2 + 3x - 1) - (x^3 - 3x^2 + 3x - 9) = 8$.

31. If $f'(x) = g'(x)$, then $f(x) = g(x) + k$. Let $x = 1$,
$f(1) = g(1) + k = (1)^3 - 4(1) + 6 + k = 3 + k = 2$, so $k = -1$. $f(x) = x^3 - 4x + 5$.

33. Let $h = f - g$, then h is continuous on $[a, b]$, differentiable on (a, b), and $h(a) = f(a) - g(a) = 0$, $h(b) = f(b) - g(b) = 0$. By Rolle's Theorem there is some c in (a, b) where $h'(c) = 0$. But $h'(c) = f'(c) - g'(c)$ so $f'(c) - g'(c) = 0$, $f'(c) = g'(c)$.

35. Similar to the proof of part (a) with $f'(c) = 0$.

37. From the Mean-Value Theorem there is a point c in (a, b) where
$$f'(c) = \frac{f(b) - f(a)}{b - a} = \frac{0}{b - a} = 0.$$

39. Similar to proof given in text; assume that $f(x) < 0$ and replace the word "maximum" by "minimum".

41.   Let $s(t)$ be the position function of the automobile for $0 \leq t \leq 5$, then by the Mean-Value Theorem there is at least one point c in $(0,5)$ where

$$s'(c) = v(c) = [s(5) - s(0)]/(5-0) = 4/5 = 0.8 \text{ mi/min} = 48 \text{ mi/hr}.$$

EXERCISE SET 4.11

1.   **(a)** positive, negative, slowing down   **(b)** positive, positive, speeding up
       (c) negative, positive, slowing down

3.   **(a)** left because $v = ds/dt < 0$ at t_0.
       (b) negative because $a = d^2s/dt^2$ and the curve is concave down at $t_0(d^2s/dt^2 < 0)$.
       (c) speeding up because v and a have the same sign.
       (d) $v < 0$ and $a > 0$ at t_1 so the particle is slowing down because v and a have opposite signs.

5.   **(a)** At 60 mi/hr the slope of the estimated tangent line is about 4.6 mi/hr per sec. Use 1 mi $= 5,280$ ft and 1 hr $= 3600$ sec to get $a = dv/dt \approx 4.6(5,280)/(3600) \approx 6.7$ ft/sec^2.
       (b) The slope of the tangent to the curve is maximum at $t = 0$.

7. $v = 3t^2 - 12t$, $a = 6t - 12$

t	s	v	$|v|$	a	direction; motion
1	-5	-9	9	-6	left; speeding up
2	-16	-12	12	0	left; neither
3	-27	-9	9	6	left; slowing down
4	-32	0	0	12	stopped
5	-25	15	15	18	right; speeding up

9.   **(a)** $v = 10t - 22$, speed $= |v| = |10t - 22|$. $d|v|/dt$ does not exist at $t = 2.2$ which is the only critical point. If $t = 1, 2.2, 3$ then $|v| = 12, 0, 8$. The maximum speed is 12.
       (b) the distance from the origin is $|s| = |5t^2 - 22t| = |t(5t - 22)|$, but $t(5t - 22) < 0$ for $1 \leq t \leq 3$ so $|s| = -(5t^2 - 22t) = 22t - 5t^2$, $d|s|/dt = 22 - 10t$, thus the only critical point is $t = 2.2$. $d^2|s|/dt^2 < 0$ so the particle is farthest from the origin when $t = 2.2$. Its position is $s = 5(2.2)^2 - 22(2.2) = -24.2$.

11. $s = 1 + 6t - t^2$
     $v = 2(3 - t)$
     $a = -2$

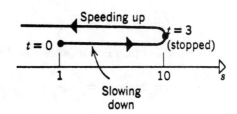

13. $s = t^3 - 9t^2 + 24t$
     $v = 3(t - 2)(t - 4)$
     $a = 6(t - 3)$

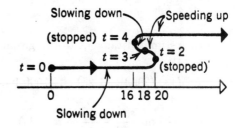

15. $s = \begin{cases} \cos t, & 0 \le t \le 2\pi \\ 1, & t > 2\pi \end{cases}$

     $v = \begin{cases} -\sin t, & 0 \le t \le 2\pi \\ 0, & t > 2\pi \end{cases}$

     $a = \begin{cases} -\cos t, & 0 \le t < 2\pi \\ 0, & t > 2\pi \end{cases}$

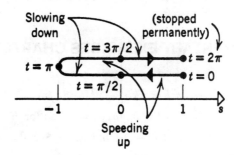

17. $s = 4t^{3/2} - 3t^2$, $v = 6t^{1/2} - 6t$, $a = 3t^{-1/2} - 6$.
     (a) $a = 0$ when $3t^{-1/2} = 6$, $t = 1/4$; $s = 5/16$, $v = 3/2$.
     (b) $v = 0$ when $6t^{1/2}(1 - t^{1/2}) = 0$ for $t > 0$, $t = 1$; $s = 1$, $a = -3$.

19. **(a)** $a = \dfrac{dv}{dt} = \dfrac{dv}{ds}\dfrac{ds}{dt} = v\dfrac{dv}{ds}$ because $v = \dfrac{ds}{dt}$.

     (b) $v = \dfrac{3}{2\sqrt{3t + 7}} = \dfrac{3}{2s}$; $\dfrac{dv}{ds} = -\dfrac{3}{2s^2}$; $a = -\dfrac{9}{4s^3} = -9/500$.

21. **(a)** $s_1 = s_2$ if they collide, so $\dfrac{1}{2}t^2 - t + 3 = -\dfrac{1}{4}t^2 + t + 1$, $\dfrac{3}{4}t^2 - 2t + 2 = 0$ which has no real solution.

(b) Find the minimum value of $D = |s_1 - s_2| = \left|\frac{3}{4}t^2 - 2t + 2\right|$. From part (a), $\frac{3}{4}t^2 - 2t + 2$

is never zero, and for $t = 0$ it is positive, hence it is always positive, so $D = \frac{3}{4}t^2 - 2t + 2$.

$\frac{dD}{dt} = \frac{3}{2}t - 2 = 0$ when $t = \frac{4}{3}$. $\frac{d^2D}{dt^2} > 0$ so D is minimum when $t = \frac{4}{3}$, $D = \frac{2}{3}$.

(c) $v_1 = t - 1$, $v_2 = -\frac{1}{2}t + 1$. $v_1 < 0$ if $0 \leq t < 1$, $v_1 > 0$ if $t > 1$; $v_2 < 0$ if $t > 2$, $v_2 > 0$ if $0 \leq t < 2$. They are moving in opposite directions during the intervals $0 \leq t < 1$ and $t > 2$.

23. **(a)** From the estimated tangent to the graph at the point where $v = 2000$, $dv/ds \approx -1.25$ ft/sec per ft.

(b) $a = v\,dv/ds \approx (2000)(-1.25) = -2500$ ft/sec^2.

25. **(a)** $\displaystyle\lim_{t_1 \to t_0} v_{ave} = \lim_{t_1 \to t_0} \frac{s(t_1) - s(t_0)}{t_1 - t_0} = s'(t_0) = v(t_0)$

(b) similar to part (a) with a in place of v.

SUPPLEMENTARY EXERCISES CHAPTER 4

1. $V = \pi R^2 h - \pi r^2 h = \pi(R^2 - r^2)h$, $dV/dt = \pi[(R^2 - r^2)dh/dt + h(2R\,dR/dt - 2r\,dr/dt)]$.
    But $dR/dt = dr/dt = 2$, $dh/dt = -3$ so for $R = 7$, $r = 4$, and $h = 5$
    $dV/dt = \pi[(49 - 16)(-3) + 5(14(2) - 8(2))] = -39\pi$. The volume is decreasing at the rate of 39π m^3/sec.

3. By similar triangles
    $x/48 = 10/y$, $x = 480/y$ so
    $dx/dt = -(480/y^2)dy/dt$. But
    $dy/dt = 32$ when $y = 16$ thus
    $dx/dt = -(480/16^2)(32) = -60$.
    The shadow is moving toward the pole
    at the rate of 60 ft/sec.

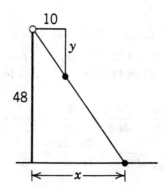

5. $f'(x) = x^2(3 - 4x)$, critical points $x = 0$, $3/4$; $f(-1) = -2$, $f(0) = 0$, $f(3/4) = 27/256$, $f(3/2) = -27/16$. $m = -2$ at $x = -1$, $M = 27/256$ at $x = 3/4$.

7. $f'(x) = 2(3 - x^2)/(x^2 + 3)^2$, critical point $x = \sqrt{3}$; $f(\sqrt{3}) = \sqrt{3}/3$, $f(2) = 4/7$, $\lim\limits_{x \to 0+} f(x) = 0$.
    No minimum on $(0, 2]$, $M = \sqrt{3}/3$ at $x = \sqrt{3}$.

9. $x^2 - 2x \geq 0$ when $x \leq 0$ or $x \geq 2$, $x^2 - 2x < 0$ when $0 < x < 2$

    $$f'(x) = \begin{cases} -2x + 2, & x < 0 \text{ or } x > 2 \\ 2x - 2, & 0 < x < 2 \end{cases}$$

    and $f'(x)$ does not exist when $x = 0, 2$. The only critical point in $(1, 3)$ is $x = 2$; $f(1) = -1$, $f(2) = 0$, $f(3) = -3$, $m = -3$ at $x = 3$, $M = 0$ at $x = 2$.

11. $f(x) = x^3 - 4x + 1$, $f'(x) = 3x^2 - 4$, $x_{n+1} = x_n - \dfrac{x_n^3 - 4x_n + 1}{3x_n^2 - 4}$

    $x_1 = -2$, $x_2 = -2.125$, $x_3 = -2.114975450, \cdots, x_5 = x_6 = -2.114907541$

    $x_1 = 0$, $x_2 = 0.25$, $x_3 = 0.254098361$, $x_4 = x_5 = 0.254101688$

    $x_1 = 2$, $x_2 = 1.875$, $x_3 = 1.860978520, \cdots, x_5 = x_6 = 1.860805853$.

13. $f'(x) = -\dfrac{2x}{(1 + x^2)^2}$

    $f''(x) = \dfrac{2(3x^2 - 1)}{(1 + x^2)^3}$

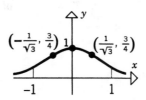

15. $f'(x) = \dfrac{2(x^3 + 1)}{x^2}$

    $f''(x) = \dfrac{2(x^3 - 2)}{x^3}$

    $f(x) = x^2 - \dfrac{2}{x}$ so $f(x)$ is

    asymptotic to $y = x^2$ for $|x|$ large.

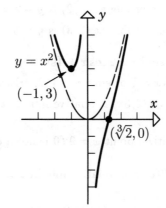

17. $f'(x) = -4\cos x \sin x$
  $\qquad = -2\sin 2x$
  $f''(x) = -4\cos 2x$

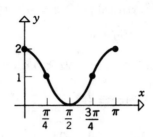

19. $f'(x) = \dfrac{3(8 - x)}{(x + 8)^3}$

  $f''(x) = \dfrac{6(x - 16)}{(x + 8)^4}$

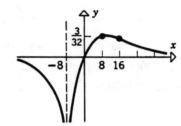

21. $f'(x) = 4x^3 - 18x^2 + 24x - 8$

  $f''(x) = 12x^2 - 36x + 24 = 12(x - 1)(x - 2)$

  $f''(x) = 0$ when $x = 1, 2$; $f(1) = 2$, $f(2) = 3$. The inflection points are $(1, 2)$ and $(2, 3)$ because the concavity changes at these points.

  $f'(1) = 2$ so the tangent line at $(1, 2)$ is $y - 2 = 2(x - 1)$, $y = 2x$.

  $f'(2) = 0$ so the tangent line at $(2, 3)$ is $y = 3$.

23. $f'(x) = 2\cos x + 2\sin 2x = 2\cos x + 4\sin x \cos x = 2\cos x(1 + 2\sin x)$;

  $f'(x) = 0$ when $\cos x = 0$ or $\sin x = -1/2$; critical points $x = \pi/2, 3\pi/2, 7\pi/6, 11\pi/6$

  relative max at $x = \pi/2, 3\pi/2$, relative min at $x = 7\pi/6, 11\pi/6$

25. $f'(x) = \dfrac{x - 9}{18x^{3/2}}$; critical point $x = 9$ (0 is not a critical point, it is not in the domain of f)

  $f''(x) = \dfrac{27 - x}{36x^{5/2}}$; $f''(9) > 0$, relative min at $x = 9$

27. $f'(x) = \sin x(2\cos x + 1)$; $f'(x) = 0$ when $\sin x = 0$ or $\cos x = -1/2$, in $(0, 2\pi)$ the critical points are $x = \pi, 2\pi/3, 4\pi/3$

$f''(x) = 2\cos 2x + \cos x;\ f''(\pi) > 0,\ f''(2\pi/3) < 0,\ f''(4\pi/3) < 0$

relative max at $x = 2\pi/3, 4\pi/3$, relative min at $x = \pi$

29. Let (x, y) be a point in the first quadrant
that is on the ellipse, then $A = (2x)(2y) = 4xy$.
But, from the equation of the ellipse,

$y^2 = \dfrac{9}{16}(16 - x^2)$ so with $S = A^2 = 16x^2 y^2$,

$S = 9x^2(16 - x^2) = 9(16x^2 - x^4)$ for $0 < x < 4$,

$dS/dx = 36x(8 - x^2)$, critical point at

$x = \sqrt{8} = 2\sqrt{2}$. $d^2 S/dx^2 > 0$ at $x = 2\sqrt{2}$ thus

S and hence A is maximum there. If $x = 2\sqrt{2}$

then $y = 3\sqrt{2}/2$. The dimensions of the

rectangle are $4\sqrt{2}$ by $3\sqrt{2}$.

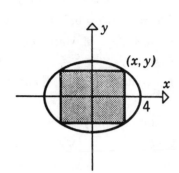

31. Let k be the amount of light admitted per unit area of clear glass. The total amount of light
admitted by the entire window is

$T = k \cdot \text{(area of clear glass)} + \dfrac{1}{2}k \cdot \text{(area of blue glass)} = 2krh + \dfrac{1}{4}\pi kr^2$.

But $P = 2h + 2r + \pi r$ which gives $h = \dfrac{1}{2}(P - 2r - \pi r)$ so

$$T = kr(P - 2r - \pi r) + \frac{1}{4}\pi k r^2 = k\left[Pr - \left(2 + \pi - \frac{\pi}{4}\right)r^2\right]$$

$$= k\left[Pr - \frac{8 + 3\pi}{4}r^2\right] \text{ for } 0 < r < \frac{P}{2 + \pi},$$

$$\frac{dT}{dr} = k\left(P - \frac{8 + 3\pi}{2}r\right), \frac{dT}{dr} = 0 \text{ when } r = \frac{2P}{8 + 3\pi}.$$

This is the only critical point and $d^2 T/dr^2 < 0$ there so the most light is admitted when
$r = 2P/(8 + 3\pi)$ ft.

33. The total area of material used is

$A = A_{\text{top}} + A_{\text{bottom}} + A_{\text{side}} = (2r)^2 + (2r)^2 + 2\pi rh = 8r^2 + 2\pi rh$.

The volume is $V = \pi r^2 h$ thus $h = V/(\pi r^2)$ so $A = 8r^2 + 2V/r$ for $r > 0$,

$dA/dr = 16r - 2V/r^2 = 2(8r^3 - V)/r^2$, $dA/dr = 0$ when $r = \sqrt[3]{V}/2$. This is the only critical

point, $d^2 A/dr^2 > 0$ there so the least material is used when $r = \sqrt[3]{V}/2$, $\dfrac{r}{h} = \dfrac{r}{V/(\pi r^2)} = \dfrac{\pi}{V}r^3$

and, for $r = \sqrt[3]{V}/2$, $\dfrac{r}{h} = \dfrac{\pi}{V}\dfrac{V}{8} = \dfrac{\pi}{8}$.

35. f is continuous on $[-2, 2]$, $f'(x) = -x/\sqrt{4 - x^2}$ so f is differentiable on $(-2, 2)$, $f(-2) = f(2) = 0$; hypotheses are satisfied. $f'(c) = 0$ for $c = 0$.

37. f is continuous on $[0, \sqrt{\pi}]$, $f'(x) = 2x \cos(x^2)$ so f is differentiable on $(0, \sqrt{\pi})$, $f(0) = f(\sqrt{\pi}) = 0$; hypotheses are satisfied. $f'(c) = 0$ when $2c \cos(c^2) = 0$ which yields $c = 0, \pm\sqrt{\pi/2}$ of which only $c = \sqrt{\pi/2}$ is in $(0, \sqrt{\pi})$.

39. f is continuous on $[0, 4]$ and differentiable on $(0, 4)$. $f'(c) = \dfrac{f(4) - f(0)}{4 - 0}$, $\dfrac{1}{2\sqrt{c}} = \dfrac{1}{2}$, $c = 1$

41. By inspection, f is continuous on $[0, 2]$ and differentiable on $(0, 2)$ except perhaps at $x = 1$. For $x = 1$, $\lim\limits_{x \to 1^-} f(x) = \lim\limits_{x \to 1^+} f(x) = f(1)$ so f is continuous at $x = 1$.

$\lim\limits_{x \to 1^-} f'(x) = \lim\limits_{x \to 1^-} (-2x) = -2$, $\lim\limits_{x \to 1^+} f'(x) = \lim\limits_{x \to 1^+} (-2/x^2) = -2$ so f is differentiable at $x = 1$ (see theorem preceding Exercise 71, Section 3.3). $f'(c) = \dfrac{f(2) - f(0)}{2 - 0} = \dfrac{1 - 3}{2} = -1$ so $c \ne 1$. If $x < 1$ then $f'(x) = -2x$ thus $f'(c) = -1$ for $c = 1/2$. If $x > 1$ then $f'(x) = -2/x^2$ thus $f'(c) = -1$ for $c = \sqrt{2}$. The values of c are $1/2$, $\sqrt{2}$.

CHAPTER 5
Integration

EXERCISE SET 5.2

1. $x^9/9 + C$

3. $\dfrac{7}{12}x^{12/7} + C$

5. $4\displaystyle\int t^{-1/2}dt = 8\sqrt{t} + C$

7. $\displaystyle\int x^{7/2}dx = \dfrac{2}{9}x^{9/2} + C$

9. $\displaystyle\int (x^{-3} + x^{1/2} - 3x^{1/4} + x^2)dx = -\dfrac{1}{2}x^{-2} + \dfrac{2}{3}x^{3/2} - \dfrac{12}{5}x^{5/4} + \dfrac{1}{3}x^3 + C$

11. $\displaystyle\int (7y^{-3/4} - y^{1/3} + 4y^{1/2})dy = 28y^{1/4} - \dfrac{3}{4}y^{4/3} + \dfrac{8}{3}y^{3/2} + C$

13. $\displaystyle\int (x + x^4)dx = x^2/2 + x^5/5 + C$

15. $\displaystyle\int x^{1/3}(4 - 4x + x^2)dx = \int (4x^{1/3} - 4x^{4/3} + x^{7/3})dx = 3x^{4/3} - \dfrac{12}{7}x^{7/3} + \dfrac{3}{10}x^{10/3} + C$

17. $\displaystyle\int (x + 2x^{-2} - x^{-4})dx = x^2/2 - 2/x + 1/(3x^3) + C$

19. $-4\cos x + 2\sin x + C$

21. $\displaystyle\int (\sec^2 x + \sec x \tan x)dx = \tan x + \sec x + C$

23. $\displaystyle\int (\sec x \tan x + 1)dx = \sec x + x + C$

25. $\displaystyle\int \sec x \tan x \, dx = \sec x + C$

27. $\displaystyle\int (1 + \sin\theta)d\theta = \theta - \cos\theta + C$

29 $\displaystyle\int (\cos\theta - 5\sec^2\theta)d\theta = \sin\theta - 5\tan\theta + C$

31. $F(x) = \displaystyle\int x^{1/3}dx = \dfrac{3}{4}x^{4/3} + C,\ F(1) = \dfrac{3}{4} + C = 2,\ C = 5/4;\ F(x) = \dfrac{3}{4}x^{4/3} + \dfrac{5}{4}$

33. $f'(x) = \dfrac{2}{3}x^{3/2} + C_1;\ f(x) = \dfrac{4}{15}x^{5/2} + C_1 x + C_2$

35. $dy/dx = 2x + 1, y = \int(2x+1)dx = x^2 + x + C; y = 0$ when $x = -3$

so $(-3)^2 + (-3) + C = 0, C = -6$ thus $y = x^2 + x - 6$.

37. $dy/dx = \int 6x\,dx = 3x^2 + C_1$. The slope of the tangent line is -3 so $dy/dx = -3$ when $x = 1$.

Thus $3(1)^2 + C_1 = -3$, $C_1 = -6$ so $dy/dx = 3x^2 - 6$, $y = \int(3x^2 - 6)dx = x^3 - 6x + C_2$; If

$x = 1$, then $y = 5 - 3(1) = 2$ so $(1)^2 - 6(1) + C_2 = 2, C_2 = 7$ thus $y = x^3 - 6x + 7$.

39. $\dfrac{d}{dx}\left[\sqrt{x^3+5}\right] = \dfrac{3x^2}{2\sqrt{x^3+5}}$ so $\int \dfrac{3x^2}{2\sqrt{x^3+5}}dx = \sqrt{x^3+5} + C$.

41. $\dfrac{d}{dx}\left[\sin\left(2\sqrt{x}\right)\right] = \dfrac{\cos\left(2\sqrt{x}\right)}{\sqrt{x}}$ so $\int \dfrac{\cos\left(2\sqrt{x}\right)}{\sqrt{x}}dx = \sin\left(2\sqrt{x}\right) + C$

43. (a) $F'(x) = G'(x) = 3x + 4$.

(b) $F(x) = (9x^2 + 24x + 16)/6 = 3x^2/2 + 4x + 8/3 = G(x) + 8/3$

45. $f(x) = \dfrac{d}{dx}(5x^3 - 3x + C) = 15x^2 - 3$.   **47.** $\int(\sec^2 x - 1)dx = \tan x - x + C$

49. $\dfrac{d}{dx}\left[\int f(x)dx - \int g(x)dx\right] = \dfrac{d}{dx}\left[\int f(x)dx\right] - \dfrac{d}{dx}\left[\int g(x)dx\right] = f(x) - g(x)$

EXERCISE SET 5.3

1. (a) $\displaystyle\int u^{23}du = u^{24}/24 + C = (x^2+1)^{24}/24 + C$

(b) $-\displaystyle\int u^3 du = -u^4/4 + C = -(\cos^4 x)/4 + C$

(c) $2\displaystyle\int \sin u\,du = -2\cos u + C = -2\cos\sqrt{x} + C$

(d) $\dfrac{3}{8}\displaystyle\int u^{-1/2}du = \dfrac{3}{4}u^{1/2} + C = \dfrac{3}{4}\sqrt{4x^2+5} + C$

3. (a) $-\displaystyle\int u\,du = -\dfrac{1}{2}u^2 + C = -\dfrac{1}{2}\cot^2 x + C$

(b) $\displaystyle\int u^9 du = \dfrac{1}{10}u^{10} + C = \dfrac{1}{10}(1+\sin t)^{10} + C$

(c) $\int (u-1)^2 u^{1/2} du = \int (u^{5/2} - 2u^{3/2} + u^{1/2})du = \frac{2}{7}u^{7/2} - \frac{4}{5}u^{5/2} + \frac{2}{3}u^{3/2} + C$

$$= \frac{2}{7}(1+x)^{7/2} - \frac{4}{5}(1+x)^{5/2} + \frac{2}{3}(1+x)^{3/2} + C$$

(d) $\int \csc^2 u\, du = -\cot u + C = -\cot(\sin x) + C$

5. $u = 2 - x^2$, $du = -2x\,dx$; $-\frac{1}{2}\int u^3 du = -u^4/8 + C = -(2-x^2)^4/8 + C$

7. $u = 8x$, $du = 8dx$; $\frac{1}{8}\int \cos u\, du = \frac{1}{8}\sin u + C = \frac{1}{8}\sin 8x + C$

9. $u = 4x$, $du = 4dx$; $\frac{1}{4}\int \sec u \tan u\, du = \frac{1}{4}\sec u + C = \frac{1}{4}\sec 4x + C$

11. $u = 7t^2 + 12$, $du = 14t\,dt$; $\frac{1}{14}\int u^{1/2}du = \frac{1}{21}u^{3/2} + C = \frac{1}{21}(7t^2 + 12)^{3/2} + C$

13. $u = x^3 + 1$, $du = 3x^2 dx$; $\frac{1}{3}\int u^{-1/2}du = \frac{2}{3}u^{1/2} + C = \frac{2}{3}\sqrt{x^3 + 1} + C$

15. $u = 4x^2 + 1$, $du = 8x\,dx$; $\frac{1}{8}\int u^{-3}du = -\frac{1}{16}u^{-2} + C = -\frac{1}{16}(4x^2 + 1)^{-2} + C$

17. $u = 5/x$, $du = -(5/x^2)dx$; $-\frac{1}{5}\int \sin u\, du = \frac{1}{5}\cos u + C = \frac{1}{5}\cos(5/x) + C$

19. $u = x^3$, $du = 3x^2 dx$; $\frac{1}{3}\int \sec^2 u\, du = \frac{1}{3}\tan u + C = \frac{1}{3}\tan(x^3) + C$

21. $u = \sin 3t$, $du = 3\cos 3t\,dt$; $\frac{1}{3}\int u^5 du = \frac{1}{18}u^6 + C = \frac{1}{18}\sin^6 3t + C$

23. $u = 2 - \sin 4\theta$, $du = -4\cos 4\theta\,d\theta$; $-\frac{1}{4}\int u^{1/2}du = -\frac{1}{6}u^{3/2} + C = -\frac{1}{6}(2 - \sin 4\theta)^{3/2} + C$

25. $u = \sec 2x$, $du = 2\sec 2x \tan 2x\,dx$; $\frac{1}{2}\int u^2 du = \frac{1}{6}u^3 + C = \frac{1}{6}\sec^3 2x + C$

27. $u = \cos 3\theta$, $du = -3\sin 3\theta\,d\theta$; $-\frac{1}{3}\int \sec^2 u\, du = -\frac{1}{3}\tan u + C = -\frac{1}{3}\tan(\cos 3\theta) + C$

29. $u = \sin(a + bx)$, $du = b\cos(a + bx)dx$

$$\frac{1}{b}\int u^n\,du = \frac{1}{b(n+1)}u^{n+1} + C = \frac{1}{b(n+1)}\sin^{n+1}(a+bx) + C$$

31. $u = x - 3$, $x = u + 3$, $dx = du$

$$\int(u+3)u^{1/2}du = \int(u^{3/2} + 3u^{1/2})du = \frac{2}{5}u^{5/2} + 2u^{3/2} + C = \frac{2}{5}(x-3)^{5/2} + 2(x-3)^{3/2} + C$$

33. $u = y + 1$, $y = u - 1$, $dy = du$

$$\int\frac{u-1}{u^{1/2}}du = \int(u^{1/2} - u^{-1/2})du = \frac{2}{3}u^{3/2} - 2u^{1/2} + C = \frac{2}{3}(y+1)^{3/2} - 2(y+1)^{1/2} + C$$

35. $u = 3\theta$, $du = 3\,d\theta$

$$\frac{1}{3}\int\tan^2 u\,du = \frac{1}{3}\int(\sec^2 u - 1)du = \frac{1}{3}(\tan u - u) + C = \frac{1}{3}(\tan 3\theta - 3\theta) + C$$

37. $u = \sqrt{x-1}$, $u^2 = x - 1$, $x = u^2 + 1$, $dx = 2u\,du$

$$\int(u^2+1)^2 u(2u)du = 2\int(u^4 + 2u^2 + 1)u^2\,du = 2\int(u^6 + 2u^4 + u^2)du$$

$$= \frac{2}{7}u^7 + \frac{4}{5}u^5 + \frac{2}{3}u^3 + C$$

$$= \frac{2}{7}(x-1)^{7/2} + \frac{4}{5}(x-1)^{5/2} + \frac{2}{3}(x-1)^{3/2} + C$$

39. **(a)** First method: $\displaystyle\int(25x^2 - 10x + 1)dx = \frac{25}{3}x^3 - 5x^2 + x + C$;

second method: $\displaystyle\frac{1}{5}\int u^2\,du = \frac{1}{15}u^3 + C = \frac{1}{15}(5x-1)^3 + C$

(b) $\dfrac{1}{15}(5x-1)^3 + C = \dfrac{1}{15}(125x^3 - 75x^2 + 15x - 1) + C = \dfrac{25}{3}x^3 - 5x^2 + x - \dfrac{1}{15} + C$;

the answers differ by a constant.

41. $f(x) = \displaystyle\int(6 - 5\,\sin 2x)dx = 6x + \frac{5}{2}\cos 2x + C$,

$f(0) = \dfrac{5}{2} + C = 3$, $C = \dfrac{1}{2}$ so $f(x) = 6x + \dfrac{5}{2}\cos 2x + \dfrac{1}{2}$

43. $u = 3x + 2$, $du = 3\,dx$; $\dfrac{1}{3}\displaystyle\int f'(u)du = \dfrac{1}{3}f(u) + C = \dfrac{1}{3}f(3x+2) + C$

45. $u = 2/x$, $du = -(2/x^2)dx$; $-\dfrac{1}{2}\displaystyle\int f'(u)du = -\dfrac{1}{2}f(u) + C = -\dfrac{1}{2}f(2/x) + C$

EXERCISE SET 5.4

1. (a) $1 + 8 + 27 = 36$ (b) $5 + 8 + 11 + 14 + 17 = 55$
(c) $20 + 12 + 6 + 2 + 0 + 0 = 40$ (d) $1 + 1 + 1 + 1 + 1 + 1 = 6$

3. $\displaystyle\sum_{k=1}^{10} k$ **5.** $\displaystyle\sum_{k=1}^{49} k(k+1)$ **7.** $\displaystyle\sum_{k=1}^{10} 2k$ **9.** $\displaystyle\sum_{k=1}^{6} (-1)^{k+1}(2k-1)$

11. $\displaystyle\sum_{k=1}^{5} (-1)^k \frac{1}{k}$ **13.** $\displaystyle\sum_{k=1}^{4} \sin\frac{(2k-1)\pi}{8}$ **15.** $\displaystyle\sum_{k=1}^{5} \frac{k}{k+1}$

17. (a) $\displaystyle\sum_{k=1}^{5} (-1)^{k+1} a_k$ (b) $\displaystyle\sum_{k=0}^{5} (-1)^{k+1} b_k$ (c) $\displaystyle\sum_{k=0}^{n} a_k x^k$ (d) $\displaystyle\sum_{k=0}^{5} a^{5-k} b^k$

19. $\displaystyle\sum_{k=1}^{100} k - \sum_{k=1}^{2} k = \frac{1}{2}(100)(100+1) - (1+2) = 5050 - 3 = 5047$

21. $\dfrac{1}{6}(20)(21)(41) = 2,870$

23. $4\displaystyle\sum_{k=1}^{6} k^3 - 2\sum_{k=1}^{6} k + \sum_{k=1}^{6} 1 = 4\left[\frac{1}{4}(6)^2(7)^2\right] - 2\left[\frac{1}{2}(6)(7)\right] + 6 = 1728$

25. $\displaystyle\sum_{k=1}^{30} k(k^2 - 4) = \sum_{k=1}^{30}(k^3 - 4k) = \sum_{k=1}^{30} k^3 - 4\sum_{k=1}^{30} k = \frac{1}{4}(30)^2(31)^2 - 4 \times \frac{1}{2}(30)(31) = 214,365$

27. $\displaystyle\sum_{k=1}^{n} \frac{3k}{n} = \frac{3}{n}\sum_{k=1}^{n} k = \frac{3}{n} \times \frac{1}{2}n(n+1) = \frac{3}{2}(n+1)$

29. $\displaystyle\sum_{k=1}^{n-1} \frac{k^3}{n^2} = \frac{1}{n^2}\sum_{k=1}^{n-1} k^3 = \frac{1}{n^2} \times \frac{1}{4}(n-1)^2 n^2 = \frac{1}{4}(n-1)^2$

31. $\dfrac{1 + 2 + 3 + \cdots + n}{n^2} = \displaystyle\sum_{k=1}^{n} \frac{k}{n^2} = \frac{1}{n^2}\sum_{k=1}^{n} k = \frac{1}{n^2} \times \frac{1}{2}n(n+1) = \frac{n+1}{2n}; \ \lim_{n \to +\infty} \frac{n+1}{2n} = \frac{1}{2}.$

33. $\displaystyle\sum_{k=1}^{n} \frac{5k}{n^2} = \frac{5}{n^2} \sum_{k=1}^{n} k = \frac{5}{n^2} \times \frac{1}{2}n(n+1) = \frac{5(n+1)}{2n}; \quad \lim_{n \to +\infty} \frac{5(n+1)}{2n} = \frac{5}{2}.$

35. $\displaystyle\sum_{k=1}^{n-1} \left(\frac{9}{n} - \frac{k}{n^2}\right) = \frac{9}{n} \sum_{k=1}^{n-1} 1 - \frac{1}{n^2} \sum_{k=1}^{n-1} k = \frac{9}{n}(n-1) - \frac{1}{n^2} \times \frac{1}{2}(n-1)(n) = \frac{17}{2}\left(\frac{n-1}{n}\right);$

$\displaystyle\lim_{n \to +\infty} \frac{17}{2}\left(\frac{n-1}{n}\right) = \frac{17}{2}.$

37. $\displaystyle 1 + 3 + 5 + \cdots + (2n-1) = \sum_{k=1}^{n}(2k-1) = 2\sum_{k=1}^{n} k - \sum_{k=1}^{n} 1 = 2 \times \frac{1}{2}n(n+1) - n = n^2.$

39. $\displaystyle (3^5 - 3^4) + (3^6 - 3^5) + \cdots + (3^{17} - 3^{16}) = 3^{17} - 3^4$

41. $\displaystyle \left(\frac{1}{2^2} - \frac{1}{1^2}\right) + \left(\frac{1}{3^2} - \frac{1}{2^2}\right) + \cdots + \left(\frac{1}{20^2} - \frac{1}{19^2}\right) = \frac{1}{20^2} - 1 = -\frac{399}{400}$

43. $\displaystyle (a_1 - a_0) + (a_2 - a_1) + \cdots + (a_n - a_{n-1}) = a_n - a_0$

45. (a) $\displaystyle\sum_{k=1}^{n} \frac{1}{(2k-1)(2k+1)} = \frac{1}{2} \sum_{k=1}^{n} \left(\frac{1}{2k-1} - \frac{1}{2k+1}\right)$

$\displaystyle = \frac{1}{2}\left[\left(1 - \frac{1}{3}\right) + \left(\frac{1}{3} - \frac{1}{5}\right) + \left(\frac{1}{5} - \frac{1}{7}\right) + \cdots + \left(\frac{1}{2n-1} - \frac{1}{2n+1}\right)\right]$

$\displaystyle = \frac{1}{2}\left[1 - \frac{1}{2n+1}\right] = \frac{n}{2n+1}$

(b) $\displaystyle\lim_{n \to +\infty} \frac{n}{2n+1} = \frac{1}{2}$

47. (a) $n + n + \cdots + n = n^2$
(n terms)

(b) -3

(c) $\displaystyle x\sum_{k=1}^{n} k = \frac{1}{2}n(n+1)x$

(d) $c + c + \cdots + c = (n - m + 1)c$
($n - m + 1$ terms)

49. (a) $\displaystyle\sum_{k=0}^{14}(k+4)(k+1)$

(b) $\displaystyle\sum_{k=5}^{19}(k-1)(k-4)$

51. $\displaystyle\sum_{k=1}^{18} k \sin\frac{\pi}{k}$ **53.**   both are valid

55.   (a) $\displaystyle\sum_{k=0}^{19} 3^{k+1} = \sum_{k=0}^{19} 3(3^k) = \frac{3(1-3^{20})}{1-3} = \frac{3}{2}(3^{20}-1)$

(b) $\displaystyle\sum_{k=0}^{25} 2^{k+5} = \sum_{k=0}^{25} 2^5 2^k = \frac{2^5(1-2^{26})}{1-2} = 2^{31} - 2^5$

(c) $\displaystyle\sum_{k=0}^{100} (-1)\left(\frac{-1}{2}\right)^k = \frac{(-1)(1-(-1/2)^{101})}{1-(-1/2)} = -\frac{2}{3}(1+1/2^{101})$

57. $\displaystyle\sum_{i=1}^{4}\left[\sum_{j=1}^{5} i + \sum_{j=1}^{5} j\right] \sum_{i=1}^{4}\left[5i + \frac{1}{2}(5)(6)\right] = 5\sum_{i=1}^{4} i + \sum_{i=1}^{4} 15 = 5\cdot\frac{1}{2}(4)(5) + (4)(15) = 110$

59. $\displaystyle\sum_{k=1}^{n}(a_k - b_k) = (a_1 - b_1) + (a_2 - b_2) + \cdots + (a_n - b_n)$

$$= (a_1 + a_2 + \cdots + a_n) - (b_1 + b_2 + \cdots + b_n) = \sum_{k=1}^{n} a_k - \sum_{k=1}^{n} b_k$$

EXERCISE SET 5.5

1. $\Delta x = \dfrac{6-2}{4} = 1,\ f(x) = 3x + 1$

(a) $c_k = 2, 3, 4, 5;\ \displaystyle\sum_{k=1}^{4} f(c_k)\Delta x = (7 + 10 + 13 + 16)(1) = 46$

(b) $d_k = 3, 4, 5, 6;\ \displaystyle\sum_{k=1}^{4} f(d_k)\Delta x = (10 + 13 + 16 + 19)(1) = 58$

3. $\Delta x = \pi/4,\ f(x) = \cos x$

(a) $c_k = -\pi/2, -\pi/4, \pi/4, \pi/2$

$$\sum_{k=1}^{4} f(c_k)\Delta x = (0 + \sqrt{2}/2 + \sqrt{2}/2 + 0)(\pi/4) = \sqrt{2}\,\pi/4 \approx 1.111$$

(b) $d_k = -\pi/4, 0, 0, \pi/4$

$$\sum_{k=1}^{4} f(d_k)\Delta x = (\sqrt{2}/2 + 1 + 1 + \sqrt{2}/2)(\pi/4) = (2 + \sqrt{2})\pi/4 \approx 2.682$$

5. $\Delta x = \dfrac{3}{n}$, $c_k = 1 + (k-1)\dfrac{3}{n}$

$$f(c_k)\Delta x = \frac{1}{2}c_k\Delta x = \frac{1}{2}\left[1 + (k-1)\frac{3}{n}\right]\frac{3}{n} = \frac{1}{2}\left[\frac{3}{n} + (k-1)\frac{9}{n^2}\right]$$

$$\sum_{k=1}^{n} f(c_k)\Delta x = \frac{1}{2}\left[\sum_{k=1}^{n}\frac{3}{n} + \frac{9}{n^2}\sum_{k=1}^{n}(k-1)\right] = \frac{1}{2}\left[3 + \frac{9}{n^2}\cdot\frac{1}{2}(n-1)n\right] = \frac{3}{2} + \frac{9}{4}\frac{n-1}{n}$$

$$A = \lim_{n\to+\infty}\left[\frac{3}{2} + \frac{9}{4}\left(1 - \frac{1}{n}\right)\right] = \frac{3}{2} + \frac{9}{4} = \frac{15}{4}$$

7. $\Delta x = \dfrac{1}{n}$, $c_k = (k-1)\dfrac{1}{n}$; $f(c_k)\Delta x = c_k^2\Delta x = (k-1)^2\dfrac{1}{n^2}\cdot\dfrac{1}{n} = \dfrac{1}{n^3}(k-1)^2$

$$\sum_{k=1}^{n} f(c_k)\Delta x = \frac{1}{n^3}\sum_{k=1}^{n}(k-1)^2 = \frac{1}{n^3}\sum_{k=1}^{n-1}k^2 = \frac{1}{n^3}\cdot\frac{1}{6}(n-1)(n)(2n-1) = \frac{1}{6}\frac{(n-1)(2n-1)}{n^2}$$

$$A = \lim_{n\to+\infty}\frac{1}{6}\left(1 - \frac{1}{n}\right)\left(2 - \frac{1}{n}\right) = \frac{1}{6}(1)(2) = \frac{1}{3}$$

9. $\Delta x = \dfrac{4}{n}$, $c_k = 2 + (k-1)\dfrac{4}{n}$

$$f(c_k)\Delta x = c_k^3\Delta x = \left[2 + \frac{4}{n}(k-1)\right]^3\frac{4}{n} = \frac{32}{n}\left[1 + \frac{2}{n}(k-1)\right]^3$$

$$= \frac{32}{n}\left[1 + \frac{6}{n}(k-1) + \frac{12}{n^2}(k-1)^2 + \frac{8}{n^3}(k-1)^3\right]$$

$$\sum_{k=1}^{n} f(c_k)\Delta x = \frac{32}{n}\left[\sum_{k=1}^{n}1 + \frac{6}{n}\sum_{k=1}^{n}(k-1) + \frac{12}{n^2}\sum_{k=1}^{n}(k-1)^2 + \frac{8}{n^3}\sum_{k=1}^{n}(k-1)^3\right]$$

$$= \frac{32}{n}\left[n + \frac{6}{n}\sum_{k=1}^{n-1}k + \frac{12}{n^2}\sum_{k=1}^{n-1}k^2 + \frac{8}{n^3}\sum_{k=1}^{n-1}k^3\right]$$

$$= \frac{32}{n}\left[n + \frac{6}{n}\cdot\frac{1}{2}(n-1)(n) + \frac{12}{n^2}\cdot\frac{1}{6}(n-1)(n)(2n-1) + \frac{8}{n^3}\cdot\frac{1}{4}(n-1)^2 n^2\right]$$

$$= 32\left[1 + 3\frac{n-1}{n} + 2\frac{(n-1)(2n-1)}{n^2} + 2\frac{(n-1)^2}{n^2}\right]$$

$$A = \lim_{n \to +\infty} 32 \left[1 + 3\left(1 - \frac{1}{n}\right) + 2\left(1 - \frac{1}{n}\right)\left(2 - \frac{1}{n}\right) + 2\left(1 - \frac{1}{n}\right)^2 \right]$$

$$= 32[1 + 3(1) + 2(1)(2) + 2(1)^2] = 320$$

11. $\Delta x = \dfrac{3}{n},\ d_k = 1 + \dfrac{3}{n}k;\ f(d_k)\Delta x = \dfrac{1}{2}d_k\Delta x = \dfrac{1}{2}\left(1 + \dfrac{3}{n}k\right)\dfrac{3}{n} = \dfrac{3}{2}\left[\dfrac{1}{n} + \dfrac{3}{n^2}k\right]$

$$\sum_{k=1}^{n} f(d_k)\Delta x = \frac{3}{2}\left[\sum_{k=1}^{n}\frac{1}{n} + \sum_{k=1}^{n}\frac{3}{n^2}k\right] = \frac{3}{2}\left[1 + \frac{3}{n^2}\cdot\frac{1}{2}n(n+1)\right] = \frac{3}{2}\left[1 + \frac{3}{2}\frac{n+1}{n}\right]$$

$$A = \lim_{n \to +\infty} \frac{3}{2}\left[1 + \frac{3}{2}\left(1 + \frac{1}{n}\right)\right] = \frac{3}{2}\left(1 + \frac{3}{2}\right) = \frac{15}{4}$$

13. $\Delta x = \dfrac{1}{n},\ d_k = \dfrac{k}{n};\ f(d_k)\Delta x = d_k^2\Delta x = \dfrac{k^2}{n^2}\dfrac{1}{n} = \dfrac{k^2}{n^3}$

$$\sum_{k=1}^{n} f(d_k)\Delta x = \sum_{k=1}^{n}\frac{k^2}{n^3} = \frac{1}{n^3}\frac{1}{6}n(n+1)(2n+1) = \frac{1}{6}\frac{(n+1)(2n+1)}{n^2}$$

$$A = \lim_{n \to +\infty} \frac{1}{6}\left(1 + \frac{1}{n}\right)\left(2 + \frac{1}{n}\right) = \frac{1}{3}$$

15. (a) Using inscribed rectangles, $\Delta x = \dfrac{b}{n},\ c_k = \dfrac{b}{n}(k - 1)$

$$f(c_k)\Delta x = c_k^3\Delta x = \frac{b^4}{n^4}(k - 1)^3$$

$$\sum_{k=1}^{n} f(c_k)\Delta x = \frac{b^4}{n^4}\sum_{k=1}^{n}(k - 1)^3 = \frac{b^4}{n^4}\sum_{k=1}^{n-1}k^3 = \frac{b^4}{4}\frac{(n - 1)^2}{n^2}$$

$$A = \lim_{n \to +\infty} \frac{b^4}{4}\left(1 - \frac{1}{n}\right)^2 = b^4/4$$

(b) $\Delta x = \dfrac{b - a}{n},\ c_k = a + \dfrac{b - a}{n}(k - 1)$

$$f(c_k)\Delta x = c_k^3\Delta x = \left[a + \frac{b - a}{n}(k - 1)\right]^3\frac{b - a}{n}$$

$$= \frac{b - a}{n}\left[a^3 + \frac{3a^2(b - a)}{n}(k - 1) + \frac{3a(b - a)^2}{n^2}(k - 1)^2 + \frac{(b - a)^3}{n^3}(k - 1)^3\right]$$

$$\sum_{k=1}^{n} f(c_k)\Delta x = (b - a)\left[a^3 + \frac{3}{2}a^2(b - a)\frac{n - 1}{n} + \frac{1}{2}a(b - a)^2\frac{(n - 1)(2n - 1)}{n^2}\right.$$

$$\left. + \frac{1}{4}(b - a)^3\frac{(n - 1)^2}{n^2}\right]$$

$$A = \lim_{n \to +\infty} \sum_{k=1}^{n} f(c_k)\Delta x$$

$$= (b-a)\left[a^3 + \frac{3}{2}a^2(b-a) + a(b-a)^2 + \frac{1}{4}(b-a)^3\right] = \frac{1}{4}(b^4 - a^4).$$

17. **(a)** $\Delta x = \dfrac{b-a}{n}$, $c_k = a + \dfrac{b-a}{n}(k-1)$; $f(c_k)\Delta x = \dfrac{b-a}{n}\left[a + \dfrac{b-a}{n}(k-1)\right]$

$$\sum_{k=1}^{n} f(c_k)\Delta x = (b-a)\left[a + \frac{b-a}{2} \cdot \frac{n-1}{n}\right]$$

$$A = \lim_{n \to +\infty}(b-a)\left[a + \frac{b-a}{2}\left(1 - \frac{1}{n}\right)\right] = \frac{1}{2}(b^2 - a^2)$$

(b) $\Delta x = \dfrac{b-a}{n}$, $d_k = a + \dfrac{b-a}{n}k$; $f(d_k)\Delta x = \dfrac{b-a}{n}\left[a + \dfrac{b-a}{n}k\right]$

$$\sum_{k=1}^{n} f(d_k)\Delta x = (b-a)\left[a + \frac{b-a}{2} \cdot \frac{n+1}{n}\right]$$

$$A = \lim_{n \to +\infty}(b-a)\left[a + \frac{b-a}{2}\left(1 + \frac{1}{n}\right)\right] = \frac{1}{2}(b^2 - a^2)$$

(c) The region is enclosed by a trapezoid so its area is

$$A = \frac{1}{2}(b-a)(b+a)$$

$$= \frac{1}{2}(b^2 - a^2).$$

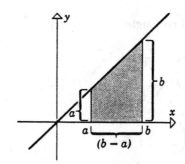

19. **(a)** 0.584145862, 0.623823864, 0.649145594

    (b) 0.761923639, 0.712712753, 0.684701150

21. **(a)** 0.919403170, 0.960215997, 0.984209789

    (b) 1.076482803, 1.038755813, 1.015625715

EXERCISE SET 5.6

1. **(a)** $(4/3)(1) + (5/2)(1) + (4)(2) = 71/6$
   (b) 2

3. **(a)** $(-9/4)(1) + (3)(2) + (63/16)(1) + (-5)(3) = -117/16$
   (b) 3

5. $\displaystyle\sum_{k=1}^{4} f(c_k)\Delta x = (1/3 + 1/5 + 1/7 + 1/9)(2) = 496/315$

   $\displaystyle\sum_{k=1}^{4} f(x_k^*)\Delta x = (1/2 + 1/4 + 1/6 + 1/8)(2) = 25/12$

   $\displaystyle\sum_{k=1}^{4} f(d_k)\Delta x = (1 + 1/3 + 1/5 + 1/7)(2) = 352/105$

   $496/315 = 1984/1260$, $25/12 = 2625/1260$, $352/105 = 4224/1260$,
   $1984/1260 \leq 2625/1260 \leq 4224/1260$

7. $\displaystyle\int_{-3}^{3} 4x(1 - 3x)\,dx$

9. $\displaystyle\lim_{\max \Delta x_k \to 0} \sum_{k=1}^{n} 2x_k^* \Delta x_k$; $a = 1$, $b = 2$

11. $\displaystyle\lim_{\max \Delta x_k \to 0} \sum_{k=1}^{n} \frac{x_k^*}{x_k^* + 1}\Delta x_k$; $a = 0$, $b = 1$

13. **(a)** 0.8   **(b)** -2.6 **(c)** -1.8 **(d)** -0.3

15. **(a)** $A = \dfrac{1}{2}(1)(2) = 1$ **(b)** $A = \dfrac{1}{2}(2)(3/2 + 1/2) = 2$

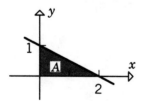

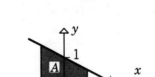

(c) $-A = -\dfrac{1}{2}(1/2)(1) = -1/4$ **(d)** $A_1 - A_2 = 1 - 1/4 = 3/4$

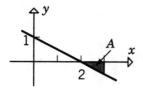   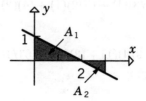

17. $A_1 - A_2 = \dfrac{1}{2}(2)(1/2) - \dfrac{1}{2}(6)(3/2)$ **19.** $A_1 + A_2 = \dfrac{1}{2}(5)(5/2) + \dfrac{1}{2}(1)(1/2)$

$\qquad\qquad = -4$ $\qquad\qquad = 13/2$

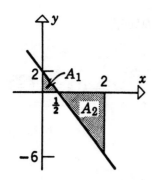   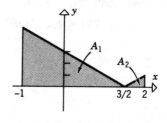

21. $\dfrac{1}{2}[\pi(1)^2] = \pi/2$

23. $\sqrt{10x - x^2} = \sqrt{25 - (x - 5)^2}; \ \displaystyle\int_0^{10} \sqrt{10x - x^2}\, dx = \dfrac{1}{2}[\pi(5)^2] = 25\pi/2$

25. $A_1 - A_2 = 0$ because
$A_1 = A_2$ by symmetry

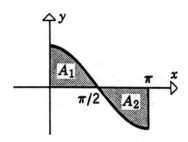

27. $A_1 + A_2 = (3)(2) + \frac{1}{2}(2)(5+3) = 14$

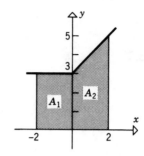

29. 0.692835360, 0.693069098, 0.693134682

31. 3.142425985, 3.141800987, 3.141625987

33. f is not bounded on [0,1] because $\lim\limits_{x \to 0+} f(x) = +\infty$, so f is not integrable on [0,1].

35. Let $S_n = \sum\limits_{k=1}^{n} f(x_k^*)\Delta x_k$. From definition 5.6.7, for any $\epsilon > 0$ there are numbers $\delta_1 > 0$ and $\delta_2 > 0$ such that $|S_n - L_1| < \epsilon$ for max $\Delta x_k < \delta_1$ and $|S_n - L_2| < \epsilon$ for max $\Delta x_k < \delta_2$.
If $\delta = \min(\delta_1, \delta_2)$ then $|S_n - L_1| < \epsilon$ and $|S_n - L_2| < \epsilon$ for max $\Delta x_k < \delta$ thus
$|L_1 - L_2| = |L_1 - S_n + S_n - L_2| = |(L_1 - S_n) + (S_n - L_2)| \le |S_n - L_1| + |S_n - L_2| < \epsilon + \epsilon = 2\epsilon$
so $|L_1 - L_2| < 2\epsilon$ for max $\Delta x_k < \delta$. Suppose $L_1 \ne L_2$ and let $\epsilon = \frac{1}{2}|L_1 - L_2|$ then $|L_1 - L_2| < 2\epsilon$
yields $|L_1 - L_2| < |L_1 - L_2|$ which is false so $L_1 \ne L_2$ is impossible.

37. Let $R_n = \sum\limits_{k=1}^{n} f(x_k^*)\Delta x_k$, $S_n = \sum\limits_{k=1}^{n} g(x_k^*)\Delta x_k$, $T_n = \sum\limits_{k=1}^{n}[f(x_k^*) + g(x_k^*)]\Delta x_k$, $R = \int_a^b f(x)dx$,
and $S = \int_a^b g(x)dx$ then $T_n = R_n + S_n$ and we want to prove that $\lim\limits_{\max \Delta x_k \to 0} T_n = R + S$.
$|T_n - (R + S)| = |(R_n - R) + (S_n - S)| \le |R_n - R| + |S_n - S|$
so for any $\epsilon > 0$ $|T_n - (R + S)| < \epsilon$ if $|R_n - R| + |S_n - S| < \epsilon$.
Because f and g are integrable on $[a, b]$, there are numbers δ_1 and δ_2 such that
$|R_n - R| < \epsilon/2$ for max $\Delta x_k < \delta_1$ and $|S_n - S| < \epsilon/2$ for max $\Delta x_k < \delta_2$.
If $\delta = \min(\delta_1, \delta_2)$ then $|R_n - R| < \epsilon/2$ and $|S_n - S| < \epsilon/2$ for max $\Delta x_k < \delta$ thus

$|R_n - R| + |S_n - S| < \epsilon$ and so $|T_n - (R + S)| < \epsilon$ for $\max \Delta x_k < \delta$ which shows that

$$\lim_{\max \Delta x_k \to 0} T_n = R + S.$$

EXERCISE SET 5.7

1. $\left. \frac{1}{4}x^4 \right]_2^3 = 65/4$

3. $\left. \frac{1}{2}x^2 + \frac{1}{5}x^5 \right]_{-1}^2 = 81/10$

5. $\left. \frac{1}{3}t^3 - t^2 + 8t \right]_1^2 = 22/3$

7. $\int_1^3 x^{-2} dx = \left. -\frac{1}{x} \right]_1^3 = 2/3$

9. $\left. -\frac{1}{2x^2} + \frac{2}{x} - \frac{1}{3x^3} \right]_1^2 = -1/3$

11. $\left. \frac{2}{3}x^{3/2} \right]_1^9 = 52/3$

13. $\left. \frac{4}{5}y^{5/2} \right]_4^9 = 844/5$

15. $\left. 6\sqrt{x} - \frac{10}{3}x^{3/2} + \frac{2}{\sqrt{x}} \right]_1^4 = -55/3$

17. $\left. -\cos\theta \right]_{-\pi/2}^{\pi/2} = 0$

19. $\int_{-\pi/4}^{\pi/4} \cos x \, dx = \left. \sin x \right]_{-\pi/4}^{\pi/4} = \sqrt{2}$

21. $\left. \frac{1}{2}x^2 - 2\cot x \right]_{\pi/6}^{\pi/2} = \pi^2/9 + 2\sqrt{3}$

23. $\int_0^{3/2} (3 - 2x)dx + \int_{3/2}^2 (2x - 3)dx = \left. (3x - x^2) \right]_0^{3/2} + \left. (x^2 - 3x) \right]_{3/2}^2 = 9/4 + 1/4 = 5/2$

25. $\int_0^{\pi/2} \cos x \, dx + \int_{\pi/2}^{3\pi/4} (-\cos x)dx = \left. \sin x \right]_0^{\pi/2} - \left. \sin x \right]_{\pi/2}^{3\pi/4} = 2 - \sqrt{2}/2$

27. $\int_{-2}^0 x^2 dx + \int_0^3 (-x)dx = \left. \frac{1}{3}x^3 \right]_{-2}^0 - \left. \frac{1}{2}x^2 \right]_0^3 = -11/6$

29. $\int_{-1}^2 x \, dx + 2\int_{-1}^2 f(x)dx = \left. \frac{1}{2}x^2 \right]_{-1}^2 + 2(3) = 15/2$

31. $\displaystyle\int_1^5 f(x)dx = \int_0^5 f(x)dx - \int_0^1 f(x)dx = 1 - (-2) = 3.$

33. negative, because $\sqrt{x}/(1-x) < 0$ for $2 \le x \le 3$.

35. positive, because $x^2/(3 - \cos x) > 0$ for $0 < x \le 4$.

37. negative, because $\displaystyle\int_2^0 x^2 \sin\sqrt{x}\,dx = -\int_0^2 x^2 \sin\sqrt{x}\,dx$ and $x^2 \sin\sqrt{x} > 0$ for $0 < x \le 2$.

39. $\displaystyle\int_1^2 \frac{x}{x^4+1}\,dx < \int_1^2 \frac{1}{x^3}\,dx = -\frac{1}{2x^2}\Big]_1^2 = 3/8.$

41. $\displaystyle m = \int_1^3 \frac{1}{x^{1.01}}\,dx = -\frac{100}{x^{0.01}}\Big]_1^3 = 1.092599583;$

$\displaystyle M = \int_1^3 \frac{1}{x^{0.99}}\,dx = 100x^{0.01}\Big]_1^3 = 1.104669194$

43. $\displaystyle 0.665867079; \int_1^3 \frac{1}{x^2}\,dx = -\frac{1}{x}\Big]_1^3 = 2/3$

45. $\displaystyle\frac{\sqrt{1} + \sqrt{2} + \sqrt{3} + \cdots + \sqrt{n}}{n^{3/2}} = \sum_{k=1}^n \frac{\sqrt{k}}{n^{3/2}} = \sum_{k=1}^n \sqrt{\frac{k}{n}}\frac{1}{n} = \sum_{k=1}^n f(x_k^*)\Delta x,$ where

$f(x) = \sqrt{x}, x_k^* = k/n$, and $\Delta x = 1/n$, so $\displaystyle\lim_{n \to +\infty}\sum_{k=1}^n f(x_k^*)\Delta x = \int_0^1 \sqrt{x}\,dx = 2/3.$

47. $\displaystyle A = \int_0^3 (x^2 + 1)dx = \frac{1}{3}x^3 + x\Big]_0^3 = 12$

49. $\displaystyle A = \int_0^{2\pi/3} 3\sin x\,dx = -3\cos x\Big]_0^{2\pi/3} = 9/2$

51. $A_1 = \displaystyle\int_{-3}^{-2}(x^2 - 3x - 10)dx$

$\qquad = \dfrac{1}{3}x^3 - \dfrac{3}{2}x^2 - 10x\Big]_{-3}^{-2} = 23/6,$

$\qquad A_2 = -\displaystyle\int_{-2}^{5}(x^2 - 3x - 10)dx = 343/6,$

$\qquad A_3 = \displaystyle\int_{5}^{8}(x^2 - 3x - 10)dx = 243/6,$

$\qquad A = A_1 + A_2 + A_3 = 203/2$

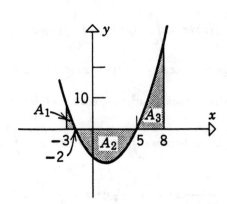

53. (b), (c) are always valid. Use $f(x) = g(x) = 1$ to show that (a), (d), and (e) are false.

EXERCISE SET 5.8

1. (a) $\displaystyle\int_{1}^{3}u^7 du$ **(b)** $-\dfrac{1}{2}\displaystyle\int_{7}^{4}u^{1/2}du$ **(c)** $\dfrac{1}{\pi}\displaystyle\int_{-\pi}^{\pi}\sin u\,du$

$\quad$ **(d)** $\displaystyle\int_{0}^{1}u^2 du$ **(e)** $\dfrac{1}{2}\displaystyle\int_{3}^{4}(u-3)u^{1/2}du$ **(f)** $\displaystyle\int_{-3}^{0}(u+5)u^{20}du$

3. $u = 2x + 1,\ \dfrac{1}{2}\displaystyle\int_{1}^{3}u^4 du = \dfrac{1}{10}u^5\Big]_{1}^{3} = 121/5,\ \text{or}\ \dfrac{1}{10}(2x+1)^5\Big]_{0}^{1} = 121/5$

5. $u = 1 - 2x,\ -\dfrac{1}{2}\displaystyle\int_{3}^{1}u^3 du = -\dfrac{1}{8}u^4\Big]_{3}^{1} = 10,\ \text{or}\ -\dfrac{1}{8}(1-2x)^4\Big]_{-1}^{0} = 10$

7. $u = 1 + x,\ \displaystyle\int_{1}^{9}(u-1)u^{1/2}du = \displaystyle\int_{1}^{9}(u^{3/2} - u^{1/2})du = \dfrac{2}{5}u^{5/2} - \dfrac{2}{3}u^{3/2}\Big]_{1}^{9} = 1192/15,$

$\quad$ or $\dfrac{2}{5}(1+x)^{5/2} - \dfrac{2}{3}(1+x)^{3/2}\Big]_{0}^{8} = 1192/15$

9. $u = x/2,\ 8\displaystyle\int_{0}^{\pi/4}\sin u\,du\ = -8\cos u\Big]_{0}^{\pi/4} = 8 - 4\sqrt{2},\ \text{or}\ -8\cos(x/2)\Big]_{0}^{\pi/2} = 8 - 4\sqrt{2}$

11. $u = x^2 + 2,\ \dfrac{1}{2}\displaystyle\int_6^3 u^{-3}du = -\dfrac{1}{4u^2}\Big]_6^3 = -1/48,$ or $-\dfrac{1}{4}\dfrac{1}{(x^2+2)^2}\Big]_{-2}^{-1} = -1/48$

13. $\dfrac{2}{3}(3u+1)^{1/2}\Big]_0^1 = 2/3$

15. $\dfrac{2}{3}(x^3+9)^{1/2}\Big]_{-1}^1 = \dfrac{2}{3}(\sqrt{10} - 2\sqrt{2})$

17. $u = x^2 + 4x + 7,\ \dfrac{1}{2}\displaystyle\int_{12}^{28} u^{-1/2}du = u^{1/2}\Big]_{12}^{28} = \sqrt{28} - \sqrt{12} = 2(\sqrt{7} - \sqrt{3})$

19. $\dfrac{1}{2}\sin^2 x\Big]_{-3\pi/4}^{-\pi/4} = 0$

21. $\dfrac{5}{2}\sin(x^2)\Big]_0^{\sqrt{\pi}} = 0$

23. $u = \sqrt{x},\ 2\displaystyle\int_\pi^{2\pi}\sin u\,du = -2\cos u\Big]_\pi^{2\pi} = -4$

25. $u = \sin 3x,\ \dfrac{1}{3}\displaystyle\int_0^{-1} u^2 du = \dfrac{1}{9}u^3\Big]_0^{-1} = -1/9$

27. $u = 3\theta,\ \dfrac{1}{3}\displaystyle\int_{\pi/4}^{\pi/3}\sec^2 u\,du = \dfrac{1}{3}\tan u\Big]_{\pi/4}^{\pi/3} = (\sqrt{3} - 1)/3$

29. $u = 4 - 3y,\ y = \dfrac{1}{3}(4 - u),\ dy = -\dfrac{1}{3}du$

$-\dfrac{1}{27}\displaystyle\int_4^1 \dfrac{16 - 8u + u^2}{u^{1/2}}du = \dfrac{1}{27}\displaystyle\int_1^4 (16u^{-1/2} - 8u^{1/2} + u^{3/2})du$

$= \dfrac{1}{27}\left[32u^{1/2} - \dfrac{16}{3}u^{3/2} + \dfrac{2}{5}u^{5/2}\right]_1^4 = 106/405$

31. $A = \displaystyle\int_0^{\pi/8} 3\cos 2x\,dx = \dfrac{3}{2}\sin 2x\Big]_0^{\pi/8} = 3\sqrt{2}/4$

33. $\displaystyle\int_{-2}^2 \sqrt{4 - u^2}\,du = \dfrac{1}{2}[\pi(2)^2] = 2\pi$

35. $-\dfrac{1}{2}\displaystyle\int_1^0 \sqrt{1 - u^2}\,du = \dfrac{1}{2}\displaystyle\int_0^1 \sqrt{1 - u^2}\,du = \dfrac{1}{2}\cdot\dfrac{1}{4}[\pi(1)^2] = \pi/8$

37. $u = 1/x$, $-\displaystyle\int_2^1 f(u)\,du = \int_1^2 f(u)\,du = 3$

39. $u = 3x + 1$, $\dfrac{1}{3}\displaystyle\int_1^4 f(u)\,du = \dfrac{5}{3}$

41. $\sin x = \cos(\pi/2 - x)$,

$$\int_0^{\pi/2} \sin^n x\,dx = \int_0^{\pi/2} \cos^n(\pi/2 - x)\,dx = -\int_{\pi/2}^0 \cos^n u\,du\,(u = \pi/2 - x)$$

$$= \int_0^{\pi/2} \cos^n u\,du = \int_0^{\pi/2} \cos^n x\,dx \text{ (by replacing } u \text{ by } x)$$

43. $2k$, because $f(-x) = f(x)$ on $[-2, 2]$.

45. 0, because $f(-x) = -f(x)$ on $[-1, 1]$.

47. $\displaystyle\sum_{k=1}^n \sin[\pi(k/n)](1/n) = \sum_{k=1}^n f(x_k^*)\Delta x$, where $f(x) = \sin(\pi x)$, $x_k^* = k/n$, and $\Delta x = 1/n$ so

$$\lim_{n \to +\infty} \sum_{k=1}^n f(x_k^*)\Delta x = \int_0^1 \sin(\pi x)\,dx = -\frac{1}{\pi}\cos(\pi x)\Big]_0^1 = 2/\pi.$$

49. (a) $I = -\displaystyle\int_a^0 \frac{f(a - u)}{f(a - u) + f(u)}\,du = \int_0^a \frac{f(a - u) + f(u) - f(u)}{f(a - u) + f(u)}\,du$

$$= \int_0^a du - \int_0^a \frac{f(u)}{f(a - u) + f(u)}\,du, I = a - I \text{ so } 2I = a, I = a/2$$

(b) $3/2$ 　　　　　　　　　　　　　　　　**(c)** $\pi/4$

EXERCISE SET 5.9

1. $f_{\text{ave}} = \dfrac{1}{3 - 1}\displaystyle\int_1^3 3x\,dx = \dfrac{3}{4}x^2\Big]_1^3 = 6.$

3. $f_{\text{ave}} = \dfrac{1}{\pi - 0}\displaystyle\int_0^\pi \sin x\,dx = -\dfrac{1}{\pi}\cos x\Big]_0^\pi = 2/\pi.$

5. $f_{\text{ave}} = \dfrac{1}{5 - 0}\displaystyle\int_0^5 \sqrt{3x + 1}\,dx = \dfrac{2}{45}(3x + 1)^{3/2}\Big]_0^5 = 14/5.$

7. $f_{\text{ave}} = \dfrac{1}{2-(-2)} \displaystyle\int_{-2}^{2} \sqrt{4-x^2}\, dx = \dfrac{1}{4} \times \dfrac{1}{2}\pi(2)^2 = \pi/2.$

9. **(a)** $f_{\text{ave}} = \dfrac{1}{2-0} \displaystyle\int_{0}^{2} x^2 dx = 4/3$ **(b)** $(x^*)^2 = 4/3, x^* = \pm 2/\sqrt{3},$ but only $2/\sqrt{3}$ is in $[0,2]$.

(c)

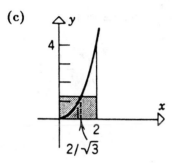

11. $f_{\text{ave}} = \dfrac{1}{9} \displaystyle\int_{0}^{9} x^{1/2} dx = 2; \ \sqrt{x^*} = 2, \ x^* = 4$

13. $f_{\text{ave}} = \dfrac{1}{x_1 - x_0} \displaystyle\int_{x_0}^{x_1} (\alpha x + \beta) dx = \dfrac{1}{2}\alpha(x_0 + x_1) + \beta;$

$\alpha x^* + \beta = \dfrac{1}{2}\alpha(x_0 + x_1) + \beta, \ x^* = (x_0 + x_1)/2.$

15. $V_p/\sqrt{2} = 120, V_p = 120\sqrt{2} \approx 169.7$ volts

17. $a_{\text{ave}} = \dfrac{v(t_1) - v(t_0)}{t_1 - t_0} = \dfrac{1}{t_1 - t_0}v(t)\Big]_{t_0}^{t_1} = \dfrac{1}{t_1 - t_0}\displaystyle\int_{t_0}^{t_1} a(t) dt$

19. time to fill tank $=$ (volume of tank)/(rate of filling) $= [\pi(3)^2 5]/(1) = 45\pi$, force on bottom at time $t =$ weight of water in tank at time $t = (62.4)$ (rate of filling)(time) $= 62.4t,$

force$_{\text{ave}} = \dfrac{1}{45\pi} \displaystyle\int_{0}^{45\pi} 62.4t\, dt = 1404\pi$ lb.

21. **(a)** $\displaystyle\int_{a}^{b} [f(x) - f_{\text{ave}}]\, dx = \int_{a}^{b} f(x) dx - \int_{a}^{b} f_{\text{ave}} dx = \int_{a}^{b} f(x) dx - f_{\text{ave}}(b-a) = 0$

because $f_{\text{ave}}(b-a) = \displaystyle\int_{a}^{b} f(x) dx.$

(b) No, because if $\int_a^b [f(x) - c]dx = 0$ then $\int_a^b f(x)dx - c(b-a) = 0$ so

$c = \dfrac{1}{b-a}\displaystyle\int_a^b f(x)dx = f_{\text{ave}}$ is the only value.

23. **(a)** $x^3 + 1$ **(b)** $F(x) = \dfrac{1}{4}t^4 + t\Big]_1^x = \dfrac{1}{4}x^4 + x - \dfrac{5}{4}$; $F'(x) = x^3 + 1$

25. $\sin\sqrt{x}$ **27.** $|x|$ **29.** $\displaystyle\int_2^x \dfrac{1}{t-1}dt$ **31.** $\displaystyle\int_0^x \dfrac{1}{t-1}dt$

33. **(a)** $(0, +\infty)$ because f is continuous there and 1 is in $(0, +\infty)$.
   (b) at $x = 1$ because $F(1) = 0$

35. $F'(x) = \dfrac{\cos x}{x^2 + 3}$, $F''(x) = \dfrac{-(x^2 + 3)\sin x - 2x\cos x}{(x^2 + 3)^2}$
   (a) 0 **(b)** $1/3$ **(c)** 0

37. The domain is $(-\infty, +\infty)$; $F(x)$ is 0 if $x = 1$, positive if $x > 1$, and negative if $x < 1$.

39. The domain is $[-2, 2]$; $F(x)$ is 0 if $x = -1$, positive if $-1 < x \le 2$, and negative if $-2 \le x < -1$.

41. $x < 0 : F(x) = \displaystyle\int_{-1}^x (-t)dt = -\dfrac{1}{2}t^2\Big]_{-1}^x = \dfrac{1}{2}(1 - x^2)$,

$x \ge 0 : F(x) = \displaystyle\int_{-1}^0 (-t)dt + \int_0^x t\,dt = \dfrac{1}{2} + \dfrac{1}{2}x^2$; $F(x) = \begin{cases} (1 - x^2)/2, & x < 0 \\ (1 + x^2)/2, & x \ge 0 \end{cases}$

43. $x \le 0 : F(x) = \displaystyle\int_{-1}^x t^2 dt = \dfrac{1}{3}t^3\Big]_{-1}^x = \dfrac{1}{3}(x^3 + 1)$,

$x > 0 : F(x) = \displaystyle\int_{-1}^0 t^2 dt + \int_0^x 2t\,dt = \dfrac{1}{3} + x^2$; $F(x) = \begin{cases} (x^3 + 1)/3, & x \le 0 \\ x^2 + 1/3, & x > 0 \end{cases}$

45. $\dfrac{1}{x^3}(3x^2) = \dfrac{3}{x}$

47. $F'(x) = \dfrac{1}{1 + x^2} + \dfrac{1}{1 + (1/x)^2}(-1/x^2) = 0$ so F is constant on $(0, +\infty)$.

49. **(a)** $\sin^2(x^3)(3x^2) - \sin^2(x^2)(2x) = 3x^2\sin^2(x^3) - 2x\sin^2(x^2)$

(b) $\dfrac{1}{1+x}(1) - \dfrac{1}{1-x}(-1) = \dfrac{2}{1-x^2}$

51. $\displaystyle\int_x^b f(t)dt = -\int_b^x f(t)dt$ so $\dfrac{d}{dx}\int_x^b f(t)dt = -\dfrac{d}{dx}\int_b^x f(t)dt = -f(x)$

SUPPLEMENTARY EXERCISES CHAPTER 5

1. $-x^{-2}/2 + 2\sqrt{x} + 5\cos x + C$

3. $u = \sqrt{x} + 2,\ 2\displaystyle\int u^8 du = \dfrac{2}{9}u^9 + C = \dfrac{2}{9}(\sqrt{x}+2)^9 + C$

5. $u = \sqrt{2x^2 - 5},\ du = 2x/\sqrt{2x^2-5}\,dx,\ \dfrac{1}{2}\displaystyle\int \sin u\,du = -\dfrac{1}{2}\cos\sqrt{2x^2-5} + C$

7. $\displaystyle\int (3x^{1/2} + x^{11/6})dx = 2x^{3/2} + \dfrac{6}{17}x^{17/6} + C$

9. $u = \sin 5t,\ \dfrac{1}{5}\displaystyle\int \sec^2 u\,du = \dfrac{1}{5}\tan(\sin 5t) + C$

11. (a) $\displaystyle\int (y^5 + 4y^3 + 4y)dy = \dfrac{1}{6}y^6 + y^4 + 2y^2 + C$

    (b) $\dfrac{1}{6}(y^2 + 2)^3 + C$

    [answer to (b)] − [answer to (a)]

$$= \dfrac{1}{6}(y^6 + 6y^4 + 12y^2 + 8) + C - \left(\dfrac{1}{6}y^6 + y^4 + 2y^2 + C\right) = 4/3$$

13. $\displaystyle\int_0^1 u^4 du = 1/5$

15. $u = x - 1,\ x = u + 1,\ \displaystyle\int_1^4 \dfrac{u-1}{\sqrt{u}}du = \int_1^4 (u^{1/2} - u^{-1/2})du = \dfrac{2}{3}u^{3/2} - 2u^{1/2}\Big]_1^4 = 8/3$

17. $\dfrac{4}{\pi}\displaystyle\int_{\pi/2}^{\pi} \cos u\,du = \dfrac{4}{\pi}\sin u\Big]_{\pi/2}^{\pi} = -4/\pi$

19. $\displaystyle\int_{-2}^{1/2} -(2x-1)dx + \int_{1/2}^{2} (2x-1)dx = (-x^2 + x)\Big]_{-2}^{1/2} + (x^2 - x)\Big]_{1/2}^{2} = 17/2$

21. $\int_0^x \frac{1}{(3t+1)^2}\,dt = -\frac{1}{3(3t+1)}\Big]_0^x = -\frac{1}{3(3x+1)} + \frac{1}{3} = \frac{1}{6}$, $3x+1 = 2$, $x = 1/3$

23. **(a)** $5+5+5+5 = 20$ $\qquad$ **(b)** $2+2+2+2 = 8$

(c) $n+n+n+n = 4n$ $\qquad$ **(d)** $0+1/5+2/6 = 8/15$

(e) $6/4 + 6/9 + 6/16 = 61/24$ $\qquad$ **(f)** 9

(g) $\sin(0) + \sin(\pi/4) + \sin(\pi/2) + \sin(3\pi/4) + \sin(\pi) = 0 + \sqrt{2}/2 + 1 + \sqrt{2}/2 + 0 = 1 + \sqrt{2}$

(h) $\sqrt{2}/2 + (\sqrt{2}/2)^2 + (\sqrt{2}/2)^3 + (\sqrt{2}/2)^4 = 3\sqrt{2}/4 + 3/4$

25. **(a)** $\displaystyle\sum_{k=1}^{9}(-1)^{k+1}\left(\frac{k}{k+1}\right)^2 = \sum_{k=2}^{10}(-1)^k\left(\frac{k-1}{k}\right)^2$

(b) $\displaystyle\sum_{k=1}^{11}(-1)^{k+1}\frac{\pi^{k+1}}{k} = \sum_{k=2}^{12}(-1)^k\frac{\pi^k}{k-1}$

27. **(a)** $\Delta x = 4/n$, $c_k = 4k/n$

$$\sum_{k=1}^{n} f(c_k)\Delta x = \sum_{k=1}^{n}(16 - 16k^2/n^2)(4/n)$$

$$= \frac{64}{n}\sum_{k=1}^{n}1 - \frac{64}{n^3}\sum_{k=1}^{n}k^2 = 64 - \frac{32}{3}\frac{(n+1)(2n+1)}{n^2}$$

(b) $d_k = 4(k-1)/n$

$$\sum_{k=1}^{n} f(d_k)\Delta x = \sum_{k=1}^{n}(16 - 16(k-1)^2/n^2)(4/n) = \frac{64}{n}\sum_{k=1}^{n}1 - \frac{64}{n^3}\sum_{k=1}^{n}(k-1)^2$$

$$= 64 - \frac{32}{3}\frac{(n-1)(2n-1)}{n^2}$$

(c) area $= \displaystyle\lim_{n\to+\infty}\left[64 - \frac{32}{3}\left(1 + \frac{1}{n}\right)\left(2 + \frac{1}{n}\right)\right] = 128/3$; $\displaystyle\int_0^4(16 - x^2)\,dx = 128/3$

29. **(a)** $\Delta x = 2/n$, because f is constant c_k can be chosen anywhere in the k-th subinterval so

$f(c_k) = 6$ and $\displaystyle\sum_{k=1}^{n} f(c_k)\Delta x = \sum_{k=1}^{n}(6)(2/n) = 12$

(b) same as for (a)

(c) area $= \displaystyle\lim_{n\to+\infty} 12 = 12$; $\displaystyle\int_{-1}^{1} 6\,dx = 12$

31. **(a)** $2\displaystyle\int_3^5 P(x)\,dx + \int_3^5 Q(x)\,dx = 2(3) + (4) = 10$

(b) $-\int_1^5 P(x)dx = -(-1) = 1$

(c) $-\int_3^5 Q(u)du = -\int_3^5 Q(x)dx = -4$

(d) $\int_3^5 P(x)dx + \int_5^1 P(x)dx = \int_3^5 P(x)dx - \int_1^5 P(x)dx = (3) - (-1) = 4$

33. $f_{ave} = \int_{-2}^{-1} 3x^2 dx = 7;\ 3(x^*)^2 = 7,\ x^* = \pm\sqrt{7/3}$ but only $-\sqrt{7/3}$ is in $[-2, -1]$

35. $f_{ave} = \dfrac{1}{4}\int_{-3}^1 (2 + |x|)dx = \dfrac{1}{4}\left[\int_{-3}^0 (2 - x)dx + \int_0^1 (2 + x)dx\right] = \dfrac{1}{4}[21/2 + 5/2] = 13/4;$

$2 + |x^*| = 13/4,\ |x^*| = 5/4,\ x^* = \pm 5/4$ but only $-5/4$ is in $[-3, 1]$

CHAPTER 6
Applications of the Definite Integral

EXERCISE SET 6.1

1. $A = \int_{-1}^{2} (x^2 + 1 - x)dx = (x^3/3 + x - x^2/2)\Big]_{-1}^{2} = 9/2$

3. $A = \int_{1}^{2} (y - 1/y^2)dy = (y^2/2 + 1/y)\Big]_{1}^{2} = 1$

5. (a) $A = \int_{0}^{4} (4x - x^2)dx = 32/3$

(b) $A = \int_{0}^{16} (\sqrt{y} - y/4)dy = 32/3$

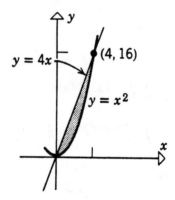

7. Eliminate x to get $y^2 = y + 2$,
$y^2 - y - 2 = 0, (y + 1)(y - 2) = 0$,
$y = -1$ and 2 with corresponding values
of $x = 1/2$ and 2.

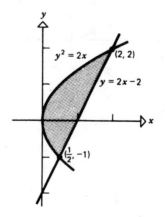

(a) $A = \int_{0}^{1/2} [\sqrt{2x} - (-\sqrt{2x})]dx$

$+ \int_{1/2}^{2} [\sqrt{2x} - (2x - 2)]dx$

$= \int_{0}^{1/2} 2\sqrt{2}x^{1/2}dx + \int_{1/2}^{2} (\sqrt{2}x^{1/2} - 2x + 2)dx$

$= 2/3 + 19/12 = 9/4$

(b) $A = \int_{-1}^{2} [(y/2 + 1) - y^2/2]dy = 9/4$

9. $A = \int_{1/4}^{1} (\sqrt{x} - x^2)\,dx = 49/192$

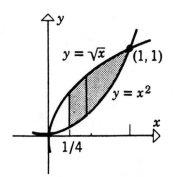

11. $A = \int_{\pi/4}^{\pi/2} (0 - \cos 2x)\,dx$

$= -\int_{\pi/4}^{\pi/2} \cos 2x\,dx = 1/2$

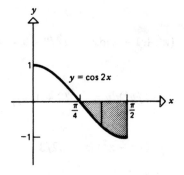

13. $A = \int_{0}^{4} [0 - (y^2 - 4y)]\,dy$

$= \int_{0}^{4} (4y - y^2)\,dy = 32/3$

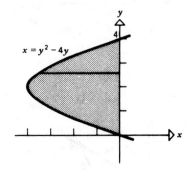

15. Equate $\sec^2 x$ and 2 to get $\sec^2 x = 2$,
$\sec x = \pm\sqrt{2}$, $x = \pm\pi/4$
$A = \int_{-\pi/4}^{\pi/4} (2 - \sec^2 x)\,dx = \pi - 2$

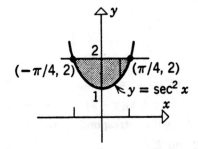

17. Eliminate y to get $6 - x = x^2 + 4$,
$x^2 + x - 2 = 0$, $(x + 2)(x - 1) = 0$,
$x = -2, 1$ with corresponding values of $y = 8, 5$.

$$A = \int_{-2}^{1} [(6 - x) - (x^2 + 4)]dx$$

$$= \int_{-2}^{1} (2 - x - x^2)dx = 9/2$$

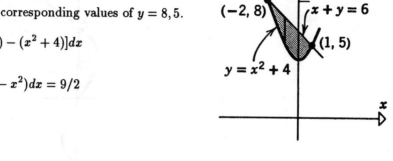

19. $A = \int_{-1}^{4} [(y + 6) - (-y^2)]dy$

$$= \int_{-1}^{4} (y + 6 + y^2)dy$$

$$= 355/6$$

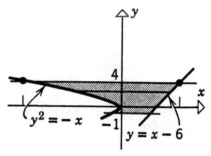

21. $y = 2 + |x - 1| = \begin{cases} 3 - x, & x \leq 1 \\ 1 + x, & x \geq 1 \end{cases}$,

$$A = \int_{-5}^{1} \left[\left(-\frac{1}{5}x + 7\right) - (3 - x)\right] dx$$

$$+ \int_{1}^{5} \left[\left(-\frac{1}{5}x + 7\right) - (1 + x)\right] dx$$

$$= \int_{-5}^{1} \left(\frac{4}{5}x + 4\right) dx + \int_{1}^{5} \left(6 - \frac{6}{5}x\right) dx$$

$$= 72/5 + 48/5 = 24$$

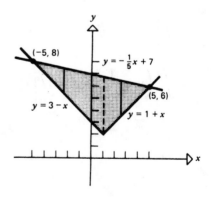

23. $A = \displaystyle\int_{-1}^{0} (y^3 - y)dy + \int_{0}^{1} -(y^3 - y)dy$

$\qquad = 1/2$

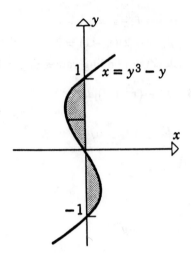

25. From the symmetry of the region

$$A = 2\int_{\pi/4}^{5\pi/4} (\sin x - \cos x)dx = 4\sqrt{2}$$

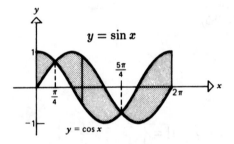

27. $A = \displaystyle\int_{1}^{4} \left(y - \frac{1}{\sqrt{y}}\right) dy = 11/2$

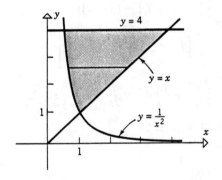

29.   The tangent line at $(4, 2)$ is

$$y = \frac{1}{4}x + 1;$$

$$A = \int_0^4 \left[\left(\frac{1}{4}x + 1 \right) - \sqrt{x} \right] dx = 2/3$$

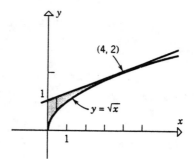

31.   **(a)** $A = \int_1^b x^{-1/2} dx = 2(\sqrt{b} - 1)$ **(b)** $\lim\limits_{b \to +\infty} A = +\infty$

33. $\int_0^k 2\sqrt{y}\,dy = \int_k^9 2\sqrt{y}\,dy$

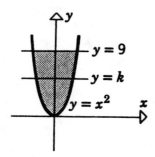

$$\int_0^k y^{1/2}\,dy = \int_k^9 y^{1/2}\,dy$$

$$\frac{2}{3}k^{3/2} = \frac{2}{3}(27 - k^{3/2})$$

$$k^{3/2} = 27/2$$

$$k = (27/2)^{2/3} = 9/\sqrt[3]{4}$$

35.   Solve for y to get $y = (b/a)\sqrt{a^2 - x^2}$ for the upper half of the ellipse; make use of symmetry
to get $A = 4 \int_0^a \frac{b}{a}\sqrt{a^2 - x^2}\,dx = \frac{4b}{a} \int_0^a \sqrt{a^2 - x^2}\,dx = \frac{4b}{a} \times \frac{1}{4}\pi a^2 = \pi ab.$

37.   **(a)** It gives the area of the region that is between f and g when $f(x) > g(x)$ <u>minus</u> the area
of the region between f and g when $f(x) < g(x)$, for $a \leq x \leq b$.
   (b) It gives the area of the region that is between f and g for $a \leq x \leq b$.

39.   The curves intersect at $x = 0$ and, by Newton's Method, at $x \approx 2.595739080 = b$, so

$$A \approx \int_0^b (\sin x - 0.2x)dx = -\cos x - 0.1x^2 \Big]_0^b \approx 1.180898334$$

EXERCISE SET 6.2

1. $V = \pi \int_{-1}^{3} (3-x)dx = 8\pi$

3. $V = \pi \int_{0}^{2} \frac{1}{4}(3-y)^2 dy = 13\pi/6$

5. $V = \pi \int_{0}^{2} x^4 dx = 32\pi/5$

7. $V = \pi \int_{1}^{2} (1+x^3)^2 dx$

$= \pi \int_{1}^{2} (1+2x^3+x^6)dx$

$= 373\pi/14$

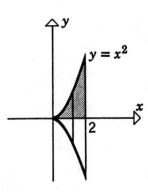

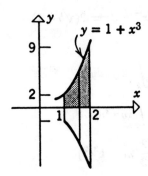

9. $V = \pi \int_{-3}^{3} (9-x^2)^2 dx$

$= \pi \int_{-3}^{3} (81-18x^2+x^4)dx$

$= 1296\pi/5$

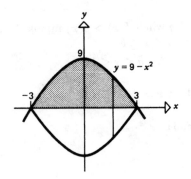

11. $V = \pi \int_0^4 [(4x)^2 - (x^2)^2] dx$

$= \pi \int_0^4 (16x^2 - x^4) dx = 2048\pi/15$

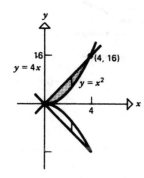

13. $V = \pi \int_0^{\pi/4} (\cos^2 x - \sin^2 x) dx$

$= \pi \int_0^{\pi/4} \cos 2x \, dx = \pi/2$

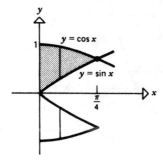

15. $V = \pi \int_0^1 [(\sqrt{x})^2 - x^2] dx$

$= \pi \int_0^1 (x - x^2) dx = \pi/6$

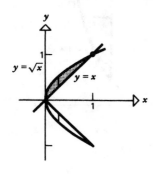

17. $V = \pi \int_0^1 y^{2/3} dy = 3\pi/5$

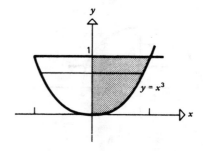

19. $V = \pi \int_{-1}^{3} (1 + y) dy = 8\pi$

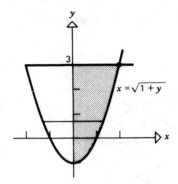

$x = \sqrt{1 + y}$

21. $V = \pi \int_{\pi/4}^{3\pi/4} \csc^2 y \, dy = 2\pi$

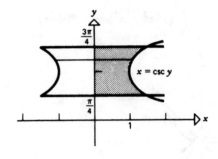

$x = \csc y$

23. $V = \pi \int_{1}^{3} (9 - y^2) dy = 28\pi/3$

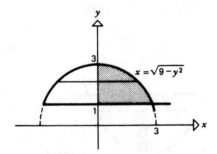

$x = \sqrt{9 - y^2}$

25. $V = \pi \int_{2}^{9} [(y - 1)^{2/3} - 1] dy$

$= 58\pi/5$

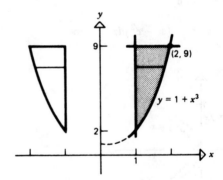

$(2, 9)$

$y = 1 + x^3$

27. $V = \pi \int_{-1}^{2} [(y+2)^2 - y^4]dy$

$= 72\pi/5$

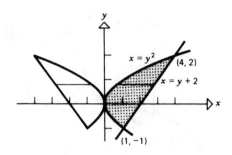

29. $V = \pi \int_{-4}^{4} [(25 - x^2) - 9]dx$

$= 2\pi \int_{0}^{4} (16 - x^2)dx$

$= 256\pi/3$

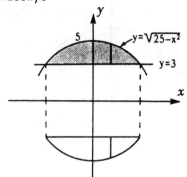

31. (a) $V(b) = \pi \int_{1}^{b} \frac{1}{x^2}dx = \pi(1 - 1/b)$

(b) $\lim\limits_{b \to +\infty} \pi(1 - 1/b) = \pi$

33. $V = \pi \int_{0}^{b} x dx = \pi b^2/2; \pi b^2/2 = 2, b^2 = 4/\pi, b = 2/\sqrt{\pi}$

35. $V = \pi \int_{0}^{3} (9 - y^2)^2 dy$

$= \pi \int_{0}^{3} (81 - 18y^2 + y^4)dy$

$= 648\pi/5$

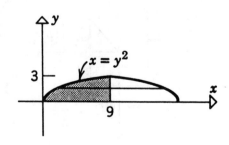

37. $V = \pi \int_{0}^{1} [(\sqrt{x}+1)^2 - (x+1)^2]dx$

$= \pi \int_{0}^{1} (2\sqrt{x} - x - x^2)dx = \pi/2$

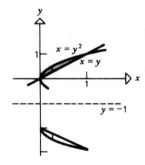

39. $V = \pi \int_{-a}^{a} \frac{b^2}{a^2}(a^2 - x^2)dx$

$= 4\pi a b^2 / 3$

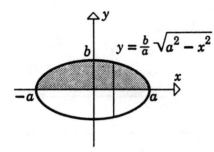

$y = \frac{b}{a}\sqrt{a^2 - x^2}$

41. $V = \pi \int_{-1}^{0} (x + 1)dx$

$+\pi \int_{0}^{1} [(x + 1) - 2x]dx$

$= \pi/2 + \pi/2 = \pi$

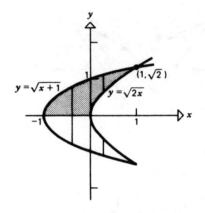

$y = \sqrt{x + 1}$

$(1, \sqrt{2})$

$y = \sqrt{2x}$

43. By similar triangles, $R/r = y/h$ so
$R = ry/h$ and $A(y) = \pi r^2 y^2 / h^2$.

$V = (\pi r^2 / h^2) \int_{0}^{h} y^2 dy = \pi r^2 h / 3$

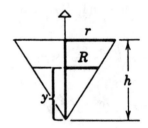

45. $V = 2\pi \int_{0}^{L/2} [(r^2 - y^2) - (r^2 - L^2/4)]dy$

$= 2\pi \int_{0}^{L/2} (L^2/4 - y^2)dy$

$= \pi L^3 / 6$

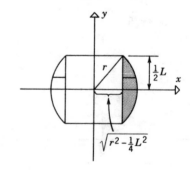

$\frac{1}{2}L$

$\sqrt{r^2 - \frac{1}{4}L^2}$

47. (a)   **(b)**

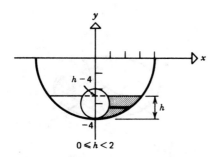

$0 \leqslant h < 2$ $2 \leqslant h \leqslant 4$

If the cherry is partially submerged then $0 \leq h < 2$ as shown in Figure (a); if it is totally submerged then $2 \leq h \leq 4$ as shown in Figure (b). The radius of the glass is 4 cm and that of the cherry is 1 cm so points on the sections shown in the figures satisfy the equations $x^2 + y^2 = 16$ and $x^2 + (y+3)^2 = 1$. We will find the volumes of the solids that are generated when the shaded regions are revolved about the y-axis.

For $0 \leq h < 2$,

$$V = \pi \int_{-4}^{h-4} [(16 - y^2) - (1 - (y+3)^2)]dy = 6\pi \int_{-4}^{h-4} (y+4)dy = 3\pi h^2;$$

for $2 \leq h \leq 4$,

$$V = \pi \int_{-4}^{-2} [(16 - y^2) - (1 - (y+3)^2)]dy + \pi \int_{-2}^{h-4} (16 - y^2)dy$$

$$= 6\pi \int_{-4}^{-2} (y+4)dy + \pi \int_{-2}^{h-4} (16 - y^2)dy = 12\pi + \frac{1}{3}\pi(12h^2 - h^3 - 40)$$

$$= \frac{1}{3}\pi(12h^2 - h^3 - 4)$$

so

$$V = \begin{cases} 3\pi h^2 & \text{if } 0 \leq h < 2 \\ \dfrac{1}{3}\pi(12h^2 - h^3 - 4) & \text{if } 2 \leq h \leq 4 \end{cases}.$$

49. $A(x) = \pi(x^2/4)^2 = \pi x^4/16$, $V = \displaystyle\int_0^{20} (\pi x^4/16)dx = 40{,}000\pi \text{ ft}^3$

51. With $y = \sqrt{9 - x^2}$, which is the upper half of the circle, $A(x)$ is the area of an equilateral triangle whose sides are each of length $2y$ so

$$A(x) = \frac{\sqrt{3}}{4}(2y)^2 = \sqrt{3}y^2 = \sqrt{3}(9 - x^2),$$

$$V = \int_{-3}^{3} \sqrt{3}(9 - x^2)dx$$

$$= 2\sqrt{3}\int_{0}^{3}(9 - x^2)dx = 36\sqrt{3}.$$

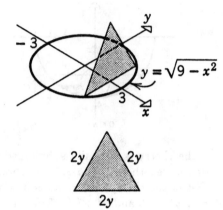

53. $V = \int_{\pi/4}^{3\pi/4} \sin^2 x \, dx$

$$= \frac{1}{2}\int_{\pi/4}^{3\pi/4}(1 - \cos 2x)dx$$

$$= (\pi + 2)/4$$

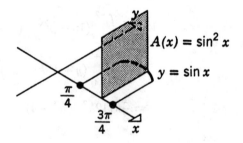

55. $\tan\theta = h/x$ so $h = x\tan\theta$,

$$A(y) = \frac{1}{2}hx = \frac{1}{2}x^2\tan\theta = \frac{1}{2}(r^2 - y^2)\tan\theta$$

because $x^2 = r^2 - y^2$,

$$V = \frac{1}{2}\tan\theta\int_{-r}^{r}(r^2 - y^2)dy = \tan\theta\int_{0}^{r}(r^2 - y^2)dy = \frac{2}{3}r^3\tan\theta$$

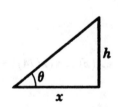

57. Each cross section perpendicular to
the y-axis is a square so

$$A(y) = x^2 = r^2 - y^2,$$
$$\frac{1}{8}V = \int_0^r (r^2 - y^2)dy$$
$$V = 8(2r^3/3) = 16r^3/3$$

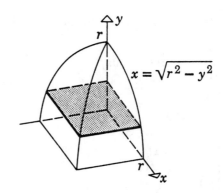

59. **(a)** If $V = \int_0^h A(x)dx$, then $\dfrac{dV}{dt} = \dfrac{dV}{dh}\dfrac{dh}{dt} = A(h)\dfrac{dh}{dt}$ so $A(h)\dfrac{dh}{dt} = -kA(h)$, $\dfrac{dh}{dt} = -k$.

(b) If $dh/dt = -k$, then $h = -kt + C$. But $h = h_0$ when $t = 0$ so $C = h_0$, thus $h = h_0 - kt$; $h = 0$ when $t = h_0/k$.

EXERCISE SET 6.3

1. $V = \displaystyle\int_1^2 2\pi x(x^2)dx = 2\pi \int_1^2 x^3 dx = 15\pi/2$

3. $V = \displaystyle\int_0^1 2\pi y(2y - 2y^2)dy = 4\pi \int_0^1 (y^2 - y^3)dy = \pi/3$

5. $V = \displaystyle\int_0^1 2\pi(x)(x^3)dx$

$= 2\pi \displaystyle\int_0^1 x^4 dx = 2\pi/5$

7. $V = \displaystyle\int_1^3 2\pi x(1/x)dx$

$= 2\pi \displaystyle\int_1^3 dx = 4\pi$

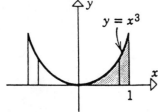

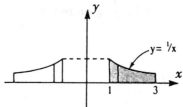

9. $V = \displaystyle\int_{1}^{2} 2\pi x[(2x-1)-(-2x+3)]dx$

$= 8\pi \displaystyle\int_{1}^{2}(x^2-x)dx = 20\pi/3$

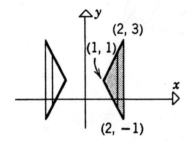

11. $V = \displaystyle\int_{0}^{2} 2\pi x(4-x^2)^{1/3}dx$

$+ \displaystyle\int_{2}^{4} 2\pi x[-(4-x^2)^{1/3}]dx$

$= 3\pi\sqrt[3]{4}(1+3\sqrt[3]{3})$

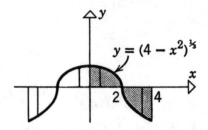

13. $V = \displaystyle\int_{0}^{1} 2\pi y^3 dy = \pi/2$

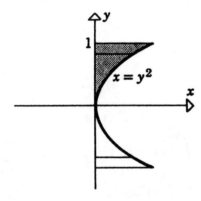

15. $V = \displaystyle\int_{0}^{1} 2\pi y(1-\sqrt{y})dy$

$= 2\pi \displaystyle\int_{0}^{1}(y-y^{3/2})dy = \pi/5$

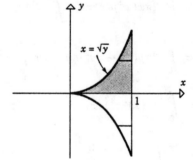

17. **(b)** $V = 2\pi \displaystyle\int_{0}^{\pi} x\sin x\,dx = 2\pi(\sin x - x\cos x)\Big]_{0}^{\pi} = 2\pi^2$

19. (a) $V = \int_0^1 2\pi x(x^3 - 3x^2 + 2x)dx = 7\pi/30$

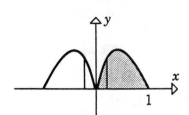

(b) much easier; the method of slicing would require that x be expressed in terms of y.

21. $V = \int_0^1 2\pi(1-y)y^{1/3}dy$

$\quad = 2\pi \int_0^1 (y^{1/3} - y^{4/3})dy = 9\pi/14$

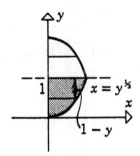

23. $x = \dfrac{h}{r}(r - y)$ is an equation of line through $(0, r)$ and $(h, 0)$ so

$V = \int_0^r 2\pi y \left[\dfrac{h}{r}(r - y)\right] dy$

$\quad = \dfrac{2\pi h}{r} \int_0^r (ry - y^2)dy = \pi r^2 h/3$

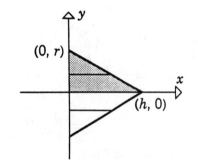

25. $V = \int_0^a 2\pi x(2\sqrt{r^2 - x^2})dx$

$\quad = 4\pi \int_0^a x(r^2 - x^2)^{1/2}dx$

$\quad = -\dfrac{4\pi}{3}(r^2 - x^2)^{3/2}\Big]_0^a$

$\quad = \dfrac{4\pi}{3}\left[r^3 - (r^2 - a^2)^{3/2}\right]$

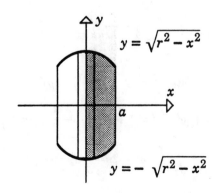

27. $V_x = \pi \int_{1/2}^{b} \frac{1}{x^2} dx = \pi(2 - 1/b)$, $V_y = 2\pi \int_{1/2}^{b} dx = \pi(2b - 1)$;

$V_x = V_y$ if $2 - 1/b = 2b - 1$, $2b^2 - 3b + 1 = 0$, solve to get $b = 1/2$ (reject) or $b = 1$.

EXERCISE SET 6.4

1. (a) $L = \int_{1}^{2} \sqrt{1 + 2^2} dx = \sqrt{5} \int_{1}^{2} dx = \sqrt{5}$

(b) $L = \int_{2}^{4} \sqrt{1 + (1/2)^2} dy = \frac{1}{2}\sqrt{5} \int_{2}^{4} dy = \sqrt{5}$

(c) $L = \sqrt{(2-1)^2 + (4-2)^2} = \sqrt{5}$ ($y = 2x$ is a line).

3. $f'(x) = \frac{9}{2}x^{1/2}$, $1 + [f'(x)]^2 = 1 + \frac{81}{4}x$,

$$L = \int_{0}^{1} \sqrt{1 + 81x/4}\,dx = \frac{8}{243}\left(1 + \frac{81}{4}x\right)^{3/2}\Bigg]_{0}^{1} = (85\sqrt{85} - 8)/243$$

5. $\frac{dy}{dx} = \frac{2}{3}x^{-1/3}$, $1 + \left(\frac{dy}{dx}\right)^2 = 1 + \frac{4}{9}x^{-2/3} = \frac{9x^{2/3} + 4}{9x^{2/3}}$,

$$L = \int_{1}^{8} \frac{\sqrt{9x^{2/3} + 4}}{3x^{1/3}}\,dx = \frac{1}{18}\int_{13}^{40} u^{1/2}\,du, \quad u = 9x^{2/3} + 4$$

$$= \frac{1}{27}u^{3/2}\Bigg]_{13}^{40} = \frac{1}{27}(40\sqrt{40} - 13\sqrt{13}) = \frac{1}{27}(80\sqrt{10} - 13\sqrt{13})$$

or (alternate solution)

$x = y^{3/2}$, $\frac{dx}{dy} = \frac{3}{2}y^{1/2}$, $1 + \left(\frac{dx}{dy}\right)^2 = 1 + \frac{9}{4}y = \frac{4 + 9y}{4}$,

$$L = \frac{1}{2}\int_{1}^{4} \sqrt{4 + 9y}\,dy = \frac{1}{18}\int_{13}^{40} u^{1/2}\,du = \frac{1}{27}(80\sqrt{10} - 13\sqrt{13}).$$

7. $x = g(y) = \frac{1}{24}y^3 + 2y^{-1}$, $g'(y) = \frac{1}{8}y^2 - 2y^{-2}$,

$1 + [g'(y)]^2 = 1 + \left(\frac{1}{64}y^4 - \frac{1}{2} + 4y^{-4}\right) = \frac{1}{64}y^4 + \frac{1}{2} + 4y^{-4} = \left(\frac{1}{8}y^2 + 2y^{-2}\right)^2$,

$$L = \int_{2}^{4} \left(\frac{1}{8}y^2 + 2y^{-2}\right) dy = 17/6$$

9. **(a)**

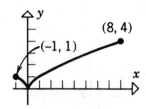

(b) dy/dx does not exist at $x = 0$

(c) $x = g(y) = y^{3/2}$, $g'(y) = \dfrac{3}{2}y^{1/2}$,

$$L = \int_0^1 \sqrt{1 + 9y/4}\,dy \quad \text{(portion for } -1 \le x \le 0)$$

$$+ \int_0^4 \sqrt{1 + 9y/4}\,dy \quad \text{(portion for } 0 \le x \le 8)$$

$$= \frac{8}{27}\left(\frac{13}{8}\sqrt{13} - 1\right) + \frac{8}{27}(10\sqrt{10} - 1) = (13\sqrt{13} + 80\sqrt{10} - 16)/27$$

11. **(a)** f' is continuous on $[a, b]$ because f is smooth so by Theorem 4.6.4 f' has a maximum value M and a minimum value m on $[a, b]$.

(b) From part (a)

$$m \le f'(x) \le M$$
$$m^2 \le [f'(x)]^2 \le M^2$$
$$1 + m^2 \le 1 + [f'(x)]^2 \le 1 + M^2$$
$$\sqrt{1 + m^2} \le \sqrt{1 + [f'(x)]^2} \le \sqrt{1 + M^2}$$
$$\int_a^b \sqrt{1 + m^2}\,dx \le \int_a^b \sqrt{1 + [f'(x)]^2}\,dx \le \int_a^b \sqrt{1 + M^2}\,dx$$
$$(b - a)\sqrt{1 + m^2} \le L \le (b - a)\sqrt{1 + M^2}$$

13. 4.645975301

EXERCISE SET 6.5

1. $S = \displaystyle\int_0^1 2\pi(7x)\sqrt{1 + 49}\,dx = 70\pi\sqrt{2}\int_0^1 x\,dx = 35\pi\sqrt{2}$

3. $f'(x) = -x/\sqrt{4-x^2}$, $1 + [f'(x)]^2 = 1 + \dfrac{x^2}{4-x^2} = \dfrac{4}{4-x^2}$,

$S = \displaystyle\int_{-1}^{1} 2\pi\sqrt{4-x^2}(2/\sqrt{4-x^2})dx = 4\pi\int_{-1}^{1} dx = 8\pi$

5. $f'(x) = \dfrac{1}{2}x^{-1/2} - \dfrac{1}{2}x^{1/2}$, $1 + [f'(x)]^2 = 1 + \dfrac{1}{4}x^{-1} - \dfrac{1}{2} + \dfrac{1}{4}x = \left(\dfrac{1}{2}x^{-1} + \dfrac{1}{2}x\right)^2$,

$S = \displaystyle\int_{1}^{3} 2\pi \left(x^{1/2} - \dfrac{1}{3}x^{3/2}\right)\left(\dfrac{1}{2}x^{-1} + \dfrac{1}{2}x\right) dx = \dfrac{\pi}{3}\int_{1}^{3}(3 + 2x - x^2)dx = 16\pi/9$

7. $S = \displaystyle\int_{0}^{2} 2\pi(9y+1)\sqrt{82}dy = 2\pi\sqrt{82}\int_{0}^{2}(9y+1)dy = 40\pi\sqrt{82}$

9. $g'(y) = -y/\sqrt{9-y^2}$, $1 + [g'(y)]^2 = \dfrac{9}{9-y^2}$,

$S = \displaystyle\int_{-2}^{2} 2\pi\sqrt{9-y^2} \cdot \dfrac{3}{\sqrt{9-y^2}}dy = 6\pi\int_{-2}^{2} dy = 24\pi$

11. $x = g(y) = \dfrac{1}{4}y^4 + \dfrac{1}{8}y^{-2}$, $g'(y) = y^3 - \dfrac{1}{4}y^{-3}$,

$1 + [g'(y)]^2 = 1 + \left(y^6 - \dfrac{1}{2} + \dfrac{1}{16}y^{-6}\right) = \left(y^3 + \dfrac{1}{4}y^{-3}\right)^2$

$S = \displaystyle\int_{1}^{2} 2\pi\left(\dfrac{1}{4}y^4 + \dfrac{1}{8}y^{-2}\right)\left(y^3 + \dfrac{1}{4}y^{-3}\right) dy = \dfrac{\pi}{16}\int_{1}^{2}(8y^7 + 6y + y^{-5})dy = 16{,}911\pi/1024$

13. Revolve the line segment joining the points $(0,0)$ and (h,r) about the x-axis. An equation of the line segment is $y = (r/h)x$ for $0 \le x \le h$ so

$S = \displaystyle\int_{0}^{h} 2\pi(r/h)x\sqrt{1 + r^2/h^2}dx = \dfrac{2\pi r}{h^2}\sqrt{r^2+h^2}\int_{0}^{h} x\,dx = \pi r\sqrt{r^2+h^2}$

15. $f(x) = \sqrt{r^2 - x^2}$, $f'(x) = -x/\sqrt{r^2-x^2}$, $1 + [f'(x)]^2 = r^2/(r^2 - x^2)$,

$S = \displaystyle\int_{a}^{a+h} 2\pi\sqrt{r^2 - x^2}(r/\sqrt{r^2-x^2})dx = 2\pi r\int_{a}^{a+h} dx = 2\pi rh$

17. **(a)** length of arc of sector $=$ circumference of base of cone, $\ell\theta = 2\pi r$, $\theta = 2\pi r/\ell$;

$S = $ area of sector $= \dfrac{1}{2}\ell^2(2\pi r/\ell) = \pi r\ell$

(b) $S = \pi r_2 \ell_2 - \pi r_1 \ell_1$
$= \pi r_2(\ell_1 + \ell) - \pi r_1 \ell_1$
$= \pi[(r_2 - r_1)\ell_1 + r_2\ell]$

Using similar triangles
$$\ell_2/r_2 = \ell_1/r_1$$
$$r_1\ell_2 = r_2\ell_1$$
$$r_1(\ell_1 + \ell) = r_2\ell_1$$
$$(r_2 - r_1)\ell_1 = r_1\ell$$

so $S = \pi(r_1\ell + r_2\ell) = \pi(r_1 + r_2)\ell$

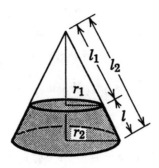

19. **(a)** $1 \le \sqrt{1 + [f'(x)]^2}$ so $2\pi f(x) \le 2\pi f(x)\sqrt{1 + [f'(x)]^2}$. By part (b) of Theorem 5.7.4

$$\int_a^b 2\pi f(x)dx \le \int_a^b 2\pi f(x)\sqrt{1 + [f'(x)]^2}dx$$

$$2\pi \int_a^b f(x)dx \le S, 2\pi A \le S$$

(b) $2\pi A = S$ if $f'(x) = 0$ for all x in $[a, b]$ so $f(x)$ is constant on $[a, b]$.

EXERCISE SET 6.6

1. $s(t) = \int (2t - 3)dt = t^2 - 3t + C$, $s(1) = (1)^2 - 3(1) + C = 5$, $C = 7$, $s(t) = t^2 - 3t + 7$.

3. $s(t) = \int (t^3 - 2t^2 + 1)dt = \frac{1}{4}t^4 - \frac{2}{3}t^3 + t + C$,

   $s(0) = \frac{1}{4}(0)^4 - \frac{2}{3}(0)^3 + 0 + C = 1$, $C = 1$, $s(t) = \frac{1}{4}t^4 - \frac{2}{3}t^3 + t + 1$.

5. $v(t) = \int 4\,dt = 4t + C_1$, $v(0) = 4(0) + C_1 = 1$, $C_1 = 1$, $v(t) = 4t + 1$,

   $s(t) = \int (4t + 1)dt = 2t^2 + t + C_2$, $s(0) = 2(0)^2 + 0 + C_2 = 0$, $C_2 = 0$, $s(t) = 2t^2 + t$.

7. $v(t) = \int 4\cos 2t\,dt = 2\sin 2t + C_1$, $v(0) = 2\sin 0 + C_1 = -1$, $C_1 = -1$,

   $v(t) = 2\sin 2t - 1$, $s(t) = \int (2\sin 2t - 1)dt = -\cos 2t - t + C_2$,

   $s(0) = -\cos 0 - 0 + C_2 = -3$, $C_2 = -2$, $s(t) = -\cos 2t - t - 2$.

9. (a) $s = \displaystyle\int \sin \frac{1}{2}\pi t \, dt = -\frac{2}{\pi}\cos\frac{1}{2}\pi t + C$

$s = 0$ when $t = 0$ which gives $C = \dfrac{2}{\pi}$ so $s = -\dfrac{2}{\pi}\cos\dfrac{1}{2}\pi t + \dfrac{2}{\pi}$.

$a = \dfrac{dv}{dt} = \dfrac{\pi}{2}\cos\dfrac{1}{2}\pi t$. When $t = 1 : s = 2/\pi$, $v = 1$, $|v| = 1$, $a = 0$.

(b) $v = -3\displaystyle\int t\, dt = -\frac{3}{2}t^2 + C_1$, $v = 0$ when $t = 0$ which gives $C_1 = 0$ so $v = -\dfrac{3}{2}t^2$.

$s = -\dfrac{3}{2}\displaystyle\int t^2 dt = -\frac{1}{2}t^3 + C_2$, $s = 1$ when $t = 0$ which gives $C_2 = 1$ so $s = -\dfrac{1}{2}t^3 + 1$.

When $t = 1 : s = 1/2$, $v = -3/2$, $|v| = 3/2$, $a = -3$.

11. If $a(t) = k$ then $v(t) = \displaystyle\int k\, dt = kt + C_1$, but $v(0) = 60$ mph $= 88$ ft/sec so $k(0) + C_1 = 88$,

$C_1 = 88$, $v(t) = kt + 88$; $s(t) = \displaystyle\int (kt + 88)dt = \frac{1}{2}kt^2 + 88t + C_2$, $s(0) = 0$ so $C_2 = 0$,

$s(t) = \dfrac{1}{2}kt^2 + 88t$. $v(t) = 0$ at the instant when the car comes to a stop so $kt + 88 = 0$,

$t = -88/k$; $s(t) = 180$ at this instant so $\dfrac{1}{2}k(-88/k)^2 + 88(-88/k) = 180$, $-3872/k = 180$,

$k \approx -21.5$ ft/sec^2.

13. $s = 0$ and $v = 112$ when $t = 0$ so $v(t) = -32t + 112$, $s(t) = -16t^2 + 112t$.

(a) $v(3) = 16$ ft/sec, $v(5) = -48$ ft/sec.

(b) $v = 0$ when the projectile is at its maximum height so $-32t + 112 = 0$, $t = 7/2$ sec, $s(7/2) = -16(7/2)^2 + 112(7/2) = 196$ ft.

(c) $s = 0$ when it reaches the ground so $-16t^2 + 112t = 0$, $-16t(t - 7) = 0$, $t = 0, 7$ of which $t = 7$ is when it is at ground level on its way down. $v(7) = -112$, $|v| = 112$ ft/sec.

15. (a) $s(t) = 0$ when it hits the ground, $s(t) = -16t^2 + 16t = -16t(t - 1) = 0$ when $t = 1$ sec.

(b) The projectile moves upward until it gets to its highest point where $v(t) = 0$, $v(t) = -32t + 16 = 0$ when $t = 1/2$ sec.

17. (a) $s(t) = 0$ when the package hits the ground, $s(t) = -16t^2 + 20t + 200 = 0$ when (use the quadratic formula) $t = (5 + 5\sqrt{33})/8$ sec.

(b) $v(t) = -32t + 20$, $v[(5 + 5\sqrt{33})/8] = -20\sqrt{33}$, the speed at impact is $20\sqrt{33}$ ft/sec.

19. $s(t) = -4.9t^2 + 49t + 150$ and $v(t) = -9.8t + 49$.

(a) The projectile reaches its maximum height when $v(t) = 0$, $-9.8t + 49 = 0$, $t = 5$ sec.

(b) $s(5) = -4.9(5)^2 + 49(5) + 150 = 272.5$ m.

(c) The projectile reaches its starting point when $s(t) = 150$, $-4.9t^2 + 49t + 150 = 150$, $-4.9t(t - 10) = 0$, $t = 10$ sec.

    (d) $v(10) = -9.8(10) + 49 = -49$ m/sec.

    (e) $s(t) = 0$ when the projectile hits the ground, $-4.9t^2 + 49t + 150 = 0$ when (use the quadratic formula) $t \approx 12.46$ sec.

    (f) $v(12.46) = -9.8(12.46) + 49 \approx -73.1$, the speed at impact is about 73.1 m/sec.

21. $s(t) = -16t^2 + v_0 t$, $v(t) = -32t + v_0$. $v = 0$ when it reaches maximum height so $-32t + v_0 = 0$, $t = v_0/32$ is the time it takes to get there, thus $s(v_0/32) = -16(v_0/32)^2 + v_0(v_0/32) = 1,000$, $v_0^2 = 64,000$, $v_0 = 80\sqrt{10}$ ft/sec (positive because fired upward).

23. $s = -16t^2 + v_0 t + s_0$, but $s_0 = 0$ so $s = -16t^2 + v_0 t$. $s = 0$ when $t = 8$ so $0 = -16(8)^2 + v_0(8)$, $v_0 = 128$ ft/sec. $v = 0$ at its highest point so $-32t + 128 = 0$, $t = 4$, $s = -16(4)^2 + 128(4) = 256$ ft.

25. **(a)** negative, because v is decreasing

    (b) increasing, because the graph of $v(t)$ is concave up

    (c) negative, because the area between the graph of $v(t)$ and the t-axis appears to be greater where $v < 0$ compared to where $v > 0$.

27. displacement $= \displaystyle\int_0^2 (t^2 + t - 2)dt = 2/3$

    distance $= \displaystyle\int_0^2 |t^2 + t - 2|dt = \int_0^1 -(t^2 + t - 2)dt + \int_1^2 (t^2 + t - 2)dt = 7/6 + 11/6 = 3$

29. displacement $= \displaystyle\int_0^\pi \cos t \, dt = 0$

    distance $= \displaystyle\int_0^\pi |\cos t|dt = \int_0^{\pi/2} \cos t \, dt + \int_{\pi/2}^\pi -\cos t \, dt = 1 + 1 = 2$

31. $\qquad v(t) = t^3 - 3t^2 + 2t = t(t-1)(t-2)$

    displacement $= \displaystyle\int_0^3 (t^3 - 3t^2 + 2t)dt = 9/4$

    distance $= \displaystyle\int_0^3 |v(t)|dt = \int_0^1 v(t)dt + \int_1^2 -v(t)dt + \int_2^3 v(t)dt = 11/4$

33. $\qquad v(t) = \dfrac{1}{2}t^2 - 2t$

    displacement $= \displaystyle\int_1^5 (\frac{1}{2}t^2 - 2t)dt = -10/3$

    distance $= \displaystyle\int_1^5 |\frac{1}{2}t^2 - 2t|dt = \int_1^4 -(\frac{1}{2}t^2 - 2t)dt + \int_4^5 (\frac{1}{2}t^2 - 2t)dt = 17/3$

35. $$v(t) = \frac{2}{5}\sqrt{5t+1} + \frac{8}{5}$$

$$\text{displacement} = \int_0^3 \left(\frac{2}{5}\sqrt{5t+1} + \frac{8}{5}\right) dt = \frac{4}{75}(5t+1)^{3/2} + \frac{8}{5}t\Big]_0^3 = 204/25$$

$$\text{distance} = \int_0^3 |v(t)|dt = \int_0^3 v(t)dt = 204/25$$

EXERCISE SET 6.7

1. (a) $W = 30[5 - (-2)] = 210$ ft·lb   (b) $W = \int_1^6 x^{-2}dx = 5/6$ ft·lb

3. $F(x) = kx$, $F(0.2) = 0.2k = 100$, $k = 500$ N/m, $W = \int_0^{0.8} 500x\,dx = 160$ J

5. $W = \int_0^1 kx\,dx = k/2 = 10$, $k = 20$ lb/ft   **7.** $W = \int_0^6 (9-x)\rho(25\pi)dx = 900\pi\rho$ ft·lb

9. $w/4 = x/3$, $w = 4x/3$

$$W = \int_0^2 (3-x)(9810)(4x/3)(6)dx$$

$$= 78480 \int_0^2 (3x - x^2)dx$$

$$= 261,600 \text{ J}$$

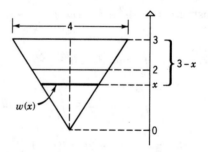

11. (a) $W = \int_0^9 (10 - x)62.4(300)dx$

$$= 18,720 \int_0^9 (10 - x)dx$$

$$= 926,640 \text{ ft·lb}$$

(b) to empty the pool in one hour
would require $926,640/3600 = 257.4$
ft·lb of work per second so
hp of motor $= 257.4/550 = 0.468$

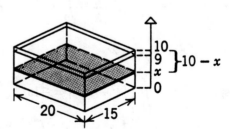

13. $W = \displaystyle\int_0^{100} 15(100 - x)dx$

$\qquad = 75,000 \text{ ft·lb}$

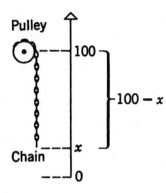

15. **(a)** $F(4000) = k/(4000)^2 = 6000,\ k = 9.6 \times 10^{10}$

$\qquad$ **(b)** $W = \displaystyle\int_{4000}^{5000} 9.6 \times 10^{10} x^{-2} dx = 4,800,000 \text{ mi·lb}$

EXERCISE SET 6.8

1. **(a)** $F = \rho h A = (62.4)(5)(9) = 2,808 \text{ lb}$
$\qquad$ **(b)** $F = (40)(10)(9) = 3,600 \text{ lb}$

3. $F = \displaystyle\int_1^3 9810x(4)dx$

$\qquad = 39240 \displaystyle\int_1^3 x\,dx$

$\qquad = 156,960 \text{ N}$

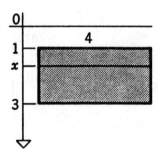

5. by similar triangles

$$\frac{w(x)}{6} = \frac{10-x}{8}$$

$$w(x) = \frac{3}{4}(10-x),$$

$$F = \int_2^{10} 62.4x \left[\frac{3}{4}(10-x)\right] dx$$

$$= 46.8 \int_2^{10} (10x - x^2)dx = 6988.8\,\text{lb}$$

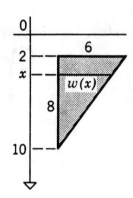

7. $F = \int_0^5 9810x(2\sqrt{25-x^2})dx$

$$= 19,620 \int_0^5 x(25-x^2)^{1/2}dx$$

$$= 8.175 \times 10^5\,\text{N}$$

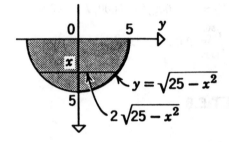

9. Find the forces on the upper and
lower halves and add them:

$$\frac{w_1(x)}{\sqrt{2}a} = \frac{x}{\sqrt{2}a/2}, \; w_1(x) = 2x$$

$$F_1 = \int_0^{\sqrt{2}a/2} \rho x(2x)dx$$

$$= 2\rho \int_0^{\sqrt{2}a/2} x^2 dx = \sqrt{2}\rho a^3/6,$$

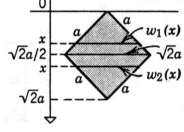

$$\frac{w_2(x)}{\sqrt{2}a} = \frac{\sqrt{2}a - x}{\sqrt{2}a/2}, \; w_2(x) = 2(\sqrt{2}a - x)$$

$$F_2 = \int_{\sqrt{2}a/2}^{\sqrt{2}a} \rho x[2(\sqrt{2}a - x)]dx = 2\rho \int_{\sqrt{2}a/2}^{\sqrt{2}a} (\sqrt{2}ax - x^2)dx = \sqrt{2}\rho a^3/3,$$

$$F = F_1 + F_2 = \sqrt{2}\rho a^3/6 + \sqrt{2}\rho a^3/3 = \rho a^3/\sqrt{2}$$

11. $\sqrt{16^2 + 4^2} = \sqrt{272} = 4\sqrt{17}$ is the

other dimension of the bottom.

$(h(x) - 4)/4 = x/(4\sqrt{17})$

$h(x) = x/\sqrt{17} + 4,$

$F = \displaystyle\int_0^{4\sqrt{17}} 62.4(x/\sqrt{17} + 4)10dx$

$= 624\displaystyle\int_0^{4\sqrt{17}} (x/\sqrt{17} + 4)dx$

$= 14,976\sqrt{17}\,\text{lb}$

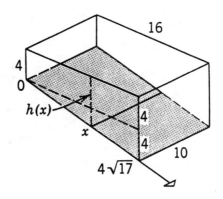

13.   **(a)**   From Exercise 12, $F = 4\rho_0(h+1)$ so (assuming that ρ_0 is constant) $dF/dt = 4\rho_0(dh/dt)$ which is a positive constant if dh/dt is a positive constant.

      (b)   If $dh/dt = 20$ then $dF/dt = 80\rho_0$ lb/min from part (a).

SUPPLEMENTARY EXERCISES CHAPTER 6

1.   **(a)** $\displaystyle\int_0^2 (x + 2 - x^2)dx$

     (b) $\displaystyle\int_0^2 \sqrt{y}dy + \int_2^4 (\sqrt{y} - y + 2)dy$

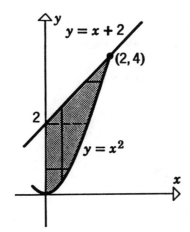

3.   **(a)** $\displaystyle\int_0^9 [\sqrt{x} - (-\sqrt{x})]dx = \int_0^9 2\sqrt{x}dx$

(b) $\int_{-3}^{3}(9-y^2)dy$

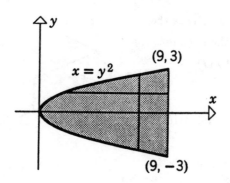

5. (a) $\int_{0}^{2} 2\pi x(x+2-x^2)dx$ **(b)** $\int_{0}^{2} \pi y\,dy + \int_{2}^{4} \pi[y-(y-2)^2]dy$

7. (a) $\int_{0}^{4} 2\pi x[x/2-(2-\sqrt{4-x})]dx$ **(b)** $\int_{0}^{2} \pi[(4y-y^2)^2-4y^2]dy$

9. (a) $\int_{0}^{9} 2\pi x(2\sqrt{x})dx = \int_{0}^{9} 4\pi x^{3/2}dx$ **(b)** $\int_{-3}^{3} \pi(81-y^4)dy$

11. (a) $A = \int_{0}^{4} \sqrt{4-y}\,dy = 16/3$

(b) $V = \int_{0}^{4} \pi(4-y)dy = 8\pi$

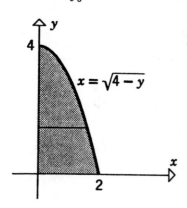

13. $A = \int_{-1}^{0} (x^3 - x)dx + \int_{0}^{1} (x - x^3)dx$

$\qquad + \int_{1}^{2} (x^3 - x)dx$

$\qquad = 1/4 + 1/4 + 9/4 = 11/4$

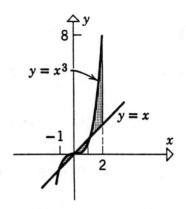

15. Let r be the radius of the semicircle shown in the figure, then by similar triangles

$2r/a = (b - y)/b$, $r = \dfrac{a}{2b}(b - y)$ so

$A(y) = \dfrac{1}{2}\pi r^2 = \dfrac{1}{2}\pi \dfrac{a^2}{4b^2}(b - y)^2 = \dfrac{\pi a^2}{8b^2}(b - y)^2,$

$V = \int_{0}^{b} A(y)dy = \int_{0}^{b} \dfrac{\pi a^2}{8b^2}(b - y)^2 dy = \pi a^2 b/24$

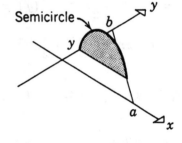

17. **(a)** $V = \int_{0}^{4} 2\pi(4 - x)\sqrt{x}\,dx$

$\qquad = 2\pi \int_{0}^{4} (4x^{1/2} - x^{3/2})dx$

$\qquad = 256\pi/15$

(b) $V = \int_{0}^{4} \pi[4 - (2 - \sqrt{x})^2]dx$

$\qquad = \pi \int_{0}^{4} (4x^{1/2} - x)dx = 40\pi/3$

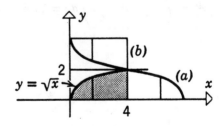

19. $y = \dfrac{x^{3/2}}{\sqrt{8}}$, $0 \leq x \leq 2$; $y' = \dfrac{3x^{1/2}}{2\sqrt{8}}$, $L = \int_{0}^{2} \sqrt{1 + \dfrac{9}{32}x}\,dx = 61/27$

21. $y' = \dfrac{1}{2}x^4 - \dfrac{1}{2}x^{-4}$, $1 + (y')^2 = 1 + \left(\dfrac{1}{4}x^{16} - \dfrac{1}{2} + \dfrac{1}{4}x^{-16}\right) = \left(\dfrac{1}{2}x^4 + \dfrac{1}{2}x^{-4}\right)^2$,

$L = \displaystyle\int_1^2 \left(\dfrac{1}{2}x^4 + \dfrac{1}{2}x^{-4}\right) dx = 779/240$

23. $y' = 3x^2$, $1 + (y')^2 = 1 + 9x^4$, $S = \displaystyle\int_1^2 2\pi x^3\sqrt{1 + 9x^4}\, dx = \pi(145^{3/2} - 10^{3/2})/27$

25. $y' = x^{1/2} - \dfrac{1}{4}x^{-1/2}$, $1 + (y')^2 = 1 + \left(x - \dfrac{1}{2} + \dfrac{1}{16}x^{-1}\right) = \left(x^{1/2} + \dfrac{1}{4}x^{-1/2}\right)^2$,

$S = \displaystyle\int_0^9 2\pi x\left(x^{1/2} + \dfrac{1}{4}x^{-1/2}\right) dx = 2\pi\int_0^9 \left(x^{3/2} + \dfrac{1}{4}x^{1/2}\right) dx = 1017\pi/5$

27. $y' = \dfrac{1}{2}x^{-1/2} - \dfrac{1}{2}x^{1/2}$, $1 + (y')^2 = \left(\dfrac{1}{2}x^{-1/2} + \dfrac{1}{2}x^{1/2}\right)^2$,

$S = \displaystyle\int_0^3 2\pi x\left(\dfrac{1}{2}x^{-1/2} + \dfrac{1}{2}x^{1/2}\right) dx = \pi\int_0^3 (x^{1/2} + x^{3/2})dx = 28\pi\sqrt{3}/5$

29. $F(x) = kx$, $F(4) = 4k = 2$, $k = 1/2$, $W = \displaystyle\int_2^4 \dfrac{1}{2}x\, dx = 3$ in·lb

31. $F(x) = 250 + \dfrac{3}{4}(40 - x) = 280 - \dfrac{3}{4}x$,

$W = \displaystyle\int_0^{40} \left(280 - \dfrac{3}{4}x\right) dx = 10,600$ ft·lb

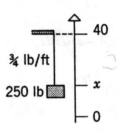

33. $A(y) = \pi x^2 = \pi(y/2 + 4)$,

$W = \displaystyle\int_{-8}^0 62.4(4 - y)[\pi(y/2 + 4)]dy$

$= 31.2\pi \displaystyle\int_{-8}^0 (32 - 4y - y^2)dy$

$= 6656\pi$ ft·lb

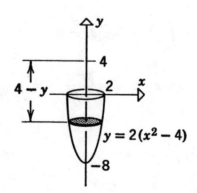

35. By similar triangles

$$w(x)/4 = (x-1)/2$$
$$w(x) = 2(x-1)$$

$$F = \int_1^3 \rho x [2(x-1)]dx$$

$$= 2\rho \int_1^3 (x^2 - x)dx = 28\rho/3 \text{ lb}$$

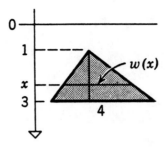

CHAPTER 7
Logarithm and Exponential Functions

EXERCISE SET 7.1

1. (a) $f(g(x)) = 4(x/4) = x$, $g(f(x)) = (4x)/4 = x$, f and g are inverse functions
    (b) $f(g(x)) = 3(3x - 1) + 1 = 9x - 2 \neq x$ so f and g are not inverse functions
    (c) $f(g(x)) = \sqrt[3]{(x^3 + 2)} - 2 = x$, $g(f(x)) = (x - 2) + 2 = x$, f and g are inverse functions
    (d) $f(g(x)) = (x^{1/4})^4 = x$, $g(f(x)) = (x^4)^{1/4} = |x| \neq x$, f and g are not inverse functions

3. $f'(x) = 3$; f is increasing on $(-\infty, +\infty)$ so f has an inverse.

5. $f(x) = (2 + x)(1 - x)$; f does not have an inverse because f is not one-to-one, for example $f(-2) = f(1) = 0$.

7. f does not have an inverse because f is not one-to-one, for example $f(0) = f(1) = -1$.

9. $f(x) = (x - 1)^3$; f has an inverse because two different numbers cannot have the same cube so f is one-to-one.

11. $f'(x) = 10x^4 + 3x^2 + 3 \geq 3$ for $-\infty < x < +\infty$; f is increasing on $(-\infty, +\infty)$ so f has an inverse.

13. $f'(x) = \cos x > 0$ for $-\pi/2 < x < \pi/2$; f is increasing on $(-\pi/2, \pi/2)$ so f has an inverse.

15. $y = f^{-1}(x)$, $x = f(y) = y^5$, $y = x^{1/5} = f^{-1}(x)$

17. $y = f^{-1}(x)$, $x = f(y) = 7y - 6$, $y = \frac{1}{7}(x + 6) = f^{-1}(x)$

19. $y = f^{-1}(x)$, $x = f(y) = 3y^3 - 5$, $y = \sqrt[3]{(x + 5)/3} = f^{-1}(x)$

21. $y = f^{-1}(x)$, $x = f(y) = \sqrt[3]{2y - 1}$, $y = (x^3 + 1)/2 = f^{-1}(x)$

23. $y = f^{-1}(x)$, $x = f(y) = 3/y^2$, $y = -\sqrt{3/x} = f^{-1}(x)$

25. $y = f^{-1}(x), x = f(y) = \begin{cases} 5/2 - y, & y < 2 \\ 1/y, & y \geq 2 \end{cases}, y = f^{-1}(x) = \begin{cases} 5/2 - x, & x > 1/2 \\ 1/x, & x \leq 1/2 \end{cases}$

27. $y = f^{-1}(x)$, $x = f(y) = 5y^3 + y - 7$, $\dfrac{dx}{dy} = 15y^2 + 1$, $\dfrac{dy}{dx} = \dfrac{1}{15y^2 + 1}$;

   check: $1 = 15y^2 \dfrac{dy}{dx} + \dfrac{dy}{dx}$, $\dfrac{dy}{dx} = \dfrac{1}{15y^2 + 1}$.

29. $y = f^{-1}(x)$, $x = f(y) = \tan 2y$, $\dfrac{dx}{dy} = 2\sec^2 2y$, $\dfrac{dy}{dx} = \dfrac{1}{2\sec^2 2y}$;

check: $1 = (2\sec^2 2y)\dfrac{dy}{dx}$, $\dfrac{dy}{dx} = \dfrac{1}{2\sec^2 2y}$.

31. $y = f^{-1}(x)$, $x = f(y) = 2y^5 + y^3 + 1$, $\dfrac{dx}{dy} = 10y^4 + 3y^2$, $\dfrac{dy}{dx} = \dfrac{1}{10y^4 + 3y^2}$;

check: $1 = 10y^4\dfrac{dy}{dx} + 3y^2\dfrac{dy}{dx}$, $\dfrac{dy}{dx} = \dfrac{1}{10y^4 + 3y^2}$.

33. **(a)** $f(g(x)) = f(\sqrt{x})$
$= (\sqrt{x})^2 = x, x > 1$;
$g(f(x)) = g(x^2)$
$= \sqrt{x^2} = x, x > 1$.

(b)

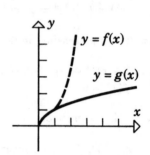

(c) No, because $f(g(x)) = x$ for every x in the domain of g is not satisfied (the domain of g is $x > 0$).

35. $y = f^{-1}(x)$, $x = f(y) = (y+2)^4$ for $y \geq 0$, $y = f^{-1}(x) = x^{1/4} - 2$ for $x \geq 16$.

37. $y = f^{-1}(x)$, $x = f(y) = -\sqrt{3 - 2y}$ for $y \leq 3/2$, $y = f^{-1}(x) = (3 - x^2)/2$ for $x \leq 0$.

39. $y = f^{-1}(x)$, $x = f(y) = y - 5y^2$ for $y \geq 1$, $5y^2 - y + x = 0$ for $y \geq 1$,
$y = f^{-1}(x) = (1 + \sqrt{1 - 20x})/10$ for $x \leq -4$.

41. **(a)** $f(f(x)) = \dfrac{3 - \dfrac{3 - x}{1 - x}}{1 - \dfrac{3 - x}{1 - x}} = \dfrac{3 - 3x - 3 + x}{1 - x - 3 + x} = x$ so $f = f^{-1}$

(b) symmetric about the line $y = x$

43. **(a)** $f(x) = x^3 - 3x^2 + 2x = x(x-1)(x-2)$ so $f(0) = f(1) = f(2) = 0$ thus f is not one-to-one.

(b) $f'(x) = 3x^2 - 6x + 2$, $f'(x) = 0$ when $x = \dfrac{6 \pm \sqrt{36 - 24}}{6} = 1 \pm \sqrt{3}/3$. $f'(x) > 0$ (f is increasing) if $x < 1 - \sqrt{3}/3$, $f'(x) < 0$ (f is decreasing) if $1 - \sqrt{3}/3 < x < 1 + \sqrt{3}/3$, so $f(x)$ takes on values less than $f(1 - \sqrt{3}/3)$ on both sides of $1 - \sqrt{3}/3$ thus $1 - \sqrt{3}/3$ is the largest value of k.

45. If $f^{-1}(x) = 1$, then $x = f(1) = 2(1)^3 + 5(1) + 3 = 10$.

47. $f'(x) = 3x^2 + 1, f'(2) = 13$ so $(f^{-1})'(10) = 1/13$.

49. $f'(x) = 2\cos 2x, f'(\pi/12) = \sqrt{3}$ so $(f^{-1})'(1/2) = 1/\sqrt{3}$.

51. (a) $f'(x) = \sqrt[3]{1 + x^2} > 0$ on $(-\infty, +\infty)$ so f is one-to-one there because f is increasing.
  (b) $f(1) = 0, f'(1) = \sqrt[3]{2}$ so $(f^{-1})'(0) = 1/\sqrt[3]{2}$.

53.

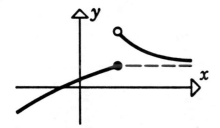

55. $F'(x) = 2f'(2g(x))g'(x)$ so $F'(3) = 2f'(2g(3))g'(3)$. By inspection $f(1) = 3$, so
$g(3) = f^{-1}(3) = 1$ and $g'(3) = (f^{-1})'(3) = 1/f'(f^{-1}(3)) = 1/f'(1) = 1/7$ because
$f'(x) = 4x^3 + 3x^2$. Thus $F'(3) = 2f'(2)(1/7) = 2(44)(1/7) = 88/7$.

EXERCISE SET 7.2

1. (a) -4 (b) 4 (c) $1/4$

3. (a) 2.9690 (b) 0.0341

5. (a) $\log_2 16 = \log_2(2^4) = 4$

(b) $\log_2\left(\dfrac{1}{32}\right) = \log_2(2^{-5}) = -5$

(c) $\log_4 4 = 1$

(d) $\log_9 3 = \log_9(9^{1/2}) = 1/2$

7. (a) 1.3655

(b) -0.3011

9. (a) $2\ln a + \dfrac{1}{2}\ln b + \dfrac{1}{2}\ln c = 2r + s/2 + t/2$.

(b) $\ln b - 3\ln a - \ln c = s - 3r - t$.

11. (a) $1 + \log x + \dfrac{1}{2}\log(x - 3)$

(b) $2\ln|x| + 3\ln\sin x - \dfrac{1}{2}\ln(x^2 + 1)$

13. $\log \dfrac{2^4(16)}{3} = \log(256/3)$

15. $\ln \dfrac{\sqrt[3]{x}(x+1)^2}{\cos x}$

17. $\sqrt{x} = 10^{-1} = 0.1, \ x = 0.01$

19. $1/x = e^{-2}, \ x = e^2$

21. $2x = 8, \ x = 4$

23. $\log_{10} x = 5, \ x = 10^5$

25. $\ln 2x^2 = \ln 3, \ 2x^2 = 3, \ x^2 = 3/2, \ x = \sqrt{3/2}$ (we discard $-\sqrt{3/2}$ because it does not satisfy the original equation).

27. $\ln 5^{-2x} = \ln 3, \ -2x \ln 5 = \ln 3, \ x = -\dfrac{\ln 3}{2 \ln 5}$

29. $e^{3x} = 7/2, \ 3x = \ln(7/2), \ x = \dfrac{1}{3} \ln(7/2)$

31. $e^{-x}(x+2) = 0$ so $e^{-x} = 0$ (impossible) or $x + 2 = 0, \ x = -2$

33. $e^{-2x} - 3e^{-x} + 2 = 0, \ (e^{-x} - 1)(e^{-x} - 2) = 0$; if $e^{-x} - 1 = 0$, then $x = 0$, if $e^{-x} - 2 = 0$, then $x = -\ln 2$.

35. $4 - 12e^{2x} = 1, \ 12e^{2x} = 3, \ e^{2x} = 1/4, \ x = \frac{1}{2} \ln(1/4) = -\ln 2$

37. $\log(1/2) < 0$ so $3 \log(1/2) < 2 \log(1/2)$

39. (a) Let $x = \log_a c$, then $a^x = c$ so $\log_b a^x = \log_b c, \ x \log_b a = \log_b c, \ x = (\log_b c)/(\log_b a)$.

   (b) $\log_2 7.35 = (\log 7.35)/(\log 2) = (\ln 7.35)/(\ln 2) \approx 2.8777$;
   $\log_5 0.6 = (\log 0.6)/(\log 5) = (\ln 0.6)/(\ln 5) \approx -0.3174$

41. $75e^{-t/125} = 15, t = -125 \ln(1/5) = 125 \ln 5 \approx 201$ days.

43. (a) 7.4; basic   (b) 4.2; acidic   (c) 6.4; acidic   (d) 5.9; basic

45. (a) 140 dB; damage   (b) 120 dB; damage
   (c) 80 dB; no damage   (d) 75 dB; no damage

47. Let I_A and I_B be the intensities of the automobile and blender, respectively. Then $\log_{10} I_A/I_0 = 7$ and $\log_{10} I_B/I_0 = 9.3, \ I_A = 10^7 I_0$ and $I_B = 10^{9.3} I_0$, so $I_B/I_A = 10^{2.3} \approx 200$.

49. (a) $\log E = 4.4 + 1.5(8.2) = 16.7, E = 10^{16.7} \approx 5 \times 10^{16}$ J

   (b) Let M_1 and M_2 be the magnitudes of earthquakes with energies of E and $10E$, respectively. Then $1.5(M_2 - M_1) = \log(10E) - \log E = \log 10 = 1$,
   $M_2 - M_1 = 1/1.5 = 2/3 \approx 0.67$.

51. If $t = -2x$, then $x = -t/2$ and $\lim\limits_{x \to 0}(1 - 2x)^{1/x} = \lim\limits_{t \to 0}(1 + t)^{-2/t} = \lim\limits_{t \to 0}[(1 + t)^{1/t}]^{-2} = e^{-2}$.

EXERCISE SET 7.3

1. $3x + 2 > 0, x > -2/3$

3. $4 - x^2 > 0, x^2 < 4, -2 < x < 2$

5. $1 + \ln x \geq 0, \ln x \geq -1, x \geq e^{-1}$

7. (a) $x^2 > 0$; all $x \neq 0$

    (b) $x > 0$

9. $\dfrac{1}{2x}(2) = 1/x$

11. $2(\ln x)\left(\dfrac{1}{x}\right) = \dfrac{2\ln x}{x}$

13. $\dfrac{1}{\tan x}(\sec^2 x) = \dfrac{\sec^2 x}{\tan x}$

15. $\dfrac{1}{x/(1 + x^2)}\left[\dfrac{(1 + x^2)(1) - x(2x)}{(1 + x^2)^2}\right] = \dfrac{1 - x^2}{x(1 + x^2)}$

17. $\dfrac{3x^2 - 14x}{x^3 - 7x^2 - 3}$

19. $\dfrac{1}{2}(\ln x)^{-1/2}\left(\dfrac{1}{x}\right) = \dfrac{1}{2x\sqrt{\ln x}}$

21. $\cos(5/\ln x)\dfrac{d}{dx}[5(\ln x)^{-1}] = \cos(5/\ln x)[-5(\ln x)^{-2}(1/x)] = -\dfrac{5\cos(5/\ln x)}{x(\ln x)^2}$

23. $-\dfrac{2x^3}{3 - 2x} + 3x^2\ln(3 - 2x)$

25. $2(x^2 + 1)[\ln(x^2 + 1)]\dfrac{2x}{x^2 + 1} + 2x[\ln(x^2 + 1)]^2 = 4x\ln(x^2 + 1) + 2x[\ln(x^2 + 1)]^2$

27. $\dfrac{(1 + \ln x)(2x) - x^2(0 + 1/x)}{(1 + \ln x)^2} = \dfrac{x(1 + 2\ln x)}{(1 + \ln x)^2}$

29. $\dfrac{dy}{dx} + \dfrac{1}{xy}\left(x\dfrac{dy}{dx} + y\right) = 0, \dfrac{dy}{dx} = -\dfrac{y}{x(y + 1)}$

31. $\dfrac{1}{2}\ln|x| + C$

33. $u = x^3 - 4, du = 3x^2 dx, \dfrac{1}{3}\int\dfrac{1}{u}du = \dfrac{1}{3}\ln|x^3 - 4| + C$

35. $u = \tan x$, $du = \sec^2 x \, dx$, $\displaystyle\int \frac{1}{u} du = \ln|\tan x| + C$

37. $u = 1 + \cos 3\theta$, $du = -3\sin 3\theta \, d\theta$

$$-\frac{1}{3}\int \frac{1}{u} du = -\frac{1}{3}\ln|1 + \cos 3\theta| + C = -\frac{1}{3}\ln(1 + \cos 3\theta) + C \text{ because } 1 + \cos 3\theta \geq 0$$

39. divide $x^2 + 1$ into x^3 to get

$$\int \frac{x^3}{x^2+1} dx = \int \left[x - \frac{x}{x^2+1}\right] dx = \int x \, dx - \int \frac{x}{x^2+1} dx = \frac{1}{2}x^2 - \frac{1}{2}\ln(x^2+1) + C$$

41. $u = \ln y$, $du = \dfrac{1}{y} dy$, $\displaystyle\int u^3 du = \frac{1}{4}(\ln y)^4 + C$

43. $u = 3x + 2$, $\dfrac{1}{3}\displaystyle\int_2^5 \frac{1}{u} du = \frac{1}{3}\ln|u|\Big]_2^5 = \frac{1}{3}(\ln 5 - \ln 2) = \frac{1}{3}\ln\frac{5}{2}$

45. $u = x^2 + 5$, $\dfrac{1}{2}\displaystyle\int_6^5 \frac{1}{u} du = \frac{1}{2}\ln|u|\Big]_6^5 = \frac{1}{2}(\ln 5 - \ln 6) = \frac{1}{2}\ln\frac{5}{6}$

47. $\dfrac{d}{dx}\left[\ln\cos x - \dfrac{1}{2}\ln(4 - 3x^2)\right] = -\tan x + \dfrac{3x}{4 - 3x^2}$

49. $\dfrac{d}{dx}\left[\dfrac{1}{2}\ln x + \dfrac{1}{3}\ln(x+3) + \dfrac{1}{5}\ln(3x-2)\right] = \dfrac{1}{2x} + \dfrac{1}{3(x+3)} + \dfrac{3}{5(3x-2)}$

51. $\ln|y| = \ln|x| + \dfrac{1}{3}\ln|1 + x^2|$, $\dfrac{dy}{dx} = x^3\sqrt{1+x^2}\left[\dfrac{1}{x} + \dfrac{2x}{3(1+x^2)}\right]$

53. $\ln|y| = \dfrac{1}{3}\ln|x^2 - 8| + \dfrac{1}{2}\ln|x^3 + 1| - \ln|x^6 - 7x + 5|$

$$\dfrac{dy}{dx} = \dfrac{(x^2-8)^{1/3}\sqrt{x^3+1}}{x^6 - 7x + 5}\left[\dfrac{2x}{3(x^2-8)} + \dfrac{3x^2}{2(x^3+1)} - \dfrac{6x^5 - 7}{x^6 - 7x + 5}\right]$$

55. **(a)** $u = \dfrac{ab}{t}$, $t = \dfrac{ab}{u}$, $dt = -\dfrac{ab}{u^2}du$, $\displaystyle\int_a^{ab} \frac{1}{t} dt = \int_b^1 \frac{u}{ab}\left(-\frac{ab}{u^2}\right) du = \int_1^b \frac{1}{u} du = \int_1^b \frac{1}{t} dt$

(b) $\ln(ab) = \displaystyle\int_1^{ab} \frac{1}{t} dt = \int_1^a \frac{1}{t} dt + \int_a^{ab} \frac{1}{t} dt = \int_1^a \frac{1}{t} dt + \int_1^b \frac{1}{t} dt = \ln a + \ln b$

57. $\log_b x = \dfrac{\ln x}{\ln b}$, $\dfrac{d}{dx}[\log_b x] = \dfrac{1}{\ln b} \cdot \dfrac{1}{x}$

59. $\log_x 2 = \dfrac{\ln 2}{\ln x}, \dfrac{d}{dx}[\log_x 2] = -\dfrac{\ln 2}{x(\ln x)^2}$

61. $\dfrac{k}{n^2 + k^2} = \dfrac{k/n}{1 + k^2/n^2}\dfrac{1}{n}$ so $\displaystyle\sum_{k=1}^{n} \dfrac{k}{n^2 + k^2} = \sum_{k=1}^{n} f(x_k^*)\Delta x$ where $f(x) = \dfrac{x}{1 + x^2}, x_k^* = \dfrac{k}{n}$, and

$\Delta x = \dfrac{1}{n}, 0 \le x \le 1; \displaystyle\lim_{n \to +\infty} \sum_{k=1}^{n} \dfrac{k}{n^2 + k^2} = \lim_{n \to +\infty} \sum_{k=1}^{n} f(x_k^*)\Delta x = \int_0^1 \dfrac{x}{1 + x^2}dx = \dfrac{1}{2}\ln 2.$

63. Let $f(x) = x^2 - \ln x$, then $f'(x) = 2x - 1/x = (2x^2 - 1)/x$; $f'(x) = 0$ if $x = 1/\sqrt{2}$ at which there is a relative minimum and hence the absolute minimum on $(0, +\infty)$. The minimum value is $1/2 - \ln(1/\sqrt{2}) = (1 + \ln 2)/2$.

65. Let $f(x) = (\ln^2 x)/x$, then $f'(x) = (2\ln x - \ln^2 x)/x^2 = (2 - \ln x)(\ln x)/x^2$; $f'(x) = 0$ if $x = 1$ or $x = e^2$, at which there is a relative minimum and a relative maximum, respectively. The relative minimum value is $f(1) = 0$; the relative maximum value is $f(e^2) = 4/e^2$.

67. Let $f(x) = x - 1 - \ln x$, then $f'(x) = 1 - 1/x = (x - 1)/x$, $f'(x) = 0$ when $x = 1$ where the minimum value occurs, so $f(x) \ge f(1) = 0, x - 1 - \ln x \ge 0, \ln x \le x - 1$.

69. $p = c/v$ so the average pressure with respect to volume is

$$p_{\text{ave}} = \dfrac{1}{v_1 - v_0}\int_{v_0}^{v_1} \dfrac{c}{v}dv = \dfrac{c}{v_1 - v_0}(\ln v_1 - \ln v_0) = \dfrac{c}{v_1 - v_0}\ln(v_1/v_0)$$

71. $A = \displaystyle\int_0^{\pi/3} \tan x\, dx = \int_0^{\pi/3} \dfrac{\sin x}{\cos x}dx = -\ln|\cos x|\Big]_0^{\pi/3} = \ln 2$

73. $V = \pi\displaystyle\int_1^4 \dfrac{1}{x}dx = \pi\ln|x|\Big]_1^4 = \pi\ln 4$

75. $f(x) = \ln x - x + 2, x_{n+1} = x_n - \dfrac{\ln x_n - x_n + 2}{1/x_n - 1}$;

$x_1 = 0.2, x_2 = 0.152359478, x_3 = 0.158447821, \cdots, x_5 = x_6 = 0.158594340;$

$x_1 = 3, x_2 = 3.147918433, x_3 = 3.146193441, x_4 = x_5 = 3.146193221.$

EXERCISE SET 7.4

1. (a) $x^{-1}, x > 0$ (b) $x^2, x \ne 0$

    (c) $-x^2, -\infty < x < +\infty$ (d) $-x, -\infty < x < +\infty$

(e) $x^3, x > 0$

(f) $\ln x + x, x > 0$

(g) $x - \sqrt[3]{x}, -\infty < x < +\infty$

(h) $\dfrac{e^x}{x}, x > 0$

3. $f(\ln 2) = e^{\ln 2} + 3e^{-\ln 2} = 2 + 3e^{\ln(1/2)} = 2 + 3/2 = 7/2$

5. (a) $\pi^{-x} = e^{-x \ln \pi}$

(b) $x^{2x} = e^{2x \ln x}$

7. $-10xe^{-5x^2}$

9. $x^3 e^x + 3x^2 e^x = x^2 e^x (x + 3)$

11. $\dfrac{dy}{dx} = \dfrac{(e^x + e^{-x})(e^x + e^{-x}) - (e^x - e^{-x})(e^x - e^{-x})}{(e^x + e^{-x})^2}$

$= \dfrac{(e^{2x} + 2 + e^{-2x}) - (e^{2x} - 2 + e^{-2x})}{(e^x + e^{-x})^2} = 4/(e^x + e^{-x})^2$

13. $(x \sec^2 x + \tan x)e^{x \tan x}$

15. $(1 - 3e^{3x})e^{(x - e^{3x})}$

17. $\dfrac{(x-1)e^{-x}}{1 - xe^{-x}} = \dfrac{x - 1}{e^x - x}$

19. $e^{ax}(a \cos bx - b \sin bx)$

21. $y = e^{\ln(x^3 + 1)} = x^3 + 1, dy/dx = 3x^2$

23. $f'(x) = -3^{-x} \ln 3; y = 3^{-x}, \ln y = -x \ln 3, \dfrac{1}{y}y' = -\ln 3, y' = -y \ln 3 = -3^{-x} \ln 3$

25. $f'(x) = \pi^{x \tan x}(\ln \pi)(x \sec^2 x + \tan x);$

$y = \pi^{x \tan x}, \ln y = (x \tan x) \ln \pi, \dfrac{1}{y}y' = (\ln \pi)(x \sec^2 x + \tan x)$

$y' = \pi^{x \tan x}(\ln \pi)(x \sec^2 x + \tan x)$

27. (a) Because x^x is not of the form a^x where a is constant.

(b) $y = x^x, \ln y = x \ln x, \dfrac{1}{y}y' = 1 + \ln x, y' = x^x(1 + \ln x)$

29. $\ln y = (\ln x) \ln(x^3 - 2x), \dfrac{1}{y}\dfrac{dy}{dx} = \dfrac{3x^2 - 2}{x^3 - 2x} \ln x + \dfrac{1}{x} \ln(x^3 - 2x),$

$\dfrac{dy}{dx} = (x^3 - 2x)^{\ln x}\left[\dfrac{3x^2 - 2}{x^3 - 2x} \ln x + \dfrac{1}{x} \ln(x^3 - 2x)\right]$

31. $\ln y = (\tan x)\ln(\ln x),\ \dfrac{1}{y}\dfrac{dy}{dx} = \dfrac{1}{x\ln x}\tan x + (\sec^2 x)\ln(\ln x),$

$\dfrac{dy}{dx} = (\ln x)^{\tan x}\left[\dfrac{\tan x}{x\ln x} + (\sec^2 x)\ln(\ln x)\right]$

33. $\ln y = e^x \ln x,\ \dfrac{1}{y}\dfrac{dy}{dx} = \dfrac{e^x}{x} + e^x \ln x,\ \dfrac{dy}{dx} = x^{(e^x)}\left[\dfrac{e^x}{x} + e^x \ln x\right]$

35. $y = Ae^{2x} + Be^{-4x},\ y' = 2Ae^{2x} - 4Be^{-4x},\ y'' = 4Ae^{2x} + 16Be^{-4x}$ so
$y'' + 2y' - 8y = (4Ae^{2x} + 16Be^{-4x}) + 2(2Ae^{2x} - 4Be^{-4x}) - 8(Ae^{2x} + Be^{-4x}) = 0.$

37. **(a)** $f'(x) = ke^{kx},\ f''(x) = k^2 e^{kx},\ f'''(x) = k^3 e^{kx}, \ldots, f^{(n)}(x) = k^n e^{kx}$
(b) $f'(x) = -ke^{-kx},\ f''(x) = k^2 e^{-kx},\ f'''(x) = -k^3 e^{-kx}, \ldots, f^{(n)}(x) = (-1)^n k^n e^{-kx}$

39. $f'(x) = \dfrac{1}{\sqrt{2\pi}\sigma}\exp\left[-\dfrac{1}{2}\left(\dfrac{x-\mu}{\sigma}\right)^2\right]\dfrac{d}{dx}\left[-\dfrac{1}{2}\left(\dfrac{x-\mu}{\sigma}\right)^2\right]$

$= \dfrac{1}{\sqrt{2\pi}\sigma}\exp\left[-\dfrac{1}{2}\left(\dfrac{x-\mu}{\sigma}\right)^2\right]\left[-\left(\dfrac{x-\mu}{\sigma}\right)\left(\dfrac{1}{\sigma}\right)\right]$

$= -\dfrac{1}{\sqrt{2\pi}\sigma^3}(x-\mu)\exp\left[-\dfrac{1}{2}\left(\dfrac{x-\mu}{\sigma}\right)^2\right]$

41. $-\dfrac{1}{5}e^{-5x} + C$ **43.** $e^{\sin x} + C$

45. $-\dfrac{1}{6}\displaystyle\int e^{-2x^3}(-6x^2)dx = -\dfrac{1}{6}e^{-2x^3} + C$

47. $u = 1 + e^x,\ du = e^x dx,\ \displaystyle\int \dfrac{1}{u}du = \ln(1 + e^x) + C$

49. $u = 1 + e^{2t},\ du = 2e^{2t}dt,\ \dfrac{1}{2}\displaystyle\int u^{1/2}du = \dfrac{1}{3}(1 + e^{2t})^{3/2} + C$

51. $\displaystyle\int e^{\sin x}\cos x\, dx = e^{\sin x} + C = \exp(\sin x) + C$

53. $u = 2 - e^{-x},\ du = e^{-x}dx,\ \displaystyle\int \sec^2 u\, du = \tan(2 - e^{-x}) + C$

55. $\dfrac{\pi^{\sin x}}{\ln \pi} + C$ **57.** $\dfrac{1}{2}x^2 \ln 3 - 4\pi e^2 \sin x + C$

59. $\ln(e^x) + \ln(e^{-x}) = \ln(e^x e^{-x}) = \ln 1 = 0$ so $\displaystyle\int [\ln(e^x) + \ln(e^{-x})]dx = C$

61. $u = \sqrt{y}$, $du = \dfrac{1}{2\sqrt{y}}dy$, $2\displaystyle\int e^u du = 2e^{\sqrt{y}} + C$

63. $u = 3 - 4e^x$, $du = -4e^x dx$, $u = -1$ when $x = 0$, $u = -17$ when $x = \ln 5$

$$-\frac{1}{4}\int_{-1}^{-17} u\, du = -\frac{1}{8}u^2\Big]_{-1}^{-17} = -36$$

65. $3x - e^x\big]_1^2 = 3 + e - e^2$

67. $u = e^x + 4$, $du = e^x dx$, $u = e^{-\ln 3} + 4 = \dfrac{1}{3} + 4 = \dfrac{13}{3}$ when $x = -\ln 3$,

$u = e^{\ln 3} + 4 = 3 + 4 = 7$ when $x = \ln 3$, $\displaystyle\int_{13/3}^{7} \frac{1}{u}du = \ln u\Big]_{13/3}^{7} = \ln(7) - \ln(13/3) = \ln(21/13)$

69. $g(x) = e^{-x}f(x)$, $g'(x) = e^{-x}f'(x) - e^{-x}f(x) = e^{-x}[f'(x) - f(x)] = 0$ if $f'(x) = f(x)$ thus $g(x) = k$ because $g'(x) = 0$ so $e^{-x}f(x) = k$, $f(x) = ke^x$.

71. $2^x = 3^{x+1}$, $\ln(2^x) = \ln(3^{x+1})$, $x \ln 2 = (x + 1)\ln 3$, $x \ln 2 = x \ln 3 + \ln 3$,

$x(\ln 2 - \ln 3) = \ln 3$, $x = \dfrac{\ln 3}{\ln 2 - \ln 3} = \dfrac{\ln 3}{\ln(2/3)}$

73. $f'(x) = ex^{e-1}$

75. $\dfrac{dk}{dT} = k_0 \exp\left[-\dfrac{q}{2}\dfrac{T - T_0}{T_0 T}\right]\left(-\dfrac{q}{2T^2}\right) = -\dfrac{qk_0}{2T^2}\exp\left[-\dfrac{q}{2}\dfrac{T - T_0}{T_0 T}\right]$

77. Divide $e^x + 3$ into e^{2x} to get $\dfrac{e^{2x}}{e^x + 3} = e^x - \dfrac{3e^x}{e^x + 3}$ so

$$\int \frac{e^{2x}}{e^x + 3}dx = \int e^x dx - 3\int \frac{e^x}{e^x + 3}dx = e^x - 3\ln(e^x + 3) + C$$

79. $y = f^{-1}(x)$, $x = f(y) = e^{1/y}$, $1/y = \ln x$, $y = \dfrac{1}{\ln x} = f^{-1}(x)$

81. $y = f^{-1}(x)$, $x = f(y) = 1 - \ln(3y)$, $\ln(3y) = 1 - x$, $y = \dfrac{1}{3}e^{1-x} = f^{-1}(x)$

83. Let $f(x) = x^3 e^{-2x}$, then $f'(x) = -2x^3 e^{-2x} + 3x^2 e^{-2x} = x^2 e^{-2x}(3 - 2x)$; $f'(x) = 0$ if $x = 0$ or $x = 3/2$. The maximum value occurs at $x = 3/2$ so the maximum value is $f(3/2) = (27/8)e^{-3}$.

85. $A = \displaystyle\int_0^{\ln 3} (3 - e^x)\,dx$

$= (3x - e^x)]_0^{\ln 3} = 3\ln 3 - 2$

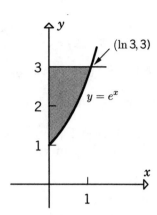

87. $V = \pi \displaystyle\int_0^{\ln 3} e^{2x}\,dx = \dfrac{\pi}{2}e^{2x}\bigg]_0^{\ln 3} = 4\pi.$

89. The area of the region shown in

the diagram is $A = \displaystyle\int_1^5 \ln x\,dx.$

If $y = \ln x$, then $x = e^y$ so

$A = \displaystyle\int_0^{\ln 5} (5 - e^y)\,dy = (5y - e^y)\bigg]_0^{\ln 5}$

$= 5\ln 5 - 4.$

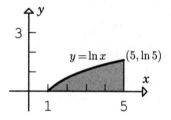

91. $f(x) = e^x + x^2 - 2, x_{n+1} = x_n - \dfrac{e^{x_n} + x_n^2 - 2}{e^{x_n} + 2x_n}$

$x_1 = -1.5, \cdots, x_5 = x_6 = -1.315973778;$

$x_1 = 0.5, \cdots, x_4 = x_5 = 0.537274449$ so

$-1.315973778 < x < 0.537274449.$

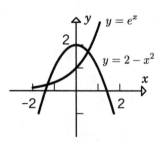

93. (a) $\dfrac{dS}{df} = \dfrac{(e^{bf} - 1)(3af^2) - af^3(be^{bf})}{(e^{bf} - 1)^2} = \dfrac{af^2(3e^{bf} - 3 - bfe^{bf})}{(e^{bf} - 1)^2}$, $dS/df = 0$ if $f = 0$ (where

the minimum value occurs), and if $3e^{bf} - 3 - bfe^{bf} = 0$, or $3e^{-bf} + bf - 3 = 0$ (where

the maximum value occurs).

(b) $f(x) = 3e^{-x} + x - 3$, $x_{n+1} = x_n - \dfrac{3e^{-x_n} + x_n - 3}{1 - 3e^{-x_n}}$, $x_1 = 3, \cdots, x_4 = x_5 = 2.821439372$.

(c) $f \approx 2.821439372/b = (2.821439372)T/(4.8043 \times 10^{-11}) \approx 5.87 \times 10^{10}T$.

95. For any $\epsilon > 0$, $0 < e^x < \epsilon$ if $\ln e^x < \ln \epsilon$, $x < \ln \epsilon$; choose $x_0 = \ln \epsilon$.

97. $y = x^x$, $\ln y = x \ln x$, $y'/y = 1 + \ln x$, $y' = x^x(1 + \ln x)$; $y' = 0$ when $\ln x = -1$, $x = e^{-1}$ at which there is a relative minimum and hence the absolute minimum for $x > 0$. The minimum value is $e^{-1/e}$.

99. **(a)** The area under $1/t$ for $x \leq t \leq x + 1$ is less than the area of the rectangle with altitude $1/x$ and base 1, but greater than the area of the rectangle with altitude $1/(x+1)$ and base 1.

(b) $\displaystyle\int_x^{x+1} \frac{1}{t} dt = \ln t \Big]_x^{x+1} = \ln(x+1) - \ln x = \ln(1 + 1/x)$, so

$1/(x+1) < \ln(1 + 1/x) < 1/x$ for $x > 0$.

(c) From part (b), $e^{1/(x+1)} < e^{\ln(1+1/x)} < e^{1/x}$, $e^{1/(x+1)} < 1 + 1/x < e^{1/x}$,

$e^{x/(x+1)} < (1 + 1/x)^x < e$; by the Squeezing Theorem, $\displaystyle\lim_{x \to +\infty} (1 + 1/x)^x = e$.

(d) Use the inequality $e^{x/(x+1)} < (1 + 1/x)^x$ to get $e < (1 + 1/x)^{x+1}$ so

$(1 + 1/x)^x < e < (1 + 1/x)^{x+1}$.

EXERCISE SET 7.5

1. **(a)** $+\infty$ **(b)** 0 **3.** **(a)** $+\infty$ **(b)** $+\infty$

5. **(a)** 1 **(b)** 1 **7.** **(a)** $+\infty$ **(b)** 0

9.

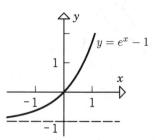

11.

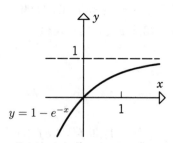

13. **(a)** yes, because $\lim\limits_{x\to 0} f(x) = f(0)$

   (b) no, because $\lim\limits_{x\to 0^+} f'(x) = \lim\limits_{x\to 0^+} e^x = 1$ and $\lim\limits_{x\to 0^-} f'(x) = \lim\limits_{x\to 0^-} (-e^{-x}) = -1$ so

   $\lim\limits_{x\to 0^+} f'(x) \neq \lim\limits_{x\to 0^-} f'(x)$

   (c) $f(x) = \begin{cases} e^x, & x \geq 0 \\ e^{-x}, & x < 0 \end{cases}$

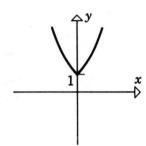

15. $e^x \cos x = e^x$ when $\cos x = 1$
   $x = 0, \pm 2\pi, \pm 4\pi, \ldots$

   $e^x \cos x = -e^x$ when $\cos x = -1$
   $x = \pm\pi, \pm 3\pi, \pm 5\pi, \ldots$

   $e^x \cos x = 0$ when $\cos x = 0$
   $x = \pm\pi/2, \pm 3\pi/2, \pm 5\pi/2, \ldots$

   $-1 \leq \cos x \leq 1$ thus

   $-e^x \leq e^x \cos x \leq e^x$ and so

   $\lim\limits_{x\to -\infty} e^x \cos x = 0$ by the Squeezing Theorem.

   $f(x) = e^x \cos x$, $f'(x) = e^x(\cos x - \sin x)$,

   $f'(x) = 0$ when $\sin x = \cos x$,

   $\tan x = 1$, $x = \pi/4 + n\pi$, $n = 0, \pm 1, \pm 2, \ldots$

   $f''(x) = -2e^x \sin x$, $f''(x) = 0$ when $x = 0, \pm\pi, \pm 2\pi, \ldots$

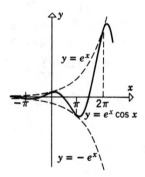

17. $\lim\limits_{x\to -\infty} \dfrac{e^x + e^{-x}}{e^x - e^{-x}} = \lim\limits_{x\to -\infty} \dfrac{e^{2x} + 1}{e^{2x} - 1} = -1$ **19.** 0

21. $\dfrac{d}{dx}[e^x]\Big|_{x=0} = \lim\limits_{h\to 0} \dfrac{e^{(0+h)} - e^0}{h} = \lim\limits_{h\to 0} \dfrac{e^h - 1}{h} = 1$

23. $\lim\limits_{x\to 0} \dfrac{1 - e^{-x}}{x} = \lim\limits_{x\to 0} -\dfrac{e^{-x} - e^0}{x - 0} = -\dfrac{d}{dx}[e^{-x}]\Big|_{x=0} = e^{-x}\Big|_{x=0} = 1$

25. let $h = 1/x$ then $x = 1/h$ and $\displaystyle\lim_{x \to +\infty} x(e^{1/x} - 1) = \lim_{h \to 0+} \dfrac{e^h - 1}{h} = \dfrac{d}{dx}[e^x]\Big|_{x=0} = 1$

27. **(a)** $\displaystyle\lim_{x \to +\infty} xe^x = +\infty, \quad \lim_{x \to -\infty} xe^x = 0.$

(b) $y = xe^x$
$y' = (x + 1)e^x$
$y'' = (x + 2)e^x$

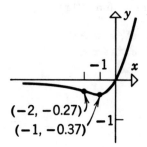

$(-2, -0.27)$
$(-1, -0.37)$

29. **(a)** $\displaystyle\lim_{x \to +\infty} \dfrac{x^2}{e^{2x}} = 0, \quad \lim_{x \to -\infty} \dfrac{x^2}{e^{2x}} = +\infty.$

(b) $y = x^2/e^{2x} = x^2 e^{-2x}$
$y' = 2x(1 - x)e^{-2x}$
$y'' = 2(2x^2 - 4x + 1)e^{-2x}$
$y'' = 0$ if $2x^2 - 4x + 1 = 0$
$x = \dfrac{4 \pm \sqrt{16 - 8}}{2},$
$\quad = 1 \pm \sqrt{2}/2 \approx 0.29, 1.71$

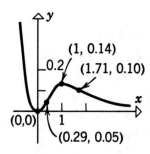

$(1, 0.14)$
$(1.71, 0.10)$
$(0,0)$
$(0.29, 0.05)$

31. **(a)**

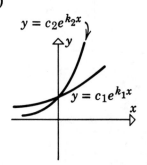

(b)

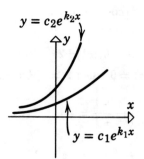

(c)

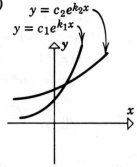

33. $m_{\text{line}} = \dfrac{1/e - 1}{1 - 0} = \dfrac{1 - e}{e}$, an equation

of the line is $y = \dfrac{1 - e}{e}x + 1$ so

$$A = \int_0^1 \left(\frac{1-e}{e}x + 1 - e^{-x} \right) dx$$

$$= \frac{1-e}{2e}x^2 + x + e^{-x} \Big]_0^1 = \frac{3-e}{2e}$$

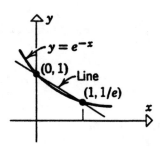

35.   **(a)** If $\ln x < \sqrt{x}$, then $x < e^{\sqrt{x}}, 1/x > 1/e^{\sqrt{x}} = e^{-\sqrt{x}}$.

   (b) $\displaystyle\lim_{x \to +\infty} e^{x - n\sqrt{x}} = \lim_{x \to +\infty} e^{\sqrt{x}(\sqrt{x} - n)} = +\infty$ so $\displaystyle\lim_{x \to +\infty} \frac{e^x}{x^n} = +\infty$ because $\dfrac{e^x}{x^n} > e^{x - n\sqrt{x}}$.

37.   If $x > 1$, then $0 < \ln x < \sqrt{x}, 0 < \dfrac{\ln x}{x^n} < \dfrac{\sqrt{x}}{x^n} = \dfrac{1}{x^{(n-1/2)}}$, but $\displaystyle\lim_{x \to +\infty} \frac{1}{x^{(n-1/2)}} = 0$ so

$\displaystyle\lim_{x \to +\infty} \frac{\ln x}{x^n} = 0.$

39. $\displaystyle\lim_{x \to 0^+} x^n \ln x = \lim_{t \to +\infty} \frac{\ln(1/t)}{t^n} = \lim_{t \to +\infty} \frac{-\ln t}{t^n} = 0.$

41. $y = -\ln x$ **43.** $y = \ln(x - 1)$

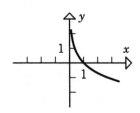   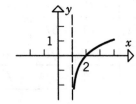

45. $y = \dfrac{\ln x}{x^2}$

$y' = \dfrac{1 - 2\ln x}{x^3}$

$y'' = \dfrac{6\ln x - 5}{x^4}$

$y' = 0$ if $x = e^{1/2}$

$y'' = 0$ if $x = e^{5/6}$

$\lim\limits_{x \to +\infty} y = 0$, $\lim\limits_{x \to 0^+} y = -\infty$

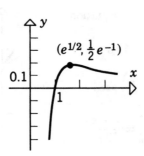

47. $(\ln x)/x \le 1/e$ so $e\ln x \le x, \ln(x^e) \le x, x^e \le e^x$ with equality only for $x = e$ because the maximum value of $(\ln x)/x$ occurs only at $x = e$.

EXERCISE SET 7.6

1.

	(a)	(b)	(c)	(d)	(e)	(f)
$\sinh x_0$	-2	$-3/4$	$-4/3$	$1/\sqrt{3}$	$8/15$	-1
$\cosh x_0$	$\sqrt{5}$	$5/4$	$5/3$	$2/\sqrt{3}$	$17/15$	$\sqrt{2}$
$\tanh x_0$	$-2/\sqrt{5}$	$-3/5$	$-4/5$	$1/2$	$8/17$	$-1/\sqrt{2}$
$\coth x_0$	$-\sqrt{5}/2$	$-5/3$	$-5/4$	2	$17/8$	$-\sqrt{2}$
$\operatorname{sech} x_0$	$1/\sqrt{5}$	$4/5$	$3/5$	$\sqrt{3}/2$	$15/17$	$1/\sqrt{2}$
$\operatorname{csch} x_0$	$-1/2$	$-4/3$	$-3/4$	$\sqrt{3}$	$15/8$	-1

(a) $\cosh^2 x_0 = 1 + \sinh^2 x_0 = 1 + (-2)^2 = 5$, $\cosh x_0 = \sqrt{5}$

(b) $\sinh^2 x_0 = \cosh^2 x_0 - 1 = \dfrac{25}{16} - 1 = \dfrac{9}{16}$, $\sinh x_0 = -\dfrac{3}{4}$ (because $x_0 < 0$)

(c) $\operatorname{sech}^2 x_0 = 1 - \tanh^2 x_0 = 1 - \left(-\dfrac{4}{5}\right)^2 = 1 - \dfrac{16}{25} = \dfrac{9}{25}$, $\operatorname{sech} x_0 = \dfrac{3}{5}$, $\cosh x_0 = \dfrac{1}{\operatorname{sech} x_0} = \dfrac{5}{3}$,

     from $\dfrac{\sinh x_0}{\cosh x_0} = \tanh x_0$ we get $\sinh x_0 = \left(\dfrac{5}{3}\right)\left(-\dfrac{4}{5}\right) = -\dfrac{4}{3}$

(d) $\operatorname{csch}^2 x_0 = \coth^2 x_0 - 1 = 4 - 1 = 3$, $\operatorname{csch} x_0 = \sqrt{3}$, $\sinh x_0 = \dfrac{1}{\operatorname{csch} x_0} = \dfrac{1}{\sqrt{3}}$, from

     $\dfrac{\cosh x_0}{\sinh x_0} = \coth x_0$ we get $\cosh x_0 = \left(\dfrac{1}{\sqrt{3}}\right)(2) = \dfrac{2}{\sqrt{3}}$

(e) $\cosh x_0 = \dfrac{1}{\operatorname{sech} x_0} = \dfrac{17}{15},\ \sinh^2 x_0 = \cosh^2 x_0 - 1 = \dfrac{289}{225} - 1 = \dfrac{64}{255},\ \sinh x_0 = \dfrac{8}{15}$ (because $x_0 > 0$)

(f) $\sinh x_0 = \dfrac{1}{\operatorname{csch} x_0} = -1,\ \cosh^2 x_0 = 1 + \sinh^2 x_0 = 2,\ \cosh x_0 = \sqrt{2}$

3. from (6b) and (1), $\cosh 2x = \cosh^2 x + \sinh^2 x$ and $\cosh^2 x = 1 + \sinh^2 x$ so $\cosh 2x = 2\sinh^2 x + 1$

5. $\cosh(-x) = \dfrac{1}{2}[e^{(-x)} + e^{-(-x)}] = \dfrac{1}{2}(e^{-x} + e^{x}) = \cosh x$

7. $\tanh(x + y) = \dfrac{\sinh(x+y)}{\cosh(x+y)} = \dfrac{\sinh x \cosh y + \cosh x \sinh y}{\cosh x \cosh y + \sinh x \sinh y}$

$= \dfrac{\dfrac{\sinh x \cosh y}{\cosh x \cosh y} + \dfrac{\cosh x \sinh y}{\cosh x \cosh y}}{\dfrac{\cosh x \cosh y}{\cosh x \cosh y} + \dfrac{\sinh x \sinh y}{\cosh x \cosh y}} = \dfrac{\tanh x + \tanh y}{1 + \tanh x \tanh y}$

9. $\tanh 2x = \dfrac{\sinh 2x}{\cosh 2x} = \dfrac{2\sinh x \cosh x}{\cosh^2 x + \sinh^2 x}$ (from (6a) and (6b))

$= \dfrac{2\tanh x}{1 + \tanh^2 x}$ (after dividing numerator and denominator by $\cosh^2 x$)

11. from (7a) with x replaced by $\dfrac{x}{2}$: $\cosh x = 2\sinh^2 \dfrac{x}{2} + 1,$

$2\sinh^2 \dfrac{x}{2} = \cosh x - 1,\ \sinh^2 \dfrac{x}{2} = \dfrac{1}{2}(\cosh x - 1),\ \sinh \dfrac{x}{2} = \pm\sqrt{\dfrac{1}{2}(\cosh x - 1)}$

13. add (4b) to (9b) then let $x = \dfrac{a+b}{2}$ and $y = \dfrac{a-b}{2}$.

15. (a) $\dfrac{d}{dx}(\sinh x) = \dfrac{d}{dx}\left[\dfrac{1}{2}(e^x - e^{-x})\right] = \dfrac{1}{2}(e^x + e^{-x}) = \cosh x$

(b) $\dfrac{d}{dx}(\coth x) = \dfrac{d}{dx}\left[\dfrac{e^x + e^{-x}}{e^x - e^{-x}}\right] = \dfrac{(e^x - e^{-x})(e^x - e^{-x}) - (e^x + e^{-x})(e^x + e^{-x})}{(e^x - e^{-x})^2}$

$= \dfrac{(e^{2x} - 2 + e^{-2x}) - (e^{2x} + 2 + e^{-2x})}{(e^x - e^{-x})^2} = -\dfrac{4}{(e^x - e^{-x})^2} = -\operatorname{csch}^2 x$

(c) $\dfrac{d}{dx}(\text{sech } x) = \dfrac{d}{dx}\left[\dfrac{2}{e^x + e^{-x}}\right]$

$\qquad = \dfrac{d}{dx}[2(e^x + e^{-x})^{-1}] = -2(e^x + e^{-x})^{-2}(e^x - e^{-x})$

$\qquad = -\dfrac{2}{(e^x + e^{-x})}\dfrac{e^x - e^{-x}}{e^x + e^{-x}} = -\text{sech } x \tanh x$

(d) proceed as in (c) using $\dfrac{2}{e^x - e^{-x}}$

17. $4\cosh(4x - 8)$

19. $-\dfrac{1}{x}\text{csch}^2(\ln x)$

21. $\dfrac{1}{x^2}\text{csch}(1/x)\coth(1/x)$

23. $\dfrac{2 + 5\cosh(5x)\sinh(5x)}{\sqrt{4x + \cosh^2(5x)}}$

25. $x^{5/2}\tanh(\sqrt{x})\text{sech}^2(\sqrt{x}) + 3x^2\tanh^2(\sqrt{x})$

27. $\dfrac{1}{7}\sinh^7 x + C$

29. $\dfrac{2}{3}(\tanh x)^{3/2} + C$

31. $\ln(\cosh x) + C$

33. $-\dfrac{1}{3}\text{sech}^3 x + C$

35. $2\sinh(\sqrt{x}) + C$

37. (a) $\dfrac{1}{2}(e^{\ln x} + e^{-\ln x}) = \dfrac{1}{2}\left(x + \dfrac{1}{x}\right) = \dfrac{x^2 + 1}{2x},\ x > 0$

$\quad$ (b) $\dfrac{1}{2}(e^{\ln x} - e^{-\ln x}) = \dfrac{1}{2}\left(x - \dfrac{1}{x}\right) = \dfrac{x^2 - 1}{2x},\ x > 0$

$\quad$ (c) $\dfrac{e^{2\ln x} - e^{-2\ln x}}{e^{2\ln x} + e^{-2\ln x}} = \dfrac{x^2 - 1/x^2}{x^2 + 1/x^2} = \dfrac{x^4 - 1}{x^4 + 1},\ x > 0$

$\quad$ (d) $\dfrac{1}{2}(e^{-\ln x} + e^{\ln x}) = \dfrac{1}{2}\left(\dfrac{1}{x} + x\right) = \dfrac{1 + x^2}{2x},\ x > 0$

39. positive on $(0, +\infty)$, negative on $(-\infty, 0)$, increasing on $(-\infty, +\infty)(dy/dx = \text{sech}^2 x > 0)$ concave up on $(-\infty, 0)$, concave down on $(0, +\infty)(d^2y/dx^2 = -2\text{sech}^2 x \tanh x)$

41. (a) $\cosh x = \sqrt{1 + \sinh^2 x} \geq \sqrt{1} = 1$

$\quad$ (b) $\cosh x \geq 1$ (part (a)) so $0 < \dfrac{1}{\cosh x} \leq 1,\ 0 < \text{sech } x \leq 1$

43. using $\sinh x + \cosh x = e^x$ (5a), $(\sinh x + \cosh x)^n = (e^x)^n = e^{nx} = \sinh nx + \cosh nx$

45. **(a)** $\displaystyle\lim_{x\to+\infty}\frac{\cosh x}{e^x} = \lim_{x\to+\infty}\frac{e^x+e^{-x}}{2e^x} = \lim_{x\to+\infty}\frac{1}{2}(1+e^{-2x}) = 1/2$

(b) $\displaystyle\lim_{x\to+\infty}\frac{\sinh ax}{e^x} = \lim_{x\to+\infty}\frac{e^{ax}-e^{-ax}}{2e^x} = \lim_{x\to+\infty}\frac{1}{2}[e^{(a-1)x}-e^{-(a+1)x}]$

$$= \begin{cases} +\infty, a>1 \\ 1/2, a=1 \\ 0, 0<a<1 \end{cases}$$

47. $\displaystyle A = \int_0^{\ln 3}\sinh 2x\,dx = \frac{1}{2}\cosh 2x\Big]_0^{\ln 3} = \frac{1}{2}[\cosh(2\ln 3)-1],$

but $\cosh(2\ln 3) = \cosh(\ln 9) = \dfrac{1}{2}(e^{\ln 9}+e^{-\ln 9}) = \dfrac{1}{2}(9+1/9) = 41/9$ so $A = \dfrac{1}{2}[41/9-1] = 16/9$.

49. $\displaystyle V = \pi\int_0^5(\cosh^2 2x - \sinh^2 2x)dx = \pi\int_0^5 dx = 5\pi$

51. $y' = \sinh x,\ 1+(y')^2 = 1+\sinh^2 x = \cosh^2 x$

$\displaystyle L = \int_0^{\ln 2}\cosh x\,dx = \sinh x\Big]_0^{\ln 2} = \sinh(\ln 2) = \frac{1}{2}(e^{\ln 2}-e^{-\ln 2}) = \frac{1}{2}\left(2-\frac{1}{2}\right) = \frac{3}{4}$

53. **(a)** $y' = \sinh(x/a),\ 1+(y')^2 = 1+\sinh^2(x/a) = \cosh^2(x/a)$

$\displaystyle L = 2\int_0^b\cosh(x/a)\,dx = 2a\sinh(x/a)\Big]_0^b = 2a\sinh(b/a)$

(b) The highest point is at $x=b$, the lowest at $x=0$,
so $S = a\cosh(b/a) - a\cosh(0) = a\cosh(b/a) - a$.

55. From part (b) of Exercise 53, $S = a\cosh(b/a) - a$ so $30 = a\cosh(200/a) - a$. Let $u = 200/a$,
then $a = 200/u$ so $30 = (200/u)[\cosh u - 1]$, $\cosh u - 1 = 0.15u$. If $f(u) = \cosh u - 0.15u - 1$,
then $u_{n+1} = u_n - \dfrac{\cosh u_n - 0.15u_n - 1}{\sinh u_n - 0.15}$; $u_1 = 0.3, \cdots, u_4 = u_5 = 0.297792782 \approx 200/a$ so
$a \approx 671.6079505$. From part (a),
$L = 2a\sinh(b/a) \approx 2(671.6079505)\sinh(0.297792782) \approx 405.9\,\text{ft}.$

EXERCISE SET 7.7

1. $\dfrac{1}{y}dy = \dfrac{1}{x}dx,\ \ln|y| = \ln|x| + C_1,\ \ln\left|\dfrac{y}{x}\right| = C_1,\ \dfrac{y}{x} = \pm e^{C_1} = C,\ y = Cx$

3. $\dfrac{1}{1+y}dy = -\dfrac{x}{\sqrt{1+x^2}}dx$, $\ln|1+y| = -\sqrt{1+x^2} + C_1$,

$1 + y = \pm e^{-\sqrt{1+x^2}+C_1} = \pm e^{C_1}e^{-\sqrt{1+x^2}} = Ce^{-\sqrt{1+x^2}}$, $y = Ce^{-\sqrt{1+x^2}} - 1$

5. $e^y\,dy = \dfrac{\sin x}{\cos^2 x}dx = \sec x \tan x\,dx$, $e^y = \sec x + C$, $y = \ln(\sec x + C)$

7. $\rho = e^{\int 3dx} = e^{3x}$, $e^{3x}y = \displaystyle\int e^x\,dx = e^x + C$, $y = e^{-2x} + Ce^{-3x}$

9. $\rho = e^{\int dx} = e^x$, $e^x y = \displaystyle\int e^x \cos(e^x)dx = \sin(e^x) + C$, $y = e^{-x}\sin(e^x) + Ce^{-x}$

11. $y' + \dfrac{3}{x}y = -2x^3$, $\rho = e^{\int \frac{3}{x}dx} = e^{3\ln x} = x^3$,

$x^3 y = \displaystyle\int -2x^6\,dx = -\dfrac{2}{7}x^7 + C$, $y = -\dfrac{2}{7}x^4 + Cx^{-3}$

13. $\rho = e^{\int -x\,dx} = e^{-x^2/2}$, $e^{-x^2/2}y = \displaystyle\int xe^{-x^2/2}dx = -e^{-x^2/2} + C$,

$y = -1 + Ce^{x^2/2}$, $3 = -1 + C$, $C = 4$, $y = -1 + 4e^{x^2/2}$

15. $\rho = e^{\int dt} = e^t$, $e^t y = \displaystyle\int 2e^t\,dt = 2e^t + C$, $y = 2 + Ce^{-t}$, $1 = 2 + C$, $C = -1$, $y = 2 - e^{-t}$

17. $y^2 t\dfrac{dy}{dt} = t - 1$, $y^2\,dy = \left(1 - \dfrac{1}{t}\right)dt$, $\dfrac{1}{3}y^3 = t - \ln t + C$,

$9 = 1 + C$, $C = 8$, $\dfrac{1}{3}y^3 = t - \ln t + 8$, $y = \sqrt[3]{3t - 3\ln t + 24}$

19. $\dfrac{dy}{dx} = \dfrac{y^2}{3\sqrt{x}}$, $\dfrac{1}{y^2}dy = \dfrac{1}{3\sqrt{x}}dx$, $-\dfrac{1}{y} = \dfrac{2}{3}\sqrt{x} + C$; $y = -1$ when $x = 1$ so $1 = \dfrac{2}{3} + C$,

$C = \dfrac{1}{3}$, $-\dfrac{1}{y} = \dfrac{2}{3}\sqrt{x} + \dfrac{1}{3}$, $y = -\dfrac{3}{2\sqrt{x} + 1}$

21. $\dfrac{dy}{dx} = xe^y$, $e^{-y}\,dy = x\,dx$, $-e^{-y} = \dfrac{1}{2}x^2 + C$; $y = 0$ when $x = 2$ so $-1 = 2 + C$, $C = -3$,

$-e^{-y} = \dfrac{1}{2}x^2 - 3$, $y = -\ln(3 - \dfrac{1}{2}x^2)$

23. (a) $A(h) = \pi(1)^2 = \pi$, $\pi\dfrac{dh}{dt} = -0.025\sqrt{h}$, $\dfrac{\pi}{\sqrt{h}}dh = -0.025dt$, $2\pi\sqrt{h} = -0.025t + C$; $h = 4$

when $t = 0$ so $4\pi = C$, $2\pi\sqrt{h} = -0.025t + 4\pi$, $\sqrt{h} = 2 - \dfrac{0.025}{2\pi}t$, $h \approx (2 - 0.003979t)^2$.

(b) $h = 0$ when $t \approx 2/0.003979 \approx 502.6\,\text{sec} \approx 8.4\,\text{min}$

25. $\dfrac{dv}{dt} = -0.04v^2,\ \dfrac{1}{v^2}dv = -0.04dt,\ -\dfrac{1}{v} = -0.04t + C;\ v = 50$ when $t = 0$ so $-\dfrac{1}{50} = C,$

$-\dfrac{1}{v} = -0.04t - \dfrac{1}{50},\ v = \dfrac{50}{2t+1}.$ But $v = \dfrac{dx}{dt}$ so $\dfrac{dx}{dt} = \dfrac{50}{2t+1},\ x = 25\ln(2t+1) + C_1;\ x = 0$

when $t = 0$ so $C_1 = 0,\ x = 25\ln(2t+1).$

27. (a) $\dfrac{dv}{dt} = \dfrac{ck}{m_0 - kt} - g,\ v = -c\ln(m_0 - kt) - gt + C;\ v = 0$ when $t = 0$ so $0 = -c\ln m_0 + C,$

$C = c\ln m_0,\ v = c\ln m_0 - c\ln(m_0 - kt) - gt = c\ln\dfrac{m_0}{m_0 - kt} - gt.$

(b) $m_0 - kt = 0.2m_0$ when $t = 100$ so

$v = 2500\ln\dfrac{m_0}{0.2m_0} - 9.8(100) = 2500\ln 5 - 980 \approx 3044\,\text{m/sec}.$

29. (a) $vdv = -\dfrac{gR^2}{x^2}dx,\ \dfrac{1}{2}v^2 = \dfrac{gR^2}{x} + C;\ v = v_0$ when $x = R$ so $\dfrac{1}{2}v_0^2 = gR + C,$

$C = \dfrac{1}{2}v_0^2 - gR,\ \dfrac{1}{2}v^2 = \dfrac{gR^2}{x} + \dfrac{1}{2}v_0^2 - gR,\ v^2 = \dfrac{2gR^2}{x} + v_0^2 - 2gR.$

(b) From the result in part (a), $v^2 > 0$ for all $x \geq R$ if $v_0^2 - 2gR \geq 0,\ v_0 \geq \sqrt{2gR}.$

(c) 1 mi = 5280 ft so 32 ft/sec^2 = 32/5280 mi/sec^2,

$v_0 = \sqrt{2gR} = \sqrt{2(32/5280)(3960)} \approx 6.9\,\text{mi/sec}.$

31. $\dfrac{dy}{dt}$ = rate in $-$ rate out, where y is the amount of salt at time $t,$

$\dfrac{dy}{dt} = (4)(2) - \left(\dfrac{y}{50}\right)(2) = 8 - \dfrac{1}{25}y$ so $\dfrac{dy}{dt} + \dfrac{1}{25}y = 8$ and $y(0) = 25.$

$\rho = e^{\int \frac{1}{25}dt} = e^{t/25},\ e^{t/25}y = \int 8e^{t/25}dt = 200e^{t/25} + C,$

$y = 200 + Ce^{-t/25},\ 25 = 200 + C,\ C = -175,$

(a) $y = 200 - 175e^{-t/25}$

(b) when $t = 25,\ y = 200 - 175e^{-1} \approx 136$ lb

33. At time t there are $500 + (20 - 10)t = 500 + 10t$ gallons of brine in the tank so

$\dfrac{dy}{dt} = 0 - \dfrac{y}{500 + 10t}(10) = -\dfrac{y}{50 + t},\ \dfrac{dy}{dt} + \dfrac{1}{50 + t}y = 0$ and $y(0) = 50,$

$\rho = e^{\int 1/(50+t)dt} = e^{\ln(50+t)} = 50 + t,\ (50 + t)y = C,\ y = \dfrac{C}{50 + t},$

$50 = C/50$, $C = 2500$, $y = \dfrac{2500}{50 + t}$

The tank reaches the point of overflowing when $500 + 10t = 1000$, $t = 50$ min so
$y = 2500/(50 + 50) = 25$ lb.

35. (a) From (28), $k = -\dfrac{1}{T} \ln 2 = -\dfrac{1}{140} \ln 2 \approx -0.005$ so $y = 10e^{-0.005t}$ (approximately).

 (b) 10 weeks = 70 days so $y = 10e^{-0.35} \approx 7$ mg

37. $100e^{0.02t} = 5000$, $e^{0.02t} = 50$, $t = \dfrac{1}{0.02} \ln 50 \approx 196$ days

39. $y = y_0 e^{kt}$, but $y = 0.6y_0$ when $t = 5$ so $0.6y_0 = y_0 e^{5k}$, $e^{5k} = 0.6$

 $k = \dfrac{1}{5} \ln 0.6$ so $T = -5\dfrac{\ln 2}{\ln 0.6} \approx 6.8$ years

41. (a) $y = 10,000e^{kt}$, but $y = 12,000$ when $t = 10$ so $10,000e^{10k} = 12,000$, $k = \dfrac{1}{10} \ln 1.2$.
 When $t = 20$, $y = 10,000e^{20k} = 10,000e^{2\ln 1.2} = 10,000(1.44) = 14,400$.

 (b) From (27), $T = 10\dfrac{\ln 2}{\ln 1.2} \approx 38$ years

43. $\dfrac{dT}{dt} = k(T - C)$, $k < 0$

 $\dfrac{dT}{dt} - kT = -kC$, $\rho = e^{\int -k\,dt} = e^{-kt}$, $e^{-kt}T = \int -kCe^{-kt}\,dt = Ce^{-kt} + K$,

 $T = C + Ke^{kt}$. But $T = T_0$ when $t = 0$ so $T_0 = C + K$, $K = T_0 - C$, $T = C + (T_0 - C)e^{kt}$

45. (a) In t years the interest will be compounded nt times at an interest rate of r/n each time. The value at the end of 1 interval is $P + (r/n)P = P(1 + r/n)$, at the end of 2 intervals it is $P(1 + r/n) + (r/n)P(1 + r/n) = P(1 + r/n)^2$, and continuing in this fashion the value at the end of nt intervals is $P(1 + r/n)^{nt}$.

 (b) Let $x = r/n$, then $n = r/x$ and
 $$\lim_{n \to +\infty} P(1 + r/n)^{nt} = \lim_{x \to 0^+} P(1 + x)^{rt/x} = \lim_{x \to 0^+} P[(1 + x)^{1/x}]^{rt} = Pe^{rt}.$$

 (c) The rate of increase is $dA/dt = rPe^{rt} = rA$.

47. Let $y = y_0 e^{kt}$ with $y = y_1$ when $t = t_1$ and $y = y_1/2$ when $t = t_1 + T$ then $y_0 e^{kt_1} = y_1$ (i) and
 $y_0 e^{k(t_1 + T)} = y_1/2$ (ii). Divide (i) by (ii) to get $e^{-kT} = 2$, $T = -\dfrac{1}{k} \ln 2$.

SUPPLEMENTARY EXERCISES CHAPTER 7

1. (a) $f(g(x)) = m\left(\dfrac{1}{mx}\right) = 1/x \neq x$; f and g are not inverses.

    (b) $f(g(x)) = \dfrac{3}{(3-x)/x + 1} = x$, $g(f(x)) = \dfrac{3 - 3/(x+1)}{3/(x+1)} = x$; f and g are inverses.

    (c) $f(g(x)) = (x^{1/3} + 2)^3 - 8 = x + 6x^{2/3} + 12x^{1/3} \neq x$; f and g are not inverses.

    (d) $f(g(x)) = x + 1 - 1 = x$, $g(f(x)) = \sqrt[3]{x^3 - 1 + 1} = x$; f and g are inverses.

    (e) $f(g(x)) = \sqrt{e^{2\ln x}} = \sqrt{x^2} = x$ where $x > 0$, $g(f(x)) = 2\ln\sqrt{e^x} = \ln e^x = x$; f and g are inverses.

3. $f(0) = f(2)$; f is not one-to-one so $f^{-1}(x)$ does not exist.

5. $y = f^{-1}(x)$, $x = f(y) = e^{2y} + 1$, $e^{2y} = x - 1$, $y = \dfrac{1}{2}\ln(x-1) = f^{-1}(x)$.

7. f^{-1} will exist if and only if f is one-to-one. Let x_1, x_2 be any two distinct points in the domain of $y = f(x) = (ax + b)/(cx + d)$.

    $y_1 = f(x_1) = (ax_1 + b)/(cx_1 + d)$, $y_2 = f(x_2) = (ax_2 + b)/(cx_2 + d)$,

    $y_2 - y_1 = \dfrac{(ax_2 + b)(cx_1 + d) - (ax_1 + b)(cx_2 + d)}{(cx_2 + d)(cx_1 + d)}$

    $\qquad = \dfrac{ad(x_2 - x_1) - bc(x_2 - x_1)}{(cx_2 + d)(cx_1 + d)} = \dfrac{(ad - bc)(x_2 - x_1)}{(cx_2 + d)(cx_2 + d)}$

    f will be one-to-one if $y_1 \neq y_2$ (or equivalently $y_2 - y_1 \neq 0$) whenever $x_1 \neq x_2$, which occurs when $ad - bc \neq 0$. To find $f^{-1}(x)$ in this case, solve $y = (ax + b)/(cx + d)$ for x to get $x = (-dy + b)/(cy - a) = f^{-1}(y)$ so $f^{-1}(x) = (-dx + b)/(cx - a)$.

9. (a) $f(x) = \begin{cases} 2x - 5, & x \geq 5/2 \\ -2x + 5, & x < 5/2 \end{cases}$, $f'(x) = \begin{cases} 2, & x > 5/2 \\ -2, & x < 5/2 \end{cases}$

    and $f'(x)$ does not exist at $x = 5/2$. $f(x)$ is minimum when $x = 5/2$ and f is decreasing for $x < 5/2$ because $f'(x) < 0$, so f is one-to-one for x in the interval $(-\infty, 5/2)$

    (b) $f'(x) = 2(x + 2)$, so f is decreasing for $x < -2$ and increasing for $x > -2$. f is one-to-one for x in $(-2, +\infty)$

    (c) $f'(x) = -\sin(x - 2\pi/3)$, $f'(x) = 0$ when $x - 2\pi/3 = n\pi$, $x = 2\pi/3 + n\pi$ where n is an integer. $f'(-\pi/3) = f'(2\pi/3) = 0$ and $f'(x) > 0$ if $-\pi/3 < x < 2\pi/3$ so f is one-to-one for x in $(-\pi/3, 2\pi/3)$.

11. $y = f^{-1}(x)$, $x = f(y) = \dfrac{3}{y+1}$, $y = \dfrac{3}{x} - 1 = f^{-1}(x)$; $f'(x) = -\dfrac{3}{(x+1)^2}$,

$f'(f^{-1}(x)) = -\dfrac{3}{(3/x)^2} = -\dfrac{x^2}{3}$, $(f^{-1})'(x) = -\dfrac{3}{x^2}$.

13. $y = f^{-1}(x)$, $x = f(y) = e^{y/2}$, $y = 2\ln x = f^{-1}(x)$; $f'(x) = \dfrac{1}{2}e^{x/2}$, $f'(f^{-1}(x)) = \dfrac{1}{2}e^{\ln x} = \dfrac{x}{2}$,

$(f^{-1})'(x) = \dfrac{2}{x}$.

15. **(a)** $\ln(1/12) = -\ln 12 = -\ln(2^2 \cdot 3) = -(2\ln 2 + \ln 3) = -(2r + s)$

 (b) $\ln(9/\sqrt{8}) = \ln(3^2 \cdot 2^{-3/2}) = 2\ln 3 - \dfrac{3}{2}\ln 2 = 2s - 3r/2$

 (c) $\ln(\sqrt[4]{8/3}) = \dfrac{1}{4}\ln(2^3/3) = \dfrac{1}{4}(3\ln 2 - \ln 3) = (3r - s)/4$

17. **(a)** $25^x = 3^{1-x}$, $(5^2)^x = 3^{1-x}$, $5^{2x} = 3^{1-x}$, $\ln 5^{2x} = \ln 3^{1-x}$,

 $2x\ln 5 = (1-x)\ln 3$, $x = (\ln 3)/(2\ln 5 + \ln 3)$

 (b) $\sinh x = \dfrac{1}{4}\cosh x$, $\dfrac{1}{2}(e^x - e^{-x}) = \dfrac{1}{8}(e^x + e^{-x})$, $3e^x = 5e^{-x}$, $e^{2x} = 5/3$, $x = (\ln 5 - \ln 3)/2$

19. **(a)** $\cosh x = (1 + \sinh^2 x)^{1/2} = (1 + 9/25)^{1/2} = \sqrt{34}/5$
 (b) $\tanh x = \sinh x / \cosh x = -3/\sqrt{34}$
 (c) $\sinh 2x = 2\sinh x \cosh x = -6\sqrt{34}/25$

21. $y = e^{-x/2}$, $dy/dx = -\dfrac{1}{2}e^{-x/2} = -1/(2\sqrt{e^x})$

23. $dy/dx = (\ln x - 1)/(\ln x)^2$ **25.** $y = x/x = 1$, $dy/dx = 0$

27. $y = x\ln 10 - \ln \sin x$, $dy/dx = \ln 10 - \cot x$

29. $y = x^4 e^{\tan x}$, $dy/dx = x^3 e^{\tan x}(x \sec^2 x + 4)$

31. $dy/dx = \dfrac{1 + x/\sqrt{x^2 + a^2}}{x + \sqrt{x^2 + a^2}} = 1/\sqrt{x^2 + a^2}$

33. $y = \exp(3x^2)$, $dy/dx = 6x \exp(3x^2)$

35. $y = (\ln \sqrt{x})^{1/2}$, $dy/dx = \dfrac{1}{2}(\ln \sqrt{x})^{-1/2}\left(\dfrac{1}{2x}\right) = 1/(4x\sqrt{\ln \sqrt{x}})$

37. $dy/dx = \pi^x(\pi x^{\pi-1}) + x^\pi(\pi^x \ln \pi) = \pi^x x^{\pi-1}(\pi + x\ln \pi)$

39. $dy/dx = 5\cosh[\tanh(5x)]\operatorname{sech}^2(5x)$

41. $y = e^{3x}(1 + 2e^{-x} + e^{-2x}) = e^{3x} + 2e^{2x} + e^x$, $dy/dx = 3e^{3x} + 4e^{2x} + e^x$.

43. $y = e^{ax}\sin bx$, $y' = e^{ax}[b\cos bx + a\sin bx]$,
$y'' = e^{ax}[2ab\cos bx + (a^2 - b^2)\sin bx]$ so $y'' - 2ay' + (a^2 + b^2)y = 0$

45. (a) $dy = -e^{-x}dx$ (b) $dy = \dfrac{1}{1+x}dx$ (c) $dy = 2x(\ln 2)2^{x^2}dx$

47. (a) $\dfrac{1}{\sqrt{4 + e^{\ln x}}}\left(\dfrac{1}{x}\right) = \dfrac{1}{x\sqrt{4+x}}$ (b) $\sqrt{\ln e^{5x} + e^{5x}}(5e^{5x}) = 5e^{5x}\sqrt{5x + e^{5x}}$

49. $Y = \ln y = \ln(Ce^{kt}) = \ln C + kt$ which is linear in t and Y so the graph is a straight line.

51. $u = 1 + e^x$, $\displaystyle\int \dfrac{1}{u}du = \ln(1 + e^x) + C$ **53.** $\dfrac{x^{e+1}}{e+1} + C$

55. 55. $\displaystyle\int \left(\dfrac{4}{x} - \dfrac{3}{x^2}\right)dx = 4\ln|x| + 3/x + C$

57. $u = 2\sec x - 1$, $\dfrac{1}{2}\displaystyle\int \dfrac{1}{u}du = \dfrac{1}{2}\ln|2\sec x - 1| + C$

59. $u = \sin 2x$, $\dfrac{1}{2}\displaystyle\int \exp(u)du = \dfrac{1}{2}\exp(\sin 2x) + C$

61. $u = \tanh x$, $\displaystyle\int u\,du = \dfrac{1}{2}\tanh^2 x + C$ **63.** $u = \ln x$, $\displaystyle\int_1^2 \dfrac{1}{u}du = \ln 2$

65. $u = \tan x$, $\displaystyle\int_0^1 2^u du = \dfrac{2^u}{\ln 2}\Big]_0^1 = \dfrac{1}{\ln 2}$ **67.** $\dfrac{d}{dx}\left[\dfrac{e^{kx}}{k} + C\right] = e^{kx}$

69. $y = x^3 e^{-x}$
$y' = x^2(3 - x)e^{-x}$
$y' = 0$ when $x = 0, 3$
$y'' = x(x^2 - 6x + 6)e^{-x}$
$y'' = 0$ when $x = 0, 3 \pm \sqrt{3}$

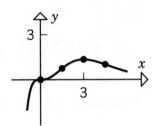

71. $y'' = e^y y' + 2y' + 1$. If x_0 is a critical point of y then when $x = x_0$, $y' = 0$ and
$y'' = e^{y_0}(0) + 2(0) + 1 = 1 > 0$ so a relative minimum occurs at x_0.

73. $A = \int_0^b e^{-2x} dx = -\dfrac{1}{2}(e^{-2b} - 1) = \dfrac{1}{2} - \dfrac{1}{2}e^{-2b}$. If $A = 1/4$ then $1/2 - e^{-2b}/2 = 1/4$, $e^{-2b} = 1/2$,
$b = (1/2)\ln 2$. $\lim\limits_{b \to +\infty} A = 1/2$

75. $y' = \sinh x$, $1 + (y')^2 = 1 + \sinh^2 x = \cosh^2 x$,
$$S = \int_0^1 2\pi \cosh x(\cosh x)dx = 2\pi \int_0^1 \cosh^2 x\, dx = \pi \int_0^1 (1 + \cosh 2x)dx$$
$$= \pi \left(x + \frac{1}{2}\sinh 2x \right)\Big]_0^1 = \pi(2 + \sinh 2)/2$$

77.   Let y = population (in millions) t years after 1970, then $y = 205e^{0.018t}$.

(a) $t = 2000 - 1970 = 30$ for the year 2000 so $y = 205e^{(0.018)(30)} = 205e^{0.54} \approx 352$ million.

(b)   1 billion = 1000 million, $205e^{0.018t} = 1000$ when $t = (1/0.018)\ln(1000/205) \approx 88$. The
population will reach one billion in the year $1970 + 88 = 2058$.

CHAPTER 8
Inverse Trigonometric and Hyperbolic Functions

EXERCISE SET 8.1

1. (a) $-\pi/2$ (b) π (c) $-\pi/4$ (d) $\pi/4$ (e) 0 (f) $\pi/2$

3. $\theta = -\pi/3$; $\cos\theta = 1/2$, $\tan\theta = -\sqrt{3}$, $\cot\theta = -1/\sqrt{3}$, $\sec\theta = 2$, $\csc\theta = -2/\sqrt{3}$

5. $\tan\theta = 4/3$, $0 < \theta < \pi/2$; use the triangle shown to get $\sin\theta = 4/5$, $\cos\theta = 3/5$, $\cot\theta = 3/4$, $\sec\theta = 5/3$, $\csc\theta = 5/4$.

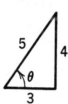

7. (a) $\pi/7$ (b) $\sin^{-1}(\sin\pi) = \sin^{-1}(\sin 0) = 0$
    (c) $\sin^{-1}(\sin(5\pi/7)) = \sin^{-1}(\sin(2\pi/7)) = 2\pi/7$
    (d) Note that $\pi/2 < 630 - 200\pi < \pi$ so
        $\sin(630) = \sin(630 - 200\pi) = \sin(\pi - (630 - 200\pi)) = \sin(201\pi - 630)$
        where $0 < 201\pi - 630 < \pi/2$; $\sin^{-1}(\sin 630) = \sin^{-1}(\sin(201\pi - 630)) = 201\pi - 630$.

9. (a) $0 \le x \le \pi$ (b) $-1 \le x \le 1$
    (c) $-\pi/2 < x < \pi/2$ (d) $-\infty < x < +\infty$
    (e) $0 < x \le \pi/2$ or $-\pi < x \le -\pi/2$ (f) $|x| \ge 1$

11. Let $\theta = \cos^{-1}(3/5)$,
    $\sin 2\theta = 2\sin\theta\cos\theta$
          $= 2(4/5)(3/5) = 24/25$

13. $\sin^{-1}(1) = \pi/2$

15. Let $\theta = \sec^{-1}(3/2)$,

$$\tan 2\theta = \frac{2\tan\theta}{1-\tan^2\theta}$$

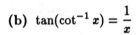

$$= \frac{2(\sqrt{5}/2)}{1-5/4} = -4\sqrt{5}$$

17. **(a)** $\tan^{-1}\dfrac{1}{2} + \tan^{-1}\dfrac{1}{3} = \tan^{-1}\dfrac{1/2+1/3}{1-(1/2)\,(1/3)} = \tan^{-1}1 = \pi/4$

(b) $2\tan^{-1}\dfrac{1}{3} = \tan^{-1}\dfrac{1}{3} + \tan^{-1}\dfrac{1}{3} = \tan^{-1}\dfrac{1/3+1/3}{1-(1/3)\,(1/3)} = \tan^{-1}\dfrac{3}{4}$,

$2\tan^{-1}\dfrac{1}{3} + \tan^{-1}\dfrac{1}{7} = \tan^{-1}\dfrac{3}{4} + \tan^{-1}\dfrac{1}{7} = \tan^{-1}\dfrac{3/4+1/7}{1-(3/4)\,(1/7)} = \tan^{-1}1 = \pi/4$

19. **(a)** $\cos(\tan^{-1}x) = \dfrac{1}{\sqrt{1+x^2}}$ **(b)** $\tan(\cot^{-1}x) = \dfrac{1}{x}$

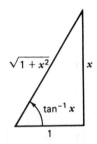   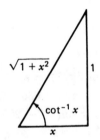

(c) $\sin(\sec^{-1}x) = \dfrac{\sqrt{x^2-1}}{x}$ **(d)** $\cot(\csc^{-1}x) = \sqrt{x^2-1}$

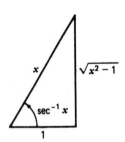   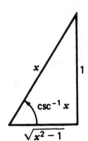

21. **(a)**

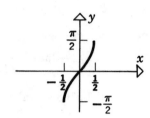

(b)

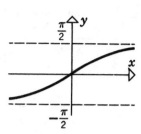

23. **(a)** let $\theta = \sin^{-1}(-x)$ then $\sin\theta = -x$, $-\pi/2 \le \theta \le \pi/2$. But $\sin(-\theta) = -\sin\theta$ and $-\pi/2 \le -\theta \le \pi/2$ so $\sin(-\theta) = -(-x) = x$, $-\theta = \sin^{-1}x$, $\theta = -\sin^{-1}x$.

(b) proof is similar to that in part (a).

25. If $-1 \le x < 0$ then $0 < -x \le 1$ so $\sin^{-1}(-x) + \cos^{-1}(-x) = \pi/2$, but $\sin^{-1}(-x) = -\sin^{-1}x$ and $\cos^{-1}(-x) = \pi - \cos^{-1}x$ thus $-\sin^{-1}x + (\pi - \cos^{-1}x) = \pi/2$, $\sin^{-1}x + \cos^{-1}x = \pi/2$.

27. **(a)** $55.0°$ **(b)** $33.6°$ **(c)** $25.8°$

29. $x = \pi + \tan^{-1}k$ **31.** $x = \pi - \sin^{-1}(0.37) \approx 2.7626$

33. $x = \tan^{-1}(3.16) - \pi \approx -1.8773$ **35.** $\theta = -\cos^{-1}(0.23) \approx -76.7°$

37. **(b)** $\theta = \sin^{-1}\frac{R}{R+h} = \sin^{-1}\frac{6378}{16,378} \approx 23°$

39. $\sin 2\theta = gR/v^2 = (9.8)(18)/(14)^2 = 0.9$, $2\theta = \sin^{-1}(0.9)$ or $2\theta = 180° - \sin^{-1}(0.9)$ so $\theta = \frac{1}{2}\sin^{-1}(0.9) \approx 32°$ or $\theta = 90° - \frac{1}{2}\sin^{-1}(0.9) \approx 58°$. The ball will have a lower parabolic trajectory for $\theta = 32°$ and hence will result in the shorter time of flight.

41. $y = 0$ when $x^2 = 6000v^2/g$, $x = 10v\sqrt{60/g} = 1000\sqrt{30}$ for $v = 400$ and $g = 32$; $\tan\theta = 3000/x = 3/\sqrt{30}$, $\theta = \tan^{-1}(3/\sqrt{30}) \approx 29°$.

43. **(a)** $\sin^{-1} X = \tan^{-1} \dfrac{X}{\sqrt{1 - X^2}}$

$\qquad\quad = ATN \dfrac{X}{\sqrt{1 - X^2}}$

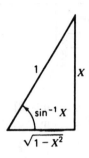

(b) $\sin^{-1} X + \cos^{-1} X = \pi/2$

$\qquad \cos^{-1} X = \pi/2 - \sin^{-1} X$

$\qquad\qquad = \pi/2 - \tan^{-1} \dfrac{X}{\sqrt{1 - X^2}}$

$\qquad\qquad \approx 1.5708 - \tan^{-1} \dfrac{X}{\sqrt{1 - X^2}}$

$\qquad\qquad = 1.5708 - ATN \dfrac{X}{\sqrt{1 - X^2}}$

EXERCISE SET 8.2

1. **(a)** $\dfrac{1}{\sqrt{1 - x^2/9}}(1/3) = 1/\sqrt{9 - x^2}$

   (b) $-2/\sqrt{1 - (2x + 1)^2}$

3. **(a)** $\dfrac{1}{x^7\sqrt{x^{14} - 1}}(7x^6) = \dfrac{7}{x\sqrt{x^{14} - 1}}$

   (b) $-1/\sqrt{e^{2x} - 1}$

5. **(a)** $\dfrac{1}{\sqrt{1 - 1/x^2}}(-1/x^2) = -\dfrac{1}{|x|\sqrt{x^2 - 1}}$

   (b) $\dfrac{\sin x}{\sqrt{1 - \cos^2 x}} = \dfrac{\sin x}{|\sin x|} = \begin{cases} 1, & \sin x > 0 \\ -1, & \sin x < 0 \end{cases}$

7. **(a)** $\dfrac{e^x}{x\sqrt{x^2 - 1}} + e^x \sec^{-1} x$

   (b) $\dfrac{3x^2(\sin^{-1} x)^2}{\sqrt{1 - x^2}} + 2x(\sin^{-1} x)^3$

9. **(a)** $\dfrac{1}{1 + (1 - x)^2/(1 + x)^2}\left[\dfrac{(1 + x)(-1) - (1 - x)(1)}{(1 + x)^2}\right] = -\dfrac{2}{(1 + x)^2 + (1 - x)^2} = -1/(x^2 + 1)$

   (b) $10(1 + x \csc^{-1} x)^9(-1/\sqrt{x^2 - 1} + \csc^{-1} x)$

11. **(a)** $\dfrac{1}{1 + (1 - x)/(1 + x)} \dfrac{1}{2}\left(\dfrac{1 - x}{1 + x}\right)^{-1/2} \dfrac{(1 + x)(-1) - (1 - x)(1)}{(1 + x)^2} = -\dfrac{1}{2\sqrt{1 - x^2}}$

   (b) $\dfrac{x + 2x \ln x}{\sqrt{1 - x^4 \ln^2 x}}$

13. $\sin^{-1}(xy) = \cos^{-1}(x-y)$, $\dfrac{1}{\sqrt{1-x^2y^2}}(xy'+y) = -\dfrac{1}{\sqrt{1-(x-y)^2}}(1-y')$,

$$y' = \frac{y\sqrt{1-(x-y)^2}+\sqrt{1-x^2y^2}}{\sqrt{1-x^2y^2}-x\sqrt{1-(x-y)^2}}$$

15. $\tan^{-1}x\big]_{-1}^{1} = \tan^{-1}1 - \tan^{-1}(-1) = \pi/4 - (-\pi/4) = \pi/2$

17. $\sec^{-1}x\big]_{-\sqrt{2}}^{-2/\sqrt{3}} = \sec^{-1}(-2/\sqrt{3}) - \sec^{-1}(-\sqrt{2}) = 7\pi/6 - 5\pi/4 = -\pi/12$

19. $u = 4x$, $\dfrac{1}{4}\displaystyle\int \dfrac{1}{1+u^2}du = \dfrac{1}{4}\tan^{-1}4x + C$ **21.** $u = e^x$, $\displaystyle\int \dfrac{1}{1+u^2}du = \tan^{-1}(e^x) + C$

23. $u = \sqrt{x}$, $2\displaystyle\int_1^{\sqrt{3}} \dfrac{1}{u^2+1}du = 2\tan^{-1}u\bigg]_1^{\sqrt{3}} = 2(\tan^{-1}\sqrt{3} - \tan^{-1}1) = 2(\pi/3 - \pi/4) = \pi/6$

25. $u = \tan x$, $\displaystyle\int \dfrac{1}{\sqrt{1-u^2}}du = \sin^{-1}(\tan x) + C$

27. $u = \ln x$, $\displaystyle\int \dfrac{1}{\sqrt{1-u^2}}du = \sin^{-1}(\ln x) + C$

29. (a) $\sin^{-1}(x/3) + C$ (b) $(1/\sqrt{5})\tan^{-1}(x/\sqrt{5}) + C$
(c) $(1/\sqrt{\pi})\sec^{-1}(x/\sqrt{\pi}) + C$

31. $u = \sqrt{3}x^2$, $\dfrac{1}{2\sqrt{3}}\displaystyle\int_0^{\sqrt{3}} \dfrac{1}{\sqrt{4-u^2}}du = \dfrac{1}{2\sqrt{3}}\sin^{-1}\dfrac{u}{2}\bigg]_0^{\sqrt{3}} = \dfrac{1}{2\sqrt{3}}\left(\dfrac{\pi}{3}\right) = \dfrac{\pi}{6\sqrt{3}}$

33. $u = 3x$, $\dfrac{1}{3}\displaystyle\int_0^{2\sqrt{3}} \dfrac{1}{4+u^2}du = \dfrac{1}{6}\tan^{-1}\dfrac{u}{2}\bigg]_0^{2\sqrt{3}} = \dfrac{1}{6}(\pi/3) = \pi/18$

35. $A = \displaystyle\int_0^{1/6} \dfrac{1}{\sqrt{1-9x^2}}dx = \dfrac{1}{3}\displaystyle\int_0^{1/2} \dfrac{1}{\sqrt{1-u^2}}du = \dfrac{1}{3}\sin^{-1}u\bigg]_0^{1/2} = \pi/18$

37. $V = \displaystyle\int_{-2}^{2} \pi\dfrac{1}{4+x^2}dx = \dfrac{\pi}{2}\tan^{-1}(x/2)\bigg]_{-2}^{2} = \pi^2/4$

39. $A = \displaystyle\int_0^{\pi/2} (1 - \sin y)\,dy$

$\qquad = (y + \cos y)]_0^{\pi/2} = \pi/2 - 1$

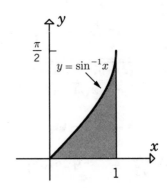

41. $\theta = \pi - (\alpha + \beta)$

$\qquad = \pi - \cot^{-1}(x - 2) - \cot^{-1}\dfrac{5 - x}{4},$

$\dfrac{d\theta}{dx} = \dfrac{1}{1 + (x - 2)^2} + \dfrac{-1/4}{1 + (5 - x)^2/16}$

$\qquad = -\dfrac{3(x^2 - 2x - 7)}{[1 + (x - 2)^2][16 + (5 - x)^2]}$

$d\theta/dx = 0$ when $x = \dfrac{2 \pm \sqrt{4 + 28}}{2} = 1 \pm 2\sqrt{2}$,

only $1 + 2\sqrt{2}$ is in $[2, 5]$; $d\theta/dx > 0$ for x in $[2, 1 + 2\sqrt{2})$,

$d\theta/dx < 0$ for x in $(1 + 2\sqrt{2}, 5]$, θ is maximum when $x = 1 + 2\sqrt{2}$.

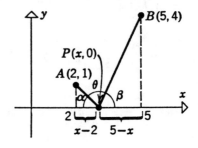

43. $\theta = \tan^{-1}(x/3)$

$\dfrac{d\theta}{dt} = \dfrac{3}{9 + x^2}\dfrac{dx}{dt}, \dfrac{dx}{dt} = \dfrac{9 + x^2}{3}\dfrac{d\theta}{dt}$

$\dfrac{dx}{dt}\bigg|_{x=2} = \dfrac{9 + 4}{3}(4\pi) = 52\pi/3 \text{ mi/min}$

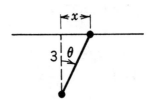

45. $\theta = \alpha - \beta$

$\qquad = \cot^{-1}(x/12) - \cot^{-1}(x/2)$

$\dfrac{d\theta}{dx} = -\dfrac{12}{144 + x^2} + \dfrac{2}{4 + x^2}$

$\qquad = \dfrac{10(24 - x^2)}{(144 + x^2)(4 + x^2)}$

$d\theta/dx = 0$ when $x = \sqrt{24} = 2\sqrt{6}$, by

the first derivative test θ is

maximum there.

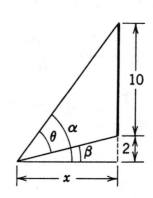

47. By the Mean-Value Theorem on the interval $[0, x]$,

$$\frac{\tan^{-1} x - \tan^{-1} 0}{x - 0} = \frac{\tan^{-1} x}{x} = \frac{1}{1 + c^2} \text{ for } c \text{ in } (0, x), \text{ but}$$

$$\frac{1}{1 + x^2} < \frac{1}{1 + c^2} < 1 \text{ for } c \text{ in } (0, x) \text{ so } \frac{1}{1 + x^2} < \frac{\tan^{-1} x}{x} < 1, \frac{x}{1 + x^2} < \tan^{-1} x < x.$$

49. (a) $A = \displaystyle\int_0^{0.8} \frac{1}{\sqrt{1 - x^2}} dx = \sin^{-1} x \Big]_0^{0.8} = \sin^{-1}(0.8)$

(b) The calculator was in degree mode instead of radian mode; the correct answer is 0.93.

EXERCISE SET 8.3

1. (a) let $y = \cosh^{-1} x$, then $x = \cosh y = \dfrac{1}{2}(e^y + e^{-y})$, $e^y - 2x + e^{-y} = 0$, $e^{2y} - 2xe^y + 1 = 0$,

$e^y = \dfrac{2x \pm \sqrt{4x^2 - 4}}{2} = x \pm \sqrt{x^2 - 1}$. To determine which sign to take, note that $y \geq 0$
so $e^{-y} \leq e^y$, $x = (e^y + e^{-y})/2 \leq (e^y + e^y)/2 = e^y$, hence $e^y \geq x$ thus $e^y = x + \sqrt{x^2 - 1}$,
$y = \cosh^{-1} x = \ln(x + \sqrt{x^2 - 1})$.

(b) $\dfrac{d}{dx}(\cosh^{-1} x) = \dfrac{1 + x/\sqrt{x^2 - 1}}{x + \sqrt{x^2 - 1}} = 1/\sqrt{x^2 - 1}$

3. (a) let $y = \text{sech}^{-1} x$ then $x = \text{sech } y = 1/\cosh y$, $\cosh y = 1/x$, $y = \cosh^{-1}(1/x)$; the proofs
for the remaining two are similar.

(b) $\dfrac{d}{dx}(\text{sech}^{-1} x) = \dfrac{d}{dx}(\cosh^{-1}(1/x)) = \dfrac{(-1/x^2)}{\sqrt{1/x^2 - 1}} = -\dfrac{1}{x\sqrt{1 - x^2}}$; the remaining two are
done similarly.

(c) $\text{sech}^{-1} x = \cosh^{-1}(1/x) = \ln\left[\dfrac{1}{x} + \sqrt{\dfrac{1}{x^2} - 1}\right] = \ln\left[\dfrac{1 + \sqrt{1 - x^2}}{x}\right]$; the remaining two
are done similarly.

5. (a) $\ln(3 + \sqrt{8})$ **(b)** $\ln(\sqrt{5} - 2)$

7. (a) $\dfrac{1}{\sqrt{1 + x^2/9}}\left(\dfrac{1}{3}\right) = 1/\sqrt{9 + x^2}$ **(b)** $2/\sqrt{(2x + 1)^2 - 1}$

9. (a) $-\dfrac{7x^6}{x^7\sqrt{1 - x^{14}}} = -\dfrac{7}{x\sqrt{1 - x^{14}}}$ **(b)** $-1/\sqrt{1 + e^{2x}}$

11. (a) $\dfrac{1}{\sqrt{1 + 1/x^2}}(-1/x^2) = -\dfrac{1}{|x|\sqrt{x^2 + 1}}$

(b) $\dfrac{\sinh x}{\sqrt{\cosh^2 x - 1}} = \dfrac{\sinh x}{|\sinh x|} = \begin{cases} 1, & x > 0 \\ -1, & x < 0 \end{cases}$

13. (a) $-\dfrac{e^x}{x\sqrt{1 - x^2}} + e^x \operatorname{sech}^{-1} x$

(b) $3x^2(\sinh^{-1} x)^2/\sqrt{1 + x^2} + 2x(\sinh^{-1} x)^3$

15. (a) $\dfrac{1}{1 - (1 - x)^2/(1 + x)^2}\left[-\dfrac{2}{(1 + x)^2}\right] = -1/(2x)$

(b) $10(1 + x\operatorname{csch}^{-1} x)^9\left(-\dfrac{x}{|x|\sqrt{1 + x^2}} + \operatorname{csch}^{-1} x\right)$

17. $x = \sqrt{2}u, \displaystyle\int \dfrac{\sqrt{2}}{\sqrt{2u^2 - 2}} du = \int \dfrac{1}{\sqrt{u^2 - 1}} du = \cosh^{-1}(x/\sqrt{2}) + C$

19. $u = e^x, \displaystyle\int \dfrac{1}{u\sqrt{1 - u^2}} du = -\operatorname{sech}^{-1}(e^x) + C$

21. $u = x^3, \dfrac{1}{3}\displaystyle\int \dfrac{1}{u\sqrt{1 + u^2}} du = -\dfrac{1}{3}\operatorname{csch}^{-1}|x^3| + C$

23. $\coth^{-1} x\Big]_2^3 = \coth^{-1} 3 - \coth^{-1} 2$

$\qquad = \dfrac{1}{2}\ln\dfrac{3 + 1}{3 - 1} - \dfrac{1}{2}\ln\dfrac{2 + 1}{2 - 1} = \dfrac{1}{2}\ln 2 - \dfrac{1}{2}\ln 3 \approx \dfrac{1}{2}(0.6931 - 1.0986) \approx -0.2028$

25. (a)   **(b)**

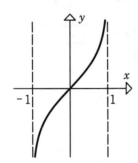

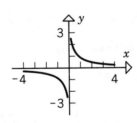

27. $\dfrac{d}{dx}(\operatorname{sech}^{-1}|x|) = \dfrac{d}{dx}(\operatorname{sech}^{-1}\sqrt{x^2}) = -\dfrac{1}{\sqrt{x^2}\sqrt{1 - x^2}}\dfrac{x}{\sqrt{x^2}} = -\dfrac{1}{x\sqrt{1 - x^2}}$

29.   If $-1 < x < 1$ then $\dfrac{1+x}{1-x} > 0$, $\tanh^{-1} x = \dfrac{1}{2}\ln\dfrac{1+x}{1-x} = \dfrac{1}{2}\ln\left|\dfrac{1+x}{1-x}\right|$; if $|x| > 1$ then

$\dfrac{x+1}{x-1} > 0$, but $\dfrac{x+1}{x-1} = \left|\dfrac{1+x}{1-x}\right|$ so $\coth^{-1} x = \dfrac{1}{2}\ln\dfrac{x+1}{x-1} = \dfrac{1}{2}\ln\left|\dfrac{1+x}{1-x}\right|$,

$\displaystyle\int \dfrac{1}{1-x^2}dx = \dfrac{1}{2}\ln\left|\dfrac{1+x}{1-x}\right| + C$

31.   Let $y = \sinh^{-1} x$ then $x = \sinh y$, $\dfrac{dy}{dx} = \dfrac{1}{dx/dy} = \dfrac{1}{\cosh y} = \dfrac{1}{\sqrt{1+\sinh^2 y}} = \dfrac{1}{\sqrt{1+x^2}}$.

33.   Let $u = -x$, $\displaystyle\int \dfrac{1}{\sqrt{x^2-1}}dx = -\int \dfrac{1}{\sqrt{u^2-1}}du = -\cosh^{-1} u + C = -\cosh^{-1}(-x) + C$.

35.   **(a)** $vdv = 16x\,dx$, $v^2/2 = 8x^2 + C$; $v = 0$ when $x = 1/4$ so $C = -1/2$, $v^2/2 = 8x^2 - 1/2$,
      $v^2 = 16x^2 - 1$, $v = \sqrt{16x^2 - 1}$.

      (b) $\dfrac{dx}{dt} = \sqrt{16x^2 - 1}$, $\dfrac{1}{\sqrt{16x^2 - 1}}dx = dt$, $\dfrac{1}{4}\cosh^{-1}(4x) = t + C$; $x = 1/4$ when $t = 0$ so
      $C = \dfrac{1}{4}\cosh^{-1}(1) = 0$, $\dfrac{1}{4}\cosh^{-1}(4x) = t$, $x = \dfrac{1}{4}\cosh(4t)$. If $x = 2$, then
      $t = \dfrac{1}{4}\cosh^{-1}(8) \approx 0.7$ sec.

SUPPLEMENTARY EXERCISES CHAPTER 8

1.   **(a)** $2\pi/3$ **(b)** $3/4$
     (c) $\cos[\sin^{-1}(4/5)] = 3/5$
     (d) $\cos[\sin^{-1}(-4/5)] = \cos[-\sin^{-1}(4/5)] = \cos[\sin^{-1}(4/5)] = 3/5$

3.   **(a)** $\pi/4$
     (b) $\sin^{-1}[\sin(5\pi/4)] = \sin^{-1}[\sin(-\pi/4)] = -\pi/4$
     (c) $\tan(\sec^{-1} 5) = 2\sqrt{6}$
     (d) $\tan^{-1}[\cot(\pi/6)] = \tan^{-1}(\sqrt{3}) = \pi/3$

5.   **(a)** let $\alpha = \cos^{-1}(4/5)$, $\beta = \sin^{-1}(5/13)$
          $\cos(\alpha + \beta) = \cos\alpha\cos\beta - \sin\alpha\sin\beta = (4/5)(12/13) - (3/5)(5/13) = 33/65$
     (b) let $\alpha = \sin^{-1}(4/5)$, $\beta = \cos^{-1}(5/13)$
          $\sin(\alpha + \beta) = \sin\alpha\cos\beta + \cos\alpha\sin\beta = (4/5)(5/13) + (3/5)(12/13) = 56/65$

(c) let $\alpha = \tan^{-1}(1/3)$, $\beta = \tan^{-1}(2)$; $\tan(\alpha + \beta) = \dfrac{\tan\alpha + \tan\beta}{1 - \tan\alpha\tan\beta} = \dfrac{1/3 + 2}{1 - (1/3)(2)} = 7$

7. $\tanh u = -3/5$, $\operatorname{sech}^2 u = 1 - \tanh^2 u = 16/25$, $\operatorname{sech} u = 4/5$, $\cosh u = 5/4$,
$\sinh u = (\tanh u)(\cosh u) = -3/4$, $\cosh 2u = \cosh^2 u + \sinh^2 u = 34/16$.

9.   **(a)**

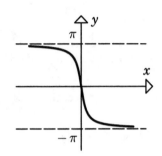

(b)

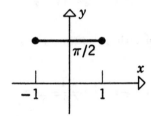

$f(x) = \cos^{-1} x + \sin^{-1} x = \pi/2$

11. $-[\sec^{-1}(x^2)]^{-2}\dfrac{1}{x^2\sqrt{x^4 - 1}}(2x) = -\dfrac{2[\sec^{-1}(x^2)]^{-2}}{x\sqrt{x^4 - 1}}$

13. $(\sec x \tan x)/\sqrt{\sec^2 x - 1} = (\sec x \tan x)/|\tan x|$

15. $1/[1 - (\ln x)^2] + \tanh^{-1}(\ln x)$ **17.** $3/[2\sqrt{\sin^{-1} 3x}\sqrt{1 - 9x^2}]$

19. $\exp(\sec^{-1} x)/(x\sqrt{x^2 - 1})$ **21.** $\pi^{\sin^{-1} x}(\ln \pi)/\sqrt{1 - x^2}$

23. $y = \tanh^{-1}(1/\coth x) = \tanh^{-1}(\tanh x)$ if $x \neq 0$, so $y = x$, $dy/dx = 1$ if $x \neq 0$.

25. $y = \tan^{-1} x$, $y' = 1/(1 + x^2)$, $y'' = -2x/(1 + x^2)^2$; $\sin y = x/\sqrt{1 + x^2}$, $\cos y = 1/\sqrt{1 + x^2}$,
$-2\sin y \cos^3 y = -2\dfrac{x}{(1 + x^2)^{1/2}}\dfrac{1}{(1 + x^2)^{3/2}} = -2x/(1 + x^2)^2 = y''$

27. $u = e^x$, $\displaystyle\int \dfrac{1}{1 - u^2}du = \dfrac{1}{2}\ln\left|\dfrac{1 + e^x}{1 - e^x}\right| + C$

29. $u = \ln x$, $\displaystyle\int \dfrac{1}{\sqrt{u^2 - 1}}du = \cosh^{-1}(\ln x) + C$, $x > e$

31. $u = 3x$, $\dfrac{1}{3}\displaystyle\int_{2/\sqrt{3}}^{2} \dfrac{1}{u\sqrt{u^2 - 1}}du = \dfrac{1}{3}\sec^{-1} u\bigg]_{2/\sqrt{3}}^{2} = \dfrac{1}{3}(\pi/3 - \pi/6) = \pi/18$

33. $\displaystyle\int \dfrac{1}{x^{1/2} + x^{3/2}}dx = \int \dfrac{1}{x^{1/2}(1 + x)}dx$, $u = x^{1/2}$, $2\displaystyle\int \dfrac{1}{1 + u^2}du = 2\tan^{-1}\sqrt{x} + C$

35. $u = \sqrt{x}$, $2\displaystyle\int_{1/2}^{1/\sqrt{2}} \frac{1}{\sqrt{1-u^2}}du = 2\sin^{-1}u\Big]_{1/2}^{1/\sqrt{2}} = 2\left(\dfrac{\pi}{4} - \dfrac{\pi}{6}\right) = \pi/6$

37. $f'(x) = \dfrac{2}{1+4x^2} - \dfrac{1}{1+x^2} = \dfrac{1-2x^2}{(1+4x^2)(1+x^2)}$, the critical point is $x = 1/\sqrt{2}$ for $x > 0$, $f'(x) > 0$ for $0 < x < 1/\sqrt{2}$ and $f'(x) < 0$ for $x > 1/\sqrt{2}$ so $f(x)$ assumes its maximum value at $x = 1/\sqrt{2}$

CHAPTER 9
Techniques of Integration

EXERCISE SET 9.1

1. $3\displaystyle\int \frac{x}{4x-1}dx = \frac{3}{16}(4x + \ln|4x-1|) + C$ (#6; $a = -1$, $b = 4$)

3. $\dfrac{1}{5}\ln\left|\dfrac{x}{2x+5}\right| + C$ (#11; $a = 5$, $b = 2$)

5. $\dfrac{1}{30}(6x+6)(2x-3)^{3/2} + C = \dfrac{1}{5}(x+1)(2x-3)^{3/2} + C$ (#14; $a = -3$, $b = 2$)

7. $\dfrac{1}{2}\ln\left|\dfrac{\sqrt{4-3x}-2}{\sqrt{4-3x}+2}\right| + C$ (#20; $a = 4$, $b = -3$)

9. $\dfrac{1}{2\sqrt{5}}\ln\left|\dfrac{x+\sqrt{5}}{x-\sqrt{5}}\right| + C$ (#25; $a = \sqrt{5}$)

11. $\dfrac{x}{2}\sqrt{x^2-3} - \dfrac{3}{2}\ln\left|x + \sqrt{x^2-3}\right| + C$ (#28; $a^2 = 3$)

13. $\dfrac{x}{2}\sqrt{x^2+4} - 2\ln\left|x + \sqrt{x^2+4}\right| + C$ (#34; $a^2 = 4$)

15. $\dfrac{x}{2}\sqrt{9-x^2} + \dfrac{9}{2}\sin^{-1}\dfrac{x}{3} + C$ (#41; $a = 3$)

17. $\sqrt{3-x^2} - \sqrt{3}\ln\left|\dfrac{\sqrt{3}+\sqrt{3-x^2}}{x}\right| + C$ (#43; $a = \sqrt{3}$)

19. $-\dfrac{\sin 5x}{10} + \dfrac{\sin x}{2} + C$ (#80; $m = 3$, $n = 2$)

21. $\dfrac{x^4}{16}[4\ln x - 1] + C$ (#104; $n = 3$)

23. $\dfrac{e^{-2x}}{13}(-2\sin 3x - 3\cos 3x) + C$ (#106; $a = -2$, $b = 3$)

25. $\dfrac{1}{2}\displaystyle\int \dfrac{u}{(4-3u)^2}\,du = \dfrac{1}{18}\left[\dfrac{4}{4-3u} + \ln|4-3u|\right] + C \quad (\#8;\ a=4, b=-3)$

$\qquad\qquad\qquad = \dfrac{1}{18}\left[\dfrac{4}{4-3e^{2x}} + \ln|4-3e^{2x}|\right] + C$

27. $\dfrac{2}{3}\displaystyle\int \dfrac{1}{u^2+4}\,du = \dfrac{1}{3}\tan^{-1}\dfrac{u}{2} + C \quad (\#24;\ a=2) = \dfrac{1}{3}\tan^{-1}\dfrac{3\sqrt{x}}{2} + C$

29. $\dfrac{1}{3}\displaystyle\int \dfrac{1}{\sqrt{u^2-4}}\,du = \dfrac{1}{3}\ln\left|u+\sqrt{u^2-4}\right| + C \quad (\#27;\ a^2=4) = \dfrac{1}{3}\ln\left|3x+\sqrt{9x^2-4}\right| + C$

31. $\dfrac{1}{54}\displaystyle\int \dfrac{u^2}{\sqrt{5-u^2}}\,du = \dfrac{1}{54}\left[-\dfrac{u}{2}\sqrt{5-u^2} + \dfrac{5}{2}\sin^{-1}\dfrac{u}{\sqrt{5}}\right] + C \quad (\#45;\ a=\sqrt{5})$

$\qquad\qquad\qquad = \dfrac{1}{108}\left[-3x^2\sqrt{5-9x^4} + 5\sin^{-1}\dfrac{3x^2}{\sqrt{5}}\right] + C$

33. $\displaystyle\int \sin^2 u\,du = \dfrac{1}{2}u - \dfrac{1}{4}\sin 2u + C \quad (\#70) = \dfrac{1}{2}\ln x - \dfrac{1}{4}\sin(2\ln x) + C$

35. $\dfrac{1}{4}\displaystyle\int ue^u\,du = \dfrac{1}{4}e^u(u-1) + C \quad (\#98) = \dfrac{1}{4}e^{-2x}(-2x-1) + C = -\dfrac{1}{4}e^{-2x}(2x+1) + C$

37. $u = \cos 3x, \quad -\dfrac{1}{3}\displaystyle\int \dfrac{1}{u(u+1)^2}\,du = -\dfrac{1}{3}\left[\dfrac{1}{u+1} + \ln\left|\dfrac{u}{u+1}\right|\right] + C \quad (\#13;\ a=1, b=1)$

$\qquad\qquad\qquad = -\dfrac{1}{3}\left[\dfrac{1}{\cos 3x+1} + \ln\left|\dfrac{\cos 3x}{\cos 3x+1}\right|\right] + C$

39. $u = 4x^2, \quad \dfrac{1}{8}\displaystyle\int \dfrac{1}{u^2-1}\,du = \dfrac{1}{16}\ln\left|\dfrac{u-1}{u+1}\right| + C \quad (\#26;\ a=1) = \dfrac{1}{16}\ln\left|\dfrac{4x^2-1}{4x^2+1}\right| + C$

41. $u = 2e^x, \quad \dfrac{1}{2}\displaystyle\int \sqrt{3-u^2}\,du = \dfrac{1}{2}\left[\dfrac{u}{2}\sqrt{3-u^2} + \dfrac{3}{2}\sin^{-1}\dfrac{u}{\sqrt{3}}\right] + C \quad (\#41;\ a=\sqrt{3})$

$\qquad\qquad\qquad = \dfrac{1}{2}e^x\sqrt{3-4e^{2x}} + \dfrac{3}{4}\sin^{-1}\dfrac{2e^x}{\sqrt{3}} + C$

43. $u = 3x,$

$\dfrac{1}{3}\displaystyle\int \sqrt{(5/3)u - u^2}\,du = \dfrac{1}{3}\left[\dfrac{u-5/6}{2}\sqrt{(5/3)u-u^2} + \dfrac{25}{72}\sin^{-1}\left(\dfrac{u-5/6}{5/6}\right)\right] + C\ (\#50;\ a=5/6)$

$\qquad\qquad\qquad = \dfrac{18x-5}{36}\sqrt{5x-9x^2} + \dfrac{25}{216}\sin^{-1}\dfrac{18x-5}{5} + C$

45. $u = 3x$, $\dfrac{1}{9}\displaystyle\int u\sin u\,du = \dfrac{1}{9}[\sin u - u\cos u] + C$ (#83) $= \dfrac{1}{9}[\sin 3x - 3x\cos 3x] + C$

47. $u = -\sqrt{x}$, $x = u^2$, $dx = 2u\,du$,

$2\displaystyle\int ue^u\,du = 2e^u(u-1) + C$ (#98) $= 2e^{-\sqrt{x}}(-\sqrt{x}-1) + C = -2e^{-\sqrt{x}}(\sqrt{x}+1) + C$

49. $\displaystyle\int \dfrac{1}{(x+2)^2 - 9}\,dx$; $u = x + 2$,

$\displaystyle\int \dfrac{1}{u^2-9}\,du = \dfrac{1}{6}\ln\left|\dfrac{u-3}{u+3}\right| + C$ (#26; $a = 3$) $= \dfrac{1}{6}\ln\left|\dfrac{x-1}{x+5}\right| + C$

51. $\displaystyle\int \dfrac{x}{\sqrt{9 - (x-2)^2}}\,dx$; $u = x - 2$, $x = u + 2$,

$\displaystyle\int \dfrac{u+2}{\sqrt{9-u^2}}\,du = \int \dfrac{u}{\sqrt{9-u^2}}\,du + 2\int \dfrac{1}{\sqrt{9-u^2}}\,du$

$\qquad\qquad = -\sqrt{9-u^2} + 2\sin^{-1}\dfrac{u}{3} + C$ (#4 and #40)

$\qquad\qquad = -\sqrt{5 + 4x - x^2} + 2\sin^{-1}\dfrac{x-2}{3} + C$

53. $\displaystyle\int_2^x \dfrac{1}{t(4-t)}\,dt = \dfrac{1}{4}\ln\dfrac{t}{4-t}\Big]_2^x$ (#11; $a = 4, b = -1$)

$\qquad\qquad = \dfrac{1}{4}\left[\ln\dfrac{x}{4-x} - \ln 1\right] = \dfrac{1}{4}\ln\dfrac{x}{4-x}, \dfrac{1}{4}\ln\dfrac{x}{4-x} = 0.5, \ln\dfrac{x}{4-x} = 2,$

$\dfrac{x}{4-x} = e^2$, $x = 4e^2 - e^2 x$, $x(1+e^2) = 4e^2$, $x = 4e^2/(1+e^2) \approx 3.523188312.$

55. $A = \displaystyle\int_0^4 \sqrt{25 - x^2}\,dx = \left(\dfrac{1}{2}x\sqrt{25-x^2} + \dfrac{25}{2}\sin^{-1}\dfrac{x}{5}\right)\Big]_0^4$ (#41; $a = 5$)

$\qquad\qquad = 6 + \dfrac{25}{2}\sin^{-1}\dfrac{4}{5} \approx 17.59119022$

57. $A = \displaystyle\int_0^1 \dfrac{1}{25 - 16x^2}\,dx$; $u = 4x$,

$A = \dfrac{1}{4}\displaystyle\int_0^4 \dfrac{1}{25-u^2}\,du = \dfrac{1}{40}\ln\left|\dfrac{u+5}{u-5}\right|\Big]_0^4$ (#25; $a = 5$) $= \dfrac{1}{40}\ln 9 \approx 0.054930614$

59. $V = 2\pi\displaystyle\int_0^{\pi/2} x\cos x\,dx = 2\pi(\cos x + x\sin x)\big]_0^{\pi/2}$ (#84) $= \pi(\pi - 2) \approx 3.586419094$

61. $V = 2\pi \int_0^3 x e^{-x} dx; \ u = -x,$

$V = 2\pi \int_0^{-3} u e^u du = 2\pi e^u (u - 1)]_0^{-3}$ (#98) $= 2\pi(1 - 4e^{-3}) \approx 5.031899801$

63. $L = \int_0^2 \sqrt{1 + 16x^2} \, dx; \ u = 4x,$

$L = \frac{1}{4} \int_0^8 \sqrt{1 + u^2} \, du = \frac{1}{4} \left(\frac{u}{2} \sqrt{1 + u^2} + \frac{1}{2} \ln \left| u + \sqrt{1 + u^2} \right| \right) \Big]_0^8$ (#28; $a^2 = 1$)

$= \sqrt{65} + \frac{1}{8} \ln(8 + \sqrt{65}) \approx 8.409316783$

65. $S = 2\pi \int_0^\pi (\sin x)\sqrt{1 + \cos^2 x} \, dx; \ u = \cos x,$

$S = -2\pi \int_1^{-1} \sqrt{1 + u^2} \, du = 4\pi \int_0^1 \sqrt{1 + u^2} \, du$

$= 4\pi \left(\frac{u}{2}\sqrt{1 + u^2} + \frac{1}{2} \ln \left| u + \sqrt{1 + u^2} \right| \right) \Big]_0^1$ (#28; $a^2 = 1$)

$= 2\pi[\sqrt{2} + \ln(1 + \sqrt{2})] \approx 14.42359944$

EXERCISE SET 9.2

1. $u = x, \ dv = e^{-x}dx, \ du = dx, \ v = -e^{-x}$

$\int x e^{-x} dx = -x e^{-x} + \int e^{-x} dx = -x e^{-x} - e^{-x} + C$

3. $u = x^2, \ dv = e^x dx, \ du = 2x \, dx, \ v = e^x; \ \int x^2 e^x dx = x^2 e^x - 2 \int x e^x dx.$

For $\int x e^x dx$ use $u = x, \ dv = e^x dx, \ du = dx, \ v = e^x$ to get

$\int x e^x dx = x e^x - e^x + C_1$ so $\int x^2 e^x dx = x^2 e^x - 2x e^x + 2e^x + C$

5. $u = x, \ dv = \sin 2x \, dx, \ du = dx, \ v = -\frac{1}{2} \cos 2x$

$\int x \sin 2x \, dx = -\frac{1}{2} x \cos 2x + \frac{1}{2} \int \cos 2x \, dx = -\frac{1}{2} x \cos 2x + \frac{1}{4} \sin 2x + C$

7. $u = x^2$, $dv = \cos x\, dx$, $du = 2x\, dx$, $v = \sin x$; $\displaystyle\int x^2 \cos x\, dx = x^2 \sin x - 2\int x \sin x\, dx$

For $\displaystyle\int x \sin x\, dx$ use $u = x$, $dv = \sin x\, dx$ to get

$$\int x \sin x\, dx = -x \cos x + \sin x + C_1 \text{ so } \int x^2 \cos x\, dx = x^2 \sin x + 2x \cos x - 2\sin x + C$$

9. $u = \ln x$, $dv = \sqrt{x}\, dx$, $du = \dfrac{1}{x}dx$, $v = \dfrac{2}{3}x^{3/2}$

$$\int \sqrt{x} \ln x\, dx = \frac{2}{3}x^{3/2} \ln x - \frac{2}{3}\int x^{1/2}dx = \frac{2}{3}x^{3/2} \ln x - \frac{4}{9}x^{3/2} + C$$

11. $u = (\ln x)^2$, $dv = dx$, $du = 2\dfrac{\ln x}{x}dx$, $v = x$; $\displaystyle\int (\ln x)^2 dx = x(\ln x)^2 - 2\int \ln x\, dx$.

Use $u = \ln x$, $dv = dx$ to get

$$\int \ln x\, dx = x \ln x - \int dx = x \ln x - x + C_1 \text{ so } \int (\ln x)^2 dx = x(\ln x)^2 - 2x \ln x + 2x + C$$

13. $u = \ln(2x + 3)$, $dv = dx$, $du = \dfrac{2}{2x+3}dx$, $v = x$

$$\int \ln(2x+3)dx = x \ln(2x+3) - \int \frac{2x}{2x+3}dx$$

but $\displaystyle\int \frac{2x}{2x+3}dx = \int \left(1 - \frac{3}{2x+3}\right)dx = x - \frac{3}{2}\ln(2x+3) + C_1$ so

$$\int \ln(2x+3)dx = x \ln(2x+3) - x + \frac{3}{2}\ln(2x+3) + C$$

15. $u = \sin^{-1} x$, $dv = dx$, $du = 1/\sqrt{1-x^2}dx$, $v = x$

$$\int \sin^{-1} x\, dx = x \sin^{-1} x - \int x/\sqrt{1-x^2}dx = x \sin^{-1} x + \sqrt{1-x^2} + C$$

17. $u = \tan^{-1}(2x)$, $dv = dx$, $du = \dfrac{2}{1+4x^2}dx$, $v = x$

$$\int \tan^{-1}(2x)dx = x \tan^{-1}(2x) - \int \frac{2x}{1+4x^2}dx = x \tan^{-1}(2x) - \frac{1}{4}\ln(1+4x^2) + C$$

19. $u = e^x$, $dv = \sin x\, dx$, $du = e^x dx$, $v = -\cos x$; $\displaystyle\int e^x \sin x\, dx = -e^x \cos x + \int e^x \cos x\, dx$.

For $\displaystyle\int e^x \cos x\, dx$ use $u = e^x$, $dv = \cos x\, dx$ to get $\displaystyle\int e^x \cos x = e^x \sin x - \int e^x \sin x\, dx$ so

$$\int e^x \sin x \, dx = -e^x \cos x + e^x \sin x - \int e^x \sin x \, dx,$$

$$2 \int e^x \sin x \, dx = e^x(\sin x - \cos x) + C_1, \quad \int e^x \sin x \, dx = \frac{1}{2}e^x(\sin x - \cos x) + C$$

21. $u = e^{ax}$, $dv = \sin bx \, dx$, $du = ae^{ax}dx$, $v = -\frac{1}{b}\cos bx$

$$\int e^{ax} \sin bx \, dx = -\frac{1}{b}e^{ax}\cos bx + \frac{a}{b}\int e^{ax}\cos bx \, dx. \text{ Use } u = e^{ax}, dv = \cos bx \, dx \text{ to get}$$

$$\int e^{ax} \cos bx \, dx = \frac{1}{b}e^{ax}\sin bx - \frac{a}{b}\int e^{ax}\sin bx \, dx \text{ so}$$

$$\int e^{ax} \sin bx \, dx = -\frac{1}{b}e^{ax}\cos bx + \frac{a}{b^2}e^{ax}\sin bx - \frac{a^2}{b^2}\int e^{ax}\sin bx \, dx,$$

$$\int e^{ax} \sin bx \, dx = \frac{e^{ax}}{a^2+b^2}(a\sin bx - b\cos bx) + C$$

23. $u = \sin(\ln x)$, $dv = dx$, $du = \dfrac{\cos(\ln x)}{x}dx$, $v = x$

$$\int \sin(\ln x)dx = x\sin(\ln x) - \int \cos(\ln x)dx. \text{ Use } u = \cos(\ln x), dv = dx \text{ to get}$$

$$\int \cos(\ln x)dx = x\cos(\ln x) + \int \sin(\ln x)dx \text{ so}$$

$$\int \sin(\ln x)dx = x\sin(\ln x) - x\cos(\ln x) - \int \sin(\ln x)dx,$$

$$\int \sin(\ln x)dx = (x/2)[\sin(\ln x) - \cos(\ln x)] + C$$

25. $u = x$, $dv = \sec^2 x \, dx$, $du = dx$, $v = \tan x$

$$\int x \sec^2 x \, dx = x\tan x - \int \tan x \, dx = x\tan x - \int \frac{\sin x}{\cos x}dx = x\tan x + \ln|\cos x| + C$$

27. $u = x^2$, $dv = xe^{x^2}dx$, $du = 2x \, dx$, $v = \frac{1}{2}e^{x^2}$

$$\int x^3 e^{x^2} \, dx = \frac{1}{2}x^2 e^{x^2} - \int xe^{x^2}dx = \frac{1}{2}x^2 e^{x^2} - \frac{1}{2}e^{x^2} + C$$

29. $u = x$, $dv = e^{-5x}dx$, $du = dx$, $v = -\frac{1}{5}e^{-5x}$

$$\int_0^1 xe^{-5x}dx = -\frac{1}{5}xe^{-5x}\Big]_0^1 + \frac{1}{5}\int_0^1 e^{-5x}dx$$

$$= -\frac{1}{5}e^{-5} - \frac{1}{25}e^{-5x}\Big]_0^1 = -\frac{1}{5}e^{-5} - \frac{1}{25}(e^{-5} - 1) = (1 - 6e^{-5})/25$$

31. $u = \ln x,\ dv = x^2 dx,\ du = \dfrac{1}{x}dx,\ v = \dfrac{1}{3}x^3$

$$\int_1^e x^2 \ln x\, dx = \dfrac{1}{3}x^3 \ln x\Big]_1^e - \dfrac{1}{3}\int_1^e x^2 dx = \dfrac{1}{3}e^3 - \dfrac{1}{9}x^3\Big]_1^e = \dfrac{1}{3}e^3 - \dfrac{1}{9}(e^3 - 1) = (2e^3 + 1)/9$$

33. $u = \ln(x+3),\ dv = dx,\ du = \dfrac{1}{x+3}dx,\ v = x$

$$\int_{-2}^2 \ln(x+3)dx = x\ln(x+3)]_{-2}^2 - \int_{-2}^2 \dfrac{x}{x+3}dx = 2\ln 5 + 2\ln 1 - \int_{-2}^2 \left[1 - \dfrac{3}{x+3}\right]dx$$

$$= 2\ln 5 - [x - 3\ln(x+3)]_{-2}^2 = 2\ln 5 - (2 - 3\ln 5) + (-2 - 3\ln 1) = 5\ln 5 - 4$$

35. $u = \sec^{-1}\sqrt{\theta},\ dv = d\theta,\ du = \dfrac{1}{2\theta\sqrt{\theta-1}}d\theta,\ v = \theta$

$$\int_2^4 \sec^{-1}\sqrt{\theta}\,d\theta = \theta\sec^{-1}\sqrt{\theta}\Big]_2^4 - \dfrac{1}{2}\int_2^4 \dfrac{1}{\sqrt{\theta-1}}d\theta = 4\sec^{-1}2 - 2\sec^{-1}\sqrt{2} - \sqrt{\theta-1}\Big]_2^4$$

$$= 4\left(\dfrac{\pi}{3}\right) - 2\left(\dfrac{\pi}{4}\right) - \sqrt{3} + 1 = \dfrac{5\pi}{6} - \sqrt{3} + 1$$

37. $u = x,\ dv = \sin 4x\, dx,\ du = dx,\ v = -\dfrac{1}{4}\cos 4x$

$$\int_0^{\pi/2} x\sin 4x\, dx = -\dfrac{1}{4}x\cos 4x\Big]_0^{\pi/2} + \dfrac{1}{4}\int_0^{\pi/2}\cos 4x\, dx = -\pi/8 + \dfrac{1}{16}\sin 4x\Big]_0^{\pi/2} = -\pi/8$$

39. $u = \tan^{-1}\sqrt{x},\ dv = \sqrt{x}dx,\ du = \dfrac{1}{2\sqrt{x}(1+x)}dx,\ v = \dfrac{2}{3}x^{3/2}$

$$\int_1^3 \sqrt{x}\tan^{-1}\sqrt{x}dx = \dfrac{2}{3}x^{3/2}\tan^{-1}\sqrt{x}\Big]_1^3 - \dfrac{1}{3}\int_1^3 \dfrac{x}{1+x}dx$$

$$= \dfrac{2}{3}x^{3/2}\tan^{-1}\sqrt{x}\Big]_1^3 - \dfrac{1}{3}\int_1^3 \left[1 - \dfrac{1}{1+x}\right]dx$$

$$= \left[\dfrac{2}{3}x^{3/2}\tan^{-1}\sqrt{x} - \dfrac{1}{3}x + \dfrac{1}{3}\ln|1+x|\right]_1^3 = (2\sqrt{3}\pi - \pi/2 - 2 + \ln 2)/3$$

41. $u = x^2,\ dv = \dfrac{x}{\sqrt{x^2+1}}dx,\ du = 2x\, dx,\ v = \sqrt{x^2+1}$

$$\int_0^1 \dfrac{x^3}{\sqrt{x^2+1}}dx = x^2\sqrt{x^2+1}\Big]_0^1 - 2\int_0^1 x(x^2+1)^{1/2}dx$$

$$= \sqrt{2} - \dfrac{2}{3}(x^2+1)^{3/2}\Big]_0^1 = \sqrt{2} - \dfrac{2}{3}[2\sqrt{2} - 1] = (2 - \sqrt{2})/3$$

43. **(a)** $A = \displaystyle\int_1^e \ln x \, dx = (x \ln x - x) \Big]_1^e = 1$

(b) $V = \pi \displaystyle\int_1^e (\ln x)^2 dx = \pi(x(\ln x)^2 - 2x \ln x + 2x) \Big]_1^e = \pi(e - 2)$

45. $V = 2\pi \displaystyle\int_0^\pi x \sin x \, dx = 2\pi(-x \cos x + \sin x) \Big]_0^\pi = 2\pi^2$

47. **(a)** $\displaystyle\int \sin^3 x \, dx = -\frac{1}{3} \sin^2 x \cos x + \frac{2}{3} \int \sin x \, dx = -\frac{1}{3} \sin^2 x \cos x - \frac{2}{3} \cos x + C$

(b) $\displaystyle\int \sin^4 x \, dx = -\frac{1}{4} \sin^3 x \cos x + \frac{3}{4} \int \sin^2 x \, dx,$

$\displaystyle\int \sin^2 x \, dx = -\frac{1}{2} \sin x \cos x + \frac{1}{2} x + C_1$ so

$\displaystyle\int_0^{\pi/4} \sin^4 x \, dx = -\frac{1}{4} \sin^3 x \cos x - \frac{3}{8} \sin x \cos x + \frac{3}{8} x \Big]_0^{\pi/4}$

$\qquad = -\frac{1}{4}(1/\sqrt{2})^3(1/\sqrt{2}) - \frac{3}{8}(1/\sqrt{2})(1/\sqrt{2}) + 3\pi/32 = 3\pi/32 - 1/4$

49. **(a)** $u = 5x,$

$\displaystyle\int \cos^3 5x \, dx = \frac{1}{5} \int \cos^3 u \, du = \frac{1}{5} \left[\frac{1}{3} \cos^2 u \sin u + \frac{2}{3} \int \cos u \, du \right]$

$\qquad = \frac{1}{15} \cos^2 u \sin u + \frac{2}{15} \sin u + C = \frac{1}{15} \cos^2 5x \sin 5x + \frac{2}{15} \sin 5x + C$

(b) $u = x^2,$

$\displaystyle\int x \cos^4(x^2) dx = \frac{1}{2} \int \cos^4 u \, du = \frac{1}{2} \left[\frac{1}{4} \cos^3 u \sin u + \frac{3}{4} \int \cos^2 u \, du \right]$

$\qquad = \frac{1}{8} \cos^3 u \sin u + \frac{3}{8} \left[\frac{1}{2} \cos u \sin u + \frac{1}{2} \int du \right]$

$\qquad = \frac{1}{8} \cos^3 u \sin u + \frac{3}{16} \cos u \sin u + \frac{3}{16} u + C$

$\qquad = \frac{1}{8} \cos^3(x^2) \sin(x^2) + \frac{3}{16} \cos(x^2) \sin(x^2) + \frac{3}{16} x^2 + C$

51. $u = \sin^{n-1} x,\ dv = \sin x\, dx,\ du = (n-1)\sin^{n-2} x \cos x\, dx,\ v = -\cos x$

$$\int \sin^n x\, dx = -\sin^{n-1} x \cos x + (n-1)\int \sin^{n-2} x \cos^2 x\, dx$$

$$= -\sin^{n-1} x \cos x + (n-1)\int \sin^{n-2} x (1-\sin^2 x)\, dx$$

$$= -\sin^{n-1} x \cos x + (n-1)\int \sin^{n-2} x\, dx - (n-1)\int \sin^n x\, dx,$$

$$n\int \sin^n x\, dx = -\sin^{n-1} x \cos x + (n-1)\int \sin^{n-2} x\, dx,$$

$$\int \sin^n x\, dx = -\frac{1}{n}\sin^{n-1} x \cos x + \frac{n-1}{n}\int \sin^{n-2} x\, dx$$

53. (a) $u = x^n,\ dv = e^x dx,\ du = nx^{n-1} dx,\ v = e^x;\ \int x^n e^x dx = x^n e^x - n\int x^{n-1} e^x dx$

(b) $\int x^3 e^x dx = x^3 e^x - 3\int x^2 e^x dx = x^3 e^x - 3\left[x^2 e^x - 2\int x e^x dx\right]$

$$= x^3 e^x - 3x^2 e^x + 6\left[x e^x - \int e^x dx\right] = x^3 e^x - 3x^2 e^x + 6x e^x - 6e^x + C$$

55. $u = x,\ dv = f''(x)dx,\ du = dx,\ v = f'(x)$

$$\int_{-1}^{1} x f''(x)dx = x f'(x)\big]_{-1}^{1} - \int_{-1}^{1} f'(x)dx$$

$$= f'(1) - f'(-1) - f(x)\big]_{-1}^{1} = f'(1) - f'(-1) + f(-1) - f(1)$$

57. $du = -(1/x^2)dx,\ v = x;\ \int \frac{1}{x}dx = 1 + \int \frac{1}{x}dx$ so $0 = 1$???

The "obvious" cancellation of the indefinite integrals causes the problem. Instead, proceed as follows:

$$\int \frac{1}{x}dx - \int \frac{1}{x}dx = 1,\ \int (0)dx = 1,\ \text{but}\ \int (0)dx = C\ \text{so}\ C = 1.$$

EXERCISE SET 9.3

1. $u = \cos x,\ -\int u^5 du = -\frac{1}{6}\cos^6 x + C$ **3.** $u = \sin ax,\ \frac{1}{a}\int u\, du = \frac{1}{2a}\sin^2 ax + C$

5. $\int \sin^2 5\theta\, d\theta = \frac{1}{2}\int (1-\cos 10\theta)d\theta = \frac{1}{2}\theta - \frac{1}{20}\sin 10\theta + C$

7. $\displaystyle\int \cos^4(x/4)dx = \frac{1}{4}\int [1 + \cos(x/2)]^2 dx = \frac{1}{4}\int [1 + 2\cos(x/2) + \cos^2(x/2)]dx$

$\displaystyle = \frac{1}{4}\int \left[1 + 2\cos(x/2) + \frac{1}{2}(1 + \cos x)\right] dx = \frac{1}{4}\int \left[\frac{3}{2} + 2\cos(x/2) + \frac{1}{2}\cos x\right] dx$

$\displaystyle = \frac{3}{8}x + \sin(x/2) + \frac{1}{8}\sin x + C$

9. $\displaystyle\int \cos^5 \theta\, d\theta = \int (1 - \sin^2 \theta)^2 \cos\theta\, d\theta = \int (1 - 2\sin^2 \theta + \sin^4 \theta)\cos\theta\, d\theta$

$\displaystyle = \sin\theta - \frac{2}{3}\sin^3 \theta + \frac{1}{5}\sin^5 \theta + C$

11. $\displaystyle\int \sin^2 2t \cos^3 2t\, dt = \int \sin^2 2t(1 - \sin^2 2t)\cos 2t\, dt = \int (\sin^2 2t - \sin^4 2t)\cos 2t\, dt$

$\displaystyle = \frac{1}{6}\sin^3 2t - \frac{1}{10}\sin^5 2t + C$

13. $\displaystyle\int \cos^4 x \sin^3 x\, dx = \int \cos^4 x(1 - \cos^2 x)\sin x\, dx$

$\displaystyle = \int (\cos^4 x - \cos^6 x)\sin x\, dx = -\frac{1}{5}\cos^5 x + \frac{1}{7}\cos^7 x + C$

15. $\displaystyle\int \sin^5 \theta \cos^4 \theta\, d\theta = \int (1 - \cos^2 \theta)^2 \cos^4 \theta \sin\theta\, d\theta$

$\displaystyle = \int (\cos^4 \theta - 2\cos^6 \theta + \cos^8 \theta)\sin\theta\, d\theta = -\frac{1}{5}\cos^5 \theta + \frac{2}{7}\cos^7 \theta - \frac{1}{9}\cos^9 \theta + C$

17. $\displaystyle\int \sin^2 x \cos^2 x\, dx = \frac{1}{4}\int (1 - \cos 2x)(1 + \cos 2x)dx = \frac{1}{4}\int (1 - \cos^2 2x)dx = \frac{1}{4}\int \sin^2 2x\, dx$

$\displaystyle = \frac{1}{8}\int (1 - \cos 4x)dx = \frac{1}{8}x - \frac{1}{32}\sin 4x + C$

19. $\displaystyle\int \sin x \cos 2x\, dx = \frac{1}{2}\int (\sin 3x - \sin x)dx = -\frac{1}{6}\cos 3x + \frac{1}{2}\cos x + C$

21. $\displaystyle\int \sin x \cos(x/2)dx = \frac{1}{2}\int [\sin(3x/2) + \sin(x/2)]dx = -\frac{1}{3}\cos(3x/2) - \cos(x/2) + C$

23. $u = \cos x, \; -\displaystyle\int u^{-8}du = 1/(7\cos^7 x) + C$

25. $\int_0^{\pi/4} \cos^3 x \, dx = \int_0^{\pi/4} (1 - \sin^2 x) \cos x \, dx$

$$= \sin x - \frac{1}{3} \sin^3 x \Big]_0^{\pi/4} = (\sqrt{2}/2) - \frac{1}{3}(\sqrt{2}/2)^3 = 5\sqrt{2}/12$$

27. $\int_0^{\pi/3} \sin^4 3x \cos^3 3x \, dx = \int_0^{\pi/3} \sin^4 3x (1 - \sin^2 3x) \cos 3x \, dx = \frac{1}{15} \sin^5 3x - \frac{1}{21} \sin^7 3x \Big]_0^{\pi/3} = 0$

29. $\int_0^{\pi/6} \sin 2x \cos 4x \, dx = \frac{1}{2} \int_0^{\pi/6} (\sin 6x - \sin 2x) dx = -\frac{1}{12} \cos 6x + \frac{1}{4} \cos 2x \Big]_0^{\pi/6}$

$$= [(-1/12)(-1) + (1/4)(1/2)] - [-1/12 + 1/4] = 1/24$$

31. **(a)** $\int_0^{2\pi} \sin mx \cos nx \, dx = \frac{1}{2} \int_0^{2\pi} [\sin(m+n)x + \sin(m-n)x] dx$

$$= -\frac{\cos(m+n)x}{2(m+n)} - \frac{\cos(m-n)x}{2(m-n)} \Big]_0^{2\pi},$$

$m + n$ and $m - n$ are integers so $\cos[(m+n)2\pi] = \cos[(m-n)2\pi] = 1$ thus

$$\int_0^{2\pi} \sin mx \cos nx \, dx = \left[\left(-\frac{1}{2(m+n)} - \frac{1}{2(m-n)}\right) + \left(\frac{1}{2(m+n)} + \frac{1}{2(m-n)}\right)\right] = 0$$

(b) $\int_0^{2\pi} \cos mx \cos nx \, dx = \frac{1}{2} \int_0^{2\pi} [\cos(m+n)x + \cos(m-n)x] dx$

$$= \frac{\sin(m+n)x}{2(m+n)} + \frac{\sin(m-n)x}{2(m-n)} \Big]_0^{2\pi} = 0$$

(c) $\int_0^{2\pi} \sin mx \sin nx \, dx = \frac{1}{2} \int_0^{2\pi} [\cos(m-n)x - \cos(m+n)x] dx$

$$= \frac{\sin(m-n)x}{2(m-n)} + \frac{\sin(m+n)x}{2(m+n)} \Big]_0^{2\pi} = 0$$

33. $V = \pi \int_0^{\pi/4} (\cos^2 x - \sin^2 x) dx = \pi \int_0^{\pi/4} \cos 2x \, dx = \frac{1}{2} \pi \sin 2x \Big]_0^{\pi/4} = \pi/2$

35. **(a)** $\int_0^{\pi/2} \sin^3 x \, dx = \frac{2}{3}$ **(b)** $\int_0^{\pi/2} \sin^4 x \, dx = \frac{1 \cdot 3}{2 \cdot 4} \cdot \frac{\pi}{2} = 3\pi/16$

(c) $\int_0^{\pi/2} \sin^5 x \, dx = \frac{2 \cdot 4}{3 \cdot 5} = 8/15$ **(d)** $\int_0^{\pi/2} \sin^6 x \, dx = \frac{1 \cdot 3 \cdot 5}{2 \cdot 4 \cdot 6} \cdot \frac{\pi}{2} = 5\pi/32$

37. For (6), $\int \sin^3 x \, dx = -\dfrac{1}{3} \sin^2 x \cos x + \dfrac{2}{3} \int \sin x \, dx = -\dfrac{1}{3} \sin^2 x \cos x - \dfrac{2}{3} \cos x + C$,

but $\sin^2 x = 1 - \cos^2 x$ so

$$\int \sin^3 x \, dx = -\frac{1}{3}(1 - \cos^2 x) \cos x - \frac{2}{3} \cos x + C = -\cos x + \frac{1}{3} \cos^3 x + C$$

The derivation of (7) is similar.

EXERCISE SET 9.4

1. $\dfrac{1}{3} \tan(3x + 1) + C$

3. $\dfrac{1}{2} \ln|\cos(e^{-2x})| + C$

5. $\dfrac{1}{2} \ln|\sec 2x + \tan 2x| + C$

7. $u = \tan x, \; \displaystyle\int u^2 \, du = \dfrac{1}{3} \tan^3 x + C$

9. $\displaystyle\int \tan^3 4x (1 + \tan^2 4x) \sec^2 4x \, dx = \int (\tan^3 4x + \tan^5 4x) \sec^2 4x \, dx$

$$= \frac{1}{16} \tan^4 4x + \frac{1}{24} \tan^6 4x + C$$

11. $\displaystyle\int \sec^4 x (\sec^2 x - 1) \sec x \tan x \, dx = \int (\sec^6 x - \sec^4 x) \sec x \tan x \, dx = \frac{1}{7} \sec^7 x - \frac{1}{5} \sec^5 x + C$

13. $\displaystyle\int (\sec^2 x - 1)^2 \sec x \, dx = \int (\sec^5 x - 2 \sec^3 x + \sec x) dx$

$$= \int \sec^5 x \, dx - 2 \int \sec^3 x \, dx + \int \sec x \, dx$$

$$= \frac{1}{4} \sec^3 x \tan x + \frac{3}{4} \int \sec^3 x \, dx - 2 \int \sec^3 x \, dx + \ln|\sec x + \tan x|$$

$$= \frac{1}{4} \sec^3 x \tan x - \frac{5}{4} \left[\frac{1}{2} \sec x \tan x + \frac{1}{2} \ln|\sec x + \tan x| \right] + \ln|\sec x + \tan x| + C$$

$$= \frac{1}{4} \sec^3 x \tan x - \frac{5}{8} \sec x \tan x + \frac{3}{8} \ln|\sec x + \tan x| + C$$

15. $\displaystyle\int \sec^2 2t (\sec 2t \tan 2t) dt = \frac{1}{6} \sec^3 2t + C$

17. $\displaystyle\int \sec^4 x \, dx = \int (1 + \tan^2 x) \sec^2 x \, dx = \int (\sec^2 x + \tan^2 x \sec^2 x) dx = \tan x + \frac{1}{3} \tan^3 x + C$

19. $u = \pi x$, use reduction formula (3) to get

$$\frac{1}{\pi} \int \sec^6 u \, du = \frac{1}{\pi} \left[\frac{1}{5} \sec^4 u \tan u + \frac{4}{5} \int \sec^4 u \, du \right]$$

$$= \frac{1}{5\pi} \sec^4 u \tan u + \frac{4}{5\pi} \left[\frac{1}{3} \sec^2 u \tan u + \frac{2}{3} \tan u \right] + C$$

$$= \frac{1}{5\pi} \sec^4 \pi x \tan \pi x + \frac{4}{15\pi} \sec^2 \pi x \tan \pi x + \frac{8}{15\pi} \tan \pi x + C$$

21. Use reduction formula (4) to get $\int \tan^4 x \, dx = \frac{1}{3} \tan^3 x - \tan x + x + C$

23. $u = \tan(x^2)$, $\frac{1}{2} \int u^2 \, du = \frac{1}{6} \tan^3(x^2) + C$

25. $\int (\csc^2 x - 1) \csc^2 x (\csc x \cot x) dx = \int (\csc^4 x - \csc^2 x)(\csc x \cot x) dx$

$$= -\frac{1}{5} \csc^5 x + \frac{1}{3} \csc^3 x + C$$

27. $\int (\csc^2 x - 1) \cot x \, dx = \int \csc x (\csc x \cot x) dt - \int \frac{\cos x}{\sin x} dx = -\frac{1}{2} \csc^2 x - \ln|\sin x| + C$

29. $\int \tan^{1/2} x (1 + \tan^2 x) \sec^2 x \, dx = \frac{2}{3} \tan^{3/2} x + \frac{2}{7} \tan^{7/2} x + C$

31. $\int_0^{\pi/6} (\sec^2 2x - 1) dx = \frac{1}{2} \tan 2x - x \Big]_0^{\pi/6} = \sqrt{3}/2 - \pi/6$

33. $u = x/2$,

$$2 \int_0^{\pi/4} \tan^5 u \, du = \frac{1}{2} \tan^4 u - \tan^2 u - 2 \ln|\cos u| \Big]_0^{\pi/4} = 1/2 - 1 - 2\ln(1/\sqrt{2}) = -1/2 + \ln 2$$

35. $y' = \tan x$, $1 + (y')^2 = 1 + \tan^2 x = \sec^2 x$,

$$L = \int_0^{\pi/4} \sqrt{\sec^2 x} \, dx = \int_0^{\pi/4} \sec x \, dx = \ln|\sec x + \tan x| \Big]_0^{\pi/4} = \ln(\sqrt{2} + 1)$$

37. (a) $\int \csc x \, dx = \int \sec(\pi/2 - x) dx = -\ln|\sec(\pi/2 - x) + \tan(\pi/2 - x)| + C$

$$= -\ln|\csc x + \cot x| + C$$

(b) $-\ln|\csc x + \cot x| = \ln\dfrac{1}{|\csc x + \cot x|} = \ln\dfrac{|\csc x - \cot x|}{|\csc^2 x - \cot^2 x|} = \ln|\csc x - \cot x|.$

$$-\ln|\csc x + \cot x| = -\ln\left|\dfrac{1}{\sin x} + \dfrac{\cos x}{\sin x}\right| = \ln\left|\dfrac{\sin x}{1 + \cos x}\right|$$

$$= \ln\left|\dfrac{2\sin(x/2)\cos(x/2)}{2\cos^2(x/2)}\right| = \ln|\tan(x/2)|$$

39. $a\sin x + b\cos x = \sqrt{a^2 + b^2}\left[\dfrac{a}{\sqrt{a^2 + b^2}}\sin x + \dfrac{b}{\sqrt{a^2 + b^2}}\cos x\right]$

$$= \sqrt{a^2 + b^2}(\sin x\cos\theta + \cos x\sin\theta)$$

where $\cos\theta = a/\sqrt{a^2 + b^2}$ and $\sin\theta = b/\sqrt{a^2 + b^2}$ so $a\sin x + b\cos x = \sqrt{a^2 + b^2}\sin(x + \theta)$

and $\displaystyle\int\dfrac{dx}{a\sin x + b\cos x} = \dfrac{1}{\sqrt{a^2 + b^2}}\int\csc(x + \theta)dx$

$$= -\dfrac{1}{\sqrt{a^2 + b^2}}\ln|\csc(x + \theta) + \cot(x + \theta)| + C$$

EXERCISE SET 9.5

1. $x = 2\sin\theta,\ dx = 2\cos\theta\,d\theta,$

$$4\int\cos^2\theta\,d\theta = 2\int(1 + \cos 2\theta)d\theta = 2\theta + \sin 2\theta + C$$

$$= 2\theta + 2\sin\theta\cos\theta + C = 2\sin^{-1}(x/2) + \dfrac{1}{2}x\sqrt{4 - x^2} + C$$

3. $x = 3\sin\theta,\ dx = 3\cos\theta\,d\theta,$

$$9\int\sin^2\theta\,d\theta = \dfrac{9}{2}\int(1 - \cos 2\theta)d\theta = \dfrac{9}{2}\theta - \dfrac{9}{4}\sin 2\theta + C = \dfrac{9}{2}\theta - \dfrac{9}{2}\sin\theta\cos\theta + C$$

$$= \dfrac{9}{2}\sin^{-1}(x/3) - \dfrac{1}{2}x\sqrt{9 - x^2} + C$$

5. $x = 2\tan\theta,\ dx = 2\sec^2\theta\,d\theta,$

$$\dfrac{1}{8}\int\dfrac{1}{\sec^2\theta}d\theta = \dfrac{1}{8}\int\cos^2\theta\,d\theta = \dfrac{1}{16}\int(1 + \cos 2\theta)d\theta = \dfrac{1}{16}\theta + \dfrac{1}{32}\sin 2\theta + C$$

$$= \dfrac{1}{16}\theta + \dfrac{1}{16}\sin\theta\cos\theta + C = \dfrac{1}{16}\tan^{-1}\dfrac{x}{2} + \dfrac{x}{8(4 + x^2)} + C$$

7. $x = 3\sec\theta$, $dx = 3\sec\theta\tan\theta\,d\theta$,

$$3\int \tan^2\theta\,d\theta = 3\int(\sec^2\theta - 1)d\theta = 3\tan\theta - 3\theta + C = \sqrt{x^2 - 9} - 3\sec^{-1}\frac{x}{3} + C$$

9. $x = \sqrt{2}\sin\theta$, $dx = \sqrt{2}\cos\theta\,d\theta$,

$$2\sqrt{2}\int \sin^3\theta\,d\theta = 2\sqrt{2}\left(-\cos\theta + \frac{1}{3}\cos^3\theta\right) + C = -2\sqrt{2 - x^2} + \frac{1}{3}(2 - x^2)^{3/2} + C$$

11. $x = \sqrt{3}\tan\theta$, $dx = \sqrt{3}\sec^2\theta\,d\theta$,

$$\frac{1}{3}\int\frac{1}{\sec\theta}d\theta = \frac{1}{3}\int\cos\theta\,d\theta = \frac{1}{3}\sin\theta + C = \frac{x}{3\sqrt{3 + x^2}} + C$$

13. $x = \frac{3}{2}\sec\theta$, $dx = \frac{3}{2}\sec\theta\tan\theta\,d\theta$,

$$\frac{2}{9}\int\frac{1}{\sec\theta}d\theta = \frac{2}{9}\int\cos\theta\,d\theta = \frac{2}{9}\sin\theta + C = \frac{\sqrt{4x^2 - 9}}{9x} + C$$

15. $x = \sin\theta$, $dx = \cos\theta\,d\theta$, $\displaystyle\int\frac{1}{\cos^2\theta}d\theta = \int\sec^2\theta\,d\theta = \tan\theta + C = x/\sqrt{1 - x^2} + C$

17. $x = \sec\theta$, $dx = \sec\theta\tan\theta\,d\theta$

$$\int\sec\theta\,d\theta = \ln|\sec\theta + \tan\theta| + C = \ln\left|x + \sqrt{x^2 - 1}\right| + C$$

19. $x = \frac{3}{2}\sin\theta$, $dx = \frac{3}{2}\cos\theta\,d\theta$,

$$\frac{2}{9}\int\frac{1}{\sin^2\theta}d\theta = \frac{2}{9}\int\csc^2\theta\,d\theta = -\frac{2}{9}\cot\theta + C = -\frac{\sqrt{9 - 4x^2}}{9x} + C$$

21. $x = \frac{1}{3}\sec\theta$, $dx = \frac{1}{3}\sec\theta\tan\theta\,d\theta$,

$$\frac{1}{3}\int\frac{\sec\theta}{\tan^2\theta}d\theta = \frac{1}{3}\int\csc\theta\cot\theta\,d\theta = -\frac{1}{3}\csc\theta + C = -x/\sqrt{9x^2 - 1} + C$$

23. $e^x = \sin\theta$, $e^x dx = \cos\theta\,d\theta$,

$$\int\cos^2\theta\,d\theta = \frac{1}{2}\int(1 + \cos 2\theta)d\theta = \frac{1}{2}\theta + \frac{1}{4}\sin 2\theta + C = \frac{1}{2}\sin^{-1}(e^x) + \frac{1}{2}e^x\sqrt{1 - e^{2x}} + C$$

25. $x = 4\sin\theta$, $dx = 4\cos\theta\,d\theta$,

$$1024\int_0^{\pi/2}\sin^3\theta\cos^2\theta\,d\theta = 1024\left[-\frac{1}{3}\cos^3\theta + \frac{1}{5}\cos^5\theta\right]_0^{\pi/2} = 1024(1/3 - 1/5) = 2048/15$$

27. $x = \sec\theta,\ dx = \sec\theta\tan\theta\,d\theta,\ \int_{\pi/4}^{\pi/3}\frac{1}{\sec\theta}d\theta = \int_{\pi/4}^{\pi/3}\cos\theta\,d\theta = \sin\theta\Big]_{\pi/4}^{\pi/3} = (\sqrt{3}-\sqrt{2})/2$

29. $x = \sqrt{3}\tan\theta,\ dx = \sqrt{3}\sec^2\theta\,d\theta,$

$$\frac{1}{9}\int_{\pi/6}^{\pi/3}\frac{\sec\theta}{\tan^4\theta}d\theta = \frac{1}{9}\int_{\pi/6}^{\pi/3}\frac{\cos^3\theta}{\sin^4\theta}d\theta = \frac{1}{9}\int_{\pi/6}^{\pi/3}\frac{1-\sin^2\theta}{\sin^4\theta}\cos\theta\,d\theta$$

$$= \frac{1}{9}\int_{1/2}^{\sqrt{3}/2}\frac{1-u^2}{u^4}du \quad (u = \sin\theta)$$

$$= \frac{1}{9}\int_{1/2}^{\sqrt{3}/2}(u^{-4}-u^{-2})du = \frac{1}{9}\left[-\frac{1}{3u^3}+\frac{1}{u}\right]_{1/2}^{\sqrt{3}/2} = \frac{10\sqrt{3}+18}{243}$$

31. $u = x^2+4,\ du = 2x\,dx,\ \frac{1}{2}\int\frac{1}{u}du = \frac{1}{2}\ln|u|+C = \frac{1}{2}\ln(x^2+4)+C.$

$x = 2\tan\theta,\ dx = 2\sec^2\theta\,d\theta,$

$$\int\tan\theta\,d\theta = \ln|\sec\theta|+C_1 = \ln\frac{\sqrt{x^2+4}}{2}+C_1 = \ln(x^2+4)^{1/2}-\ln2+C_1$$

$$= \frac{1}{2}\ln(x^2+4)+C \text{ with } C = C_1-\ln2$$

33. $y' = \frac{1}{x},\ 1+(y')^2 = 1+\frac{1}{x^2} = \frac{x^2+1}{x^2},$

$$L = \int_1^2\sqrt{\frac{x^2+1}{x^2}}dx = \int_1^2\frac{\sqrt{x^2+1}}{x}dx;\ x = \tan\theta,\ dx = \sec^2\theta\,d\theta,$$

$$L = \int_{\pi/4}^{\tan^{-1}2}\frac{\sec^3\theta}{\tan\theta}d\theta = \int_{\pi/4}^{\tan^{-1}2}\frac{\tan^2\theta+1}{\tan\theta}\sec\theta\,d\theta = \int_{\pi/4}^{\tan^{-1}2}\left[\sec\theta\tan\theta+\frac{\sec\theta}{\tan\theta}\right]d\theta$$

$$= \int_{\pi/4}^{\tan^{-1}2}(\sec\theta\tan\theta+\csc\theta)d\theta = \sec\theta-\ln|\csc\theta+\cot\theta|\Big]_{\pi/4}^{\tan^{-1}2}$$

$$= \left[\sqrt{5}-\ln\left|\frac{\sqrt{5}}{2}+\frac{1}{2}\right|\right] - \left[\sqrt{2}-\ln|\sqrt{2}+1|\right] = \sqrt{5}-\sqrt{2}+\ln\frac{2+2\sqrt{2}}{1+\sqrt{5}}$$

35. $y' = 2x,\ 1+(y')^2 = 1+4x^2,$

$$S = 2\pi\int_0^1 x^2\sqrt{1+4x^2}dx;\ x = \frac{1}{2}\tan\theta,\ dx = \frac{1}{2}\sec^2\theta\,d\theta,$$

$$S = \frac{\pi}{4} \int_0^{\tan^{-1}2} \tan^2 \theta \sec^3 \theta \, d\theta = \frac{\pi}{4} \int_0^{\tan^{-1}2} (\sec^2 \theta - 1) \sec^3 \theta \, d\theta$$

$$= \frac{\pi}{4} \int_0^{\tan^{-1}2} (\sec^5 \theta - \sec^3 \theta) d\theta$$

$$= \frac{\pi}{4} \left[\frac{1}{4} \sec^3 \theta \tan \theta - \frac{1}{8} \sec \theta \tan \theta - \frac{1}{8} \ln|\sec \theta + \tan \theta| \right]_0^{\tan^{-1}2} = \frac{\pi}{32} [18\sqrt{5} - \ln(2 + \sqrt{5})]$$

37. **(a)** $x = 3\sinh u, \; dx = 3\cosh u \, du, \; \int du = u + C = \sinh^{-1}(x/3) + C$

(b) $x = 3\tan \theta, \; dx = 3\sec^2 \theta \, d\theta,$

$$\int \sec \theta \, d\theta = \ln|\sec \theta + \tan \theta| + C = \ln(\sqrt{x^2 + 9}/3 + x/3) + C$$

but $\sinh^{-1}(x/3) = \ln(x/3 + \sqrt{x^2/9 + 1}) = \ln(x/3 + \sqrt{x^2 + 9}/3)$ so the results agree.

39. $\displaystyle\int \frac{1}{(x-2)^2 + 9} dx = \frac{1}{3} \tan^{-1}\left(\frac{x-2}{3}\right) + C$

41. $\displaystyle\int \frac{1}{\sqrt{9 - (x-1)^2}} dx = \sin^{-1}\left(\frac{x-1}{3}\right) + C$

43. $\displaystyle\int \frac{1}{\sqrt{(x-3)^2 + 1}} dx = \sinh^{-1}(x - 3) + C.$

<u>Alternate solution:</u> let $x - 3 = \tan \theta,$

$$\int \sec \theta \, d\theta = \ln|\sec \theta + \tan \theta| + C = \ln(\sqrt{x^2 - 6x + 10} + x - 3) + C.$$

45. $\displaystyle\int \sqrt{4 - (x+1)^2} dx,$ let $x + 1 = 2\sin \theta,$

$$4 \int \cos^2 \theta \, d\theta = 2\theta + \sin 2\theta + C = 2\theta + 2\sin \theta \cos \theta + C$$

$$= 2\sin^{-1}\left(\frac{x+1}{2}\right) + \frac{1}{2}(x+1)\sqrt{3 - 2x - x^2} + C$$

47. $\displaystyle\int \frac{1}{2(x+1)^2 + 5} dx = \frac{1}{2} \int \frac{1}{(x+1)^2 + 5/2} dx = \frac{1}{\sqrt{10}} \tan^{-1}\sqrt{2/5}(x+1) + C$

49. $\int \dfrac{2x+5}{(x+1)^2+4}dx$, let $u = x+1$,

$$\int \frac{2u+3}{u^2+4}du = \int \left[\frac{2u}{u^2+4} + \frac{3}{u^2+4}\right] du = \ln(u^2+4) + \frac{3}{2}\tan^{-1}(u/2) + C$$

$$= \ln(x^2+2x+5) + \frac{3}{2}\tan^{-1}\left(\frac{x+1}{2}\right) + C$$

51. $\int \dfrac{x+3}{\sqrt{(x+1)^2+1}}dx$, let $u = x+1$,

$$\int \frac{u+2}{\sqrt{u^2+1}}du = \int \left[u(u^2+1)^{-1/2} + \frac{2}{\sqrt{u^2+1}}\right] du = \sqrt{u^2+1} + 2\sinh^{-1}u + C$$

$$= \sqrt{x^2+2x+2} + 2\sinh^{-1}(x+1) + C$$

Alternate solution: let $x+1 = \tan\theta$,

$$\int (\tan\theta+2)\sec\theta\, d\theta = \int \sec\theta\tan\theta\, d\theta + 2\int \sec\theta\, d\theta = \sec\theta + 2\ln|\sec\theta + \tan\theta| + C$$

$$= \sqrt{x^2+2x+2} + 2\ln(\sqrt{x^2+2x+2} + x+1) + C.$$

53. $\displaystyle\int_0^1 \sqrt{4x-x^2}dx = \int_0^1 \sqrt{4-(x-2)^2}dx$, let $x-2 = 2\sin\theta$,

$$4\int_{-\pi/2}^{-\pi/6} \cos^2\theta\, d\theta = 2\theta + \sin 2\theta \Big]_{-\pi/2}^{-\pi/6} = \frac{2\pi}{3} - \frac{\sqrt{3}}{2}$$

EXERCISE SET 9.6

1. $\dfrac{A}{(x-2)} + \dfrac{B}{(x+5)}$

3. $\dfrac{2x-3}{x^2(x-1)} = \dfrac{A}{x} + \dfrac{B}{x^2} + \dfrac{C}{x-1}$

5. $\dfrac{A}{x} + \dfrac{B}{x^2} + \dfrac{C}{x^3} + \dfrac{Dx+E}{x^2+1}$

7. $\dfrac{Ax+B}{x^2+5} + \dfrac{Cx+D}{(x^2+5)^2}$

9. $\dfrac{1}{(x+4)(x-1)} = \dfrac{A}{x+4} + \dfrac{B}{x-1}$; $A = -\dfrac{1}{5}$, $B = \dfrac{1}{5}$

$$-\frac{1}{5}\int \frac{1}{x+4}dx + \frac{1}{5}\int \frac{1}{x-1}dx = -\frac{1}{5}\ln|x+4| + \frac{1}{5}\ln|x-1| + C = \frac{1}{5}\ln\left|\frac{x-1}{x+4}\right| + C$$

11. $\dfrac{x}{(x-2)(x-3)} = \dfrac{A}{x-2} + \dfrac{B}{x-3}$; $A = -2$, $B = 3$

$-2\displaystyle\int \dfrac{1}{x-2}dx + 3\int \dfrac{1}{x-3}dx = -2\ln|x-2| + 3\ln|x-3| + C$

13. $\dfrac{11x+17}{(2x-1)(x+4)} = \dfrac{A}{2x-1} + \dfrac{B}{x+4}$; $A = 5$, $B = 3$

$5\displaystyle\int \dfrac{1}{2x-1}dx + 3\int \dfrac{1}{x+4}dx = \dfrac{5}{2}\ln|2x-1| + 3\ln|x+4| + C$

15. $\dfrac{1}{(x-1)(x+2)(x-3)} = \dfrac{A}{x-1} + \dfrac{B}{x+2} + \dfrac{C}{x-3}$; $A = -\dfrac{1}{6}$, $B = \dfrac{1}{15}$, $C = \dfrac{1}{10}$

$-\dfrac{1}{6}\displaystyle\int \dfrac{1}{x-1}dx + \dfrac{1}{15}\int \dfrac{1}{x+2}dx + \dfrac{1}{10}\int \dfrac{1}{x-3}dx$

$\qquad = -\dfrac{1}{6}\ln|x-1| + \dfrac{1}{15}\ln|x+2| + \dfrac{1}{10}\ln|x-3| + C$

17. $\dfrac{2x^2 - 9x - 9}{x(x+3)(x-3)} = \dfrac{A}{x} + \dfrac{B}{x+3} + \dfrac{C}{x-3}$; $A = 1$, $B = 2$, $C = -1$

$\displaystyle\int \dfrac{1}{x}dx + 2\int \dfrac{1}{x+3}dx - \int \dfrac{1}{x-3}dx = \ln|x| + 2\ln|x+3| - \ln|x-3| + C = \ln\left|\dfrac{x(x+3)^2}{x-3}\right| + C$

19. $\dfrac{x^2+2}{x+2} = x - 2 + \dfrac{6}{x+2}$, $\displaystyle\int \left(x - 2 + \dfrac{6}{x+2}\right)dx = \dfrac{1}{2}x^2 - 2x + 6\ln|x+2| + C$

21. $\dfrac{3x^2 - 10}{x^2 - 4x + 4} = 3 + \dfrac{12x - 22}{x^2 - 4x + 4}$, $\dfrac{12x - 22}{(x-2)^2} = \dfrac{A}{x-2} + \dfrac{B}{(x-2)^2}$; $A = 12$, $B = 2$

$\displaystyle\int 3dx + 12\int \dfrac{1}{x-2}dx + 2\int \dfrac{1}{(x-2)^2}dx = 3x + 12\ln|x-2| - 2/(x-2) + C$

23. $\dfrac{x^3}{x^2 - 3x + 2} = x + 3 + \dfrac{7x - 6}{x^2 - 3x + 2}$, $\dfrac{7x - 6}{(x-1)(x-2)} = \dfrac{A}{x-1} + \dfrac{B}{x-2}$; $A = -1$, $B = 8$

$\displaystyle\int (x+3)dx - \int \dfrac{1}{x-1}dx + 8\int \dfrac{1}{x-2}dx = \dfrac{1}{2}x^2 + 3x - \ln|x-1| + 8\ln|x-2| + C$

25. $\dfrac{x^5 + 2x^2 + 1}{x^3 - x} = x^2 + 1 + \dfrac{2x^2 + x + 1}{x^3 - x}$,

$\dfrac{2x^2 + x + 1}{x(x+1)(x-1)} = \dfrac{A}{x} + \dfrac{B}{x+1} + \dfrac{C}{x-1}$; $A = -1$, $B = 1$, $C = 2$

$$\int (x^2 + 1)dx - \int \frac{1}{x}dx + \int \frac{1}{x+1}dx + 2\int \frac{1}{x-1}dx$$

$$= \frac{1}{3}x^3 + x - \ln|x| + \ln|x+1| + 2\ln|x-1| + C = \frac{1}{3}x^3 + x + \ln\left|\frac{(x+1)(x-1)^2}{x}\right| + C$$

27. $\dfrac{2x^2 + 3}{x(x-1)^2} = \dfrac{A}{x} + \dfrac{B}{x-1} + \dfrac{C}{(x-1)^2};\ A = 3,\ B = -1,\ C = 5$

$$3\int \frac{1}{x}dx - \int \frac{1}{x-1}dx + 5\int \frac{1}{(x-1)^2}dx = 3\ln|x| - \ln|x-1| - 5/(x-1) + C$$

29. $\dfrac{x^2 + x - 16}{(x+1)(x-3)^2} = \dfrac{A}{x+1} + \dfrac{B}{x-3} + \dfrac{C}{(x-3)^2};\ A = -1,\ B = 2,\ C = -1$

$$-\int \frac{1}{x+1}dx + 2\int \frac{1}{x-3}dx - \int \frac{1}{(x-3)^2}dx$$

$$= -\ln|x+1| + 2\ln|x-3| + \frac{1}{x-3} + C = \ln\frac{(x-3)^2}{|x+1|} + \frac{1}{x-3} + C$$

31. $\dfrac{x^2}{(x+2)^3} = \dfrac{A}{x+2} + \dfrac{B}{(x+2)^2} + \dfrac{C}{(x+2)^3};\ A = 1,\ B = -4,\ C = 4$

$$\int \frac{1}{x+2}dx - 4\int \frac{1}{(x+2)^2}dx + 4\int \frac{1}{(x+2)^3}dx = \ln|x+2| + \frac{4}{x+2} - \frac{2}{(x+2)^2} + C$$

33. $\dfrac{2x^2 - 1}{(4x-1)(x^2+1)} = \dfrac{A}{4x-1} + \dfrac{Bx+C}{x^2+1};\ A = -14/17,\ B = 12/17,\ C = 3/17$

$$\int \frac{2x^2 - 1}{4x^3 - x^2 + 4x - 1}dx = -\frac{7}{34}\ln|4x-1| + \frac{6}{17}\ln(x^2+1) + \frac{3}{17}\tan^{-1}x + C$$

35. $\dfrac{1}{(x+2)(x-2)(x^2+4)} = \dfrac{A}{x+2} + \dfrac{B}{x-2} + \dfrac{Cx+D}{x^2+4};\ A = -1/32,\ B = 1/32,\ C = 0,\ D = -1/8$

$$\int \frac{dx}{x^4 - 16} = \frac{1}{32}\ln\left|\frac{x-2}{x+2}\right| - \frac{1}{16}\tan^{-1}(x/2) + C$$

37. $\dfrac{x^3 + 3x^2 + x + 9}{(x^2+1)(x^2+3)} = \dfrac{Ax+B}{x^2+1} + \dfrac{Cx+D}{x^2+3};\ A = 0,\ B = 3,\ C = 1,\ D = 0$

$$\int \frac{x^3 + 3x^2 + x + 9}{(x^2+1)(x^2+3)}dx = 3\tan^{-1}x + \frac{1}{2}\ln(x^2+3) + C$$

39. $\dfrac{x^3 - 3x^2 + 2x - 3}{x^2+1} = x - 3 + \dfrac{x}{x^2+1}$,

$$\int \frac{x^3 - 3x^2 + 2x - 3}{x^2+1}dx = \frac{1}{2}x^2 - 3x + \frac{1}{2}\ln(x^2+1) + C$$

41. $\dfrac{x^2+1}{(x^2+2x+3)^2} = \dfrac{Ax+B}{x^2+2x+3} + \dfrac{Cx+D}{(x^2+2x+3)^2}; \ A=0, \ B=1, \ C=D=-2$

$$\int \frac{x^2+1}{(x^2+2x+3)^2}\,dx = \int \frac{1}{(x+1)^2+2}\,dx - \int \frac{2x+2}{(x^2+2x+3)^2}\,dx$$

$$= \frac{1}{\sqrt{2}}\tan^{-1}\frac{x+1}{\sqrt{2}} + 1/(x^2+2x+3) + C$$

43. let $x=\sin\theta$ to get $\displaystyle\int \frac{1}{x^2+4x-5}\,dx,$

$$\frac{1}{(x+5)(x-1)} = \frac{A}{x+5} + \frac{B}{x-1}; \ A=-1/6, \ B=1/6$$

$$-\frac{1}{6}\int \frac{1}{x+5}\,dx + \frac{1}{6}\int \frac{1}{x-1}\,dx = \frac{1}{6}\ln\left|\frac{x-1}{x+5}\right| + C = \frac{1}{6}\ln\left|\frac{\sin\theta-1}{\sin\theta+5}\right| + C$$

45. let $u=e^x$ to get $\displaystyle\int \frac{dx}{1+e^x} = \int \frac{e^x\,dx}{e^x(1+e^x)} = \int \frac{du}{u(1+u)},$

$$\frac{1}{u(1+u)} = \frac{A}{u} + \frac{B}{1+u}; \ A=1, \ B=-1$$

$$\int \frac{du}{u(1+u)} = \ln u - \ln(1+u) + C = \ln\frac{e^x}{1+e^x} + C$$

47. (a) $x^4+1 = (x^4+2x^2+1) - 2x^2 = (x^2+1)^2 - 2x^2$

$$= [(x^2+1)+\sqrt{2}x][(x^2+1)-\sqrt{2}x]$$

$$= (x^2+\sqrt{2}x+1)(x^2-\sqrt{2}x+1); a=\sqrt{2}, b=-\sqrt{2}$$

(b) $\dfrac{x}{(x^2+\sqrt{2}x+1)(x^2-\sqrt{2}x+1)} = \dfrac{Ax+B}{x^2+\sqrt{2}x+1} + \dfrac{Cx+D}{x^2-\sqrt{2}x+1};$

$$A=0, \ B=-\frac{\sqrt{2}}{4}, \ C=0, \ D=\frac{\sqrt{2}}{4} \text{ so}$$

$$\int_0^1 \frac{x}{x^4+1}\,dx = -\frac{\sqrt{2}}{4}\int_0^1 \frac{1}{x^2+\sqrt{2}x+1}\,dx + \frac{\sqrt{2}}{4}\int_0^1 \frac{1}{x^2-\sqrt{2}x+1}\,dx$$

$$= -\frac{\sqrt{2}}{4}\int_0^1 \frac{1}{(x+\sqrt{2}/2)^2+1/2}\,dx + \frac{\sqrt{2}}{4}\int_0^1 \frac{1}{(x-\sqrt{2}/2)^2+1/2}\,dx$$

$$= -\frac{\sqrt{2}}{4}\int_{\sqrt{2}/2}^{1+\sqrt{2}/2} \frac{1}{u^2+1/2}\,du + \frac{\sqrt{2}}{4}\int_{-\sqrt{2}/2}^{1-\sqrt{2}/2} \frac{1}{u^2+1/2}\,du$$

$$= -\frac{1}{2}\tan^{-1}\sqrt{2}u\,\Big]_{\sqrt{2}/2}^{1+\sqrt{2}/2} + \frac{1}{2}\tan^{-1}\sqrt{2}u\,\Big]_{-\sqrt{2}/2}^{1-\sqrt{2}/2}$$

$$= -\frac{1}{2}\tan^{-1}(\sqrt{2}+1) + \frac{1}{2}\left(\frac{\pi}{4}\right) + \frac{1}{2}\tan^{-1}(\sqrt{2}-1) - \frac{1}{2}\left(-\frac{\pi}{4}\right)$$

$$= \frac{\pi}{4} - \frac{1}{2}[\tan^{-1}(\sqrt{2}+1) - \tan^{-1}(\sqrt{2}-1)]$$

$$= \frac{\pi}{4} - \frac{1}{2}[\tan^{-1}(1+\sqrt{2}) + \tan^{-1}(1-\sqrt{2})]$$

$$= \frac{\pi}{4} - \frac{1}{2}\tan^{-1}\left[\frac{(1+\sqrt{2}) + (1-\sqrt{2})}{1 - (1+\sqrt{2})(1-\sqrt{2})}\right] \quad \text{(Exercise 16, 8.1)}$$

$$= \frac{\pi}{4} - \frac{1}{2}\tan^{-1} 1 = \frac{\pi}{4} - \frac{1}{2}\left(\frac{\pi}{4}\right) = \frac{\pi}{8}$$

49. $V = \pi \displaystyle\int_0^2 \frac{x^4}{(9-x^2)^2}dx, \quad \frac{x^4}{x^4 - 18x^2 + 81} = 1 + \frac{18x^2 - 81}{x^4 - 18x^2 + 81},$

$$\frac{18x^2 - 81}{(9-x^2)^2} = \frac{18x^2 - 81}{(x+3)^2(x-3)^2} = \frac{A}{x+3} + \frac{B}{(x+3)^2} + \frac{C}{x-3} + \frac{D}{(x-3)^2};$$

$$A = -\frac{9}{4},\ B = \frac{9}{4},\ C = \frac{9}{4},\ D = \frac{9}{4}$$

$$V = \pi\left[x - \frac{9}{4}\ln|x+3| - \frac{9/4}{x+3} + \frac{9}{4}\ln|x-3| - \frac{9/4}{x-3}\right]_0^2 = \pi\left(\frac{19}{5} - \frac{9}{4}\ln 5\right)$$

51. $\displaystyle\int \frac{1}{y^2 - 5y + 6}dy = \int dx, \quad \frac{1}{(y-3)(y-2)} = \frac{1}{y-3} - \frac{1}{y-2},$

$$\ln|y-3| - \ln|y-2| = x + C_1,\ \ln\left|\frac{y-3}{y-2}\right| = x + C_1,\ \frac{y-3}{y-2} = \pm e^{C_1}e^x = Ce^x,$$

$$y - 3 = Ce^x y - 2Ce^x,\ y(1 - Ce^x) = 3 - 2Ce^x,\ y = \frac{3 - 2Ce^x}{1 - Ce^x}.$$

53. $\displaystyle\int \frac{1}{y^2 + y}dy = \int \frac{1}{t(t-1)}dt, \quad \frac{1}{y(y+1)} = \frac{1}{y} - \frac{1}{y+1} \ \text{and}\ \frac{1}{t(t-1)} = -\frac{1}{t} + \frac{1}{t-1},$

$$\ln|y| - \ln|y+1| = -\ln|t| + \ln|t-1| + C_1,\ \ln\left|\frac{y}{y+1}\right| = \ln\left|\frac{t-1}{t}\right| + C_1,$$

$$\frac{y}{y+1} = C_2\frac{t-1}{t};\ \text{solve for } y \text{ to get } y = \frac{C_2(t-1)/t}{1 - C_2(t-1)/t} = \frac{1}{Ct/(t-1) - 1} = \frac{t-1}{Ct - t + 1}.$$

55. (a) $\int \dfrac{1}{ay - by^2} dy = \int dt,\ \dfrac{1}{y(a - by)} = \dfrac{1/a}{y} + \dfrac{b/a}{a - by},$

$\dfrac{1}{a} \ln |y| - \dfrac{1}{a} \ln |a - by| = t + C_1,\ \dfrac{1}{a} \ln \left| \dfrac{y}{a - by} \right| = t + C_1,$

$\dfrac{y}{a - by} = C_2 e^{at},\ y = aC_2 e^{at} - bC_2 e^{at} y,\ y = \dfrac{aC_2 e^{at}}{1 + bC_2 e^{at}} = \dfrac{a}{Ce^{-at} + b}.$

(b) $\lim\limits_{t \to +\infty} y = a/b,\ \lim\limits_{t \to -\infty} y = 0$

57. (a) $x^3 - 6x^2 + 11x - 6 = (x - 1)(x - 2)(x - 3)$
(b) $x^3 - 3x^2 + x - 20 = (x - 4)(x^2 + x + 5)$
(c) $x^4 - 5x^3 + 7x^2 - 5x + 6 = (x - 2)(x - 3)(x^2 + 1)$

59. $x^4 - 3x^3 - 7x^2 + 27x - 18 = (x - 1)(x - 2)(x - 3)(x + 3),$

$\dfrac{1}{(x - 1)(x - 2)(x - 3)(x + 3)} = \dfrac{A}{x - 1} + \dfrac{B}{x - 2} + \dfrac{C}{x - 3} + \dfrac{D}{x + 3};$

$A = 1/8,\ B = -1/5,\ C = 1/12,\ D = -1/120$

$\int \dfrac{dx}{x^4 - 3x^3 - 7x^2 + 27x - 18} = \dfrac{1}{8} \ln |x - 1| - \dfrac{1}{5} \ln |x - 2| + \dfrac{1}{12} \ln |x - 3| - \dfrac{1}{120} \ln |x + 3| + C$

61. (a) $\sqrt{2}$ is a positive root of $x^2 - 2 = 0$. The only possible positive rational roots of $x^2 - 2 = 0$ are the integers 1 and 2 neither of which satisfies the equation so $\sqrt{2}$ is not rational.

(b) $\sqrt{a}$ is a positive root of $x^2 - a = 0$. The only possible positive rational roots of $x^2 - a = 0$ are the positive integers that divide a. If one of these satisfies the equation then $\sqrt{a}$ is an integer, otherwise $\sqrt{a}$ cannot be rational.

EXERCISE SET 9.7

1. $u = \sqrt{x - 2},\ x = u^2 + 2,\ dx = 2u\, du$

$\int 2u^2(u^2 + 2)du = 2 \int (u^4 + 2u^2)du = \dfrac{2}{5}u^5 + \dfrac{4}{3}u^3 + C = \dfrac{2}{5}(x - 2)^{5/2} + \dfrac{4}{3}(x - 2)^{3/2} + C$

3. $u = \sqrt{x - 4},\ x = u^2 + 4,\ dx = 2u\, du$

$\int_0^2 \dfrac{2u^2}{u^2 + 4} du = 2 \int_0^2 \left[1 - \dfrac{4}{u^2 + 4} \right] du = 2u - 4\tan^{-1}(u/2) \Big]_0^2 = 4 - \pi$

5. $u = 3 + \sqrt{x},\ x = (u - 3)^2,\ dx = 2(u - 3)du$

$\int_3^5 \dfrac{2(u - 3)}{u} du = 2 \int_3^5 (1 - 3/u)du = 2u - 6\ln |u| \Big]_3^5 = 4 - 6\ln(5/3)$

7. $u = \sqrt{x^3 + 1}$, $x^3 = u^2 - 1$, $3x^2dx = 2u\,du$

$$\frac{2}{3}\int u^2(u^2-1)du = \frac{2}{3}\int (u^4 - u^2)du = \frac{2}{15}u^5 - \frac{2}{9}u^3 + C = \frac{2}{15}(x^3+1)^{5/2} - \frac{2}{9}(x^3+1)^{3/2} + C$$

9. $u = x^{1/6}$, $x = u^6$, $dx = 6u^5du$

$$\int \frac{6u^5}{u^3 + u^2}du = 6\int \frac{u^3}{u+1}du = 6\int \left[u^2 - u + 1 - \frac{1}{u+1}\right]du$$

$$= 2x^{1/2} - 3x^{1/3} + 6x^{1/6} - 6\,\ln(x^{1/6}+1) + C$$

11. $u = v^{1/4}$, $v = u^4$, $dv = 4u^3du$; $4\int \frac{1}{u(1-u)}du = 4\int \left[\frac{1}{u} + \frac{1}{1-u}\right]du = 4\ln\frac{v^{1/4}}{|1 - v^{1/4}|} + C$

13. $u = t^{1/6}$, $t = u^6$, $dt = 6u^5du$

$$6\int \frac{u^3}{u-1}du = 6\int \left[u^2 + u + 1 + \frac{1}{u-1}\right]du = 2t^{1/2} + 3t^{1/3} + 6t^{1/6} + 6\,\ln|t^{1/6} - 1| + C$$

15. $u = \sqrt{1+x^2}$, $x^2 = u^2 - 1$, $2x\,dx = 2u\,du$, $x\,dx = u\,du$

$$\int (u^2 - 1)du = \frac{1}{3}(1+x^2)^{3/2} - (1+x^2)^{1/2} + C$$

17. $z = \sqrt{x}$, $x = z^2$, $dx = 2z\,dz$, $2\int z\sin z\,dz$; use integration by parts:

$u = z$, $dv = \sin z\,dz$, $du = dz$, $v = -\cos z$

$$2\int z\sin z\,dz = 2\left(-z\cos z + \int \cos z\,dz\right) = -2z\cos z + 2\sin z + C$$

so $\int \sin\sqrt{x}dx = -2\sqrt{x}\cos\sqrt{x} + 2\sin\sqrt{x} + C$

19. $u = \sqrt{e^x + 1}$, $e^x = u^2 - 1$, $x = \ln(u^2 - 1)$, $dx = \frac{2u}{u^2-1}du$

$$\int \frac{2}{u^2-1}du = \int \left[\frac{1}{u-1} - \frac{1}{u+1}\right]du = \ln|u-1| - \ln|u+1| + C = \ln\frac{\sqrt{e^x+1}-1}{\sqrt{e^x+1}+1} + C$$

21. $\int \frac{1}{1 + \dfrac{2u}{1+u^2} + \dfrac{1-u^2}{1+u^2}} \frac{2}{1+u^2}du = \int \frac{1}{u+1}du = \ln|\tan(x/2) + 1| + C$

23. $u = \tan(\theta/2)$, $\int \frac{d\theta}{1 - \cos\theta} = \int \frac{1}{u^2}du = -\frac{1}{u} + C = -\cot(\theta/2) + C$,

$$\int_{\pi/2}^{\pi} \frac{d\theta}{1 - \cos\theta} = -\cot(\theta/2)\Big]_{\pi/2}^{\pi} = 1$$

25. $u = \tan(x/2)$, $2\displaystyle\int \frac{1 - u^2}{(3u^2 + 1)(u^2 + 1)}du$

$$\frac{1 - u^2}{(3u^2 + 1)(u^2 + 1)} = \frac{(0)u + 2}{3u^2 + 1} + \frac{(0)u - 1}{u^2 + 1} = \frac{2}{3u^2 + 1} - \frac{1}{u^2 + 1},$$

$$\int \frac{\cos x}{2 - \cos x}dx = \frac{4}{\sqrt{3}}\tan^{-1}[\sqrt{3}\tan(x/2)] - x + C$$

27. (a) $\displaystyle\int \sec x\, dx = \int \frac{1}{\cos x}dx = \int \frac{2}{1 - u^2}du = \ln\left|\frac{1 + u}{1 - u}\right| + C = \ln\left|\frac{1 + \tan(x/2)}{1 - \tan(x/2)}\right| + C$

(b) $\displaystyle\frac{1 + \tan(x/2)}{1 - \tan(x/2)} = \frac{\tan(\pi/4) + \tan(x/2)}{1 - \tan(\pi/4)\tan(x/2)} = \tan(\pi/4 + x/2)$ (trig identity)

(c) $\displaystyle\frac{1 + \tan(x/2)}{1 - \tan(x/2)} = \frac{\cos(x/2) + \sin(x/2)}{\cos(x/2) - \sin(x/2)} = \frac{[\cos(x/2) + \sin(x/2)]^2}{\cos^2(x/2) - \sin^2(x/2)}$

$$= \frac{\cos^2(x/2) + 2\sin(x/2)\cos(x/2) + \sin^2(x/2)}{\cos x}$$

$$= \frac{1 + \sin x}{\cos x} = \sec x + \tan x$$

29. Let $u = \tanh(x/2)$ then $\cosh(x/2) = 1/\operatorname{sech}(x/2) = 1/\sqrt{1 - \tanh^2(x/2)} = 1/\sqrt{1 - u^2}$,

$\sinh(x/2) = \tanh(x/2)\cosh(x/2) = u/\sqrt{1 - u^2}$

so $\sinh x = 2\sinh(x/2)\cosh(x/2) = 2u/(1 - u^2)$,

$\cosh x = \cosh^2(x/2) + \sinh^2(x/2) = (1 + u^2)/(1 - u^2)$, $x = 2\tanh^{-1}u$, $dx = [1/(1 - u^2)]du$

$$\int \frac{dx}{2\cosh x + \sinh x} = \int \frac{1}{u^2 + u + 1}du$$

$$= \frac{2}{\sqrt{3}}\tan^{-1}\frac{2u + 1}{\sqrt{3}} + C = \frac{2}{\sqrt{3}}\tan^{-1}\frac{2\tanh(x/2) + 1}{\sqrt{3}} + C$$

31. $dx = -(1/u^2)du$, $-\displaystyle\int \frac{u}{\sqrt{3u^2 - 1}}du = -\frac{1}{3}\sqrt{3u^2 - 1} + C = -\frac{\sqrt{3 - x^2}}{3x} + C$

33. $dx = -(1/u^2)du$, $-\displaystyle\int u\sqrt{1 - 5u^2}\,du = \frac{1}{15}(1 - 5u^2)^{3/2} + C = \frac{(x^2 - 5)^{3/2}}{15x^3} + C$

EXERCISE SET 9.8

1. exact value $= 14/3 \approx 4.666666667$
    (a) 4.667600663, $|E_M| \approx 0.000933996$
    (b) 4.664795679, $|E_T| \approx 0.001870988$
    (c) 4.666651630, $|E_S| \approx 0.000015037$

3. exact value $= 2$
    (a) 2.008248408, $|E_M| \approx 0.008248408$
    (b) 1.983523538, $|E_T| \approx 0.016476462$
    (c) 2.000109517, $|E_S| \approx 0.000109517$

5. exact value $= e^{-1} - e^{-3} \approx 0.318092373$
    (a) 0.317562837, $|E_M| \approx 0.000529536$
    (b) 0.319151975, $|E_T| \approx 0.001059602$
    (c) 0.318095187, $|E_S| \approx 0.000002814$

7. $f(x) = \sqrt{x+1}$, $f''(x) = -\dfrac{1}{4}(x+1)^{-3/2}$, $f^{(4)}(x) = -\dfrac{15}{16}(x+1)^{-7/2}$; $K_2 = 1/4$, $K_4 = 15/16$
    (a) $|E_M| \leq \dfrac{27}{2400}(1/4) \approx 0.002812500$
    (b) $|E_T| \leq \dfrac{27}{1200}(1/4) \approx 0.005625000$

    (c) $|E_S| \leq \dfrac{243}{180 \times 10^4}(15/16) \approx 0.000126563$

9. $f(x) = \sin x$, $f''(x) = -\sin x$, $f^{(4)}(x) = \sin x$; $K_2 = K_4 = 1$
    (a) $|E_M| \leq \dfrac{\pi^3}{2400}(1) \approx 0.012919282$
    (b) $|E_T| \leq \dfrac{\pi^3}{1200}(1) \approx 0.025838564$

    (c) $|E_S| \leq \dfrac{\pi^5}{180 \times 10^4}(1) \approx 0.000170011$

11. $f(x) = e^{-x}$, $f''(x) = f^{(4)}(x) = e^{-x}$; $K_2 = K_4 = e^{-1}$
    (a) $|E_M| \leq \dfrac{8}{2400}(e^{-1}) \approx 0.001226265$
    (b) $|E_T| \leq \dfrac{8}{1200}(e^{-1}) \approx 0.002452530$

    (c) $|E_S| \leq \dfrac{32}{180 \times 10^4}(e^{-1}) \approx 0.000006540$

13. (a) $n > \left[\dfrac{(27)(1/4)}{(24)(5 \times 10^{-4})} \right]^{1/2} \approx 23.7$; $n = 24$
    (b) $n > \left[\dfrac{(27)(1/4)}{(12)(5 \times 10^{-4})} \right]^{1/2} \approx 33.5$; $n = 34$

    (c) $n > \left[\dfrac{(243)(15/16)}{(180)(5 \times 10^{-4})} \right]^{1/4} \approx 7.1$; $n = 8$

15. **(a)** $n > \left[\dfrac{(\pi^3)(1)}{(24)(10^{-3})} \right]^{1/2} \approx 35.9; \, n = 36$ **(b)** $n > \left[\dfrac{(\pi^3)(1)}{(12)(10^{-3})} \right]^{1/2} \approx 50.8; \, n = 51$

    (c) $n > \left[\dfrac{(\pi^5)(1)}{(180)(10^{-3})} \right]^{1/4} \approx 6.4; \, n = 8$

17. **(a)** $n > \left[\dfrac{(8)(e^{-1})}{(24)(10^{-6})} \right]^{1/2} \approx 350.2; \, n = 351$ **(b)** $n > \left[\dfrac{(8)(e^{-1})}{(12)(10^{-6})} \right]^{1/2} \approx 495.2; \, n = 496$

    (c) $n > \left[\dfrac{(32)(e^{-1})}{(180)(10^{-6})} \right]^{1/4} \approx 15.99; \, n = 16$

19. **(a)** 0.747130878   **(b)** 0.746210796   **(c)** 0.746824948

21. **(a)** 2.129469966   **(b)** 2.130644002   **(c)** 2.129861595

23. **(a)** 0.809253858   **(b)** 0.795924733   **(c)** 0.805376152

25. **(a)** $3.142425985, \, |E_M| \approx 0.000833331$ **(b)** $3.139925989, \, |E_T| \approx 0.001666665$
    (c) $3.141592614, \, |E_S| \approx 0.000000040$

27. $S_{14} = 0.693147984, \, |E_S| \approx 0.000000803 = 8.03 \times 10^{-7}$; the method used in Example 5 results in a value of n which ensures that the magnitude of the error will be less than 10^{-6}, this is not necessarily the *smallest* value of n.

29. $f(x) = x \sin x, \, f''(x) = 2 \cos x - x \sin x, \, |f''(x)| \leq 2|\cos x| + |x| \, |\sin x| \leq 2 + 2 = 4$ so $K_2 \leq 4$,
$n > \left[\dfrac{(8)(4)}{(24)(10^{-4})} \right]^{1/2} \approx 115.5, \, n = 116$.

31. $f(x) = \sqrt{x}, \, f''(x) = -\dfrac{1}{4x^{3/2}}, \, \lim\limits_{x \to 0^+} |f''(x)| = +\infty$

33. $L = \displaystyle\int_0^\pi \sqrt{1 + \cos^2 x} \, dx \approx 3.820$

35.

t(sec)	0	0	10	15	20
v (mi/hr)	0	40	60	73	84
v (ft/sec)	0	58.67	88	107.07	123.2

$\displaystyle\int_0^{20} v \, dt \approx \dfrac{20}{(3)(4)} [0 + 4(58.67) + 2(88) + 4(107.07) + 123.2] \approx 1604 \text{ ft}$

37. $\int_0^{180} v\, dt \approx \dfrac{180}{(3)(6)}[0.00 + 4(0.03) + 2(0.08) + 4(0.16) + 2(0.27) + 4(0.42) + 0.65] = 37.9$ mi

39. $V = \int_0^{16} \pi r^2 dy = \pi \int_0^{16} r^2 dy \approx \pi \dfrac{16}{(3)(4)}[(8.5)^2 + 4(11.5)^2 + 2(13.8)^2 + 4(15.4)^2 + (16.8)^2]$

$\approx 9270 \text{ cm}^3 \approx 9.3 \text{ L}$

SUPPLEMENTARY EXERCISES CHAPTER 9

1. $u = x,\ dv = \cos 2x\, dx,\ du = dx,\ v = \dfrac{1}{2}\sin 2x;\ \int x \cos 2x\, dx = \dfrac{1}{2}x \sin 2x + \dfrac{1}{4}\cos 2x + C$

3. $\int (\sec^2 x - 1)\sec x \tan x\, dx = \dfrac{1}{3}\sec^3 x - \sec x + C$

5. $u = \tan 3t,\ \dfrac{1}{3}\int u^2 du = \dfrac{1}{9}\tan^3 3t + C$

7. $\int \dfrac{1 - \cos^2 x}{1 + \cos x} dx = \int (1 - \cos x)dx = x - \sin x + C$

9. $\int x^2 \cos^2 x\, dx = \dfrac{1}{2}\int x^2(1 + \cos 2x)dx = \dfrac{1}{2}\int x^2 dx + \dfrac{1}{2}\int x^2 \cos 2x\, dx,$

use integration by parts twice to get

$\int x^2 \cos 2x\, dx = \dfrac{1}{2}x^2 \sin 2x + \dfrac{1}{2}x \cos 2x - \dfrac{1}{4}\sin 2x + C_1$

so $\int x^2 \cos^2 x\, dx = \dfrac{1}{6}x^3 + \dfrac{1}{4}(x^2 - 1/2)\sin 2x + \dfrac{1}{4}x \cos 2x + C$

11. $\int \cos^{-5} x \sin x\, dx = \dfrac{1}{4}\cos^{-4} x + C = \dfrac{1}{4}\sec^4 x + C$

13. $\dfrac{1}{8}\int (1 - \cos 2x)^2(1 + \cos 2x)dx = \dfrac{1}{8}\int (1 - \cos 2x)\sin^2 2x\, dx$

$= \dfrac{1}{8}\int \sin^2 2x\, dx - \dfrac{1}{8}\int \sin^2 2x \cos 2x\, dx$

$= \dfrac{1}{16}\int (1 - \cos 4x)dx - \dfrac{1}{48}\sin^3 2x = \dfrac{1}{16}x - \dfrac{1}{64}\sin 4x - \dfrac{1}{48}\sin^3 2x + C$

15. $\int_0^{\pi/4} \sin 5x \sin 3x\, dx = \dfrac{1}{2}\int_0^{\pi/4} (\cos 2x - \cos 8x)dx = \dfrac{1}{4}\sin 2x - \dfrac{1}{16}\sin 8x \Big]_0^{\pi/4} = 1/4$

17. $\dfrac{1}{2}\displaystyle\int_0^1 (1 - \cos 2\pi x)dx = \dfrac{1}{2}x - \dfrac{1}{4\pi}\sin 2\pi x\Big]_0^1 = 1/2$

19. $\dfrac{1}{2}\tan(x^2)\Big]_0^{\sqrt{\pi/2}} = 1/2$

21. $u = \cot^{-1} x,\ -\displaystyle\int \sin u\, du = \cos(\cot^{-1} x) + C = x/\sqrt{1 + x^2} + C$

23. $\ln|\sec(e^x) + \tan(e^x)| + C$ **25.** $u = e^{2x} + 1,\ \dfrac{1}{2}\displaystyle\int u^{-1/2}du = \sqrt{e^{2x} + 1} + C$

27. Use integration by parts with $u = e^{3x}$, $dv = \sin 2x\, dx$ to get

$\displaystyle\int e^{3x}\sin 2x\, dx = -\dfrac{1}{2}e^{3x}\cos 2x + \dfrac{3}{2}\int e^{3x}\cos 2x\, dx$ and again with

$u = e^{3x}$, $dv = \cos 2x\, dx$ to get $\displaystyle\int e^{3x}\cos 2x\, dx = \dfrac{1}{2}e^{3x}\sin 2x - \dfrac{3}{2}\int e^{3x}\sin 2x\, dx$ so, with

$I = \displaystyle\int e^{3x}\sin 2x\, dx,\ I = -\dfrac{1}{2}e^{3x}\cos 2x + \dfrac{3}{4}e^{3x}\sin 2x - \dfrac{9}{4}I,\ I = \dfrac{1}{13}e^{3x}(3\sin 2x - 2\cos 2x) + C$

29. $u = \sin^{-1}(x/2),\ dv = dx,\ du = 1/\sqrt{4 - x^2}dx,\ v = x$

$\displaystyle\int_1^2 \sin^{-1}(x/2)dx = x\sin^{-1}(x/2)\Big]_1^2 - \int_1^2 x(4 - x^2)^{-1/2}dx$

$\qquad = (2)(\pi/2) - (1)(\pi/6) + (4 - x^2)^{1/2}\Big]_1^2 = 5\pi/6 - \sqrt{3}$

31. $u = \sin(3\ln x),\ dv = dx,\ du = \dfrac{3}{x}\cos(3\ln x)dx,\ v = x$

$\displaystyle\int \sin(3\ln x)dx = x\sin(3\ln x) - 3\int \cos(3\ln x)dx.$ Use $u = \cos(3\ln x),\ dv = dx$ to get

$\displaystyle\int \cos(3\ln x)dx = x\cos(3\ln x) + 3\int \sin(3\ln x)dx$ so

$\displaystyle\int \sin(3\ln x)dx = x\sin(3\ln x) - 3x\cos(3\ln x) - 9\int \sin(3\ln x)dx,$

$\displaystyle\int \sin(3\ln x)dx = \dfrac{1}{10}x[\sin(3\ln x) - 3\cos(3\ln x)] + C$

33. $\displaystyle\int x(x^2 - 9)^{-1/2}dx = \sqrt{x^2 - 9} + C$

35. $x = 3\sin\theta,\ dx = 3\cos\theta\,d\theta$

$$3\int_{\sin^{-1}(1/3)}^{\pi/2}\frac{\cos^2\theta}{\sin\theta}\,d\theta = 3\int_{\sin^{-1}(1/3)}^{\pi/2}\frac{1-\sin^2\theta}{\sin\theta}\,d\theta = 3\int_{\sin^{-1}(1/3)}^{\pi/2}(\csc\theta - \sin\theta)d\theta$$

$$= -3\ln|\csc\theta + \cot\theta| + 3\cos\theta\Big]_{\sin^{-1}(1/3)}^{\pi/2}$$

$$= -3\ln(1) + 3\ln|3 + \sqrt{8}| - 3(\sqrt{8}/3) = 3\ln(3 + \sqrt{8}) - \sqrt{8}$$

37. $u = \sqrt{2x+3},\ x = (u^2 - 3)/2,\ dx = u\,du$

$$\frac{1}{4}\int(u^2 - 3)^2 du = \frac{1}{4}\int(u^4 - 6u^2 + 9)du$$

$$= \frac{1}{4}\left(\frac{1}{5}u^5 - 2u^3 + 9u\right) + C = \frac{1}{20}u(u^4 - 10u^2 + 45) + C$$

$$= \frac{1}{20}\sqrt{2x+3}(4x^2 + 12x + 9 - 20x - 30 + 45) + C$$

$$= \frac{1}{5}(x^2 - 2x + 6)\sqrt{2x+3} + C$$

39. $\dfrac{1}{2}\displaystyle\int\dfrac{1}{\sqrt{1 - (t + 1/2)^2}}\,dt = \dfrac{1}{2}\sin^{-1}(t + 1/2) + C$

41. $x = a\sin\theta,\ dx = a\cos\theta\,d\theta;\ \dfrac{1}{a^2}\displaystyle\int\csc^2\theta\,d\theta = -\dfrac{1}{a^2}\cot\theta + C = -\dfrac{\sqrt{a^2 - x^2}}{a^2 x} + C$

43. $x = a\sin\theta,\ dx = a\cos\theta\,d\theta$

$$a^2\int\cos^2\theta\,d\theta = \frac{1}{2}a^2\theta + \frac{1}{4}a^2\sin 2\theta + C = \frac{1}{2}a^2\sin^{-1}(x/a) + \frac{1}{2}x\sqrt{a^2 - x^2} + C$$

45. $\displaystyle\int\dfrac{x-2}{\sqrt{4 - (x-2)^2}}\,dx = \int\dfrac{u}{\sqrt{4 - u^2}}\,du\quad (u = x - 2) = -\sqrt{4 - u^2} + C = -\sqrt{4x - x^2} + C$

47. $2x^2 + 3x + 1 = (2x + 1)(x + 1),\ \dfrac{1}{(2x + 1)(x + 1)} = \dfrac{2}{2x + 1} - \dfrac{1}{x + 1}$

$$\int\frac{dx}{(2x + 1)(x + 1)} = \ln\left|\frac{2x + 1}{x + 1}\right| + C$$

49. $\dfrac{x + 1}{x(x + 3)(x - 2)} = \dfrac{-1/6}{x} + \dfrac{-2/15}{x + 3} + \dfrac{3/10}{x - 2}$

$$\int\frac{x + 1}{x^3 + x^2 - 6x}\,dx = -\frac{1}{6}\ln|x| - \frac{2}{15}\ln|x + 3| + \frac{3}{10}\ln|x - 2| + C$$

51. $u = x^3 - 3x$, $\dfrac{1}{3}\displaystyle\int \dfrac{1}{u}\,du = \dfrac{1}{3}\ln|x^3 - 3x| + C$

53. $x^4 - 1 = (x+1)(x-1)(x^2+1)$

$$\int \frac{2x^2 + 5}{x^4 - 1}\,dx = \int \left[\frac{-7/4}{x+1} + \frac{7/4}{x-1} + \frac{-3/2}{x^2+1}\right]\,dx = \frac{7}{4}\ln\left|\frac{x-1}{x+1}\right| - \frac{3}{2}\tan^{-1}x + C$$

55. $\displaystyle\int \frac{dx}{(x^2+4)(x-3)} = \int \left[\frac{(-1/13)x - (3/13)}{x^2+4} + \frac{1/13}{x-3}\right]\,dx$

$$= -\frac{1}{26}\ln(x^2+4) - \frac{3}{26}\tan^{-1}\frac{x}{2} + \frac{1}{13}\ln|x-3| + C$$

57. $\dfrac{3x^2 + 12x + 2}{(x^2+4)^2} = \dfrac{3}{x^2+4} + \dfrac{12x - 10}{(x^2+4)^2} = \dfrac{3}{x^2+4} + \dfrac{12x}{(x^2+4)^2} - \dfrac{10}{(x^2+4)^2}$

$$\int \frac{1}{(x^2+4)^2} = \frac{1}{8}\int \cos^2\theta\,d\theta \quad (x = 2\tan\theta)$$

$$= \frac{1}{16}\theta + \frac{1}{32}\sin 2\theta + C_1 = \frac{1}{16}\tan^{-1}(x/2) + \frac{x}{8(x^2+4)} + C_1$$

so $\displaystyle\int \frac{3x^2 + 12x + 2}{(x^2+4)^2}\,dx = \frac{3}{2}\tan^{-1}\frac{x}{2} - \frac{6}{x^2+4} - \frac{5}{8}\tan^{-1}\frac{x}{2} - \frac{5x}{4(x^2+4)} + C$

$$= \frac{7}{8}\tan^{-1}\frac{x}{2} - \frac{5x + 24}{4(x^2+4)} + C$$

59. $\displaystyle\int \frac{x}{(x+1)^2 + 4}\,dx = \int \frac{u-1}{u^2+4}\,du \quad (u = x+1)$

$$= \frac{1}{2}\ln(u^2+4) - \frac{1}{2}\tan^{-1}\frac{u}{2} + C = \frac{1}{2}\ln(x^2 + 2x + 5) - \frac{1}{2}\tan^{-1}\frac{x+1}{2} + C$$

61. $u = \sqrt{2}x$, $\dfrac{1}{\sqrt{2}}\displaystyle\int \dfrac{1}{\sqrt{3 - u^2}}\,du = \dfrac{1}{\sqrt{2}}\sin^{-1}\sqrt{2/3}\,x + C$

63. $u = \sqrt{t}$, $t = u^2$, $dt = 2u\,du$; $\displaystyle\int \frac{2u^2}{u^2+1}\,du = 2\int \left[1 - \frac{1}{u^2+1}\right]\,du = 2\sqrt{t} - 2\tan^{-1}\sqrt{t} + C$

65. $u = 1 + x^{1/3}$, $x = (u-1)^3$, $dx = 3(u-1)^2\,du$

$$3\int_1^2 \frac{(u-1)^4}{u}\,du = 3\int_1^2 \left[u^3 - 4u^2 + 6u - 4 + \frac{1}{u}\right]\,du$$

$$= 3\left[\frac{1}{4}u^4 - \frac{4}{3}u^3 + 3u^2 - 4u + \ln u\right]_1^2 = -7/4 + 3\ln 2$$

67. $u = \tan(x/2)$, $\tan x = \sin x / \cos x = 2u/(1 - u^2)$

$$\int \frac{dx}{1 - \tan x} = \int \frac{2u^2 - 2}{(u^2 + 1)(u^2 + 2u - 1)} du$$

$$\frac{2u^2 - 2}{(u^2 + 1)(u^2 + 2u - 1)} = \frac{u + 1}{u^2 + 1} + \frac{-u - 1}{u^2 + 2u - 1}$$

$$\int \frac{dx}{1 - \tan x} = \frac{1}{2} \ln(u^2 + 1) + \tan^{-1} u - \frac{1}{2} \ln |u^2 + 2u - 1| + C$$

$$= \tan^{-1} u - \frac{1}{2} \ln \left| \frac{u^2 + 2u - 1}{u^2 + 1} \right| + C = \tan^{-1} u - \frac{1}{2} \ln \left| \frac{2u}{1 + u^2} - \frac{1 - u^2}{1 + u^2} \right| + C$$

$$= \frac{x}{2} - \frac{1}{2} \ln |\sin x - \cos x| + C$$

69. $u = \tan(x/2)$, $\tan x = 2u/(1 - u^2)$

$$\int \frac{dx}{\sin x - \tan x} = \frac{1}{2} \int \frac{u^2 - 1}{u^3} du = \frac{1}{2} \int (1/u - u^{-3}) du = \frac{1}{2} \ln \left| \tan \frac{x}{2} \right| + \frac{1}{4} \cot^2 \frac{x}{2} + C$$

71. $\int \frac{dx}{x^2 - a^2} = \int \left[\frac{1/(2a)}{x - a} + \frac{-1/(2a)}{x + a} \right] dx$

$$= \frac{1}{2a} \ln |x - a| - \frac{1}{2a} \ln |x + a| + C = \frac{1}{2a} \ln \left| \frac{x - a}{x + a} \right| + C$$

73. **(a)** $A = \int_0^2 \frac{1}{4 + x^2} dx = \frac{1}{2} \tan^{-1} \frac{x}{2} \Big]_0^2 = \pi/8$

(b) $V = \pi \int_0^2 \frac{1}{(4 + x^2)^2} dx = \frac{\pi}{8} \int_0^{\pi/4} \cos^2 \theta \, d\theta \quad (x = 2 \tan \theta)$

$$= \frac{\pi}{16} \left(\theta + \frac{1}{2} \sin 2\theta \right) \Big]_0^{\pi/4} = \pi(\pi + 2)/64$$

(c) $V = 2\pi \int_0^2 \frac{x}{4 + x^2} dx = \pi \ln(4 + x^2) \Big]_0^2 = \pi \ln 2$

75. **(a)** $\int x^3 e^{2x} dx = \frac{1}{2} x^3 e^{2x} - \frac{3}{2} \int x^2 e^{2x} dx = \frac{1}{2} x^3 e^{2x} - \frac{3}{2} \left[\frac{1}{2} x^2 e^{2x} - \int x e^{2x} dx \right]$

$$= \frac{1}{2} x^3 e^{2x} - \frac{3}{4} x^2 e^{2x} + \frac{3}{2} \left[\frac{1}{2} x e^{2x} - \frac{1}{2} \int e^{2x} dx \right]$$

$$= \frac{1}{2} x^3 e^{2x} - \frac{3}{4} x^2 e^{2x} + \frac{3}{4} x e^{2x} - \frac{3}{8} e^{2x} + C$$

(b) $\displaystyle\int_0^{\pi/10} x^2 \sin 5x\, dx = -\frac{1}{5}x^2 \cos 5x\Big]_0^{\pi/10} + \frac{2}{5}\int_0^{\pi/10} x \cos 5x\, dx$

$\displaystyle = 0 + \frac{2}{5}\left[\frac{1}{5}x \sin 5x\right]_0^{\pi/10} - \frac{2}{25}\int_0^{\pi/10} \sin 5x\, dx$

$\displaystyle = \frac{2}{25}(\pi/10) + \frac{2}{125}\cos 5x\Big]_0^{\pi/10} = \pi/125 + \frac{2}{125}(0-1) = (\pi - 2)/125$

(c) $\displaystyle\int \sin^2 x \cos^4 x\, dx = \frac{1}{6}\sin^3 x \cos^3 x + \frac{1}{2}\int \sin^2 x \cos^2 x\, dx$

$\displaystyle = \frac{1}{6}\sin^3 x \cos^3 x + \frac{1}{2}\left[\frac{1}{4}\sin^3 x \cos x + \frac{1}{4}\int \sin^2 x\, dx\right]$

$\displaystyle = \frac{1}{6}\sin^3 x \cos^3 x + \frac{1}{8}\sin^3 x \cos x + \frac{1}{8}\left[-\frac{1}{2}\sin x \cos x + \frac{1}{2}\int dx\right]$

$\displaystyle = \frac{1}{6}\sin^3 x \cos^3 x + \frac{1}{8}\sin^3 x \cos x - \frac{1}{16}\sin x \cos x + \frac{x}{16} + C$

77. **(a)** $\displaystyle\int \frac{1 - \cos^2 \theta}{\cos^5 \theta}\sin \theta\, d\theta = \int (\cos^{-5}\theta - \cos^{-3}\theta)\sin \theta\, d\theta$

$\displaystyle = \frac{1}{4}\cos^{-4}\theta - \frac{1}{2}\cos^{-2}\theta + C = \frac{1}{4}\sec^4 \theta - \frac{1}{2}\sec^2 \theta + C$

(b) $\displaystyle\int \tan^3 \theta \sec^2 \theta\, d\theta = \frac{1}{4}\tan^4 \theta + C$ but $\displaystyle\frac{1}{4}\tan^4 \theta = \frac{1}{4}(\sec^2 \theta - 1)^2 = \frac{1}{4}(\sec^4 \theta - 2\sec^2 \theta + 1)$

so the answers to (a) and (b) differ by $1/4$.

79. **(a)** 58.9275   **(b)** 54.7328   **81.** **(a)** 1.8277   **(b)** 1.8278

83. 0.36972

85. $\displaystyle\int \frac{1}{e^{ax} + 1}dx = \int \frac{e^{-ax}}{1 + e^{-ax}}dx = -\frac{1}{a}\ln(1 + e^{-ax}) + C$

87. $\displaystyle\frac{\sqrt{1+x} + \sqrt{1-x}}{\sqrt{1+x} - \sqrt{1-x}} = \frac{(\sqrt{1+x} + \sqrt{1-x})^2}{2x} = \frac{1 + \sqrt{1-x^2}}{x}$

$\displaystyle\int \left[\frac{1}{x} + \frac{\sqrt{1-x^2}}{x}\right]dx = \int \frac{1}{x}dx + \int \frac{\sqrt{1-x^2}}{x}dx = \ln|x| + \int \frac{u^2}{u^2 - 1}du \quad (u = \sqrt{1 - x^2})$

$\displaystyle = \ln|x| + \int \left[1 + \frac{1}{u^2 - 1}\right]du = \ln|x| + u + \frac{1}{2}\ln\frac{1-u}{1+u} + C$

$$= \ln|x| + \sqrt{1-x^2} + \frac{1}{2}\ln\frac{1-\sqrt{1-x^2}}{1+\sqrt{1-x^2}} + C$$

$$= \sqrt{1-x^2} + \frac{1}{2}\ln(x^2) + \frac{1}{2}\ln\frac{(1-\sqrt{1-x^2})^2}{x^2} + C$$

$$= \sqrt{1-x^2} + \ln(1-\sqrt{1-x^2}) + C$$

89. $x - 1 = 1/u$, $x = 1 + 1/u$, $dx = -(1/u^2)du$

$$\int \frac{\sqrt{x+1}}{(x-1)^{5/2}}dx = -\int \sqrt{2u+1}\,du = -\frac{1}{3}(2u+1)^{3/2} + C = -\frac{1}{3}\left[\frac{x+1}{x-1}\right]^{3/2} + C$$

91. $\displaystyle\int \frac{1}{x^6(3+2x^{-5})}dx = -\frac{1}{10}\int \frac{1}{u}du \quad (u = 3 + 2x^{-5}) = -\frac{1}{10}\ln|u| + C = -\frac{1}{10}\ln|3+2x^{-5}| + C$

93. $\displaystyle\int \sqrt{x - \sqrt{x^2-4}}\,dx = \frac{1}{\sqrt{2}}\int(\sqrt{x+2} - \sqrt{x-2})dx = \frac{\sqrt{2}}{3}[(x+2)^{3/2} - (x-2)^{3/2}] + C$

CHAPTER 10
Improper Integrals; L'Hôpital's Rule

EXERCISE SET 10.1

1. $\lim\limits_{\ell\to+\infty}(-e^{-x})\Big]_0^\ell = \lim\limits_{\ell\to+\infty}(-e^{-\ell}+1) = 1$

3. $\lim\limits_{\ell\to+\infty}2\sqrt{x}\Big]_1^\ell = \lim\limits_{\ell\to+\infty}2(\sqrt{\ell}-1) = +\infty$, divergent

5. $\lim\limits_{\ell\to+\infty}\ln\dfrac{x-1}{x+1}\Big]_4^\ell = \lim\limits_{\ell\to+\infty}\left(\ln\dfrac{\ell-1}{\ell+1}-\ln\dfrac{3}{5}\right) = -\ln\dfrac{3}{5} = \ln\dfrac{5}{3}$

7. $\lim\limits_{\ell\to+\infty}-\dfrac{1}{2\ln^2 x}\Big]_e^\ell = \lim\limits_{\ell\to+\infty}\left[-\dfrac{1}{2\ln^2\ell}+\dfrac{1}{2}\right] = \dfrac{1}{2}$

9. $\lim\limits_{\ell\to+\infty}-\dfrac{1}{2(x^2+1)}\Big]_a^\ell = \lim\limits_{\ell\to+\infty}\left[-\dfrac{1}{2(\ell^2+1)}+\dfrac{1}{2(a^2+1)}\right] = \dfrac{1}{2(a^2+1)}$

11. $\lim\limits_{\ell\to-\infty}-\dfrac{1}{4(2x-1)^2}\Big]_\ell^0 = \lim\limits_{\ell\to-\infty}\dfrac{1}{4}[-1+1/(2\ell-1)^2] = -1/4$

13. $\lim\limits_{\ell\to-\infty}\dfrac{1}{3}e^{3x}\Big]_\ell^0 = \lim\limits_{\ell\to-\infty}\left[\dfrac{1}{3}-\dfrac{1}{3}e^{3\ell}\right] = \dfrac{1}{3}$

15. $\displaystyle\int_{-\infty}^{+\infty}x^3\,dx$ converges if $\displaystyle\int_{-\infty}^{0}x^3\,dx$ and $\displaystyle\int_{0}^{+\infty}x^3\,dx$ both converge; it diverges if either (or both)

diverge. $\displaystyle\int_{0}^{+\infty}x^3\,dx = \lim\limits_{\ell\to+\infty}\dfrac{1}{4}x^4\Big]_0^\ell = \lim\limits_{\ell\to+\infty}\dfrac{1}{4}\ell^4 = +\infty$ so $\displaystyle\int_{-\infty}^{+\infty}x^3\,dx$ is divergent.

17. $\displaystyle\int_{0}^{+\infty}\dfrac{x}{(x^2+3)^2}\,dx = \lim\limits_{\ell\to+\infty}-\dfrac{1}{2(x^2+3)}\Big]_0^\ell = \lim\limits_{\ell\to+\infty}\dfrac{1}{2}[-1/(\ell^2+3)+1/3] = \dfrac{1}{6}$,

similarly $\displaystyle\int_{-\infty}^{0}\dfrac{x}{(x^2+3)^2}\,dx = -1/6$ so $\displaystyle\int_{-\infty}^{\infty}\dfrac{x}{(x^2+3)^2}\,dx = 1/6+(-1/6) = 0$

19. $\lim\limits_{\ell\to3+}-\dfrac{1}{x-3}\Big]_\ell^4 = \lim\limits_{\ell\to3+}\left[-1+\dfrac{1}{\ell-3}\right] = +\infty$, divergent

21. $\displaystyle\lim_{\ell\to\pi/2^-} -\ln(\cos x)\Big]_0^\ell = \lim_{\ell\to\pi/2^-} -\ln(\cos\ell) = +\infty$, divergent

23. $\displaystyle\lim_{\ell\to1^-} \sin^{-1}x\Big]_0^\ell = \lim_{\ell\to1^-} \sin^{-1}\ell = \pi/2$

25. $\displaystyle\lim_{\ell\to\pi/6^-} -\sqrt{1-2\sin x}\Big]_0^\ell = \lim_{\ell\to\pi/6^-} (-\sqrt{1-2\sin\ell}+1) = 1$

27. $\displaystyle\int_0^2 \frac{dx}{x-2} = \lim_{\ell\to2^-} \ln|x-2|\Big]_0^\ell = \lim_{\ell\to2^-} (\ln|\ell-2| - \ln 2) = -\infty$, divergent

29. $\displaystyle\int_0^8 x^{-1/3}dx = \lim_{\ell\to0^+} \frac{3}{2}x^{2/3}\Big]_\ell^8 = \lim_{\ell\to0^+} \frac{3}{2}(4-\ell^{2/3}) = 6,$

$\displaystyle\int_{-1}^0 x^{-1/3}dx = \lim_{\ell\to0^-} \frac{3}{2}x^{2/3}\Big]_{-1}^\ell = \lim_{\ell\to0^-} \frac{3}{2}(\ell^{2/3}-1) = -3/2$

so $\displaystyle\int_{-1}^8 x^{-1/3}dx = 6 + (-3/2) = 9/2$

31. Define $\displaystyle\int_0^{+\infty} \frac{1}{x^2}dx = \int_0^a \frac{1}{x^2}dx + \int_a^{+\infty} \frac{1}{x^2}dx$ where $a > 0$; take $a = 1$ for convenience,

$\displaystyle\int_0^1 \frac{1}{x^2}dx = \lim_{\ell\to0^+} (-1/x)\Big]_\ell^1 = \lim_{\ell\to0^+} (1/\ell - 1) = +\infty$ so $\displaystyle\int_0^{+\infty} \frac{1}{x^2}dx$ is divergent.

33. $\displaystyle\int_0^{+\infty} \frac{e^{-\sqrt{x}}}{\sqrt{x}}dx = 2\int_0^{+\infty} e^{-u}du = 2\lim_{\ell\to+\infty} (-e^{-u})\Big]_0^\ell = 2\lim_{\ell\to+\infty} (1-e^{-\ell}) = 2$

35. $\displaystyle\int_0^{+\infty} \frac{e^{-x}}{\sqrt{1-e^{-x}}}dx = \int_0^1 \frac{du}{\sqrt{u}} = \lim_{\ell\to0^+} 2\sqrt{u}\Big]_\ell^1 = \lim_{\ell\to0^+} 2(1-\sqrt{\ell}) = 2$

37. $\displaystyle\int_0^{+\infty} e^{-ax}dx = \lim_{\ell\to+\infty} -\frac{1}{a}e^{-ax}\Big]_0^\ell = \lim_{\ell\to+\infty} \left[-\frac{1}{a}e^{-a\ell} + \frac{1}{a}\right] = \frac{1}{a} = 5, a = \frac{1}{5}$

39. $u = \sqrt{x}$, $\displaystyle\int_0^{+\infty} \frac{e^{-x}}{\sqrt{x}}dx = 2\int_0^{+\infty} e^{-u^2}du = 2(\sqrt{\pi}/2) = \sqrt{\pi}$

41. $u = \sqrt{x}$, $\displaystyle\int_0^{+\infty} \frac{\sin x}{\sqrt{x}}dx = 2\int_0^{+\infty} \sin(u^2)du = \sqrt{\pi/2}$

43. **(a)** $\displaystyle\int_0^{+\infty} \cos x\, dx = \lim_{\ell\to+\infty} \sin x \Big]_0^{\ell} = \lim_{\ell\to+\infty} \sin\ell$ which does not exist and does not become infinite.

   (b) $u = \sqrt{x}$, $\displaystyle\int_0^{+\infty} \frac{\cos\sqrt{x}}{\sqrt{x}} dx = 2\int_0^{+\infty} \cos u\, du$; $\displaystyle\int_0^{+\infty} \cos u\, du$ diverges

   (c) $u = 1/x$, $\displaystyle\int_0^1 \frac{\cos(1/x)}{x^2} dx = -\int_{+\infty}^1 \cos u\, du = \int_1^{+\infty} \cos u\, du$ which diverges

45. $\displaystyle\lim_{\ell\to+\infty}\int_0^{\ell} e^{-x}\cos x\, dx = \lim_{\ell\to+\infty} \frac{1}{2} e^{-x}(\sin x - \cos x)\Big]_0^{\ell} = \lim_{\ell\to+\infty} \frac{1}{2}[e^{-\ell}(\sin\ell - \cos\ell) + 1]$

   but both $e^{-\ell}\sin\ell$ and $e^{-\ell}\cos\ell \to 0$ as $\ell \to +\infty$ (by the Squeezing Theorem because

   $-e^{-\ell} \le e^{-\ell}\sin\ell \le e^{-\ell}$ and $-e^{-\ell} \le e^{-\ell}\cos\ell \le e^{-\ell}$) so $\displaystyle\int_0^{+\infty} e^{-x}\cos x\, dx = 1/2$.

47. If $p = 1$, $\displaystyle\int_0^1 \frac{dx}{x} = \lim_{\ell\to 0^+} \ln x \Big]_{\ell}^1 = +\infty$;

   if $p \ne 1$, $\displaystyle\int_0^1 \frac{dx}{x^p} = \lim_{\ell\to 0^+} \frac{x^{1-p}}{1-p}\Big]_{\ell}^1 = \lim_{\ell\to 0^+}[(1 - \ell^{1-p})/(1-p)] = \begin{cases} 1/(1-p), & p < 1 \\ +\infty, & p > 1 \end{cases}$

49. **(a)** $\displaystyle\int_2^{+\infty} \frac{x}{x^5 + 1} dx \le \int_2^{+\infty} \frac{dx}{x^4} = \lim_{\ell\to+\infty} -\frac{1}{3x^3}\Big]_2^{\ell} = 1/24$

   (b) $\displaystyle\int_1^{+\infty} e^{-x^2} dx \le \int_1^{+\infty} xe^{-x^2} dx = \lim_{\ell\to+\infty} -\frac{1}{2}e^{-x^2}\Big]_1^{\ell} = \frac{1}{2}e^{-1}$

51. $A = \displaystyle\int_0^{+\infty} e^{-3x} dx = \lim_{\ell\to+\infty} -\frac{1}{3}e^{-3x}\Big]_0^{\ell} = \lim_{\ell\to+\infty} \frac{1}{3}(1 - e^{-3\ell}) = 1/3$.

53. **(a)** $V = \displaystyle\int_1^{+\infty} \frac{\pi}{x^2} dx = \lim_{\ell\to+\infty} -\pi/x \Big]_1^{\ell} = \pi$,

   (b) $S = \displaystyle\int_1^{+\infty} 2\pi(1/x)\sqrt{1 + 1/x^4} dx$, but $\sqrt{1 + 1/x^4} \ge 1$ if $x \ge 1$ so

   $(2\pi/x)\sqrt{1 + 1/x^4} \ge 2\pi/x$, $\displaystyle\lim_{\ell\to+\infty}\int_1^{\ell} (2\pi/x) dx = +\infty$; S is infinite.

55. $\int \dfrac{dx}{(r^2+x^2)^{3/2}} = \dfrac{1}{r^2}\int \cos\theta\, d\theta \; (x = r\tan\theta) = \dfrac{1}{r^2}\sin\theta + C = \dfrac{x}{r^2\sqrt{r^2+x^2}} + C$

so $u = \dfrac{2\pi NIr}{k} \displaystyle\lim_{\ell\to+\infty} \dfrac{x}{r^2\sqrt{r^2+x^2}}\Bigg]_a^\ell = \dfrac{2\pi NI}{kr}\displaystyle\lim_{\ell\to+\infty}(\ell/\sqrt{r^2+\ell^2} - a/\sqrt{r^2+a^2})$

$$= \dfrac{2\pi NI}{kr}(1 - a/\sqrt{r^2+a^2})$$

57. **(a)** $\displaystyle\int_{4000}^{4000+\ell} 9.6\times 10^{10}x^{-2}dx$

(b) $\displaystyle\int_{4000}^{+\infty} 9.6\times 10^{10}x^{-2}dx = \lim_{\ell\to+\infty} -9.6\times 10^{10}/x\Bigg]_{4000}^\ell = 2.4\times 10^7$

59. $\mathcal{L}\{1\} = \displaystyle\int_0^{+\infty} e^{-st}dt = \lim_{\ell\to+\infty} -\dfrac{1}{s}e^{-st}\Bigg]_0^\ell = \dfrac{1}{s}$

61. $\mathcal{L}\{\sin t\} = \displaystyle\int_0^{+\infty} e^{-st}\sin t\, dt = \lim_{\ell\to+\infty} \dfrac{e^{-st}}{s^2+1}(-s\sin t - \cos t)\Bigg]_0^\ell = \dfrac{1}{s^2+1}$

63. $2\displaystyle\int_0^1 \cos(u^2)du \approx 1.809$

65. **(a)** $\displaystyle\int_0^4 \dfrac{1}{x^6+1}dx \approx 1.047;\ \pi/3 \approx 1.047$

(b) $\displaystyle\int_0^{+\infty} \dfrac{1}{x^6+1}dx = \int_0^4 \dfrac{1}{x^6+1}dx + \int_4^{+\infty} \dfrac{1}{x^6+1}dx$ so

$$E = \int_4^{+\infty} \dfrac{1}{x^6+1}dx < \int_4^{+\infty} \dfrac{1}{x^6}dx = \dfrac{1}{5(4)^5} < 2\times 10^{-4}.$$

EXERCISE SET 10.2

1. $\displaystyle\lim_{x\to 1} \dfrac{1/x}{1} = 1$ **3.** $\displaystyle\lim_{x\to 0} \dfrac{e^x}{\cos x} = 1$ **5.** $\displaystyle\lim_{\theta\to 0} \dfrac{\sec^2\theta}{1} = 1$

7. $\displaystyle\lim_{x\to 1} \dfrac{1/x}{\pi\sec^2\pi x} = 1/\pi$ **9.** $\displaystyle\lim_{x\to\pi^+} \dfrac{\cos x}{1} = -1$

11. $\displaystyle\lim_{x\to\pi/2^-} \dfrac{-\sin x}{-(1/2)(\pi/2-x)^{-1/2}} = \lim_{x\to\pi/2^-} 2(\sin x)\sqrt{\pi/2-x} = 0$

13. $\lim\limits_{x\to 0}\dfrac{e^x - e^{-x}}{2\sin 2x} = \lim\limits_{x\to 0}\dfrac{e^x + e^{-x}}{4\cos 2x} = 1/2$ **15.** $\lim\limits_{x\to 0}\dfrac{3\cos 3x}{2\cosh 2x} = 3/2$

17. $\lim\limits_{x\to \pi/2}\dfrac{2\cos 2x}{8x} = -\dfrac{1}{2\pi}$

19. $\lim\limits_{x\to 0}\dfrac{1 - 1/(x+1)}{2\sin 2x} = \lim\limits_{x\to 0}\dfrac{x}{2(x+1)\sin 2x} = \lim\limits_{x\to 0}\dfrac{1}{4(x+1)\cos 2x + 2\sin 2x} = 1/4$

21. $\lim\limits_{x\to 0}\dfrac{-2x + 2\sin x}{4x^3} = \lim\limits_{x\to 0}\dfrac{-2 + 2\cos x}{12x^2} = \lim\limits_{x\to 0}-\dfrac{\sin x}{12x} = -1/12$

23. $\lim\limits_{x\to 0}\dfrac{x - \sin x}{x^3} = \lim\limits_{x\to 0}\dfrac{1 - \cos x}{3x^2} = \lim\limits_{x\to 0}\dfrac{\sin x}{6x} = 1/6$ **25.** $\lim\limits_{x\to 0}\dfrac{ae^{ax} - be^{bx}}{1} = a - b$

27. $\lim\limits_{x\to 0}\dfrac{1 - \sec^2 x}{\cos x - 1} = \lim\limits_{x\to 0}\dfrac{\cos^2 x - 1}{\cos^2 x(\cos x - 1)} = \lim\limits_{x\to 0}\dfrac{\cos x + 1}{\cos^2 x} = 2$

29. $\lim\limits_{x\to +\infty}\dfrac{-\dfrac{1}{1+x^2}}{\dfrac{1}{1+1/x^2}\left(-\dfrac{2}{x^3}\right)} = \lim\limits_{x\to +\infty}\dfrac{x}{2} = +\infty$

31. $\lim\limits_{x\to 0+}\dfrac{-\sin x/\cos x}{-3\sin 3x/\cos 3x} = \lim\limits_{x\to 0+}\dfrac{\tan x}{3\tan 3x} = \lim\limits_{x\to 0+}\dfrac{\sec^2 x}{9\sec^2 3x} = 1/9$

33. **(a)** L'Hôpital's rule does not apply to the problem $\lim\limits_{x\to 1}\dfrac{3x^2 - 2x + 1}{3x^2 - 2x}$ because it is not a $\dfrac{0}{0}$ form

      (b) $\lim\limits_{x\to 1}\dfrac{3x^2 - 2x + 1}{3x^2 - 2x} = 2$

35. $\lim\limits_{x\to 0}\dfrac{k + \cos \ell x}{x^2}$ does not exist if $k \neq -1$ so suppose $k = -1$, then

$\lim\limits_{x\to 0}\dfrac{-1 + \cos \ell x}{x^2} = \lim\limits_{x\to 0}\dfrac{-\ell \sin \ell x}{2x} = \lim\limits_{x\to 0}\dfrac{-\ell^2 \cos \ell x}{2} = -\ell^2/2 = -4$ if $\ell^2 = 8$, $\ell = \pm 2\sqrt{2}$

37. $\lim\limits_{x\to 1}\dfrac{\ln x}{x^4 - 1} = \lim\limits_{x\to 1}\dfrac{1/x}{4x^3} = \dfrac{1}{4}$; $\lim\limits_{x\to 1}\sqrt{\dfrac{\ln x}{x^4 - 1}} = \sqrt{\lim\limits_{x\to 1}\dfrac{\ln x}{x^4 - 1}} = \dfrac{1}{2}$

39. $\lim\limits_{x\to +\infty}\dfrac{\ln\left(\dfrac{x+1}{x-1}\right)}{1/x} = \lim\limits_{x\to +\infty}\dfrac{-2/(x^2 - 1)}{-1/x^2} = \lim\limits_{x\to +\infty}\dfrac{2x^2}{x^2 - 1} = \lim\limits_{x\to +\infty}\dfrac{4x}{2x} = 2$

41. $\lim\limits_{x \to 0^+} \dfrac{\sin(1/x)}{(\sin x)/x}$, $\lim\limits_{x \to 0^+} \dfrac{\sin x}{x} = 1$ but $\lim\limits_{x \to 0^+} \sin(1/x)$ does not exist because $\sin(1/x)$ oscillates

between -1 and 1 as $x \to +\infty$, so $\lim\limits_{x \to 0^+} \dfrac{x \sin(1/x)}{\sin x}$ does not exist.

43. $T(\theta) = \dfrac{1}{2}\sin\theta(1 - \cos\theta)$,

$S(\theta) = \dfrac{1}{2}(\theta - \sin\theta)$,

$\lim\limits_{\theta \to 0^+} \dfrac{T(\theta)}{S(\theta)} = \lim\limits_{\theta \to 0^+} \dfrac{\sin\theta - \sin\theta\cos\theta}{\theta - \sin\theta}$

$\qquad\qquad = \lim\limits_{\theta \to 0^+} \dfrac{\cos\theta + \sin^2\theta - \cos^2\theta}{1 - \cos\theta}$

$\qquad\qquad = \lim\limits_{\theta \to 0^+} \dfrac{\cos\theta - \cos 2\theta}{1 - \cos\theta}$

$\qquad\qquad = \lim\limits_{\theta \to 0^+} \dfrac{-\sin\theta + 2\sin 2\theta}{\sin\theta}$

$\qquad\qquad = \lim\limits_{\theta \to 0^+} \dfrac{-\cos\theta + 4\cos 2\theta}{\cos\theta} = 3$

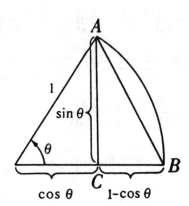

EXERCISE SET 10.3

1. $\lim\limits_{x \to +\infty} \dfrac{1/x}{1} = 0$

3. $\lim\limits_{x \to 0^+} \dfrac{-\csc^2 x}{1/x} = \lim\limits_{x \to 0^+} \dfrac{-x}{\sin^2 x} = \lim\limits_{x \to 0^+} \dfrac{-1}{2\sin x \cos x} = -\infty$

5. $\lim\limits_{x \to +\infty} \dfrac{1 + \ln x}{1 + 1/x} = +\infty$

7. $\lim\limits_{x \to +\infty} \dfrac{100x^{99}}{e^x} = \lim\limits_{x \to +\infty} \dfrac{(100)(99)x^{98}}{e^x} = \cdots = \lim\limits_{x \to +\infty} \dfrac{(100)(99)(98)\cdots(1)}{e^x} = 0$

9. $\lim\limits_{x \to +\infty} x e^{-x} = \lim\limits_{x \to +\infty} \dfrac{x}{e^x} = \lim\limits_{x \to +\infty} \dfrac{1}{e^x} = 0$

11. $\lim\limits_{x \to +\infty} x \sin(\pi/x) = \lim\limits_{x \to +\infty} \dfrac{\sin(\pi/x)}{1/x} = \lim\limits_{x \to +\infty} \dfrac{(-\pi/x^2)\cos(\pi/x)}{-1/x^2} = \lim\limits_{x \to +\infty} \pi\cos(\pi/x) = \pi$

13. $\lim\limits_{x \to +\infty} x(e^{\sin(2/x)} - 1) = \lim\limits_{x \to +\infty} \dfrac{e^{\sin(2/x)} - 1}{1/x} = \lim\limits_{x \to +\infty} \dfrac{(-2/x^2)e^{\sin(2/x)}}{-1/x^2} = \lim\limits_{x \to +\infty} 2e^{\sin(2/x)} = 2$

15. $y = (1 - 3/x)^x$, $\lim\limits_{x \to +\infty} \ln y = \lim\limits_{x \to +\infty} \dfrac{\ln(1 - 3/x)}{1/x} = \lim\limits_{x \to +\infty} \dfrac{-3}{1 - 3/x} = -3$, $\lim\limits_{x \to +\infty} y = e^{-3}$

17. $y = (e^x + x)^{1/x}$, $\lim\limits_{x \to 0} \ln y = \lim\limits_{x \to 0} \dfrac{\ln(e^x + x)}{x} = \lim\limits_{x \to 0} \dfrac{e^x + 1}{e^x + x} = 2$, $\lim\limits_{x \to 0} y = e^2$

19. $y = (1 + 1/x^2)^x$

$\lim\limits_{x \to +\infty} \ln y = \lim\limits_{x \to +\infty} \dfrac{\ln(1 + 1/x^2)}{1/x} = \lim\limits_{x \to +\infty} \dfrac{2x}{x^2 + 1} = \lim\limits_{x \to +\infty} \dfrac{1}{x} = 0$, $\lim\limits_{x \to +\infty} y = e^0 = 1$

21. $y = (1 + 1/x)^{x^2}$

$\lim\limits_{x \to +\infty} \ln y = \lim\limits_{x \to +\infty} \dfrac{\ln(1 + 1/x)}{1/x^2} = \lim\limits_{x \to +\infty} \dfrac{x^2}{2(x + 1)} = \lim\limits_{x \to +\infty} x = +\infty$, $\lim\limits_{x \to +\infty} y = +\infty$

23. $y = (2 - x)^{\tan(\pi x/2)}$, $\lim\limits_{x \to 1} \ln y = \lim\limits_{x \to 1} \dfrac{\ln(2 - x)}{\cot(\pi x/2)} = \lim\limits_{x \to 1} \dfrac{2 \sin^2(\pi x/2)}{\pi(2 - x)} = 2/\pi$, $\lim\limits_{x \to 1} y = e^{2/\pi}$

25. $y = x^{\sin x}$, $\lim\limits_{x \to 0^+} \ln y = \lim\limits_{x \to 0^+} \dfrac{\ln x}{\csc x} = \lim\limits_{x \to 0^+} \dfrac{1/x}{-\csc x \cot x} = \lim\limits_{x \to 0^+} \left(-\dfrac{\sin x}{x}\right) \tan x = 0$,

$\lim\limits_{x \to 0^+} y = e^0 = 1$

27. $y = (\sin x)^{3/\ln x}$, $\lim\limits_{x \to 0^+} \ln y = \lim\limits_{x \to 0^+} \dfrac{3 \ln \sin x}{\ln x} = \lim\limits_{x \to 0^+} (3 \cos x) \dfrac{x}{\sin x} = 3$, $\lim\limits_{x \to 0^+} y = e^3$

29. $y = (\tan x)^{\cos x}$, $\lim\limits_{x \to \pi/2^-} \ln y = \lim\limits_{x \to \pi/2^-} \dfrac{\ln \tan x}{\sec x} = \lim\limits_{x \to \pi/2^-} \dfrac{\sec^2 x / \tan x}{\sec x \tan x} = \lim\limits_{x \to \pi/2^-} \dfrac{\cos x}{\sin^2 x} = 0$,

$\lim\limits_{x \to \pi/2^-} y = e^0 = 1$

31. $y = (1 + x^2)^{1/\ln x}$

$\lim\limits_{x \to +\infty} \ln y = \lim\limits_{x \to +\infty} \dfrac{\ln(1 + x^2)}{\ln x} = \lim\limits_{x \to +\infty} \dfrac{2x^2}{1 + x^2} = \lim\limits_{x \to +\infty} 2 = 2$, $\lim\limits_{x \to +\infty} y = e^2$

33. $\lim\limits_{\theta \to 0} \left(\dfrac{1 + \cos \theta}{1 - \cos^2 \theta} - \dfrac{2}{\sin^2 \theta}\right) = \lim\limits_{\theta \to 0} \dfrac{\cos \theta - 1}{\sin^2 \theta} = \lim\limits_{\theta \to 0} \dfrac{-\sin \theta}{2 \sin \theta \cos \theta} = -1/2$

35. $\lim\limits_{x \to 0} \left(\dfrac{1}{\sin x} - \dfrac{1}{x}\right) = \lim\limits_{x \to 0} \dfrac{x - \sin x}{x \sin x} = \lim\limits_{x \to 0} \dfrac{1 - \cos x}{x \cos x + \sin x} = \lim\limits_{x \to 0} \dfrac{\sin x}{2 \cos x - x \sin x} = 0$

37. $\lim\limits_{x \to 0} \left(\dfrac{\cos x}{\sin x} - \dfrac{1}{\sin x}\right) = \lim\limits_{x \to 0} \dfrac{\cos x - 1}{\sin x} = \lim\limits_{x \to 0} \dfrac{-\sin x}{\cos x} = 0$

39. $\lim\limits_{x\to+\infty}[x-\ln(x^2+1)]=\lim\limits_{x\to+\infty}[\ln e^x-\ln(x^2+1)]=\lim\limits_{x\to+\infty}\ln\dfrac{e^x}{x^2+1},$

$\lim\limits_{x\to+\infty}\dfrac{e^x}{x^2+1}=\lim\limits_{x\to+\infty}\dfrac{e^x}{2x}=\lim\limits_{x\to+\infty}\dfrac{e^x}{2}=+\infty$ so $\lim\limits_{x\to+\infty}[x-\ln(x^2+1)]=+\infty$

41. $\lim\limits_{x\to0^+}\dfrac{\cot x}{\cot 2x}=\lim\limits_{x\to0^+}\dfrac{\tan 2x}{\tan x}=\lim\limits_{x\to0^+}\dfrac{2\sec^2 2x}{\sec^2 x}=2$

43. In each case, apply L'Hôpital's rule n times.

    (a) $\lim\limits_{x\to+\infty}\dfrac{x^n}{e^x}=\cdots=\lim\limits_{x\to+\infty}\dfrac{n(n-1)(n-2)\cdots(1)}{e^x}=0$

    (b) $\lim\limits_{x\to+\infty}\dfrac{e^x}{x^n}=\cdots=\lim\limits_{x\to+\infty}\dfrac{e^x}{n(n-1)(n-2)\cdots(1)}=+\infty$

45. **(a)** 0   **(b)** $+\infty$ **(c)** 0   **(d)** $-\infty$

      (e) $+\infty$ **(f)** $+\infty$ **(g)** $-\infty$ **(h)** $-\infty$

47. Let $y=x^x$, $\lim\limits_{x\to+\infty}y=+\infty$ and

$\lim\limits_{x\to0^+}y=1$. Use logarithmic differ-

entiation to get $dy/dx=x^x(1+\ln x)$

so $dy/dx=0$ when $x=e^{-1}$, $dy/dx<0$

if $x<e^{-1}$, $dy/dx>0$ if $x>e^{-1}$,

and $dy/dx\to-\infty$ as $x\to0^+$. Also,

$d^2y/dx^2=x^x[1/x+(1+\ln x)^2]>0$ for $x>0$ so the curve is always concave up.

49. $\displaystyle\int \ln x\,dx=x\ln x-x+C,$

$\displaystyle\int_0^1\ln x\,dx=\lim\limits_{\ell\to0^+}\int_\ell^1\ln x\,dx=\lim\limits_{\ell\to0^+}(x\ln x-x)\Big]_\ell^1=\lim\limits_{\ell\to0^+}(-1-\ell\ln\ell+\ell),$

but $\lim\limits_{\ell\to0^+}\ell\ln\ell=\lim\limits_{\ell\to0^+}\dfrac{\ln\ell}{1/\ell}=\lim\limits_{\ell\to0^+}(-\ell)=0$ so $\displaystyle\int_0^1\ln x\,dx=-1$

51. $\displaystyle\int xe^{-3x}\,dx = -\frac{1}{3}xe^{-3x} - \frac{1}{9}e^{-3x} + C,$

$$\int_0^{+\infty} xe^{-3x}\,dx = \lim_{\ell \to +\infty}\int_0^\ell xe^{-3x}\,dx = \lim_{\ell \to +\infty}\left(-\frac{1}{3}xe^{-3x} - \frac{1}{9}e^{-3x}\right)\Big]_0^\ell$$

$$= \lim_{\ell \to +\infty}\left(-\frac{1}{3}\ell e^{-3\ell} - \frac{1}{9}e^{-3\ell} + \frac{1}{9}\right),$$

but $\displaystyle\lim_{\ell \to +\infty}\ell e^{-3\ell} = \lim_{\ell \to +\infty}\frac{\ell}{e^{3\ell}} = \lim_{\ell \to +\infty}\frac{1}{3e^{3\ell}} = 0$ so $\displaystyle\int_0^{+\infty} xe^{-3x}\,dx = 1/9$

53. $\displaystyle V = 2\pi\int_0^{+\infty} xe^{-x}\,dx = 2\pi\lim_{\ell \to +\infty}-e^{-x}(x+1)\Big]_0^\ell = 2\pi\lim_{\ell \to +\infty}\left[1 - e^{-\ell}(\ell+1)\right],$

but $\displaystyle\lim_{\ell \to +\infty}e^{-\ell}(\ell+1) = \lim_{\ell \to +\infty}\frac{\ell+1}{e^\ell} = \lim_{\ell \to +\infty}\frac{1}{e^\ell} = 0$ so $V = 2\pi.$

55. **(a)** $\displaystyle\int_0^\ell \sqrt{1+t^3}\,dt \geq \int_0^\ell t^{3/2}\,dt = \frac{2}{5}t^{5/2}\Big]_0^\ell = \frac{2}{5}\ell^{5/2},$

$$\lim_{\ell \to +\infty}\int_0^\ell t^{3/2}\,dt = \lim_{\ell \to +\infty}\frac{2}{5}\ell^{5/2} = +\infty \text{ so } \int_0^{+\infty}\sqrt{1+t^3}\,dt = +\infty$$

(b) $\displaystyle\lim_{x \to +\infty}\frac{2\sqrt{1+8x^3}}{(5/2)x^{3/2}} = \lim_{x \to +\infty}\frac{4}{5}\sqrt{1/x^3+8} = 8\sqrt{2}/5$

57. $\displaystyle\lim_{x \to +\infty}\frac{1+2\cos 2x}{1}$ does not exist, nor is it $\pm\infty$; $\displaystyle\lim_{x \to +\infty}\frac{x+\sin 2x}{x} = \lim_{x \to +\infty}\left(1 + \frac{\sin 2x}{x}\right) = 1.$

59. $\displaystyle\lim_{x \to +\infty}(2 + x\cos x + \sin x)$ does not exist, nor is it $\pm\infty$; $\displaystyle\lim_{x \to +\infty}\frac{x(2+\sin x)}{x+1} = \lim_{x \to +\infty}\frac{2+\sin x}{1+1/x},$
which does not exist because $\sin x$ oscillates between -1 and 1 as $x \to +\infty.$

61. **(a)** $t = 1/x, x = 1/t,$

$$\lim_{x \to +\infty}x\left(k^{1/x}-1\right) = \lim_{t \to 0+}\frac{k^t-1}{t} = \lim_{t \to 0+} = \frac{e^{t\ln k}-1}{t} = \lim_{t \to 0+}(\ln k)e^{t\ln k} = \ln k$$

(b) $1024 = 2^{10}$ so enter the value of k, press the square root key ten times, subtract 1, and multiply the result by 1024: with $k = 0.3$ we get -1.203265293 $(\ln 0.3 \approx -1.203972804)$, with $k = 2$ we get 0.693381829 $(\ln 2 \approx 0.693147181)$

63. $u = t/\sqrt{x}, x = t^2/u^2,$

$$\lim_{x \to +\infty}xf(t/\sqrt{x}) = \lim_{u \to 0+}t^2\frac{f(u)}{u^2} = t^2\lim_{u \to 0+}\frac{f'(u)}{2u} = t^2\lim_{u \to 0+}\frac{f''(u)}{2} = \frac{1}{2}at^2$$

65. $\mathcal{L}\{t\} = \int_0^{+\infty} te^{-st}dt = \left[-\dfrac{1}{s}te^{-st} - \dfrac{1}{s^2}e^{-st}\right]_0^{+\infty}$, but

$\lim\limits_{t\to+\infty} te^{-st} = 0$ and $\lim\limits_{t\to+\infty} e^{-st} = 0$ so $\mathcal{L}\{t\} = 1/s^2$.

67. **(a)** $y = x^{\frac{\ln a}{1+\ln x}}$,

$\lim\limits_{x\to 0^+} \ln y = \lim\limits_{x\to 0^+} \dfrac{(\ln a)\ln x}{1+\ln x} = \lim\limits_{x\to 0^+} \dfrac{(\ln a)/x}{1/x} = \lim\limits_{x\to 0^+} \ln a = \ln a$, $\lim\limits_{x\to 0^+} y = e^{\ln a} = a$

(b) same as part (a) with $x \to +\infty$

(c) $y = (x+1)^{\frac{\ln a}{x}}$, $\lim\limits_{x\to 0} \ln y = \lim\limits_{x\to 0} \dfrac{(\ln a)\ln(x+1)}{x} = \lim\limits_{x\to 0} \dfrac{\ln a}{x+1} = \ln a$, $\lim\limits_{x\to 0} y = e^{\ln a} = a$

69. **(a)** $t = -\ln x, x = e^{-t}, dx = -e^{-t}dt$,

$\int_0^1 (\ln x)^n dx = -\int_{+\infty}^0 (-t)^n e^{-t}dt = (-1)^n \int_0^{+\infty} t^n e^{-t}dt = (-1)^n \Gamma(n+1)$.

(b) $t = x^n, x = t^{1/n}, dx = (1/n)t^{1/n-1}dt$,

$\int_0^{+\infty} e^{-x^n}dx = (1/n)\int_0^{+\infty} t^{1/n-1}e^{-t}dt = (1/n)\Gamma(1/n) = \Gamma(1/n+1)$.

SUPPLEMENTARY EXERCISES CHAPTER 10

1. $\int_0^{+\infty} \dfrac{dx}{x^2+4} = \lim\limits_{\ell\to+\infty} \dfrac{1}{2}\tan^{-1}(x/2)\Big]_0^{\ell} = \lim\limits_{\ell\to+\infty} \dfrac{1}{2}\tan^{-1}(\ell/2) = \pi/4$,

$\int_{\ell}^0 \dfrac{dx}{x^2+4} = \lim\limits_{\ell\to-\infty} \dfrac{1}{2}\tan^{-1}(x/2)\Big]_{\ell}^0 = \pi/4$ so $\int_{-\infty}^{+\infty} \dfrac{dx}{x^2+4} = \pi/2$

3. $\lim\limits_{\ell\to 1^-} -\sqrt{1-x^2}\Big]_0^{\ell} = \lim\limits_{\ell\to 1^-} (-\sqrt{1-\ell^2}+1) = 1$

5. $\int_0^1 x^{-2/3}dx = \lim\limits_{\ell\to 0^+} 3x^{1/3}\Big]_{\ell}^1 = 3$, $\int_{-1}^0 x^{-2/3}dx = \lim\limits_{\ell\to 0^-} 3x^{1/3}\Big]_{-1}^{\ell} = 3$, $\int_{-1}^1 x^{-2/3}dx = 6$

7. $\lim\limits_{\ell\to+\infty} -\dfrac{1}{2}e^{-x^2}\Big]_0^{\ell} = \lim\limits_{\ell\to+\infty} \dfrac{1}{2}(-e^{-\ell^2}+1) = 1/2$,

$\lim\limits_{\ell\to-\infty} -\dfrac{1}{2}e^{-x^2}\Big]_{\ell}^0 = \lim\limits_{\ell\to-\infty} \dfrac{1}{2}(-1+e^{-\ell^2}) = -1/2$, so $\int_{-\infty}^{+\infty} xe^{-x^2}dx = 1/2 - 1/2 = 0$

9. $\displaystyle\lim_{\ell\to 0^+} \ln|\sin x|\,\Big]_{\ell}^{\pi/2} = \lim_{\ell\to 0^+} -\ln|\sin\ell| = +\infty$, diverges

11. $\displaystyle\lim_{\ell\to +\infty} -1/\ln x\,\Big]_{e}^{\ell} = \lim_{\ell\to +\infty} (-1/\ln\ell + 1) = 1$

13. $\displaystyle\lim_{\ell\to +\infty} \tan^{-1}(x+1)\,\Big]_{e}^{\ell} = \lim_{\ell\to +\infty} [\tan^{-1}(\ell+1) - \tan^{-1}(1)] = \pi/2 - \pi/4 = \pi/4.$

15. If $n = -1$ then $\displaystyle\int_0^1 \frac{\ln x}{x}\,dx = \lim_{\ell\to 0^+} \frac{1}{2}(\ln x)^2\,\Big]_{\ell}^{1} = -\infty$ so the integral diverges. If $n \neq -1$ then

$$\int_0^1 x^n \ln x\,dx = \lim_{\ell\to 0^+}\left[\frac{x^{n+1}}{n+1}\ln x - \frac{x^{n+1}}{(n+1)^2}\right]_{\ell}^{1} = \lim_{\ell\to 0^+}\left[-\frac{1}{(n+1)^2} - \frac{\ell^{n+1}}{n+1}\ln\ell + \frac{\ell^{n+1}}{(n+1)^2}\right]$$

If $n < -1$ then $n+1 < 0$, $\displaystyle\lim_{\ell\to 0^+} \ell^{n+1}\ln\ell = -\infty$ and $\displaystyle\lim_{\ell\to 0^+}\ell^{n+1} = +\infty$ so the integral diverges.

If $n > -1$ then $\displaystyle\lim_{\ell\to 0^+}\ell^{n+1}\ln\ell = \lim_{\ell\to 0^+}\frac{\ln\ell}{\ell^{-(n+1)}} = \lim_{\ell\to 0^+} -\frac{\ell^{n+1}}{n+1} = 0$ and $\displaystyle\lim_{\ell\to 0^+}\ell^{n+1} = 0$ so the integral converges to $-1/(n+1)^2$.

17. $\displaystyle\lim_{x\to 0}\frac{3xe^{3x} + e^{3x} - 1}{2\sin 2x} = \lim_{x\to 0}\frac{9xe^{3x} + 6e^{3x}}{4\cos 2x} = 3/2$

19. $\displaystyle\lim_{x\to 0^+}\frac{e^{1/x}}{1/x^2} = \lim_{x\to 0^+}\frac{(-1/x^2)e^{1/x}}{-2/x^3} = \lim_{x\to 0^+}\frac{e^{1/x}}{2/x} = \lim_{x\to 0^+}\frac{(-1/x^2)e^{1/x}}{-2/x^2} = \lim_{x\to 0^+}(1/2)e^{1/x} = +\infty$

21. $\displaystyle\lim_{x\to 0^-} x^2 e^{1/x} = (0)(0) = 0$

23. $\displaystyle\lim_{\theta\to 0}\left(\frac{1}{\theta\sin\theta} - \frac{1}{\theta^2}\right) = \lim_{\theta\to 0}\frac{\theta - \sin\theta}{\theta^2\sin\theta} = \lim_{\theta\to 0}\frac{1 - \cos\theta}{\theta^2\cos\theta + 2\theta\sin\theta}$

$$= \lim_{\theta\to 0}\frac{\sin\theta}{-\theta^2\sin\theta + 4\theta\cos\theta + 2\sin\theta}$$

$$= \lim_{\theta\to 0}\frac{\cos\theta}{-\theta^2\cos\theta - 6\theta\sin\theta + 6\cos\theta} = 1/6$$

25. $\displaystyle\lim_{x\to 2}\frac{1 - e^{x-2}}{2\pi\sin 2\pi x} = \lim_{x\to 2}\frac{-e^{x-2}}{4\pi^2\cos 2\pi x} = -1/(4\pi^2)$

27. $\displaystyle\lim_{x\to 0}\frac{\sin(x^2)}{2x\cos(x^2)} = \lim_{x\to 0}\frac{2x\cos(x^2)}{-4x^2\sin(x^2) + 2\cos(x^2)} = 0$

29. $\displaystyle\lim_{x\to+\infty}\frac{3(\ln x)^2}{x}=\lim_{x\to+\infty}\frac{6\ln x}{x}=\lim_{x\to+\infty}\frac{6}{x}=0$

31. $y=(1+x)^{\ln x}$, $\displaystyle\lim_{x\to0+}\ln y=\lim_{x\to0+}\ln x\ln(1+x)=\lim_{x\to0+}\frac{\ln(1+x)}{1/\ln x}$

$$=\lim_{x\to0+}\frac{1/(1+x)}{-1/[x(\ln x)^2]}=\lim_{x\to0+}\frac{-x(\ln x)^2}{1+x},$$

but $\displaystyle\lim_{x\to0+}x(\ln x)^2=\lim_{x\to0+}\frac{(\ln x)^2}{1/x}=\lim_{x\to0+}\frac{(2\ln x)/x}{-1/x^2}$

$$=\lim_{x\to0+}\frac{2\ln x}{-1/x}=\lim_{x\to0+}\frac{2/x}{-1/x^2}=\lim_{x\to0+}(-2x)=0$$

so $\displaystyle\lim_{x\to0+}\frac{-x(\ln)^2}{1+x}=\frac{0}{1}=0$ and $\displaystyle\lim_{x\to0+}y=e^0=1$

33.   **(a)** $\displaystyle A=\int_8^{+\infty}x^{-2/3}dx=\lim_{\ell\to+\infty}3x^{1/3}\bigg]_8^\ell=+\infty$

   (b) $\displaystyle V=\int_8^{+\infty}\pi x^{-4/3}dx=\lim_{\ell\to+\infty}-3\pi x^{-1/3}\bigg]_8^\ell=3\pi/2$

35. $f(x)=\sqrt{x-x^2}-\sin^{-1}\sqrt{x}=\sqrt{x(1-x)}-\sin^{-1}\sqrt{x}$, the domain of f is $0\le x\le1$,

$$f'(x)=\frac{1}{2}(x-x^2)^{-1/2}(1-2x)-\frac{1}{\sqrt{1-x}}\Big(\frac{1}{2}x^{-1/2}\Big)=-\frac{x}{\sqrt{x-x^2}},$$

$$1+[f'(x)]^2=1+\frac{x^2}{x-x^2}=\frac{1}{1-x},$$

$$L=\int_0^1\frac{1}{\sqrt{1-x}}dx=\lim_{\ell\to1^-}-2\sqrt{1-x}\bigg]_0^\ell=\lim_{\ell\to1^-}2(-\sqrt{1-\ell}+1)=2$$

CHAPTER 11
Infinite Series

EXERCISE SET 11.1

1. $1/3, 2/4, 3/5, 4/6, 5/7, \ldots$; $\displaystyle\lim_{n \to +\infty} \frac{n}{n+2} = 1$, converges

3. $2, 2, 2, 2, 2, \ldots$; $\displaystyle\lim_{x \to +\infty} 2 = 2$, converges

5. $\dfrac{\ln 1}{1}, \dfrac{\ln 2}{2}, \dfrac{\ln 3}{3}, \dfrac{\ln 4}{4}, \dfrac{\ln 5}{5}, \ldots$; $\displaystyle\lim_{x \to +\infty} \frac{\ln n}{n} = \lim_{x \to +\infty} \frac{1}{n} = 0$, converges

7. $0, 2, 0, 2, 0, \ldots$; diverges

9. $-1, 16/9, -54/28, 128/65, -250/126, \ldots$; diverges because odd-numbered terms approach -2, even-numbered terms approach 2.

11. $6/2, 12/8, 20/18, 30/32, 42/50, \ldots$; $\displaystyle\lim_{n \to +\infty} \frac{1}{2}(1 + 1/n)(1 + 2/n) = 1/2$, converges

13. $\cos(3), \cos(3/2), \cos(1), \cos(3/4), \cos(3/5), \ldots$; $\displaystyle\lim_{n \to +\infty} \cos(3/n) = 1$, converges

15. $e^{-1}, 4e^{-2}, 9e^{-3}, 16e^{-4}, 25e^{-5}, \ldots$; $\displaystyle\lim_{x \to +\infty} x^2 e^{-x} = \lim_{x \to +\infty} \frac{x^2}{e^x} = 0$, so $\displaystyle\lim_{n \to +\infty} n^2 e^{-n} = 0$, converges

17. $2, (5/3)^2, (6/4)^3, (7/5)^4, (8/6)^5, \ldots$; let $y = \left[\dfrac{x+3}{x+1}\right]^x$, converges because

$$\lim_{x \to +\infty} \ln y = \lim_{x \to +\infty} \frac{\ln \dfrac{x+3}{x+1}}{1/x} = \lim_{x \to +\infty} \frac{2x^2}{(x+1)(x+3)} = 2, \text{ so } \lim_{n \to +\infty} \left[\frac{n+3}{n+1}\right]^n = e^2$$

19. $\left\{\dfrac{2n-1}{2n}\right\}_{n=1}^{+\infty}$; $\displaystyle\lim_{n \to +\infty} \frac{2n-1}{2n} = 1$, converges

21. $\{1/3^n\}_{n=1}^{+\infty}$; $\lim_{n \to +\infty} 1/3^n = 0$, converges

23. $\left\{\dfrac{1}{n} - \dfrac{1}{n+1}\right\}_{n=1}^{+\infty}$; $\displaystyle\lim_{n \to +\infty} \left(\frac{1}{n} - \frac{1}{n+1}\right) = 0$, converges

25. $\{\sqrt{n+1} - \sqrt{n+2}\}_{n=1}^{+\infty}$; converges because

$$\lim_{n \to +\infty} (\sqrt{n+1} - \sqrt{n+2}) = \lim_{n \to +\infty} \frac{(n+1) - (n+2)}{\sqrt{n+1} + \sqrt{n+2}} = \lim_{n \to +\infty} \frac{-1}{\sqrt{n+1} + \sqrt{n+2}} = 0$$

27. (a) $\sqrt{6}, \sqrt{6 + \sqrt{6}}, \sqrt{6 + \sqrt{6 + \sqrt{6}}}$

    (b) $\lim_{n \to +\infty} a_{n+1} = \lim_{n \to +\infty} \sqrt{6 + a_n}$, $L = \sqrt{6 + L}$, $L^2 - L - 6 = 0$, $(L - 3)(L + 2) = 0$, $L = -2$
(reject, because the terms in the sequence are positive) or $L = 3$; $\lim_{n \to +\infty} a_n = 3$.

29. (a) $1, 1, 2, 3, 5, 8, 13, 21$

    (b) $a_{n+2}/a_{n+1} = a_n/a_{n+1} + 1 = 1/(a_{n+1}/a_n) + 1$,
$\lim_{n \to +\infty} (a_{n+2}/a_{n+1}) = \lim_{n \to +\infty} [1/(a_{n+1}/a_n) + 1]$, with $L = \lim_{n \to +\infty} (a_{n+1}/a_n)$, $L = 1/L + 1$,
$L^2 - L - 1 = 0$, $L = (1 \pm \sqrt{5})/2$ so $L = (1 + \sqrt{5})/2$ because the limit cannot be negative.

31. (a) $1, \dfrac{1}{4} + \dfrac{2}{4}, \dfrac{1}{9} + \dfrac{2}{9} + \dfrac{3}{9}, \dfrac{1}{16} + \dfrac{2}{16} + \dfrac{3}{16} + \dfrac{4}{16}, \cdots = 1, 3/4, 2/3, 5/8, \cdots$

    (b) $a_n = \dfrac{1}{n^2}(1 + 2 + \cdots + n) = \dfrac{1}{n^2}\dfrac{1}{2}n(n + 1) = \dfrac{1}{2}\dfrac{n+1}{n}$, $\lim_{n \to +\infty} a_n = 1/2$

33. $\left|\dfrac{1}{n} - 0\right| = \dfrac{1}{n} < \epsilon$ if $n > 1/\epsilon$

    (a) $1/\epsilon = 1/0.5 = 2$, $N = 3$ **(b)** $1/\epsilon = 1/0.1 = 10$, $N = 11$

    (c) $1/\epsilon = 1/0.001 = 1000$, $N = 1001$

35. (a) $\left|\dfrac{1}{n} - 0\right| = \dfrac{1}{n} < \epsilon$ if $n > 1/\epsilon$, choose any $N > 1/\epsilon$.

    (b) $\left|\dfrac{n}{n+1} - 1\right| = \dfrac{1}{n+1} < \epsilon$ if $n > 1/\epsilon - 1$, choose any $N > 1/\epsilon - 1$.

37. (a) Let s_n denote the length of each
of the n sides of the polygon, then

$$\sin(\pi/n) = \frac{s_n/2}{r}$$
$$s_n = 2r\sin(\pi/n)$$
$$p_n = ns_n = 2rn\sin(\pi/n)$$

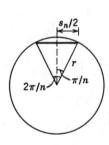

    (b) $\lim_{n \to +\infty} 2rn\sin(\pi/n) = \lim_{n \to +\infty} \dfrac{2r\sin(\pi/n)}{1/n}$

$$= \lim_{n \to +\infty} 2\pi r\cos(\pi/n) = 2\pi r$$

39. Let $y = (2^x + 3^x)^{1/x}$,

$$\lim_{x \to +\infty} \ln y = \lim_{x \to +\infty} \frac{\ln(2^x + 3^x)}{x} = \lim_{x \to +\infty} \frac{2^x \ln 2 + 3^x \ln 3}{2^x + 3^x} = \lim_{x \to +\infty} \frac{(2/3)^x \ln 2 + \ln 3}{(2/3)^x + 1} = \ln 3$$

so $\lim_{x \to +\infty} (2^n + 3^n)^{1/n} = e^{\ln 3} = 3$

EXERCISE SET 11.2

1. $a_n - a_{n+1} = \dfrac{1}{n} - \dfrac{1}{n+1} = \dfrac{1}{n(n+1)} > 0$ for $n \geq 1$, so decreasing.

3. $a_n - a_{n+1} = \dfrac{n}{2n+1} - \dfrac{n+1}{2n+3} = -\dfrac{1}{(2n+1)(2n+3)} < 0$ for $n \geq 1$, so increasing.

5. $a_n - a_{n+1} = (n - 2^n) - (n + 1 - 2^{n+1}) = 2^n - 1 > 0$ for $n \geq 1$, so decreasing.

7. $\dfrac{a_{n+1}}{a_n} = \dfrac{(n+1)/(2n+3)}{n/(2n+1)} = \dfrac{(n+1)(2n+1)}{n(2n+3)} = \dfrac{2n^2 + 3n + 1}{2n^2 + 3n} > 1$ for $n \geq 1$, so increasing.

9. $\dfrac{a_{n+1}}{a_n} = \dfrac{(n+1)e^{-(n+1)}}{ne^{-n}} = (1 + 1/n)e^{-1} < 1$ for $n \geq 1$, so decreasing.

11. $\dfrac{a_{n+1}}{a_n} = \dfrac{2^{n+1}/(n+1)!}{2^n/n!} = \dfrac{2^{n+1}}{2^n} \cdot \dfrac{n!}{(n+1)!} = \dfrac{2}{n+1} \leq 1$ for $n \geq 1$, so nonincreasing.

13. $\dfrac{a_{n+1}}{a_n} = \dfrac{(n+1)!}{3^{n+1}} \cdot \dfrac{3^n}{n!} = \dfrac{n+1}{3}$, $\dfrac{n+1}{3} \leq 1$ for $n = 1$ and 2, $\dfrac{n+1}{3} > 1$ for $n \geq 3$, so not monotone.

15. $\dfrac{a_{n+1}}{a_n} = \dfrac{10^{n+1}}{(2n+2)!} \cdot \dfrac{(2n)!}{10^n} = \dfrac{10}{(2n+2)(2n+1)} < 1$ for $n \geq 1$, so decreasing.

17. $\dfrac{a_{n+1}}{a_n} = \dfrac{(n+1)^{n+1}}{(n+1)!} \cdot \dfrac{n!}{n^n} = \dfrac{(n+1)^n}{n^n} = (1 + 1/n)^n > 1$ for $n \geq 1$, so increasing.

19. $f(x) = x/(2x+1)$, $f'(x) = 1/(2x+1)^2 > 0$ for $x \geq 1$, so increasing.

21. $f(x) = 1/(x + \ln x)$, $f'(x) = -\dfrac{1 + 1/x}{(x + \ln x)^2} < 0$ for $x \geq 1$, so decreasing.

23. $f(x) = \dfrac{\ln(x+2)}{x+2}$, $f'(x) = \dfrac{1 - \ln(x+2)}{(x+2)^2} < 0$ for $x \geq 1$, so decreasing.

25. $a_{n+1}/a_n = (n+1)/(5n) < 1$ for $n \geq 1$ so decreasing; 0 is a lower bound of $n/5^n$ for $n \geq 1$ so the sequence converges.

27. Let $f(x) = x - 1/x$, then $f'(x) = 1 + 1/x^2 > 0$ so increasing; $n - 1/n$ has no upper bound so the sequence diverges.

29. Let $f(x) = 2 + 1/x$, then $f'(x) = -1/x^2 < 0$ for $x \geq 1$ so decreasing; 2 is a lower bound because $2 + 1/n \geq 2$ for $n \geq 1$ so the sequence converges.

31. **(a)** We can discard the first two terms without affecting the convergence or the limit. This leaves the sequence $1, \dfrac{1}{2}, \dfrac{1}{3}, \ldots, \dfrac{1}{n}, \ldots$ Because $\lim\limits_{n \to +\infty} \dfrac{1}{n} = 0$, this is also the limit of the given sequence.

(b) Discard the first four terms, leaving $1, 2, 3, \ldots, n, \ldots$; $\lim\limits_{n \to +\infty} n = +\infty$ so the limit does not exist.

33. $\dfrac{a_{n+1}}{a_n} = \dfrac{3^{n+1}(1 + 3^{2n})}{3^n(1 + 3^{2n+2})} = \dfrac{3(1 + 3^{2n})}{1 + 3^{2n+2}} < \dfrac{3(1 + 3^{2n})}{3^{2n+2}} = \dfrac{1}{3^{2n+1}} + \dfrac{1}{3} < 1$ for $n \geq 1$ so decreasing.

35. **(a)** $\sqrt{2}, \sqrt{2 + \sqrt{2}}, \sqrt{2 + \sqrt{2 + \sqrt{2}}}$

(b) $a_1 = \sqrt{2} < 2$ so $a_2 = \sqrt{2 + a_1} < \sqrt{2 + 2} = 2$, $a_3 = \sqrt{2 + a_2} < \sqrt{2 + 2} = 2$, and so on indefinitely.

(c) $a_{n+1}^2 - a_n^2 = (2 + a_n) - a_n^2 = 2 + a_n - a_n^2 = (2 - a_n)(1 + a_n)$

(d) $a_n > 0$ and, from part (a), $a_n < 2$ so $2 - a_n > 0$ and $1 + a_n > 0$ thus, from part (c), $a_{n+1}^2 - a_n^2 > 0$, $a_{n+1} - a_n > 0$, $a_{n+1} > a_n$; $\{a_n\}$ is an increasing sequence.

(e) The sequence is increasing and has 2 as an upper bound so it must converge to a limit L, $\lim\limits_{n \to +\infty} a_{n+1} = \lim\limits_{n \to +\infty} \sqrt{2 + a_n}$, $L = \sqrt{2 + L}$, $L^2 - L - 2 = 0$, $(L - 2)(L + 1) = 0$ thus $\lim\limits_{n \to +\infty} a_n = 2$.

37. **(a)** If $\{a_n\}_{n=1}^{+\infty}$ is nonincreasing then $a_n \geq a_{n+1}$ for all n, thus $-a_n \leq -a_{n+1}$ so $\{-a_n\}_{n=1}^{+\infty}$ is nondecreasing.

(b) From part (a), $\{-a_n\}_{n=1}^{+\infty}$ is nondecreasing so Theorem 11.2.2 applies. Suppose there is a finite constant M_1 such that $-a_n \leq M_1$ and $\lim\limits_{n \to +\infty}(-a_n) = L_1 \leq M_1$, then $a_n \geq -M_1$, and $\lim\limits_{n \to +\infty} a_n = -L_1 \geq -M_1$; part (a) of Theorem 11.2.3 follows by letting $M = -M_1$ and $L = -L_1$. Suppose no such constant exists, then $\lim\limits_{n \to +\infty}(-a_n) = +\infty$ so $\lim\limits_{n \to +\infty} a_n = -\infty$.

39. From part (b) of Exercise 38, $\left[\dfrac{n^n}{e^{n-1}}\right]^{1/n} < \sqrt[n]{n!} < \left[\dfrac{(n+1)^{n+1}}{e^n}\right]^{1/n}$,

$$\dfrac{n}{e^{1-1/n}} < \sqrt[n]{n!} < \dfrac{(n+1)^{1+1/n}}{e}, \quad \dfrac{1}{e^{1-1/n}} < \dfrac{\sqrt[n]{n!}}{n} < \dfrac{(1+1/n)(n+1)^{1/n}}{e},$$

but $\dfrac{1}{e^{1-1/n}} \to \dfrac{1}{e}$ and $\dfrac{(1+1/n)(n+1)^{1/n}}{e} \to \dfrac{1}{e}$ as $n \to +\infty$ so $\displaystyle\lim_{n\to+\infty} \dfrac{\sqrt[n]{n!}}{n} = \dfrac{1}{e}$

41. (a) $\dfrac{a_{n+1}}{a_n} = \dfrac{(n+1)^{n+1}}{(n+1)!e^{n+1}} \cdot \dfrac{n!e^n}{n^n} = \dfrac{(1+1/n)^n}{e} < 1$ for $n \geq 1$ because $(1+1/n)^n < e$ for $n \geq 1$.

(b) Yes, because it is decreasing and bounded below by 0.

EXERCISE SET 11.3

1. (a) $s_1 = 2$, $s_2 = 12/5$, $s_3 = 62/25$, $s_4 = 312/125$

$s_n = \dfrac{2 - 2(1/5)^n}{1 - 1/5} = \dfrac{5}{2} - \dfrac{5}{2}(1/5)^n$, $\displaystyle\lim_{n\to+\infty} s_n = 5/2$, converges

(b) $\dfrac{1}{(k+1)(k+2)} = \dfrac{1}{k+1} - \dfrac{1}{k+2}$, $s_1 = 1/6$, $s_2 = 1/4$, $s_3 = 3/10$, $s_4 = 1/3$ $s_n = \dfrac{1}{2} - \dfrac{1}{n+2}$,

$\displaystyle\lim_{n\to+\infty} s_n = 1/2$, converges

(c) $s_1 = 1/4$, $s_2 = 3/4$, $s_3 = 7/4$, $s_4 = 15/4$

$s_n = \dfrac{(1/4) - (1/4)2^n}{1 - 2} = -\dfrac{1}{4} + \dfrac{1}{4}(2^n)$, $\displaystyle\lim_{n\to+\infty} s_n = +\infty$, diverges

3. geometric, $a = 1$, $r = -3/4$, sum $= \dfrac{1}{1 - (-3/4)} = 4/7$

5. geometric, $a = 7$, $r = -1/6$, sum $= \dfrac{7}{1 + 1/6} = 6$

7. geometric, $r = -3/2$, diverges

9. $s_n = \displaystyle\sum_{k=1}^{n} \left(\dfrac{1}{k+2} - \dfrac{1}{k+3}\right) = \dfrac{1}{3} - \dfrac{1}{n+3}$, $\displaystyle\lim_{n\to+\infty} s_n = 1/3$

11. $s_n = \displaystyle\sum_{k=1}^{n} \left(\dfrac{1/3}{3k-1} - \dfrac{1/3}{3k+2}\right) = \dfrac{1}{6} - \dfrac{1/3}{3n+2}$, $\displaystyle\lim_{n\to+\infty} s_n = 1/6$

13. $\displaystyle\sum_{k=1}^{\infty}\frac{4^{k+2}}{7^{k-1}}=\sum_{k=1}^{\infty}64\left(\frac{4}{7}\right)^{k-1}$; geometric, $a=64$, $r=4/7$, sum $=\dfrac{64}{1-4/7}=448/3$

15. geometric, $a=-1/2$, $r=-1/2$, sum $=\dfrac{-1/2}{1+1/2}=-1/3$

17. $0.4444\cdots=0.4+0.04+0.004+\cdots=\dfrac{0.4}{1-0.1}=4/9$

19. $5.373737\cdots=5+0.37+0.0037+0.000037+\cdots=5+\dfrac{0.37}{1-0.01}=5+37/99=532/99$

21. $0.782178217821\cdots=0.7821+0.00007821+0.000000007821+\cdots$

$$=\frac{0.7821}{1-0.0001}=7821/9999=869/1111$$

23. $s_n=\ln\dfrac{1}{2}+\ln\dfrac{2}{3}+\ln\dfrac{3}{4}+\cdots+\ln\dfrac{n}{n+1}=\ln\left(\dfrac{1}{2}\cdot\dfrac{2}{3}\cdot\dfrac{3}{4}\cdots\dfrac{n}{n+1}\right)=\ln\dfrac{1}{n+1}=-\ln(n+1)$,

$\displaystyle\lim_{n\to+\infty}s_n=-\infty$, series diverges.

25. $\ln 1-1/k^2=\ln\dfrac{k^2-1}{k^2}=\ln\dfrac{(k-1)(k+1)}{k^2}=\ln\dfrac{k-1}{k}+\ln\dfrac{k+1}{k}=\ln\dfrac{k-1}{k}-\ln\dfrac{k}{k+1}$,

$s_n=\displaystyle\sum_{k=2}^{n+1}\left[\ln\dfrac{k-1}{k}-\ln\dfrac{k}{k+1}\right]$

$=\left(\ln\dfrac{1}{2}-\ln\dfrac{2}{3}\right)+\left(\ln\dfrac{2}{3}-\ln\dfrac{3}{4}\right)+\left(\ln\dfrac{3}{4}-\ln\dfrac{4}{5}\right)+\cdots+\left(\ln\dfrac{n}{n+1}-\ln\dfrac{n+1}{n+2}\right)$

$=\ln\dfrac{1}{2}-\ln\dfrac{n+1}{n+2},\ \displaystyle\lim_{n\to+\infty}s_n=\ln\dfrac{1}{2}=-\ln 2$

27. $s_n=(1-1/3)+(1/2-1/4)+(1/3-1/5)+(1/4-1/6)+\cdots+[1/n-1/(n+2)]$

$=(1+1/2+1/3+\cdots+1/n)-(1/3+1/4+1/5+\cdots+1/(n+2))$

$=3/2-1/(n+1)-1/(n+2),\ \displaystyle\lim_{n\to+\infty}s_n=3/2$

29. $s_n=\displaystyle\sum_{k=1}^{n}\frac{1}{(2k-1)(2k+1)}=\sum_{k=1}^{n}\left[\frac{1/2}{2k-1}-\frac{1/2}{2k+1}\right]=\frac{1}{2}\left[\sum_{k=1}^{n}\frac{1}{2k-1}-\sum_{k=1}^{n}\frac{1}{2k+1}\right]$

$=\dfrac{1}{2}\left[\displaystyle\sum_{k=1}^{n}\frac{1}{2k-1}-\sum_{k=2}^{n+1}\frac{1}{2k-1}\right]=\dfrac{1}{2}\left[1-\dfrac{1}{2n+1}\right];\ \displaystyle\lim_{n\to+\infty}s_n=\dfrac{1}{2}$

31. (a) $\sum_{k=0}^{\infty}(-1)^k x^k = 1 - x + x^2 - x^3 + \cdots = \dfrac{1}{1-(-x)} = \dfrac{1}{1+x}$ if $|-x| < 1,\ |x| < 1,\ -1 < x < 1.$

(b) $\sum_{k=0}^{\infty}(x-3)^k = 1 + (x-3) + (x-3)^2 + \cdots = \dfrac{1}{1-(x-3)} = \dfrac{1}{4-x}$ if $|x-3| < 1,\ 2 < x < 4.$

(c) $\sum_{k=0}^{\infty}(-1)^k x^{2k} = 1 - x^2 + x^4 - x^6 + \cdots = \dfrac{1}{1-(-x^2)} = \dfrac{1}{1+x^2}$ if $|-x^2| < 1,\ |x| < 1,$
$-1 < x < 1.$

33. Geometric series, $a = 1/x^2$, $r = 2/x$. Converges for $|2/x| < 1,\ |x| > 2$;
$$S = \dfrac{1/x^2}{1-2/x} = \dfrac{1}{x^2-2x}.$$

35. Geometric series, $a = \sin\ x$, $r = -\dfrac{1}{2}\sin x$. Converges for $\left|-\dfrac{1}{2}\sin x\right| < 1,\ |\sin x| < 2$, so
converges for all values of x. $S = \dfrac{\sin x}{1+\dfrac{1}{2}\sin x} = \dfrac{2\sin x}{2+\sin x}.$

37. $a_2 = \dfrac{1}{2}a_1 + \dfrac{1}{2},\ a_3 = \dfrac{1}{2}a_2 + \dfrac{1}{2} = \dfrac{1}{2^2}a_1 + \dfrac{1}{2^2} + \dfrac{1}{2},\ a_4 = \dfrac{1}{2}a_3 + \dfrac{1}{2} = \dfrac{1}{2^3}a_1 + \dfrac{1}{2^3} + \dfrac{1}{2^2} + \dfrac{1}{2},$
$a_5 = \dfrac{1}{2}a_4 + \dfrac{1}{2} = \dfrac{1}{2^4}a_1 + \dfrac{1}{2^4} + \dfrac{1}{2^3} + \dfrac{1}{2^2} + \dfrac{1}{2},\cdots,a_n = \dfrac{1}{2^{n-1}}a_1 + \dfrac{1}{2^{n-1}} + \dfrac{1}{2^{n-2}} + \cdots + \dfrac{1}{2},$
$$\lim_{n\to+\infty} a_n = \lim_{n\to+\infty} \dfrac{a_1}{2^{n-1}} + \sum_{n=1}^{\infty}\left(\dfrac{1}{2}\right)^n = 0 + \dfrac{1/2}{1-1/2} = 1$$

39. By inspection, $\dfrac{\theta}{2} - \dfrac{\theta}{4} + \dfrac{\theta}{8} - \dfrac{\theta}{16} + \cdots = \dfrac{\theta/2}{1-(-1/2)} = \theta/3$

41. The series converges to $1/(1-x)$ only if $-1 < x < 1.$

EXERCISE SET 11.4

1. $\sum_{k=1}^{\infty}\dfrac{1}{2^k} = \dfrac{1/2}{1-1/2} = 1;\quad \sum_{k=1}^{\infty}\dfrac{1}{4^k} = \dfrac{1/4}{1-1/4} = 1/3;\quad \sum_{k=1}^{\infty}\left(\dfrac{1}{2^k} + \dfrac{1}{4^k}\right) = 1 + 1/3 = 4/3$

3. $\displaystyle\sum_{k=2}^{\infty} \frac{1}{k^2 - 1} = \sum_{k=2}^{\infty} \left[\frac{1/2}{k - 1} - \frac{1/2}{k + 1} \right],$

$s_n = \dfrac{1}{2} \left[\left(1 - \dfrac{1}{3} \right) + \left(\dfrac{1}{2} - \dfrac{1}{4} \right) + \left(\dfrac{1}{3} - \dfrac{1}{5} \right) + \cdots + \left(\dfrac{1}{n} - \dfrac{1}{n + 2} \right) \right]$

$\quad = \dfrac{1}{2} \left[\left(1 + \dfrac{1}{2} + \dfrac{1}{3} + \cdots + \dfrac{1}{n} \right) - \left(\dfrac{1}{3} + \dfrac{1}{4} + \dfrac{1}{5} + \cdots + \dfrac{1}{n + 2} \right) \right]$

$\quad = \dfrac{3}{4} - \dfrac{1}{2} \left(\dfrac{1}{n + 1} + \dfrac{1}{n + 2} \right),$

$\displaystyle\sum_{k=2}^{\infty} \frac{1}{k^2 - 1} = \lim_{n \to +\infty} s_n = 3/4; \quad \sum_{k=2}^{\infty} \frac{7}{10^{k-1}} = \frac{7/10}{1 - 1/10} = 7/9;$

so $\displaystyle\sum_{k=2}^{\infty} \left[\frac{1}{k^2 - 1} - \frac{7}{10^{k-1}} \right] = 3/4 - 7/9 = -1/36$

5. (a) $p = 3$, converges   (b) $p = 1/2$, diverges

    (c) $p = 1$, diverges   (d) $p = 2/3$, diverges

    (e) $p = 4/3$, converges   (f) $p = 1/4$, diverges

    (g) $p = 5/3$, converges   (h) $p = \pi$, converges

7. (a) $\displaystyle\lim_{k \to +\infty} \frac{k^2 + k + 3}{2k^2 + 1} = \frac{1}{2}$ (b) $\displaystyle\lim_{k \to +\infty} \left(1 + \frac{1}{k} \right)^k = e$

9. $\displaystyle\sum_{k=1}^{\infty} \frac{1}{k + 6} = \sum_{k=7}^{\infty} \frac{1}{k}$, diverges because the harmonic series diverges.

11. $\displaystyle\int_{1}^{+\infty} \frac{1}{5x + 2} = \lim_{\ell \to +\infty} \frac{1}{5} \ln(5x + 2) \Big]_{1}^{\ell} = +\infty$, the series diverges by the integral test.

13. $\displaystyle\int_{1}^{+\infty} \frac{1}{1 + 9x^2} dx = \lim_{\ell \to +\infty} \frac{1}{3} \tan^{-1} 3x \Big]_{1}^{\ell} = \frac{1}{3} \left(\pi/2 - \tan^{-1} 3 \right)$, the series converges by the

    integral test.

15. $\displaystyle\sum_{k=1}^{\infty} \frac{1}{\sqrt{k + 5}} = \sum_{k=6}^{\infty} \frac{1}{\sqrt{k}}$, diverges because the p-series with $p = 1/2 \leq 1$ diverges.

17. $\displaystyle\int_{1}^{+\infty} (2x - 1)^{-1/3} dx = \lim_{\ell \to +\infty} \frac{3}{4} (2x - 1)^{2/3} \Big]_{1}^{\ell} = +\infty$, the series diverges by the integral test.

19. $\displaystyle\lim_{k\to+\infty}\frac{k}{\ln(k+1)}=\lim_{k\to+\infty}\frac{1}{1/(k+1)}=+\infty$, the series diverges because $\displaystyle\lim_{k\to+\infty}u_k\neq 0$.

21. $\displaystyle\int_1^{+\infty}\frac{[\ln(x+1)]^{-2}}{x+1}\,dx=\lim_{\ell\to+\infty}\left.-1/\ln(x+1)\right]_1^\ell=1/\ln 2$, the series converges by the integral test.

23. $\displaystyle\lim_{k\to+\infty}(1+1/k)^{-k}=1/e\neq 0$, the series diverges.

25. $\displaystyle\int_1^{+\infty}\frac{\tan^{-1}x}{1+x^2}\,dx=\lim_{\ell\to+\infty}\left.\frac{1}{2}\left(\tan^{-1}x\right)^2\right]_1^\ell=3\pi^2/32$, the series converges by the integral test.

27. $\displaystyle\sum_{k=5}^{\infty}7k^{-p}=\sum_{k=5}^{\infty}7\left(\frac{1}{k^p}\right)$, converges because a p-series with $p>1$ converges.

29. $\displaystyle\lim_{k\to+\infty}k^2\sin^2(1/k)=1\neq 0$, the series diverges.

31. Use the integral test with $\displaystyle\int_2^{+\infty}\frac{dx}{x(\ln x)^p}$ to get $\displaystyle\lim_{\ell\to+\infty}\ln(\ln x)\right]_2^\ell=+\infty$ if $p=1$,

$$\lim_{\ell\to+\infty}\left.\frac{(\ln x)^{1-p}}{1-p}\right]_2^\ell=\begin{cases}+\infty & \text{if }p<1\\[2mm]\dfrac{1}{(p-1)(\ln 2)^{p-1}} & \text{if }p>1\end{cases}$$

33. Suppose $\Sigma(u_k+v_k)$ converges then so does $\Sigma[(u_k+v_k)-u_k]$, but $\Sigma[(u_k+v_k)-u_k]=\Sigma v_k$ so Σv_k converges which contradicts the assumption that Σv_k diverges. Suppose $\Sigma(u_k-v_k)$ converges then so does $\Sigma[u_k-(u_k-v_k)]=\Sigma v_k$ which leads to the same contradiction as before.

35. **(a)** diverges because $\displaystyle\sum_{k=1}^{\infty}(2/3)^{k-1}$ converges and $\displaystyle\sum_{k=1}^{\infty}1/k$ diverges.

  (b) diverges because $\displaystyle\sum_{k=1}^{\infty}\frac{k^2}{1+k^2}$ diverges and $\displaystyle\sum_{k=1}^{\infty}\frac{1}{k(k+1)}$ converges.

  (c) diverges because $\displaystyle\sum_{k=1}^{\infty}1/(3k+2)$ diverges and $\displaystyle\sum_{k=1}^{\infty}1/k^{3/2}$ converges.

  (d) converges because both $\displaystyle\sum_{k=2}^{\infty}\frac{1}{k(\ln k)^2}$ and $\displaystyle\sum_{k=2}^{\infty}1/k^2$ converge.

37. **(a)** $s_{10} \approx 1.1975$; let $f(x) = \dfrac{1}{x^3}$, then $\displaystyle\int_{10}^{+\infty} \dfrac{1}{x^3}dx = \dfrac{1}{200} = 0.0050$ and

$$\int_{11}^{+\infty} \dfrac{1}{x^3}dx = \dfrac{1}{242} \approx 0.0041 \text{ so } 1.2016 < S < 1.2026$$

(b) $\displaystyle\int_{n}^{+\infty} \dfrac{1}{x^3}dx = \dfrac{1}{2n^2} < 10^{-3}$, $n > \sqrt{500} \approx 22.4$; $n = 23$.

39. **(a)** Let $F(x) = \dfrac{1}{x}$, then $\displaystyle\int_{1}^{n} \dfrac{1}{x}dx = \ln n$ and $\displaystyle\int_{1}^{n+1} \dfrac{1}{x}dx = \ln(n+1)$, $u_1 = 1$ so

$\ln(n+1) < s_n < 1 + \ln n$; $\ln(1,000,001) < s_{1,000,000} < 1 + \ln(1,000,000)$,

$13 < s_{1,000,000} < 15$.

(b) $s_n > \ln(n+1) \geq 100$, $n \geq e^{100} - 1 \approx 2.688 \times 10^{43}$; $n = 2.69 \times 10^{43}$

EXERCISE SET 11.5

1. $\rho = \displaystyle\lim_{k \to +\infty} \dfrac{3^{k+1}/(k+1)!}{3^k/k!} = \lim_{k \to +\infty} \dfrac{3}{k+1} = 0$, the series converges.

3. $\rho = \displaystyle\lim_{k \to +\infty} \dfrac{k}{k+1} = 1$, the result is inconclusive.

5. $\rho = \displaystyle\lim_{k \to +\infty} \dfrac{(k+1)!/(k+1)^3}{k!/k^3} = \lim_{k \to +\infty} \dfrac{k^3}{(k+1)^2} = +\infty$, the series diverges.

7. $\rho = \displaystyle\lim_{k \to +\infty} \dfrac{3k+2}{2k-1} = 3/2$, the series diverges.

9. $\rho = \displaystyle\lim_{k \to +\infty} \dfrac{k^{1/k}}{5} = 1/5$, the series converges.

11. ratio test, $\rho = \displaystyle\lim_{k \to +\infty} \dfrac{2k^3}{(k+1)^3} = 2$, diverges

13. ratio test, $\rho = \displaystyle\lim_{k \to +\infty} 7/(k+1) = 0$, converges

15. ratio test, $\rho = \displaystyle\lim_{k \to +\infty} \dfrac{(k+1)^2}{5k^2} = 1/5$, converges

17. ratio test, $\rho = \displaystyle\lim_{k \to +\infty} e^{-1}(k+1)^{50}/k^{50} = e^{-1} < 1$, converges

19. root test, $\rho = \lim\limits_{k \to +\infty} k^{1/k}(2/3) = 2/3$, converges

21. diverges (integral test)

23. root test, $\rho = \lim\limits_{k \to +\infty} 4/(7k - 1) = 0$, converges

25. ratio test, $\rho = \lim\limits_{k \to +\infty} \dfrac{(k + 1)^2}{(2k + 2)(2k + 1)} = 1/4$, converges

27. integral test, $\displaystyle\int_1^{+\infty} \dfrac{dx}{1 + \sqrt{x}} = \lim\limits_{\ell \to +\infty} 2[\sqrt{x} - \ln(1 + \sqrt{x})]_1^\ell = +\infty$, diverges

29. ratio test, $\rho = \lim\limits_{k \to +\infty} \dfrac{\ln(k + 1)}{e \ln k} = \lim\limits_{k \to +\infty} \dfrac{k}{e(k + 1)} = 1/e < 1$, converges

31. ratio test, $\rho = \lim\limits_{k \to +\infty} \dfrac{k + 5}{4(k + 1)} = 1/4$, converges

33. $u_k = \dfrac{k!}{1 \cdot 3 \cdot 5 \cdots (2k - 1)}$, by the ratio test $\rho = \lim\limits_{k \to +\infty} \dfrac{k + 1}{2k + 1} = 1/2$; converges

35. $u_k = \dfrac{(k + 1)!}{1 \cdot 4 \cdot 7 \cdots (3k - 2)}$, by the ratio test $\rho = \lim\limits_{k \to +\infty} \dfrac{k + 2}{3k + 1} = \dfrac{1}{3}$; converges

37. **(a)** $\lim\limits_{x \to +\infty} \ln y = \lim\limits_{x \to +\infty} \dfrac{\ln(\ln x)}{x} = \lim\limits_{x \to +\infty} \dfrac{1}{x \ln x} = 0$ so $\lim\limits_{x \to +\infty} y = e^0 = 1 = \lim\limits_{k \to +\infty} (\ln k)^{1/k}$

    (b) $\rho = \lim\limits_{k \to +\infty} \dfrac{1}{3}(\ln k)^{1/k} = 1/3$

    (c) $\rho = \lim\limits_{k \to +\infty} \dfrac{\ln(k + 1)}{3 \ln k} = \lim\limits_{k \to +\infty} \dfrac{k}{3(k + 1)} = 1/3$

39. The result follows trivially if $a = 0$ so suppose $a \neq 0$ and consider the series $\displaystyle\sum_{k=1}^{\infty} a^k/k!$.

By the ratio test, $\rho = \lim\limits_{k \to +\infty} \dfrac{a}{k + 1} = 0$ so the series converges for every real number a and hence $\lim\limits_{k \to +\infty} a^k/k! = 0$.

41. **(a)** $s_5 \approx 1.71667$, $r_k = \dfrac{1}{k + 1}$ which is decreasing for $k \geq 6$ and $r_6 < 1$; $u_6 = 1/6!$, $r_6 = 1/7$,

    $S - s_5 < \dfrac{1/6!}{1 - 1/7} < 0.00163$.

(b) $u_{n+1} = \dfrac{1}{(n+1)!}$, $r_{n+1} = \dfrac{1}{n+2}$, $\dfrac{u_{n+1}}{1-r_{n+1}} = \dfrac{n+2}{(n+1)(n+1)!} \leq 10^{-5}$ if $n = 8$.

43. **(a)** $s_7 \approx 0.69226$, $r_k = \dfrac{k}{2(k+1)}$ which is increasing for $k \geq 8$ and $\rho = \lim\limits_{k \to +\infty} r_k = 1/2$,

$u_8 = 1/[(8)(2^8)]$, $S - s_7 < \dfrac{1/2^{11}}{1-1/2} < 0.00098$.

(b) $u_{n+1} = \dfrac{1}{(n+1)2^{n+1}}$, $\dfrac{u_{n+1}}{1-\rho} = \dfrac{1}{(n+1)2^n} \leq 10^{-5}$ if $n = 13$.

EXERCISE SET 11.6

1. $\dfrac{1}{3^k + 5} < \dfrac{1}{3^k}$, $\displaystyle\sum_{k=1}^{\infty} \dfrac{1}{3^k}$ converges

3. $\dfrac{1}{5k^2 - k} \leq \dfrac{1}{5k^2 - k^2} = \dfrac{1}{4k^2}$, $\displaystyle\sum_{k=1}^{\infty} \dfrac{1}{4k^2}$ converges

5. $\dfrac{2^k - 1}{3^k + 2k} < \dfrac{2^k}{3^k} = (2/3)^k$, $\displaystyle\sum_{k=1}^{\infty}(2/3)^k$ converges

7. $\dfrac{3}{k - 1/4} > \dfrac{3}{k}$, $\displaystyle\sum_{k=1}^{\infty} 3/k$ diverges

9. $\dfrac{9}{\sqrt{k} + 1} \geq \dfrac{9}{\sqrt{k} + \sqrt{k}} = \dfrac{9}{2\sqrt{k}}$, $\displaystyle\sum_{k=1}^{\infty} \dfrac{9}{2\sqrt{k}}$ diverges

11. $\dfrac{k^{4/3}}{8k^2 + 5k + 1} \geq \dfrac{k^{4/3}}{8k^2 + 5k^2 + k^2} = \dfrac{1}{14k^{2/3}}$, $\displaystyle\sum_{k=1}^{\infty} \dfrac{1}{14k^{2/3}}$ diverges

13. compare with the convergent series $\displaystyle\sum_{k=1}^{\infty} 1/k^5$, $\rho = \lim\limits_{k \to +\infty} \dfrac{4k^7 - 2k^6 + 6k^5}{8k^7 + k - 8} = 1/2$, converges

15. compare with the convergent series $\displaystyle\sum_{k=1}^{\infty} 5/3^k$, $\rho = \lim\limits_{k \to +\infty} \dfrac{3^k}{3^k + 1} = 1$, converges

17. compare with the divergent series $\displaystyle\sum_{k=1}^{\infty} \frac{1}{k^{2/3}}$,

$$\rho = \lim_{k \to +\infty} \frac{k^{2/3}}{(8k^2 - 3k)^{1/3}} = \lim_{k \to +\infty} \frac{1}{(8 - 3/k)^{1/3}} = 1/2, \text{ diverges}$$

19. $\displaystyle\frac{1}{k^3 + 2k + 1} < \frac{1}{k^3}, \sum_{k=1}^{\infty} 1/k^3$ converges so $\displaystyle\sum_{k=1}^{\infty} \frac{1}{k^3 + 2k + 1}$ converges by the comparison test

21. $\displaystyle\frac{1}{9k - 2} > \frac{1}{9k}, \sum_{k=1}^{\infty} \frac{1}{9k}$ diverges so $\displaystyle\sum_{k=1}^{\infty} \frac{1}{9k - 2}$ diverges by the comparison test

23. limit comparison test, compare with the convergent series $\displaystyle\sum_{k=1}^{\infty} 1/k^{5/2}$,

$$\rho = \lim_{k \to +\infty} \frac{k^3}{k^3 + 1} = 1, \text{ converges}$$

25. limit comparison test, compare with the divergent series $\displaystyle\sum_{k=1}^{\infty} 1/k$,

$$\rho = \lim_{k \to +\infty} \frac{k}{\sqrt{k^2 + k}} = 1, \text{ diverges}$$

27. limit comparison test, compare with the convergent series $\displaystyle\sum_{k=1}^{\infty} 1/k^{5/2}$,

$$\rho = \lim_{k \to +\infty} \frac{k^3 + 2k^{5/2}}{k^3 + 3k^2 + 3k} = 1, \text{ converges}$$

29. diverges because $\displaystyle\lim_{k \to +\infty} \frac{1}{4 + 2^{-k}} = 1/4 \neq 0$

31. $\displaystyle\frac{\tan^{-1} k}{k^2} < \frac{\pi/2}{k^2}, \sum_{k=1}^{\infty} \frac{\pi/2}{k^2}$ converges so $\displaystyle\sum_{k=1}^{\infty} \frac{\tan^{-1} k}{k^2}$ converges

33. $\displaystyle\sum_{k=1}^{\infty} \frac{\ln k}{k\sqrt{k}} = \sum_{k=2}^{\infty} \frac{\ln k}{k\sqrt{k}}$ because $\ln 1 = 0$,

$$\int_2^{+\infty} \frac{\ln x}{x^{3/2}} dx = \lim_{\ell \to +\infty} \left[-\frac{2\ln x}{x^{1/2}} - \frac{4}{x^{1/2}} \right]_2^{\ell} = \sqrt{2}(\ln 2 + 2) \text{ so } \sum_{k=2}^{\infty} \frac{\ln k}{k^{3/2}} \text{ converges.}$$

35. $\rho = \lim\limits_{k \to +\infty} \dfrac{1 - \cos(1/k)}{1/k^2}$, but

$$\lim_{x \to +\infty} \frac{1 - \cos(1/x)}{1/x^2} = \lim_{x \to +\infty} \frac{(-1/x^2)\sin(1/x)}{-2/x^3} = \lim_{x \to +\infty} \frac{\sin(1/x)}{2(1/x)} = 1/2, \text{ so } \rho = 1/2;$$

the series converges.

37. $\dfrac{\ln k}{k^2} < \dfrac{\sqrt{k}}{k^2} = \dfrac{1}{k^{3/2}}$, $\displaystyle\sum_{k=1}^{\infty} \dfrac{1}{k^{3/2}}$ converges so $\displaystyle\sum_{k=1}^{\infty} \dfrac{\ln k}{k^2}$ converges

39. limit comparison test, compare with $\displaystyle\sum_{k=1}^{\infty} 1/k^p$,

$$\rho = \lim_{k \to +\infty} \frac{k^p}{(a + bk)^p} = \lim_{k \to +\infty} \frac{1}{(a/k + b)^p} = 1/b^p, \text{ converges for } p > 1.$$

41. Compare with the convergent series $\displaystyle\sum_{k=1}^{\infty} 1/k!$, $\rho = \lim\limits_{k \to +\infty} \dfrac{(k+1)^2 k!}{(k+2)!} = \lim\limits_{k \to +\infty} \dfrac{k+1}{k+2} = 1$, converges

43. $k! = k(k-1)(k-2)\cdots(2)(1) \geq 2 \cdot 2 \cdot 2 \cdots 2 \cdot 1 = 2^{k-1}$, $1/k! \leq 1/2^{k-1}$, $\displaystyle\sum_{k=1}^{\infty} 1/2^{k-1}$ converges

so $\displaystyle\sum_{k=1}^{\infty} 1/k!$ converges

EXERCISE SET 11.7

1. converges

3. diverges because $\lim\limits_{k \to +\infty} a_k = \lim\limits_{k \to +\infty} \dfrac{k+1}{3k+1} = 1/3 \neq 0$

5. converges

7. $\rho = \lim\limits_{k \to +\infty} \dfrac{(3/5)^{k+1}}{(3/5)^k} = 3/5$, converges absolutely

9. $\rho = \lim\limits_{k \to +\infty} \dfrac{3k^2}{(k+1)^2} = 3$, diverges

11. $\rho = \lim\limits_{k\to+\infty} \dfrac{(k+1)^3}{ek^3} = 1/e$, converges absolutely

13. conditionally convergent, $\sum\limits_{k=1}^{\infty} \dfrac{(-1)^{k+1}}{3k}$ converges by the alternating series test but $\sum\limits_{k=1}^{\infty} \dfrac{1}{3k}$, diverges

15. divergent, $\lim\limits_{k\to+\infty} a_k \neq 0$

17. $\sum\limits_{k=1}^{\infty} \dfrac{\cos k\pi}{k} = \sum\limits_{k=1}^{\infty} \dfrac{(-1)^k}{k}$ is conditionally convergent, $\sum\limits_{k=1}^{\infty} \dfrac{(-1)^k}{k}$ converges by the alternating series test but $\sum\limits_{k=1}^{\infty} 1/k$ diverges.

19. absolutely convergent, $\sum\limits_{k=1}^{\infty} \left[\dfrac{k+2}{3k-1}\right]^k$ converges by the root test.

21. conditionally convergent, $\sum\limits_{k=1}^{\infty}(-1)^{k+1}\dfrac{k+2}{k(k+3)}$ converges by the alternating series test but $\sum\limits_{k=1}^{\infty} \dfrac{k+2}{k(k+3)}$ diverges (limit comparison test with $\sum 1/k$)

23. $\sum\limits_{k=1}^{\infty} \sin(k\pi/2) = 1 + 0 - 1 + 0 + 1 + 0 - 1 + 0 + \cdots$, divergent ($\lim\limits_{k\to+\infty} \sin(k\pi/2)$ does not exist)

25. conditionally convergent, $\sum\limits_{k=2}^{\infty} \dfrac{(-1)^k}{k \ln k}$ converges by the alternating series test but $\sum\limits_{k=2}^{\infty} \dfrac{1}{k \ln k}$ diverges (integral test)

27. absolutely convergent, $\sum\limits_{k=2}^{\infty}(1/\ln k)^k$ converges by the root test

29. conditionally convergent, let $f(x) = \dfrac{x^2+1}{x^3+2}$ then $f'(x) = \dfrac{x(4-3x-x^3)}{(x^3+2)^2} \leq 0$ for $x \geq 2$ so $\{a_k\}_{k=2}^{+\infty} = \left\{\dfrac{k^2+1}{k^3+2}\right\}_{k=2}^{+\infty}$ is nonincreasing, $\lim\limits_{k\to+\infty} a_k = 0$; the series converges by the alternating series test but $\sum\limits_{k=2}^{\infty} \dfrac{k^2+1}{k^3+2}$ diverges (limit comparison test with $\sum 1/k$)

31. $|\text{error}| < a_8 = 1/8 = 0.125$ **33.** $|\text{error}| < a_{100} = 1/\sqrt{100} = 0.1$

35. $|\text{error}| < 0.0001$ if $a_{n+1} \leq 0.0001$, $1/(n+1) \leq 0.0001$, $n+1 \geq 10,000$, $n \geq 9,999$; $n = 9,999$

37. $|\text{error}| < 0.005$ if $a_{n+1} \leq 0.005$, $1/\sqrt{n+1} \leq 0.005$, $\sqrt{n+1} \geq 200$, $n+1 \geq 40,000$, $n \geq 39,999$; $n = 39,999$

39. $a_k = \dfrac{3}{2^{k+1}}$, $|\text{error}| < a_{11} = \dfrac{3}{2^{12}} < 0.00074$; $s_{10} \approx 0.4995$; $S = \dfrac{3/4}{1-(-1/2)} = 0.5$.

41. $a_k = \dfrac{1}{(2k-1)!}$, $a_{n+1} = \dfrac{1}{(2n+1)!} \leq 10^{-4}$, $(2n+1)! \geq 10,000$, $2n+1 \geq 8$, $n \geq 3.5$; $n = 4$.
$s_4 \approx 0.84147$, $\sin(1) \approx 0.841470985$.

43. $a_k = \dfrac{1}{k2^k}$, $a_{n+1} = \dfrac{1}{(n+1)2^{n+1}} \leq 10^{-4}$, $(n+1)2^{n+1} \geq 10,000$, $n+1 \geq 10$, $n \geq 9$; $n = 9$.
$s_9 \approx 0.40553$, $\ln \dfrac{3}{2} \approx 0.405465108$.

45. **(a)** $a_k = \dfrac{1}{k^2}$, $a_{n+1} = \dfrac{1}{(n+1)^2} \leq 5 \times 10^{-3}$, $(n+1)^2 \geq 200$, $n \geq 14$; $n = 14$

     (b) $s_{10} \approx 0.817962176$, $\pi^2/12 \approx 0.822467033$, $|\text{error}| \approx 0.004504858$

47. Suppose $\Sigma|a_k|$ converges, then $\lim\limits_{k \to +\infty} |a_k| = 0$ so $|a_k| < 1$ for $k \geq K$ and thus $|a_k|^2 < |a_k|$,
$a_k^2 < |a_k|$ hence Σa_k^2 converges by the comparison test.

49. Suppose $a_1 - a_2 + a_3 - a_4 + \cdots + (-1)^{k+1}a_k + \cdots$ satisfies (a) and (b) of Theorem 11.7.1, then the
series converges and by Theorem 11.4.3b so does $-\left(a_1 - a_2 + a_3 - a_4 + \cdots + (-1)^{k+1}a_k + \cdots\right)$
which equals $-a_1 + a_2 - a_3 + a_4 - \cdots + (-1)^k a_k + \cdots$

51. $\left(1 - \dfrac{1}{2} - \dfrac{1}{4}\right) + \left(\dfrac{1}{3} - \dfrac{1}{6} - \dfrac{1}{8}\right) + \left(\dfrac{1}{5} - \dfrac{1}{10} - \dfrac{1}{12}\right) + \cdots$

     $= \left(\dfrac{1}{2} - \dfrac{1}{4}\right) + \left(\dfrac{1}{6} - \dfrac{1}{8}\right) + \left(\dfrac{1}{10} - \dfrac{1}{12}\right) + \cdots = \dfrac{1}{2}\left(1 - \dfrac{1}{2} + \dfrac{1}{3} - \dfrac{1}{4} + \dfrac{1}{5} - \dfrac{1}{6} + \cdots\right) = S/2$

53. $1 + \dfrac{1}{3^2} + \dfrac{1}{5^2} + \cdots = \left[1 + \dfrac{1}{2^2} + \dfrac{1}{3^2} + \cdots\right] - \left[\dfrac{1}{2^2} + \dfrac{1}{4^2} + \dfrac{1}{6^2} + \cdots\right]$

     $= \dfrac{\pi^2}{6} - \dfrac{1}{2^2}\left[1 + \dfrac{1}{2^2} + \dfrac{1}{3^2} + \cdots\right] = \dfrac{\pi^2}{6} - \dfrac{1}{4}\dfrac{\pi^2}{6} = \dfrac{\pi^2}{8}$

55. $1 - \dfrac{1}{2^2} + \dfrac{1}{3^2} - \cdots = \left[1 + \dfrac{1}{2^2} + \dfrac{1}{3^2} + \cdots\right] - 2\left[\dfrac{1}{2^2} + \dfrac{1}{4^2} + \dfrac{1}{6^2} + \cdots\right]$

$$= \dfrac{\pi^2}{6} - \dfrac{2}{2^2}\left[1 + \dfrac{1}{2^2} + \dfrac{1}{3^2} + \cdots\right] = \dfrac{\pi^2}{6} - \dfrac{1}{2}\dfrac{\pi^2}{6} = \dfrac{\pi^2}{12}$$

EXERCISE SET 11.8

1. $\rho = \lim\limits_{k\to+\infty} \dfrac{k+1}{k+2}|x| = |x|$, the series converges if $|x| < 1$ and diverges if $|x| > 1$. If $x = -1$, $\sum\limits_{k=0}^{\infty} \dfrac{(-1)^k}{k+1}$ converges by the alternating series test; if $x = 1$, $\sum\limits_{k=0}^{\infty} \dfrac{1}{k+1}$ diverges. The radius of convergence is 1, the interval of convergence is $[-1, 1)$.

3. $\rho = \lim\limits_{k\to+\infty} \dfrac{|x|}{k+1} = 0$, the radius of convergence is $+\infty$, the interval is $(-\infty, +\infty)$.

5. $\rho = \lim\limits_{k\to+\infty} \dfrac{5k^2|x|}{(k+1)^2} = 5|x|$, converges if $|x| < 1/5$ and diverges if $|x| > 1/5$. If $x = -1/5$, $\sum\limits_{k=1}^{\infty} \dfrac{(-1)^k}{k^2}$ converges; if $x = 1/5$, $\sum\limits_{k=1}^{\infty} 1/k^2$ converges. Radius of convergence is $1/5$, interval of convergence is $[-1/5, 1/5]$.

7. $\rho = \lim\limits_{k\to+\infty} \dfrac{k|x|}{k+2} = |x|$, converges if $|x| < 1$, diverges if $|x| > 1$. If $x = -1$, $\sum\limits_{k=1}^{\infty} \dfrac{(-1)^k}{k(k+1)}$ converges; if $x = 1$, $\sum\limits_{k=1}^{\infty} \dfrac{1}{k(k+1)}$ converges. Radius of convergence is 1, interval of convergence is $[-1, 1]$.

9. $\rho = \lim\limits_{k\to+\infty} \dfrac{\sqrt{k}}{\sqrt{k+1}}|x| = |x|$, converges if $|x| < 1$, diverges if $|x| > 1$. If $x = -1$, $\sum\limits_{k=1}^{\infty} \dfrac{-1}{\sqrt{k}}$ diverges; if $x = 1$, $\sum\limits_{k=1}^{\infty} \dfrac{(-1)^{k-1}}{\sqrt{k}}$ converges. Radius of convergence is 1, interval of convergence is $(-1, 1]$.

11. $\rho = \lim\limits_{k\to+\infty} \dfrac{|x|^2}{(2k+3)(2k+2)} = 0$, radius of convergence is $+\infty$, interval of convergence is $(-\infty, +\infty)$.

13. $\rho = \lim\limits_{k \to +\infty} \dfrac{3|x|}{k+1} = 0$, radius of convergence is $+\infty$, interval of convergence is $(-\infty, +\infty)$.

15. $\rho = \lim\limits_{k \to +\infty} \dfrac{1+k^2}{1+(k+1)^2}|x| = |x|$, converges if $|x| < 1$, diverges if $|x| > 1$. If $x = -1$, $\sum\limits_{k=0}^{\infty} \dfrac{(-1)^k}{1+k^2}$ converges; if $x = 1$, $\sum\limits_{k=0}^{\infty} \dfrac{1}{1+k^2}$ converges. Radius of convergence is 1, interval of convergence is $[-1, 1]$.

17. $\rho = \lim\limits_{k \to +\infty} \dfrac{k|x+1|}{k+1} = |x+1|$, converges if $|x+1| < 1$, diverges if $|x+1| > 1$. If $x = -2$, $\sum\limits_{k=1}^{\infty} \dfrac{-1}{k}$ diverges; if $x = 0$, $\sum\limits_{k=1}^{\infty} \dfrac{(-1)^{k+1}}{k}$ converges. Radius of convergence is 1, interval of convergence is $(-2, 0]$.

19. $\rho = \lim\limits_{k \to +\infty} (3/4)|x+5| = \dfrac{3}{4}|x+5|$, convergence if $|x+5| < 4/3$, diverges if $|x+5| > 4/3$. If $x = -19/3$, $\sum\limits_{k=0}^{\infty}(-1)^k$ diverges; if $x = -11/3$, $\sum\limits_{k=0}^{\infty} 1$ diverges. Radius of convergence is $4/3$, interval of convergence is $(-19/3, -11/3)$.

21. $\rho = \lim\limits_{k \to +\infty} \dfrac{k^2+4}{(k+1)^2+4}|x+1|^2 = |x+1|^2$, converges if $|x+1| < 1$, diverges if $|x+1| > 1$. If $x = -2$, $\sum\limits_{k=1}^{\infty} \dfrac{(-1)^{3k+1}}{k^2+4}$ converges; if $x = 0$, $\sum\limits_{k=1}^{\infty} \dfrac{(-1)^k}{k^2+4}$ converges. Radius of convergence is 1, interval of convergence is $[-2, 0]$.

23. $\rho = \lim\limits_{k \to +\infty} \dfrac{\pi|x-1|^2}{(2k+3)(2k+2)} = 0$, radius of convergence $+\infty$, interval of convergence $(-\infty, +\infty)$.

25. $x + \dfrac{1}{2}x^2 + \dfrac{3}{14}x^3 + \dfrac{3}{35}x^4 + \cdots$; $\rho = \lim\limits_{k \to +\infty} \dfrac{k+1}{3k+1}|x| = \dfrac{1}{3}|x|$, converges if $\dfrac{1}{3}|x| < 1$, $|x| < 3$ so $R = 3$.

27. $x + \dfrac{3}{2}x^2 + \dfrac{5}{8}x^3 + \dfrac{7}{48}x^4 + \cdots$; $\rho = \lim\limits_{k \to +\infty} \dfrac{2k+1}{(2k)(2k-1)}|x| = 0$ so $R = +\infty$.

29. By the ratio test for absolute convergence, $\rho = \lim\limits_{k\to+\infty} \dfrac{|x-a|}{b} = \dfrac{|x-a|}{b}$; converges if $|x-a| < b$,

diverges if $|x-a| > b$. If $x = a-b$, $\sum\limits_{k=0}^{\infty}(-1)^k$ diverges; if $x = a+b$, $\sum\limits_{k=0}^{\infty} 1$ diverges. The interval of convergence is $(a-b, a+b)$.

31. By the ratio test for absolute convergence,

$$\rho = \lim_{k\to+\infty} \frac{(k+1+p)!k!(k+q)!}{(k+p)!(k+1)!(k+1+q)!}|x| = \lim_{k\to+\infty} \frac{k+1+p}{(k+1)(k+1+q)}|x| = 0,$$

radius of convergence is $+\infty$.

33. By assumption $\sum\limits_{k=0}^{\infty} c_k x^k$ converges if $|x| < R$ so $\sum\limits_{k=0}^{\infty} c_k x^{2k} = \sum\limits_{k=0}^{\infty} c_k(x^2)^k$ converges if $|x^2| < R$,

$|x| < \sqrt{R}$. Thus $\sum\limits_{k=0}^{\infty} c_k x^{2k}$ has radius of convergence $\sqrt{R}$.

EXERCISE SET 11.9

1. $1 - 2x + 2x^2 - \dfrac{4}{3}x^3 + \dfrac{2}{3}x^4$

3. $2x - \dfrac{4}{3}x^3$

5. $x + \dfrac{1}{3}x^3$

7. $x + x^2 + \dfrac{1}{2}x^3 + \dfrac{1}{6}x^4$

9. $1 + \dfrac{1}{2}x^2 + \dfrac{5}{24}x^4$

11. $\ln 3 + \dfrac{2}{3}x - \dfrac{2}{9}x^2 + \dfrac{8}{81}x^3 - \dfrac{4}{81}x^4$

13. $e + e(x-1) + \dfrac{e}{2}(x-1)^2 + \dfrac{e}{6}(x-1)^3$

15. $2 + \dfrac{1}{4}(x-4) - \dfrac{1}{64}(x-4)^2 + \dfrac{1}{512}(x-4)^3$

17. $\dfrac{\sqrt{2}}{2} - \dfrac{\sqrt{2}}{2}(x - \pi/4) - \dfrac{\sqrt{2}}{4}(x-\pi/4)^2 + \dfrac{\sqrt{2}}{12}(x-\pi/4)^3$

19. $-\dfrac{\sqrt{3}}{2} + \dfrac{\pi}{2}(x+1/3) + \dfrac{\sqrt{3}\pi^2}{4}(x+1/3)^2 - \dfrac{\pi^3}{12}(x+1/3)^3$

21. $\dfrac{\pi}{4} + \dfrac{1}{2}(x-1) - \dfrac{1}{4}(x-1)^2 + \dfrac{1}{12}(x-1)^3$

23. $f^{(k)}(x) = (-1)^k e^{-x}$, $f^{(k)}(0) = (-1)^k$; $\sum\limits_{k=0}^{\infty} \dfrac{(-1)^k}{k!}x^k$

25. $f^{(k)}(x) = \dfrac{(-1)^k k!}{(1+x)^{k+1}}$, $f^{(k)}(0) = (-1)^k k!$; $\displaystyle\sum_{k=0}^{\infty}(-1)^k x^k$

27. $f^{(0)}(0) = 0$; for $k \geq 1$, $f^{(k)}(x) = \dfrac{(-1)^{k+1}(k-1)!}{(1+x)^k}$, $f^{(k)}(0) = (-1)^{k+1}(k-1)!$; $\displaystyle\sum_{k=1}^{\infty}\dfrac{(-1)^{k+1}}{k}x^k$

29. $f^{(k)}(0) = 0$ if k is odd, $f^{(k)}(0)$ is alternately $1/2^k$ and $-1/2^k$ if k is even.

$$\sum_{k=0}^{\infty}\frac{f^{(k)}(0)}{k!}x^k = 1 - \frac{1}{2^2 2!}x^2 + \frac{1}{2^4 4!}x^4 - \cdots = \sum_{k=0}^{\infty}\frac{(-1)^k}{2^{2k}(2k)!}x^{2k}$$

31. $f^{(k)}(0) = 0$ if k is odd, $f^{(k)}(0) = 1$ if k is even;

$$\sum_{k=0}^{\infty}\frac{f^{(k)}(0)}{k!}x^k = 1 + \frac{1}{2!}x^2 + \frac{1}{4!}x^4 + \cdots = \sum_{k=0}^{\infty}\frac{1}{(2k)!}x^{2k}$$

33. $f^{(k)}(x) = \dfrac{(-1)^k k!}{x^{k+1}}$, $f^{(k)}(-1) = -k!$; $\displaystyle\sum_{k=0}^{\infty}(-1)(x+1)^k$

35. $f^{(0)}(1) = 0$; for $k \geq 1$, $f^{(k)}(x) = \dfrac{(-1)^{k+1}(k-1)!}{x^k}$, $f^{(k)}(1) = (-1)^{k+1}(k-1)!$;

$$\sum_{k=1}^{\infty}\frac{(-1)^{k+1}}{k}(x-1)^k$$

37. $f^{(k)}(1/2) = 0$ if k is odd, $f^{(k)}(1/2)$ is alternately π^k and $-\pi^k$ if k is even;

$$\sum_{k=0}^{\infty}\frac{f^{(k)}(1/2)}{k!}(x-1/2)^k = 1 - \frac{\pi^2}{2!}(x-1/2)^2 + \frac{\pi^4}{4!}(x-1/2)^4 - \cdots$$

$$= \sum_{k=0}^{\infty}\frac{(-1)^k \pi^{2k}}{(2k)!}(x-1/2)^{2k}$$

39. $f^{(k)}(\ln 4) = 15/8$ for k even, $f^{(k)}(\ln 4) = 17/8$ for k odd, which can be written as

$$f^{(k)}(\ln 4) = \frac{16-(-1)^k}{8}; \sum_{k=0}^{\infty}\frac{16-(-1)^k}{8k!}(x-\ln 4)^k$$

EXERCISE SET 11.10

1. $f^{(6)}(x) = 2^6 e^{2x}$, $R_5(x) = \dfrac{2^6 e^{2c}}{6!}x^6$

3. $f^{(5)}(x) = -5!/(x+1)^6$, $R_4(x) = -\dfrac{1}{(c+1)^6}x^5$

5. $f^{(4)}(x) = (4+x)e^x$, $R_3(x) = \dfrac{(4+c)e^c}{4!}x^4$

7. $f^{(3)}(x) = -\dfrac{2(1-3x^2)}{(1+x^2)^3}$, $R_2(x) = -\dfrac{1-3c^2}{3(1+c^2)^3}x^3$

9. $f^{(4)}(x) = -\dfrac{15}{16}x^{-7/2}$, $R_3(x) = -\dfrac{5}{128c^{7/2}}(x-4)^4$

11. $f^{(5)}(x) = \cos x$, $R_4(x) = \dfrac{\cos c}{5!}(x-\pi/6)^5$

13. $f^{(6)}(x) = 7!/(1+x)^8$, $R_5(x) = \dfrac{7}{(1+c)^8}(x+2)^6$

15. $f^{(n+1)}(x) = (n+1)!/(1-x)^{n+2}$, $R_n(x) = \dfrac{1}{(1-c)^{n+2}}x^{n+1}$

17. $f^{(n+1)}(x) = 2^{n+1}e^{2x}$, $R_n(x) = \dfrac{2^{n+1}e^{2c}}{(n+1)!}x^{n+1}$

19. $f(x) = \cos x$, $f^{(n+1)}(x) = \pm\sin x$ or $\pm\cos x$, $|f^{(n+1)}(x)| \le 1$,

$|R_n(x)| = \dfrac{|f^{(n+1)}(c)|}{(n+1)!}|x|^{n+1} \le \dfrac{|x|^{n+1}}{(n+1)!}$, $\displaystyle\lim_{n\to+\infty}\dfrac{|x|^{n+1}}{(n+1)!} = 0$, by the Squeezing Theorem

$\displaystyle\lim_{n\to+\infty}|R_n(x)| = 0$ so $\displaystyle\lim_{n\to+\infty}R_n(x) = 0$ for all x.

21. $f(x) = e^x$, $f^{(n+1)}(x) = e^x$, $|R_n(x)| = \dfrac{e^c}{(n+1)!}|x-1|^{n+1}$. If $x \ge 1$, $e^c \le e^x$,

$|R_n(x)| \le e^x\dfrac{|x-1|^{n+1}}{(n+1)!}$, $\displaystyle\lim_{n\to+\infty}e^x\dfrac{|x-1|^{n+1}}{(n+1)!} = e^x(0) = 0$ so $\displaystyle\lim_{n\to+\infty}R_n(x) = 0$.

If $x < 1$, $e^c < e$, $|R_n(x)| < e\dfrac{|x-1|^{n+1}}{(n+1)!}$ and again $\displaystyle\lim_{n\to+\infty}R_n(x) = 0$.

23. $f(x) = e^x$, $f^{(n+1)}(x) = e^x$, $|R_n(x)| = \dfrac{e^c}{(n+1)!}|x-a|^{n+1}$. If $x \ge a$, $e^c \le e^x$,

$|R_n(x)| \le e^x\dfrac{|x-a|^{n+1}}{(n+1)!}$, $\displaystyle\lim_{n\to+\infty}e^x\dfrac{|x-a|^{n+1}}{(n+1)!} = e^x(0) = 0$ so $\displaystyle\lim_{n\to+\infty}R_n(x) = 0$.

If $x < a$, $e^c < e^a$, $|R_n(x)| < e^a\dfrac{|x-a|^{n+1}}{(n+1)!}$ and again $\displaystyle\lim_{n\to+\infty}R_n(x) = 0$.

25. $f(x) = \cos x$, $f^{(n+1)}(x) = \pm \sin x$ or $\pm \cos x$, $|f^{(n+1)}(x)| \le 1$,

$|R_n(x)| = \dfrac{|f^{(n+1)}(c)|}{(n+1)!}|x-a|^{n+1} \le \dfrac{|x-a|^{n+1}}{(n+1)!}$, $\displaystyle\lim_{n\to+\infty} \dfrac{|x-a|^{n+1}}{(n+1)!} = 0$ so $\displaystyle\lim_{n\to+\infty} R_n(x) = 0$

for all x.

27. $1 - 2x + 2x^2 - \dfrac{4}{3}x^3 + \cdots$; $(-\infty, +\infty)$

29. $x\left(1 - x + \dfrac{1}{2!}x^2 - \dfrac{1}{3!}x^3 + \cdots\right) = x - x^2 + \dfrac{1}{2!}x^3 - \dfrac{1}{3!}x^4 + \cdots$; $(-\infty, +\infty)$

31. $2x - \dfrac{2^3}{3!}x^3 + \dfrac{2^5}{5!}x^5 - \dfrac{2^7}{7!}x^7 + \cdots$; $(-\infty, +\infty)$

33. $x^2\left(1 - \dfrac{1}{2!}x^2 + \dfrac{1}{4!}x^4 - \dfrac{1}{6!}x^6 + \cdots\right) = x^2 - \dfrac{1}{2!}x^4 + \dfrac{1}{4!}x^6 - \dfrac{1}{6!}x^8 + \cdots$; $(-\infty, +\infty)$

35. $\dfrac{1}{2}\left[1 - \left(1 - \dfrac{2^2}{2!}x^2 + \dfrac{2^4}{4!}x^4 - \dfrac{2^6}{6!}x^6 + \cdots\right)\right] = x^2 - \dfrac{2^3}{4!}x^4 + \dfrac{2^5}{6!}x^6 - \dfrac{2^7}{8!}x^8 + \cdots$; $(-\infty, +\infty)$

37. $-x^2 - \dfrac{1}{2}x^4 - \dfrac{1}{3}x^6 - \dfrac{1}{4}x^8 - \cdots$; $(-1, 1)$

39. $1 + 4x^2 + 16x^4 + 64x^6 + \cdots$; $(-1/2, 1/2)$

41. $x^2\left(1 - 3x + 9x^2 - 27x^3 + \cdots\right) = x^2 - 3x^3 + 9x^4 - 27x^5 + \cdots$; $(-1/3, 1/3)$

43. $x\left(2x + \dfrac{2^3}{3!}x^3 + \dfrac{2^5}{5!}x^5 + \dfrac{2^7}{7!}x^7 + \cdots\right) = 2x^2 + \dfrac{2^3}{3!}x^4 + \dfrac{2^5}{5!}x^6 + \dfrac{2^7}{7!}x^8 + \cdots$; $(-\infty, +\infty)$

45. $1 + \dfrac{3}{2}x - \dfrac{9}{8}x^2 + \dfrac{27}{16}x^3 - \cdots$; $(-1/3, 1/3)$

47. $(1 - 2x)^{-2} = 1 + 4x + 12x^2 + 32x^3 + \cdots$; $(-1/2, 1/2)$

49. $x(1 - x^2)^{-1/2} = x + \dfrac{1}{2}x^3 + \dfrac{3}{8}x^5 + \dfrac{5}{16}x^7 + \cdots$; $(-1, 1)$

51. $\dfrac{1}{x} = \dfrac{1}{1 + (x-1)} = \dfrac{1}{1 - [-(x-1)]} = \displaystyle\sum_{k=0}^{\infty}[-(x-1)]^k = \sum_{k=0}^{\infty}(-1)^k(x-1)^k$ which is valid for

$-1 < -(x-1) < 1$, $0 < x < 2$

53. $(1 + x)^m = \dbinom{m}{0} + \displaystyle\sum_{k=1}^{\infty}\dbinom{m}{k}x^k = \sum_{k=0}^{\infty}\dbinom{m}{k}x^k$

55. $\sin \pi = 0$ $\qquad\qquad\qquad\qquad$ **57.** $e^{-\ln 3} = 1/3$

59. **(a)** $\cos 2x = 1 - \dfrac{(2x)^2}{2!} + \dfrac{(2x)^4}{4!} - \dfrac{(2x)^6}{6!} + \cdots = 1 - 2x^2 + \dfrac{2}{3}x^4 - \dfrac{4}{45}x^6 + \cdots,$

$\qquad\qquad x^2 \cos 2x = x^2 - 2x^4 + \dfrac{2}{3}x^6 - \dfrac{4}{45}x^8 + \cdots$

$\qquad$ **(b)** $\dfrac{f^{(5)}(0)}{5!} = 0$ so $f^{(5)}(0) = 0$

EXERCISE SET 11.11

1. **(a)** $3/(n+1)! < 0.5 \times 10^{-5}$, $n = 9$ $\qquad\qquad$ **(b)** $3/(n+1)! < 0.5 \times 10^{-10}$, $n = 13$

3. $|R_n(1/2)| = \dfrac{e^c}{(n+1)!}(1/2)^{n+1} < \dfrac{2}{(n+1)!}\dfrac{1}{2^{n+1}} = \dfrac{1}{2^n(n+1)!} < 0.5 \times 10^{-4}$ if $n = 5$

$\qquad$ so $\sqrt{e} = e^{0.5} \approx 1 + 0.5 + \dfrac{(0.5)^2}{2!} + \dfrac{(0.5)^3}{3!} + \dfrac{(0.5)^4}{4!} + \dfrac{(0.5)^5}{5!}$, $\sqrt{e} \approx 1.6487$

5. $|R_n(\pi/20)| \le \dfrac{(\pi/20)^{n+1}}{(n+1)!} < 0.5 \times 10^{-4}$ if $n = 3$, $\cos(\pi/20) \approx 1 - \dfrac{(\pi/20)^2}{2!} \approx 0.9877$

7. Expand about $\pi/3$ to get $\cos x = \dfrac{1}{2} - \dfrac{\sqrt{3}}{2}(x - \pi/3) - \dfrac{1}{4}(x - \pi/3)^2 + \cdots$, $58° = 29\pi/90$ radians,

$\qquad |R_n(x)| \le \dfrac{|x - \pi/3|^{n+1}}{(n+1)!}$, $|R_n(29\pi/90)| \le \dfrac{|29\pi/90 - \pi/3|^{n+1}}{(n+1)!} = \dfrac{(\pi/90)^{n+1}}{(n+1)!} < 0.5 \times 10^{-4}$

$\qquad$ if $n = 2$, $\cos 58° \approx \dfrac{1}{2} - \dfrac{\sqrt{3}}{2}(-\pi/90) - \dfrac{1}{4}(-\pi/90)^2 \approx 0.5299$

9. Let $x = 1/9$ in series (16) to get $\ln 1.25 \approx 0.223$

11. $(0.1)^3/3 < 0.5 \times 10^{-3}$ so $\tan^{-1}(0.1) \approx 0.100$ to three decimal place accuracy.

13. $|R_n(0.1)| = \dfrac{|f^{(n+1)}(c)|}{(n+1)!}(0.1)^{n+1}$ where $f^{(n+1)}(c) = \sinh c$ or $\cosh c$ for $0 < c < 0.1$, but

$\qquad \sinh c < \cosh c < \cosh 0.1$, and $\cosh 0.1 = \dfrac{1}{2}\left(e^{0.1} + e^{-0.1}\right) < \dfrac{1}{2}(2 + 1) = 1.5$

$\qquad$ so $|R_n(0.1)| < \dfrac{1.5(0.1)^{n+1}}{(n+1)!} \le 0.5 \times 10^{-4}$ if $n = 3$, $\cosh 0.1 \approx 1 + \dfrac{(0.1)^2}{2!} \approx 1.0050$

15. $\sin x = x - \dfrac{x^3}{3!} + (0)x^4 + R_4(x)$, $|R_4(x)| \le \dfrac{|x|^5}{5!} < 0.5 \times 10^{-3}$ if $|x|^5 < 0.06$, $|x| < (0.06)^{1/5} \approx 0.569$

17. $\cos x = 1 - \dfrac{x^2}{2!} + \dfrac{x^4}{4!} + (0)x^5 + R_5(x),\ |R_5(x)| \le \dfrac{|x|^6}{6!} \le \dfrac{(0.2)^6}{6!} < 9 \times 10^{-8}$

19. **(a)** $\ln 2 = 1 - 1/2 + 1/3 - 1/4 + \cdots,\ |\text{error}| < 1/(n+1) \le 0.5 \times 10^{-6}$ if $n+1 \ge 2 \times 10^6$, $n \ge 1{,}999{,}999;\ n = 1{,}999{,}999$

(b) Let $f(x) = \ln\dfrac{1+x}{1-x} = \ln(1+x) - \ln(1-x),\ f^{(n+1)}(c) = n!\left[\dfrac{(-1)^n}{(1+c)^{n+1}} + \dfrac{1}{(1-c)^{n+1}}\right]$,

$|f^{n+1}(c)| \le n!\left[\dfrac{1}{(1+c)^{n+1}} + \dfrac{1}{(1-c)^{n+1}}\right]$

$< n!\left[\dfrac{1}{(1+0)^{n+1}} + \dfrac{1}{(1-1/3)^{n+1}}\right]$ for $0 < c < 1/3$

$= n![1 + 1/(2/3)^{n+1}]$,

$|R_n(1/3)| < [1 + 1/(2/3)^{n+1}]\dfrac{(1/3)^{n+1}}{n+1} = (1/3^{n+1} + 1/2^{n+1})/(n+1) \le 0.5 \times 10^{-6}$

if $n \ge 16$ so 8 <u>terms</u> in series (16) are sufficient to assure six decimal place accuracy (even powers of x have zero coefficients).

21. $(1/2)^9/9! < 0.5 \times 10^{-3}$ and $(1/3)^7/7! < 0.5 \times 10^{-3}$ so

$\tan^{-1} 1/2 \approx 1/2 - \dfrac{(1/2)^3}{3} + \dfrac{(1/2)^5}{5} - \dfrac{(1/2)^7}{7} \approx 0.463$

$\tan^{-1} 1/3 \approx 1/3 - \dfrac{(1/3)^3}{3} + \dfrac{(1/3)^5}{5} \approx 0.322,\ \pi \approx 4(0.463 + 0.322) = 3.140$

EXERCISE SET 11.12

1. **(a)** $\dfrac{d}{dx}\left(1 + x + x^2/2! + x^3/3! + \cdots\right) = 1 + x + x^2/2! + \cdots = e^x$

(b) $\displaystyle\int \left(1 + x + x^2/2! + \cdots\right) dx = (x + x^2/2! + x^3/3! + \cdots) + C_1$
$= (1 + x + x^2/2! + x^3/3! + \cdots) + C_1 - 1 = e^x + C$

3. **(a)** $\dfrac{d}{dx}\left(x + x^3/3! + x^5/5! + \cdots\right) = 1 + x^2/2! + x^4/4! + \cdots = \cosh x$

(b) $\displaystyle\int \left(x + x^3/3! + x^5/5! + \cdots\right) dx = (x^2/2! + x^4/4! + x^6/6! + \cdots) + C_1$
$= (1 + x^2/2! + x^4/4! + x^6/6! + \cdots) + C_1 - 1$
$= \cosh x + C$

5. $1/(1+x)^2 = \dfrac{d}{dx}\left[-\dfrac{1}{1+x}\right] = \dfrac{d}{dx}\left[-\displaystyle\sum_{k=0}^{\infty}(-1)^k x^k\right]$

$\qquad\qquad = \dfrac{d}{dx}\left[\displaystyle\sum_{k=0}^{\infty}(-1)^{k+1}x^k\right] = \displaystyle\sum_{k=1}^{\infty}(-1)^{k+1}kx^{k-1}$

7. $\ln\dfrac{1}{1-x} = -\ln(1-x) = \displaystyle\int\dfrac{1}{1-x}dx - C = \int\left[\displaystyle\sum_{k=0}^{\infty}x^k\right]dx - C$

$\qquad\qquad = \displaystyle\sum_{k=0}^{\infty}\dfrac{x^{k+1}}{k+1} - C = \displaystyle\sum_{k=1}^{\infty}\dfrac{x^k}{k} - C, \ln\dfrac{1}{1-0} = 0$ so $C = 0$.

9. $x = 1/4$, $S = \ln\dfrac{1}{1-1/4} = \ln\dfrac{4}{3}$

11. $f(x) = \dfrac{1}{1-x} = \displaystyle\sum_{k=0}^{\infty}x^k$, $f'(x) = \dfrac{1}{(1-x)^2} = \displaystyle\sum_{k=1}^{\infty}kx^{k-1}$,

$\qquad f''(x) = \dfrac{2}{(1-x)^3} = \displaystyle\sum_{k=2}^{\infty}k(k-1)x^{k-2} = 2 + 6x + 12x^2 + 20x^3 + \cdots$

13. **(a)** $\rho = \displaystyle\lim_{k\to+\infty}\dfrac{|x|^{k+2}}{k+2}\cdot\dfrac{k+1}{|x|^{k+1}} = |x|\lim_{k\to+\infty}\dfrac{k+1}{k+2} = |x|$; converges if $|x| < 1$.

$\quad$ **(b)** $f'(x) = \displaystyle\sum_{k=0}^{\infty}(-1)^k x^k$, converges on $(-1,1)$

$\quad$ **(c)** $f'(x) = \dfrac{1}{1+x}$ so $f(x) = \ln(1+x) + C$ for x in $(-1,1)$, let $x = 0$ to find that $C = 0$, thus $f(x) = \ln(1+x)$.

15. $\displaystyle\int_0^1\cos\sqrt{x}\,dx = \int_0^1\left(1 - x/2! + x^2/4! - x^3/6! + \cdots\right)dx$

$\qquad\qquad = \left. x - \dfrac{1}{2\cdot 2!}x^2 + \dfrac{1}{3\cdot 4!}x^3 - \dfrac{1}{4\cdot 6!}x^4 + \cdots\right]_0^1 = 1 - \dfrac{1}{2\cdot 2!} + \dfrac{1}{3\cdot 4!} - \dfrac{1}{4\cdot 6!} + \cdots,$

$\qquad$ but $\dfrac{1}{4\cdot 6!} < 0.5\times 10^{-3}$ so $\displaystyle\int_0^1\cos\sqrt{x}\,dx \approx 1 - \dfrac{1}{2\cdot 2!} + \dfrac{1}{3\cdot 4!} \approx 0.764$

17. $\int_0^{1/2} \dfrac{dx}{1+x^4} = \int_0^{1/2} \left(1 - x^4 + x^8 - \cdots\right) dx$

$$= x - \frac{1}{5}x^5 + \frac{1}{9}x^9 - \cdots \Big]_0^{1/2} = 1/2 - \frac{(1/2)^5}{5} + \frac{(1/2)^9}{9} - \cdots,$$

but $\dfrac{(1/2)^9}{9} < 0.5 \times 10^{-3}$ so $\int_0^{1/2} \dfrac{dx}{1+x^4} \approx 1/2 - \dfrac{(1/2)^5}{5} \approx 0.494$

19. $\int_0^{0.1} e^{-x^3} dx = \int_0^{0.1} \left(1 - x^3 + x^6/2! - \cdots\right) dx$

$$= x - \frac{1}{4}x^4 + \frac{1}{7 \cdot 2!}x^7 - \cdots \Big]_0^{0.1} = 0.1 - \frac{1}{4}(0.1)^4 + \frac{1}{7 \cdot 2!}(0.1)^7 - \cdots,$$

but $\dfrac{1}{4}(0.1)^4 < 0.5 \times 10^{-3}$ so $\int_0^{0.1} e^{-x^3} dx \approx 0.100$

21. $\int_0^{1/2} (1+x^2)^{-1/4} dx = \int_0^{1/2} \left(1 - \frac{1}{4}x^2 + \frac{5}{32}x^4 - \frac{15}{128}x^6 + \cdots\right) dx$

$$= x - \frac{1}{12}x^3 + \frac{1}{32}x^5 - \frac{15}{896}x^7 + \cdots \Big]_0^{1/2}$$

$$= 1/2 - \frac{1}{12}(1/2)^3 + \frac{1}{32}(1/2)^5 - \frac{15}{896}(1/2)^7 + \cdots,$$

but $\dfrac{15}{896}(1/2)^7 < 0.5 \times 10^{-3}$ so $\int_0^{1/2} (1+x^2)^{-1/4} dx \approx 1/2 - \dfrac{1}{12}(1/2)^3 + \dfrac{1}{32}(1/2)^5 \approx 0.491$

23. $e^{-x^2} \cos x = \left(1 - x^2 + \dfrac{x^4}{2!} - \dfrac{x^6}{3!} + \cdots\right)\left(1 - \dfrac{x^2}{2!} + \dfrac{x^4}{4!} - \dfrac{x^6}{6!} + \cdots\right)$

$$= 1 - \frac{3}{2}x^2 + \frac{25}{24}x^4 - \frac{331}{720}x^6 + \cdots$$

25. $\dfrac{\sin x}{e^x} = e^{-x} \sin x = \left(1 - x + \dfrac{x^2}{2!} - \dfrac{x^3}{3!} + \dfrac{x^4}{4!} - \cdots\right)\left(x - \dfrac{x^3}{3!} + \dfrac{x^5}{5!} - \cdots\right)$

$$= x - x^2 + \frac{1}{3}x^3 - \frac{1}{30}x^5 + \cdots$$

27. $x \ln\left(1 - x^2\right) = x\left(-x^2 - \dfrac{1}{2}x^4 - \dfrac{1}{3}x^6 - \dfrac{1}{4}x^8 - \cdots\right) = -x^3 - \dfrac{1}{2}x^5 - \dfrac{1}{3}x^7 - \dfrac{1}{4}x^9 - \cdots$

29. $x^2 e^{4x}\sqrt{1+x} = x^2\left(1 + 4x + 8x^2 + \dfrac{32}{3}x^3 + \cdots\right)\left(1 + \dfrac{1}{2}x - \dfrac{1}{8}x^2 + \dfrac{1}{16}x^3 + \cdots\right)$

$$= x^2\left(1 + \dfrac{9}{2}x + \dfrac{79}{8}x^2 + \dfrac{683}{48}x^3 + \cdots\right) = x^2 + \dfrac{9}{2}x^3 + \dfrac{79}{8}x^4 + \dfrac{683}{48}x^5 + \cdots$$

31. **(a)** $\dfrac{1-\cos x}{\sin x} = \dfrac{1 - \left(1 - x^2/2! + x^4/4! - x^6/6! + \cdots\right)}{x - x^3/3! + x^5/5! - \cdots}$

$$= \dfrac{x^2/2! - x^4/4! + x^6/6! - \cdots}{x - x^3/3! + x^5/5! - \cdots} = \dfrac{x/2! - x^3/4! + x^5/6! - \cdots}{1 - x^2/3! + x^4/5! - \cdots}, x \neq 0$$

$$\lim_{x\to 0}\dfrac{1-\cos x}{\sin x} = \dfrac{0}{1} = 0$$

(b) $\ln\sqrt{1+x} - \sin 2x = \dfrac{1}{2}\ln(1+x) - \sin 2x$

$$= \dfrac{1}{2}\left(x - \dfrac{1}{2}x^2 + \dfrac{1}{3}x^3 - \cdots\right) - \left(2x - \dfrac{4}{3}x^3 + \dfrac{4}{15}x^5 - \cdots\right)$$

$$= -\dfrac{3}{2}x - \dfrac{1}{4}x^2 + \dfrac{3}{2}x^3 + \cdots,$$

$$\lim_{x\to 0}\dfrac{\ln\sqrt{1+x} - \sin 2x}{x} = \lim_{x\to 0}\left(-\dfrac{3}{2} - \dfrac{1}{4}x + \dfrac{3}{2}x^2 + \cdots\right) = -3/2$$

33. **(a)** $\sinh^{-1}x = \displaystyle\int\left(1+x^2\right)^{-1/2}dx - C = \int\left(1 - \dfrac{1}{2}x^2 + \dfrac{3}{8}x^4 - \dfrac{5}{16}x^6 + \cdots\right)dx - C$

$$= \left(x - \dfrac{1}{6}x^3 + \dfrac{3}{40}x^5 - \dfrac{5}{112}x^7 + \cdots\right) - C,$$

$\sinh^{-1}0 = 0$ so $C = 0$

(b) $\left(1+x^2\right)^{-1/2} = 1 + \displaystyle\sum_{k=1}^{\infty}\dfrac{(-1/2)(-3/2)(-5/2)\cdots(-1/2-k+1)}{k!}(x^2)^k$

$$= 1 + \sum_{k=1}^{\infty}(-1)^k\dfrac{1\cdot 3\cdot 5\cdots(2k-1)}{2^k k!}x^{2k},$$

$$\sinh^{-1}x = x + \sum_{k=1}^{\infty}(-1)^k\dfrac{1\cdot 3\cdot 5\cdots(2k-1)}{2^k k!(2k+1)}x^{2k+1}$$

(c) $R = 1$

SUPPLEMENTARY EXERCISES CHAPTER 11

1. $L = 0$ **3.** $L = 0 - 0 = 0$

5. $\sin[(2n-1)\pi/2]$ is alternately 1 and -1 so the limit does not exist

7. $a_n = (-1)^n/e^n$ is alternating; $a_n = e^{1/n}$ is decreasing;

$$a_n = \frac{1}{\sqrt{n}} - \frac{1}{\sqrt{n+1}} = \frac{\sqrt{n+1}-\sqrt{n}}{\sqrt{n}\sqrt{n+1}} = \frac{1}{\sqrt{n}\sqrt{n+1}\left(\sqrt{n+1}+\sqrt{n}\right)} \quad \text{is decreasing;}$$

$a_n = \sin \pi n = 0$ is nondecreasing; $a_n = \sin[(2n-1)\pi/2]$ is alternating;

$$a_n = \frac{n+1}{n(n+2)}, \text{ let } f(x) = \frac{x+1}{x^2+2x} \text{ then } f'(x) = -\frac{x^2+2x+2}{(x^2+2x)^2} < 0 \text{ if } x \geq 1 \text{ so } a_n \text{ is decreasing.}$$

9. (a) $\displaystyle\sum_{k=0}^{\infty} \left(\pi/q^2\right)^k$ is a geometric series which converges for $\pi/q^2 < 1$, $q^2 > \pi$, $|q| > \sqrt{\pi}$

(b) $\displaystyle\sum_{k=1}^{\infty} 1/k^{3q}$ is a p-series with $p = 3q$, converges for $3q > 1$, $q > 1/3$

(c) $\displaystyle\sum_{k=2}^{\infty} 1/(k \ln q) = \sum_{k=2}^{\infty}(1/\ln q)(1/k)$ diverges for all q because $\displaystyle\sum_{k=2}^{\infty} 1/k$ diverges

(d) $\displaystyle\sum_{k=2}^{\infty}(1/\ln q)^k$ is a geometric series which converges for $|1/\ln q| < 1$, $|\ln q| > 1$,

$q > e$ or $0 < q < e^{-1}$

11. (a) $1.3636\cdots = 1 + \displaystyle\sum_{k=1}^{\infty} 36(0.01)^k$

(b) $1.3636\cdots = 1 + \dfrac{0.36}{1-0.01} = 1 + 36/99 = 1 + 4/11 = 15/11$

13. (a) $\dfrac{1}{6}\left[\displaystyle\sum_{k=1}^{\infty}\left(\frac{1}{3}\right)^k + \sum_{k=1}^{\infty}\left(\frac{1}{2}\right)^k\right] = \dfrac{1}{6}\left[\dfrac{1/3}{1-1/3} + \dfrac{1/2}{1-1/2}\right] = 1/4$

(b) $\displaystyle\sum_{k=2}^{\infty} \ln\frac{k+1}{k} = \sum_{k=2}^{\infty}[\ln(k+1) - \ln k]$,

$s_n = [\ln 3 - \ln 2] + [\ln 4 - \ln 3] + \cdots + [\ln(n+2) - \ln(n+1)]$
$= \ln(n+2) - \ln 2, \displaystyle\lim_{n\to+\infty} s_n = +\infty$, diverges

(c) $s_n = [1^{-1/2} - 2^{-1/2}] + [2^{-1/2} - 3^{-1/2}] + \cdots + [n^{-1/2} - (n+1)^{-1/2}]$

$= 1 - (n+1)^{-1/2}, \displaystyle\lim_{n\to+\infty} s_n = 1$

15. converges (integral test, $\displaystyle\int_1^{\infty} xe^{-x^2}dx$ converges)

17. converges (comparison test, $\dfrac{\sqrt{k}}{k^2+7} < \dfrac{\sqrt{k}}{k^2} = \dfrac{1}{k^{3/2}}$)

19. converges (ratio test, $\rho = 0$)   **21.** diverges (root test, $\rho = (5/2)^3 > 1$)

23. absolutely convergent (comparison test, $2^k/(3^k+1) < 2^k/3^k = (2/3)^k$, $\sum(2/3)^k$ is a convergent geometric series)

25. diverges $\left(\lim\limits_{k\to+\infty}|u_k| = \lim\limits_{k\to+\infty}\dfrac{1}{2}(3/2)^k = +\infty\right)$

27. $\rho = \lim\limits_{k\to+\infty}\dfrac{k^{3/2}|x-1|}{(k+1)^{3/2}} = |x-1|$, converges if $|x-1| < 1$, diverges if $|x-1| > 1$.

If $x = 0$, $\sum\limits_{k=1}^{\infty}\dfrac{(-1)^k}{k^{3/2}}$ converges; if $x = 2$, $\sum\limits_{k=1}^{\infty}\dfrac{1}{k^{3/2}}$ converges. $R = 1$, interval of convergence $[0,2]$.

29. $\rho = \lim\limits_{k\to+\infty}\dfrac{k|1-x|^2}{4(k+1)} = \dfrac{1}{4}|1-x|^2$, converges if $|x-1|^2 < 4$, $|x-1| < 2$; diverges if $|x-1| > 2$.

If $x = -1$, $\sum\limits_{k=1}^{\infty}1/k$ diverges; if $x = 3$, $\sum\limits_{k=1}^{\infty}1/k$ diverges. $R = 2$, interval of convergence $(-1,3)$.

31. $\rho = \lim\limits_{k\to+\infty}\dfrac{1}{5}(k+1)|x-1| = +\infty$, $R = 0$, converges only for $x = 1$.

33. **(a)** $(x-2) - \dfrac{1}{2}(x-2)^2 + \dfrac{1}{3}(x-2)^3$ **(b)** $R_3(x) = -\dfrac{(x-2)^4}{4(c-1)^4}$, c between 2 and x

    (c) $|R_3(x)| = \dfrac{|x-2|^4}{4|c-1|^4} < \dfrac{|3/2-2|^4}{4|3/2-1|^4} = 1/4$

35. **(a)** $1 + \dfrac{1}{2}(x-1) - \dfrac{1}{8}(x-1)^2$ **(b)** $R_2(x) = \dfrac{(x-1)^3}{16c^{5/2}}$, c between 1 and x

    (c) $|R_2(x)| = \dfrac{|x-1|^3}{16c^{5/2}} < \dfrac{|4/9-1|^3}{16(4/9)^{5/2}} = \dfrac{(5/9)^3}{16(2/3)^5} < 0.0814$

37. $\ln(a+x) = \ln a(1+x/a) = \ln a + \ln(1+x/a) = \ln a + \sum\limits_{k=0}^{\infty}(-1)^k\dfrac{(x/a)^{k+1}}{k+1}$,

converges if $|x/a| < 1$, $|x| < |a| = a$ so $R = a$

39. $1/(9+x)^{1/2} = \frac{1}{3}(1+x/9)^{-1/2} = \frac{1}{3}\left[1 + \sum_{k=1}^{\infty}(-1)^k \frac{1\cdot 3\cdot 5\cdots(2k-1)}{2^k k!}(x/9)^k\right],$

converges if $|x/9| < 1$, $|x| < 9$ so $R = 9$

41. $\sec x = 1/\cos x = 1/(1 - x^2/2! + x^4/4! - \cdots) = 1 + x^2/2 + 5x^4/24 + \cdots$

43. $[\cos x]^{1/2} = \left[1 - x^2/2! + x^4/4! - \cdots\right]^{1/2} = \left[1 + (-x^2/2! + x^4/4! - \cdots)\right]^{1/2}$

$= 1 + \frac{1}{2}\left(-x^2/2! + x^4/4! - \cdots\right) - \frac{1}{8}\left(-x^2/2! + x^4/4! - \cdots\right)^2 + \cdots$

$= 1 - x^2/4 - x^4/96 + \cdots$

45. $f(x) = \ln(1 + \sin x)$, $f'(x) = \dfrac{\cos x}{1 + \sin x}$, $f''(x) = -\dfrac{1}{1 + \sin x}$, $f'''(x) = \dfrac{\cos x}{(1 + \sin x)^2}$;

$f(0) = 0$, $f'(0) = 1$, $f''(0) = -1$, $f'''(0) = 1$; $\ln(1 + \sin x) = x - \dfrac{1}{2}x^2 + \dfrac{1}{6}x^3 + \cdots$

47. $\dfrac{\ln(1-2x)}{x} = \dfrac{1}{x}\left(-2x - 2x^2 - \dfrac{8}{3}x^3 - \cdots\right) = -2 - 2x - \dfrac{8}{3}x^2 - \cdots$, $\displaystyle\lim_{x\to 0}\dfrac{\ln(1-2x)}{x} = -2$

49. $\sin x = x - x^3/3! + x^5/5! + (0)x^6 + R_6(x)$, $|R_6(x)| \leq \dfrac{|x|^7}{7!} < 6 \times 10^{-4}$ if $|x|^7 < 3.024$,
$|x| < (3.024)^{1/7} \approx 1.17$

51. $\displaystyle\int_0^1 \frac{1 - e^{-t/2}}{t}\,dt = \int_0^1 \frac{\left[1 - \left(1 - \dfrac{t}{2} + \dfrac{t^2}{8} - \dfrac{t^3}{48} + \dfrac{t^4}{384} - \dfrac{t^5}{3840} + \cdots\right)\right]}{t}\,dt$

$= \displaystyle\int_0^1 \left(\frac{1}{2} - \frac{t}{8} + \frac{t^2}{48} - \frac{t^3}{384} + \frac{t^4}{3840} - \cdots\right)dt$

$= \left.\dfrac{t}{2} - \dfrac{t^2}{16} + \dfrac{t^3}{144} - \dfrac{t^4}{1436} + \dfrac{t^5}{19200} - \cdots\right]_0^1$

$= 1/2 - 1/16 + 1/144 - 1/1436 + 1/19200 - \cdots,$

but $1/19200 < 0.5 \times 10^{-3}$ so $\displaystyle\int_0^1 \frac{1 - e^{-t/2}}{t}\,dt \approx 1/2 - 1/16 + 1/144 - 1/1436 \approx 0.444$

53. $y' = \displaystyle\sum_{n=1}^{\infty}\frac{k^n x^{n-1}}{(n-1)!} = \sum_{n=0}^{\infty}\frac{k^{n+1}x^k}{n!} = k\sum_{n=0}^{\infty}\frac{k^n x^k}{n!} = ky$, so $y' - ky = 0$

CHAPTER 12
Topics In Analytic Geometry

EXERCISE SET 12.2

1.

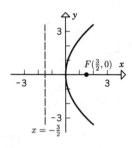

3.

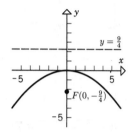

5. $y^2 = (12/5)x$

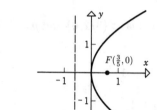

7.

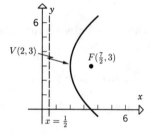

9.

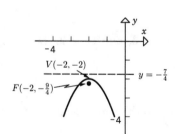

11. $(x - 2)^2 = -2(y - 5/2)$

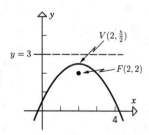

13. $(y-2)^2 = x + 2$

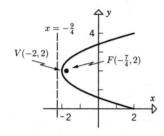

15. $(y-1)^2 = x + 1$

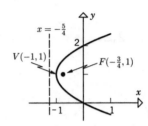

17. $y^2 = 4px$, $p = 3$, $y^2 = 12x$

19. $y^2 = -4px$, $p = 7$, $y^2 = -28x$

21. $y^2 = ax$, $2^2 = a(2)$, $a = 2$, $y^2 = 2x$

23. $x^2 = -4py$, $p = 3$, $x^2 = -12y$

25. $y^2 = a(x - h)$, $4 = a(3 - h)$ and $9 = a(2 - h)$, solve simultaneously to get $h = 19/5$, $a = -5$ so $y^2 = -5(x - 19/5)$

27. The vertex is half way between the focus and directrix so the vertex is at $(3/2, 0)$ and $p = 3/2$, $y^2 = 6(x - 3/2)$.

29. The vertex is 3 units above the directrix so $p = 3$, $(x - 1)^2 = 12(y - 1)$.

31. $(x - 5)^2 = a(y + 3)$, $(9 - 5)^2 = a(5 + 3)$ so $a = 2$, $(x - 5)^2 = 2(y + 3)$

33. **(a)** $y = Ax^2 + Bx + C$; use $(0, 3)$, $(2, 0)$ and $(3, 2)$ to get the system of equations $C = 3$, $9A + 3B + C = 2$, and $4A + 2B + C = 0$ which when solved yields $A = 7/6$, $B = -23/6$, $C = 3$ so $y = \frac{7}{6}x^2 - \frac{23}{6}x + 3$.

(b) $x = Ay^2 + By + C$; $9A + 3B + C = 0$, $4A + 2B + C = 3$, and $C = 2$. Solve to get $A = -7/6$, $B = 17/6$, $C = 2$ so $x = -\frac{7}{6}y^2 + \frac{17}{6}y + 2$

35. Complete the square to get $\left(x + \frac{B}{2A}\right)^2 = \frac{1}{A}\left(y - C + \frac{B^2}{4A}\right)$ so the vertex is at $\left(-\frac{B}{2A}, \frac{4AC - B^2}{4A}\right)$, the focus is at $\left(-\frac{B}{2A}, \frac{4AC - B^2 + 1}{4A}\right)$, and the directrix is $y = \frac{4AC - B^2 - 1}{4A}$.

37. $y = ax^2 + b$, $(20, 0)$ and $(10, 12)$
are on the curve so $400a + b = 0$
and $100a + b = 12$. Solve for b
to get $b = 16 =$ height of arch.

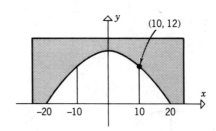

39. Let (x_0, y_0) be a point on the parabola $y^2 = 4px$, then
$$PF = \sqrt{(x_0 - p)^2 + y_0^2} = \sqrt{x_0^2 - 2px_0 + p^2 + 4px_0} = \sqrt{(x_0 + p)^2}$$
so $PF = x_0 + p$ where $x_0 \geq 0$ and PF is a minimum when $x_0 = 0$ (the vertex).

41. Use an xy-coordinate system so that $y^2 = 4px$ is an equation of the parabola, then $(1, 1/2)$ is
a point on the curve so $(1/2)^2 = 4p(1)$, $p = 1/16$. The light source should be placed at the
focus which is $1/16$ ft. from the vertex.

43. Similar to proof in text.

EXERCISE SET 12.3

1. $c^2 = 16 - 9 = 7$, $c = \sqrt{7}$

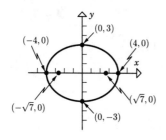

3. $\dfrac{x^2}{1} + \dfrac{y^2}{9} = 1$

$c^2 = 9 - 1 = 8$, $c = \sqrt{8}$

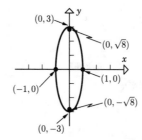

5. $\dfrac{x^2}{2} + \dfrac{y^2}{2/3} = 1$

$c^2 = 2 - 2/3 = 4/3, c = 2/\sqrt{3}$

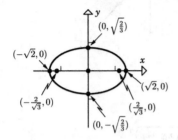

7. $\dfrac{(x-1)^2}{16} + \dfrac{(y-3)^2}{9} = 1$

$c^2 = 16 - 9 = 7, c = \sqrt{7}$

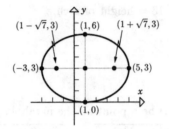

9. $\dfrac{(x+2)^2}{4} + \dfrac{(y+1)^2}{3} = 1$

$c^2 = 4 - 3 = 1, c = 1$

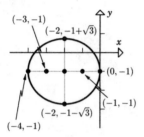

11. $\dfrac{(x+1)^2}{9} + \dfrac{(y-1)^2}{1} = 1$

$c^2 = 9 - 1 = 8, c = \sqrt{8}$

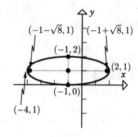

13. $\dfrac{(x+1)^2}{4} + \dfrac{(y-5)^2}{16} = 1$

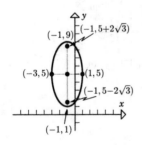

15. $x^2/9 + y^2/4 = 1$

17. $a = 26/2 = 13, \ c = 5, \ b^2 = a^2 - c^2 = 169 - 25 = 144; \ x^2/169 + y^2/144 = 1$

19. $c = 1$, $a^2 = b^2 + c^2 = 2 + 1 = 3$; $x^2/3 + y^2/2 = 1$

21. $b^2 = 16 - 12 = 4$; $x^2/16 + y^2/4 = 1$ and $x^2/4 + y^2/16 = 1$

23. $a = 6$, $(2, 3)$ satisfies $x^2/36 + y^2/b^2 = 1$ so $4/36 + 9/b^2 = 1$, $b^2 = 81/8$; $x^2/36 + y^2/(81/8) = 1$

25. The center is midway between the foci so it is at $(1, 3)$ thus $c = 1$, $b = 1$, $a^2 = 1 + 1 = 2$; $(x - 1)^2 + (y - 3)^2/2 = 1$

27. $(4, 1)$ and $(4, 5)$ are the foci so the center is at $(4, 3)$ thus $c = 2$, $a = 12/2 = 6$, $b^2 = 36 - 4 = 32$; $(x - 4)^2/32 + (y - 3)^2/36 = 1$

29. Substitute $x = 8 - 2y$ into $x^2 + 4y^2 = 40$ to get $y^2 - 4y + 3 = 0$ which yields $y = 1, 3$. Substitute these into $x = 8 - 2y$ to get $x = 6, 2$ so the points of intersection are $(2, 3)$ and $(6, 1)$.

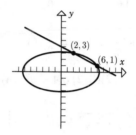

31. Substitute $x^2 = 20 - y^2$ into $x^2 + 9y^2 = 36$ to get $y^2 = 2$, $y = \pm\sqrt{2}$. Use $x^2 = 20 - y^2$ to get $x = \pm 3\sqrt{2}$ when $y = \sqrt{2}$ or $-\sqrt{2}$. The points of inter-section are $(3\sqrt{2}, \sqrt{2})$, $(3\sqrt{2}, -\sqrt{2})$, $(-3\sqrt{2}, \sqrt{2})$, and $(-3\sqrt{2}, -\sqrt{2})$.

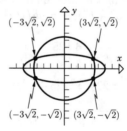

33. Use implicit differentiation on $x^2 + 4y^2 = 8$ to get $\left.\dfrac{dy}{dx}\right|_{(x_0, y_0)} = -\dfrac{x_0}{4y_0}$ where (x_0, y_0) is the point of tangency, but $-x_0/(4y_0) = -1/2$ because the slope of the line is $-1/2$ so $x_0 = 2y_0$. (x_0, y_0) is on the ellipse so $x_0^2 + 4y_0^2 = 8$ which when solved with $x_0 = 2y_0$ yields the points of tangency $(2, 1)$ and $(-2, -1)$. Substitute these into the equation of the line to get $k = \pm 4$.

35. $y = (b/a)\sqrt{a^2 - x^2}$ is the upper half of the ellipse,

$$A = 4\int_0^a \frac{b}{a}\sqrt{a^2 - x^2}\,dx = \frac{4b}{a}\int_0^a \sqrt{a^2 - x^2}\,dx = \frac{4b}{a}\left(\frac{1}{4}\pi a^2\right) = \pi ab$$

37. $y = \dfrac{b}{a}\sqrt{a^2 - x^2}$, $\dfrac{dy}{dx} = -\dfrac{b}{a}\dfrac{x}{\sqrt{a^2 - x^2}}$, $1 + \left(\dfrac{dy}{dx}\right)^2 = \dfrac{a^4 - c^2 x^2}{a^2(a^2 - x^2)}$,

$$S = 2(2\pi)\int_0^a y\sqrt{1 + (dy/dx)^2}\,dx = 4\pi\int_0^a \dfrac{b}{a}\sqrt{a^2 - x^2}\dfrac{\sqrt{a^4 - c^2 x^2}}{a\sqrt{a^2 - x^2}}\,dx$$

$$= \dfrac{4\pi b}{a^2}\int_0^a \sqrt{a^4 - c^2 x^2}\,dx = \dfrac{4\pi b}{a^2 c}\int_0^{ac}\sqrt{a^4 - u^2}\,du \quad (u = cx)$$

$$= \dfrac{4\pi b}{a^2 c}\left[\dfrac{u}{2}\sqrt{a^4 - u^2} + \dfrac{a^4}{2}\sin^{-1}\dfrac{u}{a^2}\right]_0^{ac} = \dfrac{4\pi b}{a^2 c}\left[\dfrac{1}{2}ac\sqrt{a^4 - a^2 c^2} + \dfrac{1}{2}a^4\sin^{-1}\dfrac{c}{a}\right]$$

$$= 2\pi ab\left[\dfrac{b}{a} + \dfrac{a}{c}\sin^{-1}\dfrac{c}{a}\right].$$

39. Use $\dfrac{x^2}{9} + \dfrac{y^2}{4} = 1$, $x = \dfrac{3}{2}\sqrt{4 - y^2}$,

$$V = \int_{-2}^{-2+h}(2)(3/2)\sqrt{4 - y^2}(18)dy = 54\int_{-2}^{-2+h}\sqrt{4 - y^2}\,dy$$

$$= 54\left[\dfrac{y}{2}\sqrt{4 - y^2} + 2\sin^{-1}\dfrac{y}{2}\right]_{-2}^{-2+h} = 27\left[4\sin^{-1}\dfrac{h-2}{2} + (h-2)\sqrt{4h - h^2} + 2\pi\right]$$

41. The vertex in the first quadrant is at the point where $y = x > 0$ so $x^2/a^2 + x^2/b^2 = 1$, $x^2 = a^2 b^2/(a^2 + b^2)$, $x = ab/\sqrt{a^2 + b^2}$. $A = (2x)^2 = 4x^2 = 4a^2 b^2/(a^2 + b^2)$.

43. $\sqrt{(x - 4)^2 + y^2} = \dfrac{4}{5}\left|\dfrac{25}{4} - x\right|$, $x^2 - 8x + 16 + y^2 = \dfrac{16}{25}\left(\dfrac{625}{16} - \dfrac{25}{2}x + x^2\right)$,

$9x^2 + 25y^2 = 225$, $x^2/25 + y^2/9 = 1$; center: $(0, 0)$, major axis: 10, minor axis: 6

45. $L = 2a = \sqrt{D^2 + p^2 D^2} = D\sqrt{1 + p^2}$ (see figure),

so $a = \dfrac{1}{2}D\sqrt{1 + p^2}$, but $b = \dfrac{1}{2}D$,

$T = c = \sqrt{a^2 - b^2}$

$$= \sqrt{\dfrac{1}{4}D^2(1 + p^2) - \dfrac{1}{4}D^2} = \dfrac{1}{2}pD.$$

47. **(a)** $0 < c < a$ so $0 < c/a < 1$, $0 < e < 1$.

(b) $c = 3$, $e = c/a = 3/a = 3/5$ so $a = 5$, $b^2 = a^2 - c^2 = 25 - 9 = 16$, $x^2/16 + y^2/25 = 1$.

(c) c approaches a as e approaches 1 so $b = \sqrt{a^2 - c^2}$ approaches 0; the ellipse flattens and approaches the major axis.

(d) c approaches 0 as e approaches 0 so b approaches a; the ellipse widens and approaches a circle of radius a.

49. Let R be the radius of a circle C that
is tangent to both C_1 and C_2. The distances
between the center of C and the centers
of C_1 and C_2 are, respectively,
$r_1 + R$ and $r_2 - R$ (see accompanying
diagram). Their sum is
$(r_1 + R) + (r_2 - R) = r_1 + r_2$ which is
a constant so the centers lie on an
ellipse with foci at the centers of C_1
and C_2. The length of the major axis is

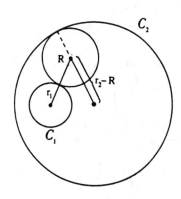

$2a = r_1 + r_2$; the center of the ellipse is at the midpoint of the line segment that joins the
centers of C_1 and C_2.

51. Similar to derivation in text.

EXERCISE SET 12.4

1. $c^2 = a^2 + b^2 = 16 + 4 = 20, c = 2\sqrt{5}$

3. $y^2/4 - x^2/9 = 1$
$c^2 = 4 + 9 = 13, c = \sqrt{13}$

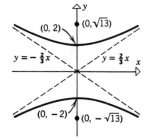

5. $x^2/1 - y^2/8 = 1$
$c^2 = 1 + 8 = 9, c = 3$

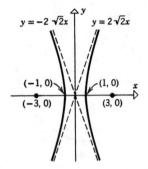

7. $c^2 = 1 + 1 = 2, c = \sqrt{2}$

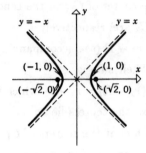

9. $c^2 = 9 + 4 = 13, c = \sqrt{13}$

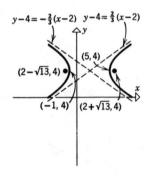

11. $(y+3)^2/36 - (x+2)^2/4 = 1$
$c^2 = 36 + 4 = 40, c = 2\sqrt{10}$

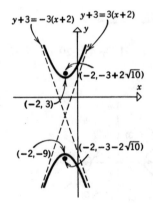

13. $(x+1)^2/4 - (y-1)^2/1 = 1$
$c^2 = 4 + 1 = 5, c = \sqrt{5}$

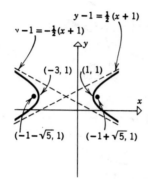

15. $(x-1)^2/4 - (y+3)^2/64 = 1$
$c^2 = 4 + 64 = 68, c = 2\sqrt{17}$

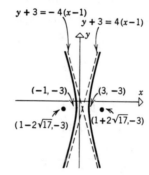

17. $a = 2, c = 3, b^2 = 9 - 4 = 5; x^2/4 - y^2/5 = 1$

19. $a = 1, b/a = 2, b = 2; x^2 - y^2/4 = 1$

21. vertices along x-axis: $b/a = 3/2$ so $a = 8/3; x^2/(64/9) - y^2/16 = 1$
vertices along y-axis: $a/b = 3/2$ so $a = 6; y^2/36 - x^2/16 = 1$

23. The form of the equation is $y^2/a^2 - x^2/b^2$ because $(5, 9)$ is above the asymptote $y = x$, $a/b = 1$ and $81/a^2 - 25/b^2 = 1$, solve to get $a^2 = b^2 = 56; y^2/56 - x^2/56 = 1$

25. $a = 2$ so $x^2/4 - y^2/b^2 = 1$, $(4, 2)$ is on the curve so $4 - 4/b^2 = 1$, $b^2 = 4/3; x^2/4 - y^2/(4/3) = 1$

27. the center is at $(2, -3)$, $a = 2, c = 3, b^2 = 9 - 4 = 5; (x-2)^2/4 - (y+3)^2/5 = 1$

29. the center is at $(6, 4)$, $a = 4, c = 5, b^2 = 25 - 16 = 9; (x-6)^2/16 - (y-4)^2/9 = 1$

31. From the definition of a hyperbola, $\left| \sqrt{(x-1)^2 + (y-1)^2} - \sqrt{x^2 + y^2} \right| = 1$,
$\sqrt{(x-1)^2 + (y-1)^2} - \sqrt{x^2 + y^2} = \pm 1$, transpose the second radical to the right hand side of the equation and square and simplify to get $\pm 2\sqrt{x^2 + y^2} = -2x - 2y + 1$, square and simplify again to get $8xy - 4x - 4y + 1 = 0$.

33. Substitute $x = 2y + 20$ into
$x^2 - 4y^2 = 36$ to get $y = -91/20$,
so $x = 2(-91/20) + 20 = 109/10$.
The curves intersect at $(109/10, -91/20)$.

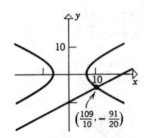

35. Eliminate x to get $y^2 = 1$, $y = \pm 1$.
Use either equation to find that
$x = \pm 2$ if $y = 1$ or if $y = -1$.
The curves intersect at $(2, 1)$,
$(2, -1)$, $(-2, 1)$, and $(-2, -1)$.

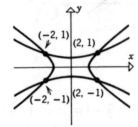

37. Let (x, y) be the coordinates of the point, use the formula for the distance between a point and a line to get $(|mx - y|/\sqrt{m^2 + 1})(|mx + y|/\sqrt{m^2 + 1}) = k^2$, $|m^2 x^2 - y^2|/(m^2 + 1) = k^2$ so $m^2 x^2 - y^2 = \pm k^2(m^2 + 1)$ which are hyperbolas with $y = \pm mx$ as asymptotes.

39. Let (x_0, y_0) be one of the points then $dy/dx\big|_{(x_0, y_0)} = 4x_0/y_0$, the tangent line is

$y = (4x_0/y_0)x + 4$, but (x_0, y_0) is on both the line and the curve which leads to
$4x_0^2 - y_0^2 + 4y_0 = 0$ and $4x_0^2 - y_0^2 = 36$, solve to get $x_0 = \pm 3\sqrt{13}/2$, $y_0 = -9$.

41. Let (x_0, y_0) be such a point. The foci are at $(-\sqrt{5}, 0)$ and $(\sqrt{5}, 0)$, the lines are perpendicular if the product of their slopes is -1 so $\dfrac{y_0}{x_0 + \sqrt{5}} \cdot \dfrac{y_0}{x_0 - \sqrt{5}} = -1$, $y_0^2 = 5 - x_0^2$ and $4x_0^2 - y_0^2 = 4$. Solve to get $x_0 = \pm 3/\sqrt{5}$, $y_0 = \pm 4/\sqrt{5}$. The coordinates are $(\pm 3/\sqrt{5}, 4/\sqrt{5})$, $(\pm 3/\sqrt{5}, -4/\sqrt{5})$.

43. **(a)** $c > a$ so $c/a > 1$, $e > 1$.

 (b) $c = 5$, $e = c/a = 5/a = 5/3$ so $a = 3$, $b^2 = c^2 - a^2 = 25 - 9 = 16$, $x^2/9 - y^2/16 = 1$.

 (c) c approaches a as e approaches 1 so $b = \sqrt{c^2 - a^2}$ approaches 0; the hyperbola flattens and approaches the focal axis, excluding the segment between the vertices.

 (d) c approaches $+\infty$ as e approaches $+\infty$ so b approaches $+\infty$; the hyperbola approaches the lines that are perpendicular to the focal axis at the vertices.

45. Let d_1 and d_2 be the distances of the first and second observers, respectively, from the point where the gun was fired, and let v be the speed of sound. Then $t = ($time for sound to reach the second observer) $-$ (time for sound to reach the first observer) $= d_2/v - d_1/v$ so $d_2 - d_1 = vt$. For constant v and t the difference of distances, d_2 and d_1 is constant so the gun was fired somewhere on a branch of a hyperbola whose foci are where the observers are.

47. Similar to the derivation in the text.

49. (a) Use $x^2/a^2 + y^2/b^2 = 1$ and $x^2/A^2 - y^2/B^2$ as the equations of the ellipse and hyperbola. If (x_0, y_0) is a point of intersection then $b^2 x_0^2 + a^2 y_0^2 = a^2 b^2$ and $B^2 x_0^2 - A^2 y_0^2 = A^2 B^2$, solve to get

$$x_0^2 = \frac{a^2 A^2 (b^2 + B^2)}{a^2 B^2 + A^2 b^2} \qquad \text{and} \qquad y_0^2 = \frac{b^2 B^2 (a^2 - A^2)}{a^2 B^2 + A^2 b^2} \qquad (1)$$

From Exercises 34 (Section 12.3) and 38 the slopes of the tangent lines to the ellipse and hyperbola at (x_0, y_0) are, respectively, $-\dfrac{b^2}{a^2}\dfrac{x_0}{y_0}$ and $\dfrac{B^2}{A^2}\dfrac{x_0}{y_0}$ so their product is $-\dfrac{b^2 B^2 x_0^2}{a^2 A^2 y_0^2}$ which, using (1), gives $-\dfrac{b^2 + B^2}{a^2 - A^2}$. The ellipse and hyperbola have the same foci so $c^2 = a^2 - b^2 = A^2 + B^2$, $a^2 - A^2 = b^2 + B^2$, $(b^2 + B^2)/(a^2 - A^2) = 1$, thus the product of the slopes of the tangent lines is -1 and the lines are perpendicular.

(b) From the figure,
$2(\alpha + \beta) = 180°$ so $\alpha + \beta = 90°$.

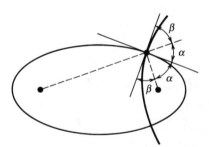

EXERCISE SET 12.5

1. (a) $\sin \theta = \sqrt{3}/2$, $\cos \theta = 1/2$
$x' = (-2)(1/2) + (6)(\sqrt{3}/2) = -1 + 3\sqrt{3}$, $y' = -(-2)(\sqrt{3}/2) + 6(1/2) = 3 + \sqrt{3}$

(b) $x = \dfrac{1}{2}x' - \dfrac{\sqrt{3}}{2}y' = \dfrac{1}{2}(x' - \sqrt{3}y')$, $y = \dfrac{\sqrt{3}}{2}x' + \dfrac{1}{2}y' = \dfrac{1}{2}(\sqrt{3}x' + y')$

$\sqrt{3}\left[\dfrac{1}{2}(x' - \sqrt{3}y')\right]\left[\dfrac{1}{2}(\sqrt{3}x' + y')\right] + \left[\dfrac{1}{2}(\sqrt{3}x' + y')\right]^2 = 6$

$$\frac{\sqrt{3}}{4}(\sqrt{3}x'^2 - 2x'y' - \sqrt{3}y'^2) + \frac{1}{4}(3x'^2 + 2\sqrt{3}x'y' + y'^2) = 6$$

$$\frac{3}{2}x'^2 - \frac{1}{2}y'^2 = 6, \quad \frac{x'^2}{4} - \frac{y'^2}{12} = 1$$

(c)

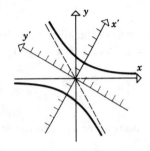

3. $\cot 2\theta = (0 - 0)/1 = 0, \; 2\theta = 90°, \; \theta = 45°$

   $x = (\sqrt{2}/2)(x' - y'), \; y = (\sqrt{2}/2)(x' + y')$

   $y'^2/18 - x'^2/18 = 1,$ hyperbola

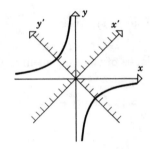

5. $\cot 2\theta = [1 - (-2)]/4 = 3/4$

   $\cos 2\theta = 3/5$

   $\sin \theta = \sqrt{(1 - 3/5)/2} = 1/\sqrt{5}$

   $\cos \theta = \sqrt{(1 + 3/5)/2} = 2/\sqrt{5}$

   $x = (1/\sqrt{5})(2x' - y')$

   $y = (1/\sqrt{5})(x' + 2y')$

   $x'^2/3 - y'^2/2 = 1,$ hyperbola

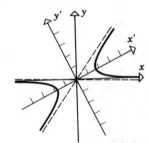

7. $\cot 2\theta = (1-3)/(2\sqrt{3}) = -1/\sqrt{3}$,
$\quad 2\theta = 120°, \theta = 60°$
$\quad\quad x = (1/2)(x' - \sqrt{3}y')$
$\quad\quad y = (1/2)(\sqrt{3}x' + y')$
$\quad\quad y' = x'^2$, parabola

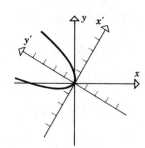

9. $\cot 2\theta = (9-16)/(-24) = 7/24$
$\quad \cos 2\theta = 7/25,$
$\quad\quad \sin\theta = 3/5, \quad\quad \cos\theta = 4/5$
$\quad\quad x = (1/5)(4x' - 3y'),$
$\quad\quad y = (1/5)(3x' + 4y')$
$\quad\quad y'^2 = 4(x' - 1)$, parabola

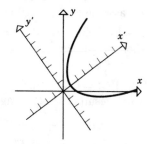

11. $\cot 2\theta = (52 - 73)/(-72) = 7/24$
$\quad \cos 2\theta = 7/25, \quad\quad \sin\theta = 3/5,$
$\quad\quad \cos\theta = 4/5$
$\quad\quad x = (1/5)(4x' - 3y'),$
$\quad\quad y = (1/5)(3x' + 4y')$
$\quad (x' + 1)^2/4 + y'^2 = 1$, ellipse

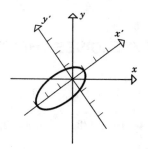

13. Let $x = x'\cos\theta - y'\sin\theta$, $y = x'\sin\theta + y'\cos\theta$ then $x^2 + y^2 = r^2$ becomes
$(\sin^2\theta + \cos^2\theta)x'^2 + (\sin^2\theta + \cos^2\theta)y'^2 = r^2$, $x'^2 + y'^2 = r^2$. Under a rotation transformation the center of the circle stays at the origin of both coordinate systems.

15. $x' = (\sqrt{2}/2)(x + y)$, $y' = (\sqrt{2}/2)(-x + y)$ which when substituted into $3x'^2 + y'^2 = 6$ yields $x^2 + xy + y^2 = 3$.

17. $\sqrt{x} + \sqrt{y} = 1$, $\sqrt{x} = 1 - \sqrt{y}$, $x = 1 - 2\sqrt{y} + y$, $2\sqrt{y} = 1 - x + y$, $4y = 1 + x^2 + y^2 - 2x + 2y - 2xy$, $x^2 - 2xy + y^2 - 2x - 2y + 1 = 0$. $\cot 2\theta = \dfrac{1-1}{2} = 0$, $2\theta = \pi/2$, $\theta = \pi/4$. Let $x = x'/\sqrt{2} - y'/\sqrt{2}$, $y = x'/\sqrt{2} + y'/\sqrt{2}$ to get $2y'^2 - 2\sqrt{2}x' + 1 = 0$, which is a parabola. From $\sqrt{x} + \sqrt{y} = 1$ we see that $0 \le x \le 1$ and $0 \le y \le 1$, so the graph is just a portion of a parabola.

19. Use (9) to express $B' - 4A'C'$ in terms of A, B, C, and θ, then simplify.

21. $\cot 2\theta = (A - C)/B = 0$ if $A = C$ so $2\theta = 90°$, $\theta = 45°$.

23. $B^2 - 4AC = (-1)^2 - 4(1)(1) = -3 < 0$; ellipse, point, or no graph

25. $B^2 - 4AC = (2\sqrt{3})^2 - 4(1)(3) = 0$; parabola, line, pair of parallel lines, or no graph

27. $B^2 - 4AC = (-24)^2 - 4(34)(41) = -5000 < 0$; ellipse, point, or no graph

29. Part (b): from (18), $A'C' < 0$ so A' and C' have opposite signs. By multiplying (19) through by -1, if necessary, assume that $A' < 0$ and $C' > 0$ so (19) can be written as
$(x' - h)^2/C' - (y' - k)^2/|A'| = K$. If $K \neq 0$ then the graph is a hyperbola (divide both sides by K), if $K = 0$ then we get the pair of intersecting lines $(x' - h)/\sqrt{C'} = \pm(y' - k)/\sqrt{|A'|}$.

Part (c): from (18), $A'C' = 0$ so either $A' = 0$ or $C' = 0$ but not both (this would imply that $A = B = C = 0$ which results in (14) being linear). Suppose $A' \neq 0$ and $C' = 0$ then complete the square to get $(x' - h)^2 = -\dfrac{E'}{A'}y' + K$. If $E' \neq 0$ the graph is a parabola, if $E' = 0$ and $K = 0$ the graph is the line $x' = h$, if $E' = 0$ and $K > 0$ the graph is the pair of parallel lines $x' = h \pm \sqrt{K}$, if $E' = 0$ and $K < 0$ there is no graph.

SUPPLEMENTARY EXERCISES CHAPTER 12

1. parabola, $(y - 3)^2 = -12(x + 2)$, $p = 3$; vertex $(-2, 3)$, focus $(-5, 3)$, directrix $x = 1$.

3. ellipse, $(x + 2)^2/4 + (y - 1)^2/9$, $a = 3$, $b = 2$, $c = \sqrt{5}$; center $(-2, 1)$, foci $(-2, 1 \pm \sqrt{5})$, major axis 6, minor axis 4.

5. hyperbola, $(x - 2)^2/9 - (y - 1)^2/1 = 1$, $a = 3$, $b = 1$, $c = \sqrt{10}$, center $(2, 1)$, foci $(2 \pm \sqrt{10}, 1)$, vertices $(-1, 1)$ and $(5, 1)$, asymptotes $y - 1 = \pm(x - 2)/3$.

7. parabola, $(y - 1)^2 = (-3/2)(x - 3)$, $p = 3/8$; vertex $(3, 1)$, focus $(21/8, 1)$, directrix $x = 27/8$.

9. $p = 4$; $(y - 3)^2 = 16(x - 1)$.

11. center $(0, 0)$, $c = 5$, $a = 6/2 = 3$, $b^2 = 16$, $y^2/9 - x^2/16 = 1$.

13. center $(0, 0)$, $c = 3$, $a = 10/2 = 5$, $b^2 = 16$; $x^2/25 + y^2/16 = 1$.

15. The curve is a parabola with focus at $(3, 4)$ and directrix $y = 2$ so the vertex is at $(3, 3)$ and $p = 1$; $(x - 3)^2 = 4(y - 3)$.

17.

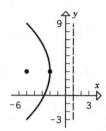

19.

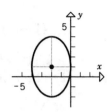

21. $\cot 2\theta = (3-3)/(-2) = 0$, $\theta = 45°$; use $x = (\sqrt{2}/2)(x'-y')$, $y = (\sqrt{2}/2)(x'+y')$ to get $x'^2/2 + y'^2/1 = 1$; ellipse.

23. $\cot 2\theta = (11-1)/(10\sqrt{3}) = 1/\sqrt{3}$, $\theta = 30°$; use $x = (1/2)(\sqrt{3}x'-y')$, $y = (1/2)(x'+\sqrt{3}y')$ to get $x'^2/(1/4) - y'^2/1 = 1$; hyperbola.

25. $\cot 2\theta = (16-9)/(-24) = -7/24$, $\cos 2\theta = -7/25$, $\sin \theta = \sqrt{(1+7/25)/2} = 4/5$,

$\cos \theta = \sqrt{(1-7/25)/2} = 3/5$, $\theta = \tan^{-1}(4/3)$; use $x = (1/5)(3x'-4y')$, $y = (1/5)(4x'+3y')$ to get $y'^2 = 4(x'-1)$; parabola.

27.

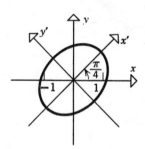

29.

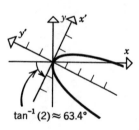

31. **(a)** $PF = \sqrt{x^2+y^2}$, $PD = |x-k|$, so $PF = ePD$ yields

$$\sqrt{x^2+y^2} = e|x-k|$$
$$x^2 + y^2 = e^2(x^2 - 2kx + k^2)$$
$$(1-e^2)x^2 + 2e^2kx + y^2 = e^2k^2$$
$$(1-e^2)\left[x^2 + \frac{2e^2k}{1-e^2}x + \frac{e^4k^2}{(1-e^2)^2}\right] + y^2 = e^2k^2 + \frac{e^4k^2}{1-e^2} = \frac{e^2k^2}{1-e^2}$$

$$(1 - e^2)\left(x + \frac{e^2 k}{1 - e^2}\right)^2 + y^2 = \frac{e^2 k}{1 - e^2}$$

$$\frac{(x + e^2 k/(1 - e^2))^2}{e^2 k^2/(1 - e^2)^2} + \frac{y^2}{e^2 k^2/(1 - e^2)} = 1; \qquad (1)$$

let $c = \dfrac{e^2 k}{1 - e^2}$, $a^2 = \dfrac{e^2 k^2}{(1 - e^2)^2}$, and $b^2 = \dfrac{e^2 k^2}{1 - e^2}$ to get $\dfrac{(x + c)^2}{a^2} + \dfrac{y^2}{b^2} = 1.$

(b) If $e > 1$, then $e^2 - 1 > 0$, rewrite (1) as $\dfrac{(x - e^2 k/(e^2 - 1))^2}{e^2 k^2/(e^2 - 1)^2} - \dfrac{y^2}{e^2 k^2/(e^2 - 1)} = 1$, then

with $c = \dfrac{e^2 k}{e^2 - 1}$, $a^2 = \dfrac{e^2 k^2}{(e^2 - 1)^2}$, and $b^2 = \dfrac{e^2 k^2}{e^2 - 1}$ we get $\dfrac{(x - c)^2}{a^2} - \dfrac{y^2}{b^2} = 1.$

CHAPTER 13
Polar Coordinates And Parametric Equations

EXERCISE SET 13.1

1.

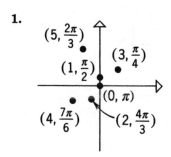

3. (a) $(3\sqrt{3}, 3)$ (b) $(-7/2, 7\sqrt{3}/2)$ (c) $(4\sqrt{2}, 4\sqrt{2})$
(d) $(5, 0)$ (e) $(-7\sqrt{3}/2, 7/2)$ (f) $(0, 0)$

5. (a) $(5, \pi)$ (b) $(4, 11\pi/6)$ (c) $(2, 3\pi/2)$
(d) $(8\sqrt{2}, 5\pi/4)$ (e) $(6, 2\pi/3)$ (f) $(\sqrt{2}, \pi/4)$

7. (a) $(-5, 0)$ (b) $(-4, 5\pi/6)$ (c) $(-2, \pi/2)$
(d) $(-8\sqrt{2}, \pi/4)$ (e) $(-6, 5\pi/3)$ (f) $(-\sqrt{2}, 5\pi/4)$

9. $x^2 + y^2 = 4$; circle   **11.** $y = 4$; horizontal line

13. $r^2 = 3r\cos\theta$, $x^2 + y^2 = 3x$, $(x - 3/2)^2 + y^2 = 9/4$; circle

15. $r^2(2\sin\theta\cos\theta) = 8$, $2(r\sin\theta)(r\cos\theta) = 8$, $xy = 4$; hyperbola

17. $r + r\sin\theta = 2$, $r = 2 - y$, $r^2 = (2 - y)^2$, $x^2 + y^2 = 4 - 4y + y^2$, $x^2 + 4y = 4$; parabola

19. $3r\cos\theta + 2r\sin\theta = 6$, $3x + 2y = 6$; line

21. $r\cos\theta = 7$ **23.** $r = 3$ **25.** $r^2 - 6r\sin\theta = 0$, $r = 6\sin\theta$

27. $r^2\cos^2\theta = 9r\sin\theta$, $r = 9\dfrac{\sin\theta}{\cos^2\theta} = 9\tan\theta\sec\theta$

29. $4(r\cos\theta)(r\sin\theta) = 9$, $4r^2\sin\theta\cos\theta = 9$, $r^2\sin 2\theta = 9/2$

31. $r^4 = 2r^2\sin\theta\cos\theta$, $r^2 = \sin 2\theta$

33. $(r-1)(r-2) = 0$, $r = 1$ or $r = 2$; the graph consists of the concentric circles $r = 1$ and $r = 2$.

35.

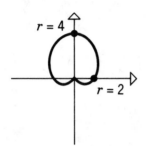

37.

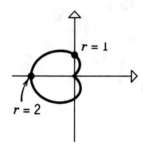

39. $r^2 = ar\sin\theta + br\cos\theta$, $x^2 + y^2 = ay + bx$ which is a circle.

41. (r, θ) and $(-r, \theta + \pi)$ are polar coordinates of the same point so if $r < 0$ then $-r > 0$ and $x = (-r)\cos(\theta + \pi) = (-r)(-\cos\theta) = r\cos\theta$, $y = (-r)\sin(\theta + \pi) = (-r)(-\sin\theta) = r\sin\theta$.

EXERCISE SET 13.2

1.

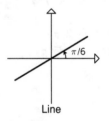

Line

3.

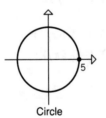

Circle

5.

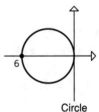

Circle

7.

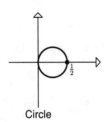

Circle

9.

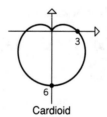

Cardioid

11.

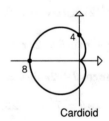

Cardioid

13.

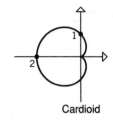

Cardioid

15.

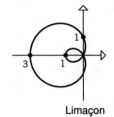

Limaçon

17.

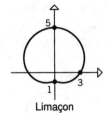

Limaçon

19.

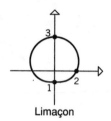

Limaçon

21.

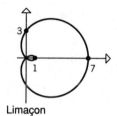

Limaçon

23.

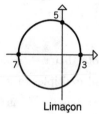

Limaçon

25.

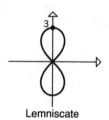

Lemniscate

27.

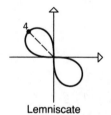

Lemniscate

29.

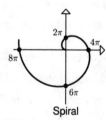

Spiral

31.

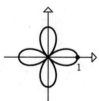

Four-petal rose

33.

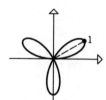

Three-petal rose

35.

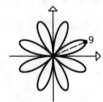

Eight-petal rose

37. $r^2 = 4\cos\theta + 4r\sin\theta,$
$x^2 + y^2 = 4x + 4y,$
$(x-2)^2 + (y-2)^2 = 8$

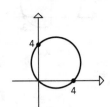

39.

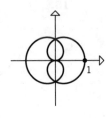

41. Note that $r \to \pm\infty$ as θ approches
odd multiples of $\pi/2$;
$x = r\cos\theta = 4\tan\theta\cos\theta = 4\sin\theta,$
$y = r\sin\theta = 4\tan\theta\sin\theta$
so $x \to \pm 4$ and $y \to \pm\infty$ as
θ approaches odd multiples of $\pi/2$.

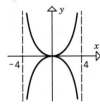

43. Note that $r \to \pm\infty$ as θ approaches
odd multiples of $\pi/2$;
$x = r\cos\theta = (2\sin\theta\tan\theta)\cos\theta = 2\sin^2\theta,$
$y = r\sin\theta = (2\sin\theta\tan\theta)\sin\theta = 2\sin^2\theta\tan\theta$
so $x \to 2$ and $y \to \pm\infty$ as
θ approaches odd multiples of $\pi/2$.

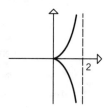

45. $\displaystyle\lim_{\theta\to 0+} y = \lim_{\theta\to 0+} \frac{\sin\theta}{\sqrt{\theta}} = 0$

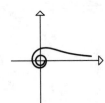

47. $y = r\sin\theta = (1+\cos\theta)\sin\theta = \sin\theta + \sin\theta\cos\theta,$
$dy/d\theta = \cos\theta - \sin^2\theta + \cos^2\theta = 2\cos^2\theta + \cos\theta - 1 = (2\cos\theta - 1)(\cos\theta + 1)$
$dy/d\theta = 0$ if $\cos\theta = 1/2$ or if $\cos\theta = -1$; $\theta = \pi/3$ or π.
If $\theta = 0, \pi/3, \pi$, then $y = 0, 3\sqrt{3}/4, 0$ so the maximum value of y is $3\sqrt{3}/4$.

49. $y = r\sin\theta = \cos 2\theta\sin\theta$, $dy/d\theta = \cos 2\theta\cos\theta - 2\sin 2\theta\sin\theta$, use the identities
$\cos 2\theta = 1 - 2\sin^2\theta$ and $\sin 2\theta = 2\sin\theta\cos\theta$ to get $dy/d\theta = \cos\theta(1 - 6\sin^2\theta)$, $dy/d\theta = 0$
for $0 \le \theta \le \pi/4$ if $\sin^2\theta = 1/6$, $\sin\theta = 1/\sqrt{6}$ where y attains its maximum value of
$y = \cos 2\theta\sin\theta = (1 - 2\sin^2\theta)\sin\theta = [1 - 2(1/6)](1/\sqrt{6}) = \sqrt{6}/9$ so the width of the
petal is $2y = 2\sqrt{6}/9$.

51. $x = r\cos\theta = (1 + \cos\theta)\cos\theta = \cos\theta + \cos^2\theta$,
$dx/d\theta = -\sin\theta - 2\sin\theta\cos\theta = -\sin\theta(1 + 2\cos\theta)$,
$dx/d\theta = 0$ if $\sin\theta = 0$ or if $\cos\theta = -1/2$; $\theta = 0$, $2\pi/3$, or π. If $\theta = 0$, $2\pi/3$, π, then
$x = 2, -1/4, 0$ so the minimum value of x is $-1/4$.

53. For the directrix to the left of the pole (see figure)
$$\frac{r}{k + r\cos\theta} = e, \quad r = ek + re\cos\theta,$$
$$r(1 - e\cos\theta) = ek \text{ so } r = \frac{ek}{1 - e\cos\theta}.$$
If the directrix is to the right of the pole,
then $r/(k - r\cos\theta) = e$ and it follows that
$r = ek/(1 + e\cos\theta)$.

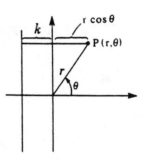

55. $r = \dfrac{3/2}{1 - \cos\theta}$, $e = 1$

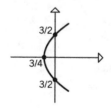

57. $r = \dfrac{3/2}{1 + \frac{1}{2}\sin\theta}$, $e = 1/2$

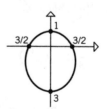

59. $r = \dfrac{2}{1 + \frac{3}{2}\cos\theta}$, $e = 3/2$

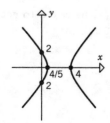

61. The graph of $r = f(\theta + \alpha)$ is the graph of $r = f(\theta)$ rotated α radians clockwise about the pole if $\alpha > 0$, $|\alpha|$ radians counterclockwise if $\alpha < 0$.

EXERCISE SET 13.3

1. $A = \displaystyle\int_{\pi/6}^{\pi/3} \frac{1}{2}\theta^2\,d\theta = 7\pi^3/1296$

3. $A = 2\displaystyle\int_{0}^{\pi} \frac{1}{2}(2 + 2\cos\theta)^2\,d\theta = 6\pi$

5. $A = 2\displaystyle\int_{\pi/6}^{\pi/2} \frac{1}{2}[25\sin^2\theta - (2 + \sin\theta)^2]\,d\theta = 8\pi/3 + \sqrt{3}$

7. $A = 2\displaystyle\int_{0}^{\pi/3} \frac{1}{2}[(2 + 2\cos\theta)^2 - 9]\,d\theta = 9\sqrt{3}/2 - \pi$

9 $A = 2\displaystyle\int_{0}^{\pi/2} \frac{1}{2}\sin 2\theta\,d\theta = 1$

11. $A = 6\displaystyle\int_{0}^{\pi/6} \frac{1}{2}(16\cos^2 3\theta)\,d\theta = 4\pi$

13. $A = 2\left[\displaystyle\int_{0}^{\pi/3} \frac{1}{2}(1 + \cos\theta)^2\,d\theta + \displaystyle\int_{\pi/3}^{\pi/2} \frac{1}{2}(9\cos^2\theta)\,d\theta\right] = 5\pi/4$

15. $A = 2\displaystyle\int_{0}^{\cos^{-1}(3/5)} \frac{1}{2}(100 - 36\sec^2\theta)\,d\theta = 100\cos^{-1}(3/5) - 48$

17. $A = \displaystyle\int_{0}^{\pi/2} \frac{1}{2}a^2\sec^4(\theta/2)\,d\theta = 4a^2/3$

19 $A = \displaystyle\int_{0}^{\pi} \frac{9}{2}e^{-4\theta}\,d\theta = \frac{9}{8}(1 - e^{-4\pi})$

21. $A = \displaystyle\int_{1/9}^{4} \frac{1}{2}\frac{1}{\theta}\,d\theta = \ln 6$

23. $A = 4\displaystyle\int_{0}^{\pi/6} \frac{1}{2}(4\cos 2\theta - 2)\,d\theta = 2\sqrt{3} - 2\pi/3$

25. **(a)** $x = r\cos\theta$ and $y = r\sin\theta$ so $r^3\cos^3\theta - r^2\sin\theta\cos\theta + r^3\sin^3\theta = 0$, $r^2[r(\cos^3\theta + \sin^3\theta) - \sin\theta\cos\theta] = 0$, $r = \sin\theta\cos\theta/(\cos^3\theta + \sin^3\theta)$.

(b) Divide numerator and denominator of the expression in part (a) by $\cos^3\theta$ to get $r = \sec\theta\tan\theta/(1 + \tan^3\theta)$. The loop is traced out for $0 \le \theta \le \pi/2$ so

$$A = \frac{1}{2} \int_0^{\pi/2} r^2 d\theta = \frac{1}{2} \int_0^{\pi/2} \frac{\sec^2\theta \tan^2\theta}{(1+\tan^3\theta)^2} d\theta, \text{ let } u = 1 + \tan^3\theta \text{ to get}$$

$$A = \frac{1}{6} \int_1^{+\infty} \frac{1}{u^2} du = -\frac{1}{6}\frac{1}{u}\Big]_1^{+\infty} = \frac{1}{6}.$$

EXERCISE SET 13.4

1. (a)

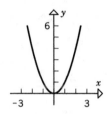

3. $\cos^2 t + \sin^2 t = 1; x^2 + y^2 = 1$

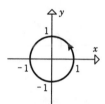

(b) $y = x^2$

5. $t = (x+4)/3;$
$y = 2x + 10$

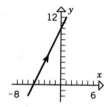

7. $\cos t = x/2, \ \sin t = y/5;$
$x^2/4 + y^2/25 = 1$

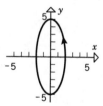

9. $\cos t = (x-3)/2, \sin t = (y-2)/4;$
$(x-3)^2/4 + (y-2)^2/16 = 1$

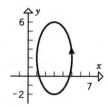

11. $\sin 2\pi t = x/4, \ \cos 2\pi t = y/4;$
$x^2/16 + y^2/16 = 1$

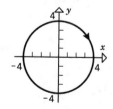

13. $\cos 2t = 1 - 2\sin^2 t;$
$x = 1 - 2y^2, -1 \le y \le 1$

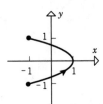

15. $y = \ln t^2 = \ln x, x \ge 1$

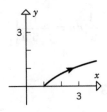

17. $x = 3y^2 - 1,$
$-1 < x \le 2, 0 < y \le 1$

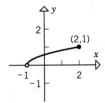

19. $x/2 + y/3 = 1,$
$0 \le x \le 2, 0 \le y \le 3$

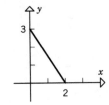

21. $y = 1 - x^2,$
$-1 \le x \le 1, 0 \le y \le 1$

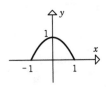

23. $x + (y - 1)^2 = 1,$
$0 \le x \le 1, 0 \le y \le 2$

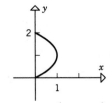

25. $x^2 + y^2 = 1,$
$\cos 1 \le x < 1, 0 < y \le \sin 1$

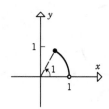

27. $(x_0, \sqrt{x_0})$ is on the curve for $x_0 \geq 0$. $y' = 1/(2\sqrt{x})$; the tangent line at $(x_0, \sqrt{x_0})$ is $y - \sqrt{x_0} = (x - x_0)/(2\sqrt{x_0})$, which crosses the x-axis at $x = -x_0 = t$ so $x_0 = -t$ hence $x = -t$, $y = \sqrt{-t}$ for $t \leq 0$.

29. $dy/dx = \dfrac{\cos t}{-\sin t} = -\cot t$, $dy/dx\big|_{t=3\pi/4} = 1$

31. $dy/dx = \dfrac{2}{1/(2\sqrt{t})} = 4\sqrt{t}$, $dy/dx\big|_{t=9} = 12$

33. $dy/dx = \dfrac{6\pi \cos 2\pi s}{-8\pi \sin 2\pi s} = -\dfrac{3}{4}\cot 2\pi s$, $dy/dx\big|_{s=-1/4} = 0$

35. $dy/dx = \dfrac{t^2}{t} = t$, $d^2y/dx^2 = \dfrac{1}{t}$, $d^2y/dx^2\big|_{t=2} = 1/2$

37. $dy/dx = \dfrac{2}{1/(2\sqrt{t})} = 4\sqrt{t}$, $d^2y/dx^2 = \dfrac{2/\sqrt{t}}{1/(2\sqrt{t})} = 4$, $d^2y/dx^2\big|_{t=1} = 4$

39. $(dx/dt)^2 + (dy/dt)^2 = (4)^2 + (3)^2 = 25$, $L = \displaystyle\int_0^2 5\,dt = 10$

41. $(dx/dt)^2 + (dy/dt)^2 = (t^2)^2 + (t)^2 = t^2(t^2 + 1)$, $L = \displaystyle\int_0^1 t(t^2 + 1)^{1/2}\,dt = (2\sqrt{2} - 1)/3$

43. $(dx/dt)^2 + (dy/dt)^2 = (-2\sin 2t)^2 + (2\cos 2t)^2 = 4$, $L = \displaystyle\int_0^{\pi/2} 2\,dt = \pi$

45. $(dx/dt)^2 + (dy/dt)^2 = [2(1 + t)]^2 + [3(1 + t)^2]^2 = (1 + t)^2[4 + 9(1 + t)^2]$,
    $L = \displaystyle\int_0^1 (1 + t)[4 + 9(1 + t)^2]^{1/2}\,dt = (80\sqrt{10} - 13\sqrt{13})/27$

47. $(dx/dt)^2 + (dy/dt)^2 = [a(1 - \cos t)]^2 + [a\sin t]^2 = 2a^2(1 - \cos t) = 4a^2\sin^2(t/2)$,
    $L = \displaystyle\int_0^{2\pi} 2a\sin(t/2)\,dt = 8a$

49. **(a)** $(dx/dt)^2 + (dy/dt)^2 = 4\sin^2 t + \cos^2 t = 4\sin^2 t + (1 - \sin^2 t) = 1 + 3\sin^2 t$,
    $L = \displaystyle\int_0^{2\pi} \sqrt{1 + 3\sin^2 t}\,dt = 4\int_0^{\pi/2} \sqrt{1 + 3\sin^2 t}\,dt$

    (b) 9.69

(c) distance traveled $= \int_{1.5}^{4.8} \sqrt{1 + 3\sin^2 t}\, dt \approx 5.16$ cm

51. $x = (2 + 3\sin\theta)\cos\theta$, $y = (2 + 3\sin\theta)\sin\theta$

53. $dy/dx = \dfrac{-e^{-t}}{e^t} = -e^{-2t}$; for $t = 2$, $dy/dx = -e^{-4}$, $(x, y) = (e^2, e^{-2})$;

$y - e^{-2} = -e^{-4}(x - e^2)$, $y = -e^{-4}x + 2e^{-2}$

55. $dy/dx = \dfrac{2t + 1}{6t^2 - 30t + 24} = \dfrac{2t + 1}{6(t - 1)(t - 4)}$

(a) $dy/dx = 0$ if $t = -1/2$

(b) $dx/dy = \dfrac{6(t - 1)(t - 4)}{2t + 1} = 0$ if $t = 1, 4$

57. If $x = 3$ then $t^2 - 3t + 5 = 3$, $t^2 - 3t + 2 = 0$, $(t - 1)(t - 2) = 0$, $t = 1$ or 2. If $t = 1$ or 2 then $y = 1$ so $(3, 1)$ is reached when $t = 1$ or 2. $dy/dx = (3t^2 + 2t - 10)/(2t - 3)$. For $t = 1$, $dy/dx = 5$, the tangent line is $y - 1 = 5(x - 3)$, $y = 5x - 14$. For $t = 2$, $dy/dx = 6$, the tangent line is $y - 1 = 6(x - 3)$, $y = 6x - 17$.

59. Assuming that $a \neq 0$ and $b \neq 0$, eliminate the parameter to get $(x - h)^2/a^2 + (y - k)^2/b^2 = 1$. If $|a| = |b|$ the curve is a circle with center (h, k) and radius $|a|$; if $|a| \neq |b|$ the curve is an ellipse with center (h, k) and major axis parallel to the x-axis when $|a| > |b|$ or major axis parallel to the y-axis when $|a| < |b|$.

61. **(a)**

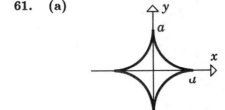

(b) Use $b = a/4$ in the equations of Exercise 60 to get $x = \dfrac{3}{4}a\cos\phi + \dfrac{1}{4}a\cos 3\phi$,

$y = \dfrac{3}{4}a\sin\phi - \dfrac{1}{4}a\sin 3\phi$; but trigonometric identities yield $\cos 3\phi = 4\cos^3\phi - 3\cos\phi$, $\sin 3\phi = 3\sin\phi - 4\sin^3\phi$ so $x = a\cos^3\phi$, $y = a\sin^3\phi$.

(c) From the result in part (b), $\cos\phi = (x/a)^{1/3}$, $\sin\phi = (y/a)^{1/3}$ so $(x/a)^{2/3} + (y/a)^{2/3} = 1$, $x^{2/3} + y^{2/3} = a^{2/3}$

63. $x' = e^t(\cos t - \sin t)$, $y' = e^t(\cos t + \sin t)$, $(x')^2 + (y')^2 = 2e^{2t}$

$$S = 2\pi \int_0^{\pi/2} (e^t \sin t)\sqrt{2e^{2t}}\,dt = 2\sqrt{2}\pi \int_0^{\pi/2} e^{2t} \sin t\,dt$$

$$= 2\sqrt{2}\pi \left[\frac{1}{5}e^{2t}(2\sin t - \cos t)\right]_0^{\pi/2} = \frac{2\sqrt{2}}{5}\pi(2e^\pi + 1)$$

65. $x' = 1$, $y' = 4t$, $(x')^2 + (y')^2 = 1 + 16t^2$, $S = 2\pi \int_0^1 t\sqrt{1 + 16t^2}\,dt = \frac{\pi}{24}(17\sqrt{17} - 1)$

67. $\dfrac{dx}{d\phi} = a(1 - \cos\phi)$, $\dfrac{dy}{d\phi} = a\sin\phi$, $\left(\dfrac{dx}{d\phi}\right)^2 + \left(\dfrac{dy}{d\phi}\right)^2 = 2a^2(1 - \cos\phi)$

$$S = 2\pi \int_0^{2\pi} a(1 - \cos\phi)\sqrt{2a^2(1 - \cos\phi)}\,d\phi = 2\sqrt{2}\pi a^2 \int_0^{2\pi} (1 - \cos\phi)^{3/2}d\phi,$$

but $1 - \cos\phi = 2\sin^2\dfrac{\phi}{2}$ so $(1 - \cos\phi)^{3/2} = 2\sqrt{2}\sin^3\dfrac{\phi}{2}$ for $0 \le \phi \le \pi$ and, taking advantage of the symmetry of the cycloid, $S = 16\pi a^2 \displaystyle\int_0^\pi \sin^3\dfrac{\phi}{2}d\phi = 64\pi a^2/3$

EXERCISE SET 13.5

1. $\theta = \pi/3$; $dr/d\theta = -\sqrt{3}$, $r = 1$, $\tan\theta = \sqrt{3}$, $m = \dfrac{1 + (\sqrt{3})(-\sqrt{3})}{-\sqrt{3} + (-\sqrt{3})} = 1/\sqrt{3}$

3. $\theta = 2$; $dr/d\theta = -1/4$, $r = 1/2$, $\tan\theta = \tan 2$, $m = \dfrac{1/2 + (\tan 2)(-1/4)}{-(1/2)\tan 2 + (-1/4)} = \dfrac{\tan 2 - 2}{2\tan 2 + 1}$

5. $\theta = 3\pi/4$; $dr/d\theta = -3\sqrt{2}/2$, $r = \sqrt{2}/2$, $\tan\theta = -1$, $m = \dfrac{\sqrt{2}/2 + (-1)(-3\sqrt{2}/2)}{-(\sqrt{2}/2)(-1) + (-3\sqrt{2}/2)} = -2$

7. $\theta = \pi/2$; $dr/d\theta = 3$, $r = 3$, $\tan\psi = 1$ **9.** $\theta = \pi$, $dr/d\theta = -5$, $r = 0$, $\tan\psi = 0$

11. $\theta = 5\pi/6$; $dr/d\theta = -2\sqrt{3}$, $r = 1$, $\tan\psi = -1/(2\sqrt{3})$

13. $r^2 + (dr/d\theta)^2 = (e^{3\theta})^2 + (3e^{3\theta})^2 = 10e^{6\theta}$, $L = \displaystyle\int_0^2 \sqrt{10}e^{3\theta}\,d\theta = \sqrt{10}(e^6 - 1)/3$

15. $r^2 + (dr/d\theta)^2 = (2a\cos\theta)^2 + (-2a\sin\theta)^2 = 4a^2$, $L = \displaystyle\int_0^\pi 2a\,d\theta = 2\pi a$

17. $r^2 + (dr/d\theta)^2 = (a\theta^2)^2 + (2a\theta)^2 = a^2\theta^2(\theta^2 + 4)$,

$$L = \int_0^\pi a\theta(\theta^2 + 4)^{1/2}d\theta = \frac{a}{3}[(\pi^2 + 4)^{3/2} - 8]$$

19. $r^2 + (dr/d\theta)^2 = [a(1 - \cos\theta)]^2 + [a\sin\theta]^2 = 4a^2\sin^2(\theta/2)$, $L = 2\int_0^\pi 2a\sin(\theta/2)d\theta = 8a$

21. **(a)** $r^2 + (dr/d\theta)^2 = (\cos n\theta)^2 + (-n\sin n\theta)^2 = \cos^2 n\theta + n^2\sin^2 n\theta$
$$= (1 - \sin^2 n\theta) + n^2\sin^2 n\theta = 1 + (n^2 - 1)\sin^2 n\theta,$$

$$L = 2\int_0^{\pi/(2n)} \sqrt{1 + (n^2 - 1)\sin^2 n\theta}\,d\theta$$

(b) $L = 2\int_0^{\pi/4} \sqrt{1 + 3\sin^2 2\theta}\,d\theta \approx 2.42$

23. **(a)** $\dfrac{dr}{dt} = 2$ and $\dfrac{d\theta}{dt} = 0.5$ so $\dfrac{dr}{d\theta} = \dfrac{dr/dt}{d\theta/dt} = \dfrac{2}{0.5} = 4$, $r = 4\theta + C$, $r = 10$ when $\theta = 0$ so
$10 = C, r = 4\theta + 10$.

(b) $r^2 + (dr/d\theta)^2 = (4\theta + 10)^2 + 16$, during the first 5 seconds the rod rotates through an
angle of $(0.5)(5) = 2.5$ radians so $L = \int_0^{2.5} \sqrt{(4\theta + 10)^2 + 16}\,d\theta$, let $u = 4\theta + 10$ to get

$$L = \frac{1}{4}\int_{10}^{20} \sqrt{u^2 + 16}\,du = \frac{1}{4}\left[\frac{u}{2}\sqrt{u^2 + 16} + 8\ln|u + \sqrt{u^2 + 16}|\right]_{10}^{20}$$

$$= \frac{1}{4}\left[10\sqrt{416} - 5\sqrt{116} + 8\ln\frac{20 + \sqrt{416}}{10 + \sqrt{116}}\right] \approx 38.9 \text{ mm}$$

25. $dx/d\theta = 4\sin^2\theta - \sin\theta - 2$, $dy/d\theta = \cos\theta(1 - 4\sin\theta)$. $dy/d\theta = 0$ when $\cos\theta = 0$ or $\sin\theta = 1/4$
so $\theta = \pi/2$, $3\pi/2$, $\sin^{-1}(1/4)$, or $\pi - \sin^{-1}(1/4)$; $dx/d\theta \neq 0$ at these points so there is a
horizontal tangent at each one.

27. $(\sqrt{3}/2, \pi/6)$ satisfies both equations so it is a point of intersection.

$$\tan\psi_1 = \frac{r}{dr/d\theta} = \frac{\sin 2\theta}{2\cos 2\theta} = \frac{1}{2}\tan 2\theta, \ \tan\psi_2 = \frac{r}{dr/d\theta} = \frac{\cos\theta}{-\sin\theta} = -\cot\theta, \ \text{at } \theta = \pi/6,$$

$$\tan\psi_1 = \sqrt{3}/2 \text{ and } \tan\psi_2 = -\sqrt{3} \text{ so } \tan\beta = \left|\frac{-\sqrt{3} - \sqrt{3}/2}{1 + (\sqrt{3}/2)(-\sqrt{3})}\right| = 3\sqrt{3}, \ \beta = \tan^{-1} 3\sqrt{3}.$$

29. $\tan\psi = r/(dr/d\theta) = e^{a\theta}/(ae^{a\theta}) = 1/a$ so ψ is a constant.

SUPPLEMENTARY EXERCISES CHAPTER 13

1.   (a) $(1, \sqrt{3})$ (b) $(0, -2)$ (c) $(0, 0)$
    (d) $(-1, 1)$ (e) $(-3, 0)$ (f) $(3/5, -4/5)$

3.   (a) 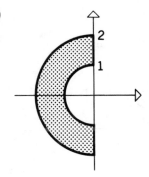   (b) 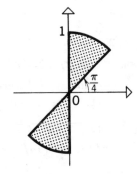

5. $r^2 \sin 2\theta = 1$, $r^2(2 \sin \theta \cos \theta) = 1$, $2(r \sin \theta)(r \cos \theta) = 1$, $2yx = 1$; hyperbola.

7. $r = -4 \csc \theta$, $r = -4/\sin \theta$, $r \sin \theta = -4$, $y = -4$, line.

9. $\theta = \pi/3$, $\tan \theta = \sqrt{3}$, $y = \sqrt{3}x$; line.   11. $r = 0$, $x = 0$ and $y = 0$; point.

13. $x = -3$, $r \cos \theta = -3$ 15. $y = 3x$, $\tan \theta = 3$, $\theta = \tan^{-1} 3$

17. 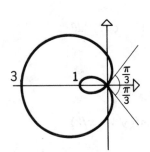   19. 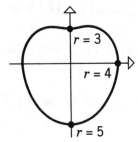

21.

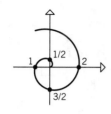

23.

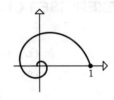

25. $a \cos 2\theta = a/2,\ \cos 2\theta = 1/2$;
one solution is $2\theta = \pi/3,\ \theta = \pi/6$
and from the symmetry of the graphs
the others are $\theta = -\pi/6,\ \pm\pi/3,\ \pm2\pi/3$,
$\pm5\pi/6$. The points of intersection are
$(a/2, \pm\pi/6),\ (a/2, \pm\pi/3),\ (a/2, \pm2\pi/3)$,
$(a/2, \pm5\pi/6)$.

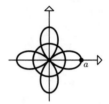

27.   **(a)** $A = 2\left[\dfrac{1}{2}\displaystyle\int_0^{\pi/3}(1+\cos\theta)^2\,d\theta + \dfrac{1}{2}\int_{\pi/3}^{\pi/2}(3\cos\theta)^2\,d\theta\right]$

$\qquad = \displaystyle\int_0^{\pi/3}(1+\cos\theta)^2\,d\theta + \int_{\pi/3}^{\pi/2}9\cos^2\theta\,d\theta$

(b) $r^2 + (dr/d\theta)^2 = (1+\cos\theta)^2 + (-\sin\theta)^2 = 2(1+\cos\theta),\ L = 2\displaystyle\int_{\pi/3}^{\pi}\sqrt{2(1+\cos\theta)}\,d\theta$

29.   **(a)** $A = \displaystyle\int_{\pi/2}^{\pi}\dfrac{1}{2}[(2\sin\theta)^2 - (2+2\cos\theta)^2]\,d\theta = 2\int_{\pi/2}^{\pi}[\sin^2\theta - (1+\cos\theta)^2]\,d\theta$

(b) $r^2 + (dr/d\theta)^2 = (2\sin\theta)^2 + (2\cos\theta)^2 = 4,\ L = \displaystyle\int_0^{\pi/2}2\,d\theta$

31. $A = 4\displaystyle\int_0^{\pi/6}\dfrac{1}{2}(2a^2\cos 2\theta - a^2)\,d\theta = 2a^2\int_0^{\pi/6}(2\cos 2\theta - 1)\,d\theta = a^2(\sqrt{3} - \pi/3)$

33. $A = 2\displaystyle\int_{\pi/2}^{\pi}[\sin^2\theta - (1+\cos\theta)^2]\,d\theta = 2\int_{\pi/2}^{\pi}(-1 - 2\cos\theta - \cos 2\theta)\,d\theta = 4 - \pi$

35. **(a)** eliminate the parameter to get
$(x-1)^2/9 + (y+1)^2/4 = 1$ for
$-2 \le x \le 4$ and $-1 \le y \le 1$.

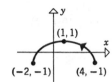

(b) $dy/dx = \dfrac{2\cos\theta}{-3\sin\theta} = -\dfrac{2}{3}\cot\theta$, $d^2y/dx^2 = \dfrac{(2/3)\csc^2\theta}{-3\sin\theta} = -\dfrac{2}{9}\csc^3\theta$; at $\theta_0 = \pi/2$,
$dy/dx = 0$ and $d^2y/dx^2 = -2/9$, $x = 1$, $y = 1$ so the tangent line is $y = 1$.

37. **(a)** Eliminate the parameter to get
$y = \ln(1/x) = -\ln x$ for
$1 \le x \le e^{-1}$ and $0 \le y \le 1$.

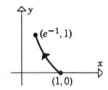

(b) $dy/dx = \dfrac{1/t}{-1/t^2} = -t$, $d^2y/dx^2 = \dfrac{-1}{-1/t^2} = t^2$; at $t_0 = 2$, $dy/dx = -2$ and $d^2y/dx^2 = 4$,
$x = 1/2$, $y = \ln 2$ so the tangent line is $y - \ln 2 = -2(x - 1/2)$, $y = -2x + 1 + \ln 2$.

39. $(dx/dt)^2 + (dy/dt)^2 = (-2\tan 2t)^2 + 2^2 = 4\sec^2 2t$, $L = \displaystyle\int_0^{\pi/6} 2\sec 2t\, dt = \ln(2 + \sqrt{3})$

41. $(dx/dt)^2 + (dy/dt)^2 = (6t)^2 + (3t^2 - 3)^2 = (3t^2 + 3)^2$, $L = \displaystyle\int_0^1 (3t^2 + 3)dt = 4$

43. $r^2 + (dr/d\theta)^2 = (e^\theta)^2 + (e^\theta)^2 = 2e^{2\theta}$, $L = \displaystyle\int_0^{2\pi} \sqrt{2}\,e^\theta d\theta = \sqrt{2}(e^{2\pi} - 1)$

45. $dx/dt = -2\cos t$, $dy/dt = 1 - 2\sin t$

(a) horizontal when $dy/dt = 0$ and $dx/dt \ne 0$; $1 - 2\sin t = 0$, $\sin t = 1/2$, $t = \pi/6, 5\pi/6$ so
(x, y) is $(0, \pi/6 + \sqrt{3})$ or $(0, 5\pi/6 - \sqrt{3})$.

(b) vertical when $dx/dt = 0$ and $dy/dt \ne 0$; $-2\cos t = 0$, $\cos t = 0$, $t = \pi/2$ so (x, y) is
$(-1, \pi/2)$.

47. $dy/dx = (2t - 4)/(1/t) = 2t^2 - 4t$, $d^2y/dx^2 = (4t - 4)/(1/t) = 4t(t - 1)$. For $t > 0$, $d^2y/dx^2 = 0$
when $t = 1$ and d^2y/dx^2 changes sign there so an inflection point occurs at $t = 1$.

49. $r = 2(1 + \cos\theta)$, $dr/d\theta = -2\sin\theta$. When $\theta = \pi/2$, $\sin\theta = 1$, $\cos\theta = 0$, $r = 2$, $dr/d\theta = -2$ so
$m = [(2)(0) + (1)(-2)]/[-(2)(1) + (0)(-2)] = 1$ and $\tan\psi = 2/(-2) = -1$, $\psi = 3\pi/4$.

CHAPTER 14
Three-Dimensional Space; Vectors

EXERCISE SET 14.1

1. **(a)** $d = \sqrt{(2-0)^2 + (1-0)^2 + (3-0)^2} = \sqrt{4+1+9} = \sqrt{14}$; midpoint $(1, 1/2, 3/2)$

    (b) $d = \sqrt{(4-5)^2 + (1-2)^2 + (6-3)^2} = \sqrt{1+1+9} = \sqrt{11}$; midpoint $(9/2, 3/2, 9/2)$

    (c) $d = \sqrt{(3+2)^2 + (0+1)^2 + (5-3)^2} = \sqrt{23+1+4} = \sqrt{30}$; midpoint $(1/2, -1/2, 4)$

    (d) $d = \sqrt{(4+1)^2 + (3+1)^2 + (-2+3)^2} = \sqrt{25+16+1} = \sqrt{42}$; midpoint $(3/2, 1, -5/2)$

3. vertices: $(4, 2, -2)$, $(4,2,1)$, $(4,1,1)$, $(4, 1, -2)$, $(-6, 1, 1)$, $(-6, 2, 1)$, $(-6, 2, -2)$, $(-6, 1, -2)$

5. **(a)** the sides have lengths 7, 14, and $7\sqrt{5}$; it is a right triangle because the sides satisfy the Pythagorean theorem, $(7\sqrt{5})^2 = 7^2 + 14^2$.

    (b) (2,1,6) is the vertex of the 90° angle because it is opposite the longest side (the hypotenuse).

    (c) area $= (1/2)(\text{altitude})(\text{base}) = (1/2)(7)(14) = 49$

7. The distance to the z-axis is the distance between (x_0, y_0, z_0) and $(0, 0, z_0)$ which is $\sqrt{x_0^2 + y_0^2}$; similarly, the distance to the x-axis is $\sqrt{y_0^2 + z_0^2}$ and the distance to the y-axis is $\sqrt{x_0^2 + z_0^2}$.

9. $(x+2)^2 + (y-4)^2 + (z+1)^2 = 36$ **11.** $x^2 + (y-1)^2 + z^2 = 9$

13. $(x-2)^2 + (y+1)^2 + (z+3)^2 = r^2$,

    (a) $r^2 = 3^2 = 9$ **(b)** $r^2 = 1^2 = 1$ **(c)** $r^2 = 2^2 = 4$

15. $r = \dfrac{1}{2}\sqrt{(-1-0)^2 + (2-2)^2 + (1-3)^2} = \dfrac{1}{2}\sqrt{5}$, center $(-1/2, 2, 2)$,

    $(x+1/2)^2 + (y-2)^2 + (z-2)^2 = 5/4$

17. $r = \sqrt{(7-3)^2 + (2+2)^2 + (1-4)^2} = \sqrt{41}$, $(x-3)^2 + (y+2)^2 + (z-4)^2 = 41$

19. $r = |[\text{distance between } (0,0,0) \text{ and } (3, -2, 4)] \pm 1| = \sqrt{29} \pm 1$,

    $x^2 + y^2 + z^2 = r^2 = \left(\sqrt{29} \pm 1\right)^2 = 30 \pm 2\sqrt{29}$

21. $(x+5)^2 + (y+2)^2 + (z+1)^2 = 49$; sphere, $C(-5, -2, -1)$, $r = 7$

23. $(x-1/2)^2 + (y-3/4)^2 + (z+5/4)^2 = 54/16$; sphere, $C(1/2, 3/4, -5/4)$, $r = 3\sqrt{6}/4$

25. $(x - 3/2)^2 + (y + 2)^2 + (z - 4)^2 = -11/4$; no graph

27. Complete the square to get $(x + 1)^2 + (y - 1)^2 + (z - 2)^2 = 9$; center $(-1, 1, 2)$, radius 3. The distance between the origin and the center is $\sqrt{6} < 3$ so the origin is inside the sphere. The largest distance is $3 + \sqrt{6}$, the smallest is $3 - \sqrt{6}$.

29. $(y + 3)^2 + (z - 2)^2 > 16$; all points outside the circular cylinder $(y + 3)^2 + (z - 2)^2 = 16$.

31. Let r be the radius of a styrofoam sphere. The distance from the origin to the center of the bowling ball is equal to the sum of the distance from the origin to the center of the styrofoam sphere nearest the origin and the distance between the center of this sphere and the center of the bowling ball so $\sqrt{3}R = \sqrt{3}r + r + R$, $(\sqrt{3} + 1)r = (\sqrt{3} - 1)R$, $r = \dfrac{\sqrt{3} - 1}{\sqrt{3} + 1}R = (2 - \sqrt{3})R$.

33. **(a)**   **(b)**   **(c)**

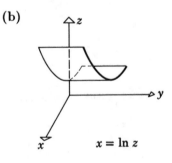

35. **(a)**   **(b)**   **(c)**

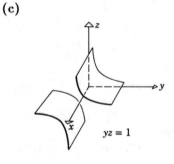

$y = e^x$ $x = \ln z$ $yz = 1$

37. **(a)**

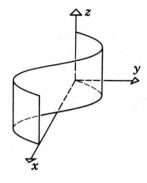

(b)

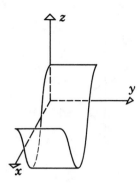

39. **(a)**

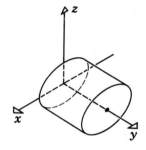

(b)

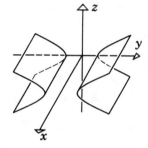

41. $(a \sin \phi \cos \theta)^2 + (a \sin \phi \sin \theta)^2 + (a \cos \phi)^2 = a^2 \sin^2 \phi \cos^2 \theta + a^2 \sin^2 \phi \sin^2 \theta + a^2 \cos^2 \phi$

$$= a^2 \sin^2 \phi (\cos^2 \theta + \sin^2 \theta) + a^2 \cos^2 \phi$$

$$= a^2 \sin^2 \phi + a^2 \cos^2 \phi = a^2 (\sin^2 \phi + \cos^2 \phi) = a^2$$

EXERCISE SET 14.2

1.

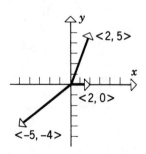

3.

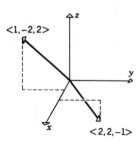

5.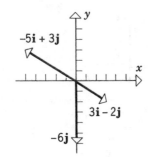

7.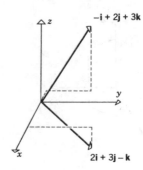

9. (a) $\langle 2-3, 8-5 \rangle = \langle -1, 3 \rangle$ (b) $\langle 0-7, 0-(-2) \rangle = \langle -7, 2 \rangle$
   (c) $\langle -4-(-6), -1-(-2) \rangle = \langle 2, 1 \rangle$ (d) $\langle -8-0, 7-0 \rangle = \langle -8, 7 \rangle$

11. (a) $\langle -3, 6, 1 \rangle$ (b) $\langle 1, -3, -5 \rangle$

13. Let (x, y) be the terminal point, then $x - 1 = 3$, $x = 4$ and $y - (-2) = -2$, $y = -4$.
     The terminal point is $(4, -4)$.

15. Let (x, y) be the initial point, then $2 - x = -2$, $x = 4$ and $0 - y = 4$, $y = -4$.
     The initial point is $(4, -4)$.

17. Let (x, y, z) be the initial point, then $5 - x = -3$, $-y = 1$, and $-1 - z = 2$ so $x = 8$,
     $y = -1$, and $z = -3$. The initial point is $(8, -1, -3)$.

19. (a) $-5\mathbf{i} - 2\mathbf{j}$ (b) $8\mathbf{i} + 10\mathbf{j}$ (c) $-2\mathbf{i} + 4\mathbf{j}$
   (d) $40\mathbf{i} + 36\mathbf{j}$ (e) $-20\mathbf{i} + 22\mathbf{j}$ (f) $-8\mathbf{i} - 8\mathbf{j}$

21. (a) $-\mathbf{i} + 4\mathbf{j} - 2\mathbf{k}$ (b) $18\mathbf{i} + 12\mathbf{j} - 6\mathbf{k}$ (c) $-\mathbf{i} - 5\mathbf{j} - 2\mathbf{k}$
   (d) $40\mathbf{i} - 4\mathbf{j} - 4\mathbf{k}$ (e) $-2\mathbf{i} - 16\mathbf{j} - 18\mathbf{k}$ (f) $-\mathbf{i} + 13\mathbf{j} - 2\mathbf{k}$

23. (a) $\|\mathbf{v}\| = \sqrt{1+1} = \sqrt{2}$ (b) $\|\mathbf{v}\| = \sqrt{4+0} = 2$ (c) $\|\mathbf{v}\| = \sqrt{2+7} = 3$

25. (a) $\|\mathbf{v}\| = \sqrt{14}$ (b) $\|\mathbf{v}\| = 3$

27. (a) $\|\langle 5, 4 \rangle\| = \sqrt{41}$ (b) $2 + 5 = 7$
   (c) $3\|\mathbf{u}\| + 4\|\mathbf{v}\| = 3\sqrt{29} + 8$ (d) $\|\langle -3, -9 \rangle\| = 3\sqrt{10}$
   (e) $\dfrac{1}{5}\langle 3, 4 \rangle = \langle 3/5, 4/5 \rangle$ (f) 1

29. (a) $\|\mathbf{u} + \mathbf{v}\| = \|2\mathbf{i} - 2\mathbf{j} + 2\mathbf{k}\| = 2\sqrt{3}$ (b) $\|\mathbf{u}\| + \|\mathbf{v}\| = \sqrt{14} + \sqrt{2}$

       (c) $\|-2\mathbf{u}\| + 2\|\mathbf{v}\| = 2\sqrt{14} + 2\sqrt{2}$ (d) $\|3\mathbf{u} - 5\mathbf{v} + \mathbf{w}\| = \|-12\mathbf{j} + 2\mathbf{k}\| = 2\sqrt{37}$

       (e) $(1/\sqrt{6})\mathbf{i} + (1/\sqrt{6})\mathbf{j} - (2/\sqrt{6})\mathbf{k}$ (f) 1

31. $6\mathbf{x} = 2\mathbf{u} - \mathbf{v} - \mathbf{w} = \langle -4, 6 \rangle, \mathbf{x} = \langle -2/3, 1 \rangle$

33. $\mathbf{u} = \dfrac{5}{7}\mathbf{i} + \dfrac{2}{7}\mathbf{j} + \dfrac{1}{7}\mathbf{k}, \ \mathbf{v} = \dfrac{8}{7}\mathbf{i} - \dfrac{1}{7}\mathbf{j} - \dfrac{4}{7}\mathbf{k}$

35. Take $\mathbf{z}$ as the diagonal of a parallelepiped with $\mathbf{u}$, $\mathbf{v}$, and $\mathbf{w}$ along its edges as shown. Then $\mathbf{z}$ can be written as the sum $c_1\mathbf{u} + c_2\mathbf{v} + c_3\mathbf{w}$.

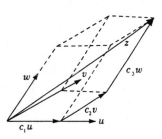

37. $c_1\mathbf{u} + c_2\mathbf{v} = \langle c_1 - 2c_2, -3c_1 + 6c_2 \rangle = \langle 3, 5 \rangle$, so $c_1 - 2c_2 = 3$ and $-3c_1 + 6c_2 = 5$ which has no solution.

39. Equate corresponding components to get the system of equations $c_1 + 3c_2 + 4c_3 = 2$, $-c_1 - c_3 = 1$, and $c_2 + c_3 = -1$. From the second and third equations, $c_1 = -1 - c_3$ and $c_2 = -1 - c_3$; substitute these into the first equation to get $-4 = 2$, which is nonsense so the system has no solution.

43. $\|3\mathbf{i} - 4\mathbf{j}\| = 5$ so the required vector is $-\dfrac{1}{5}(3\mathbf{i} - 4\mathbf{j}) = -\dfrac{3}{5}\mathbf{i} + \dfrac{4}{5}\mathbf{j}$

45. $\|6\mathbf{i} - 4\mathbf{j} + 2\mathbf{k}\| = 2\sqrt{14}$ so the required vector is $-(3\mathbf{i} - 2\mathbf{j} + \mathbf{k})/\sqrt{14}$

47. $\overrightarrow{AB} = 4\mathbf{i} + \mathbf{j} - \mathbf{k}, \ \|\overrightarrow{AB}\| = 3\sqrt{2}$ so the required vector is $(4\mathbf{i} + \mathbf{j} - \mathbf{k})/(3\sqrt{2})$

49. $-\dfrac{1}{2}\mathbf{v} = \langle -3/2, 2 \rangle$ **51.** $-2\mathbf{v} = 6\mathbf{i} - 8\mathbf{j} - 2\mathbf{k}$

53. $\|\mathbf{r} - \mathbf{r}_0\| = \|\langle x - x_0, y - y_0 \rangle\| = \sqrt{(x - x_0)^2 + (y - y_0)^2} = 1$, circle of radius 1 and center at (x_0, y_0).

55. (a) $\|\mathbf{r}\| = \sqrt{x^2 + y^2 + z^2} = 2$, sphere of radius 2 with center at $(0,0,0)$.

       (b) $\|\mathbf{r} - \mathbf{r}_0\| = \sqrt{(x - x_0)^2 + (y - y_0)^2 + (z - z_0)^2} = 3$, sphere of radius 3 with center at (x_0, y_0, z_0).

(c) $\|\mathbf{r} - \mathbf{r}_0\| = \sqrt{(x - x_0)^2 + (y - y_0)^2 + (z - z_0)^2} \le 1$, all points on or inside the sphere of radius 1 with center at (x_0, y_0, z_0).

57. **(a)** Choose two points on the line, for example $P_1(0,4)$ and $P_2(1,3)$ then $\overrightarrow{P_1 P_2} = \langle 1, -1 \rangle$ is parallel to the line, $\|\langle 1, -1 \rangle\| = \sqrt{2}$ so $\langle 1/\sqrt{2}, -1/\sqrt{2} \rangle$ and $\langle -1/\sqrt{2}, 1/\sqrt{2} \rangle$ are unit vectors parallel to the line.

(b) Pick any line that is perpendicular to the line $x + y = 4$, for example $y = x$, and proceed as in part (a) to get $\langle 1/\sqrt{2}, 1/\sqrt{2} \rangle$ and $\langle -1/\sqrt{2}, -1/\sqrt{2} \rangle$.

59. Let $R(x, y)$ be the required point then $\overrightarrow{QR} = 3/4\,\overrightarrow{QP}$,
$\langle x - 7, y + 4 \rangle = 3/4\langle -5, 7 \rangle = \langle -15/4, 21/4 \rangle$ so $x - 7 = -15/4$, $x = 13/4$ and $y + 4 = 21/4$, $y = 5/4$. The point is $(13/4, 21/4)$.

61. **(a)** $\langle \cos(\pi/3), \sin(\pi/3) \rangle = \langle 1/2, \sqrt{3}/2 \rangle$
(b) $4\langle \cos(3\pi/4), \sin(3\pi/4) \rangle = \langle -2\sqrt{2}, 2\sqrt{2} \rangle$

63. Let A, B, C be the vertices $(0,0)$, $(1,3)$, $(2,4)$ and D the fourth vertex (x,y). For the parallelogram ABCD, $\overrightarrow{AD} = \overrightarrow{BC}$, $\langle x, y \rangle = \langle 1, 1 \rangle$ so $x = 1$, $y = 1$ and D is at $(1,1)$. For the parallelogram ACBD, $\overrightarrow{AD} = \overrightarrow{CB}$, $\langle x, y \rangle = \langle -1, -1 \rangle$ so $x = -1$, $y = -1$ and D is at $(-1, -1)$.

65. Use an analytic approach as illustrated in the text.

67. Draw the triangles with sides formed by the vectors $\mathbf{u}$, $\mathbf{v}$, $\mathbf{u} + \mathbf{v}$ and $k\mathbf{u}$, $k\mathbf{v}$, $k\mathbf{u} + k\mathbf{v}$. By similar triangles, $k(\mathbf{u} + \mathbf{v}) = k\mathbf{u} + k\mathbf{v}$.

69. Let $\mathbf{a}$, $\mathbf{b}$, $\mathbf{c}$ be vectors along the sides of the triangle and A, B the midpoints of $\mathbf{a}$ and $\mathbf{b}$, then

$$\mathbf{u} = \frac{1}{2}\mathbf{a} - \frac{1}{2}\mathbf{b} = \frac{1}{2}(\mathbf{a} - \mathbf{b}) = \frac{1}{2}\mathbf{c}$$

so $\mathbf{u}$ is parallel to $\mathbf{c}$ and half as long.

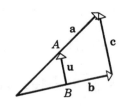

71. $\overrightarrow{AB} = \overrightarrow{AM} + \overrightarrow{MB}$,
$\overrightarrow{AC} = \overrightarrow{AM} + \overrightarrow{MC}$; add to get
$\overrightarrow{AB} + \overrightarrow{AC} = 2\,\overrightarrow{AM} + \overrightarrow{MB} + \overrightarrow{MC}$ but
$\overrightarrow{MB} = -\overrightarrow{MC}$ so $\overrightarrow{AB} + \overrightarrow{AC} = 2\,\overrightarrow{AM}$.

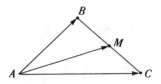

73. $\overrightarrow{AB} = \overrightarrow{AM} + \overrightarrow{MN} + \overrightarrow{NB},$

$\overrightarrow{AD} = \overrightarrow{AM} + \overrightarrow{MN} + \overrightarrow{ND},$

$\overrightarrow{CB} = \overrightarrow{CM} + \overrightarrow{MN} + \overrightarrow{NB},$

$\overrightarrow{CD} = \overrightarrow{CM} + \overrightarrow{MN} + \overrightarrow{ND},$ where

$\overrightarrow{AM} = -\overrightarrow{CM}$ and $\overrightarrow{NB} = -\overrightarrow{ND};$

add to get

$\overrightarrow{AB} + \overrightarrow{AD} + \overrightarrow{CB} + \overrightarrow{CD} = 4\,\overrightarrow{MN}$

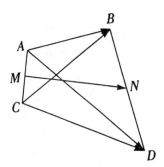

75. $\displaystyle\sum_{k=1}^{n} \overrightarrow{PP_k} = \sum_{k=1}^{n}(\mathbf{r}_k - \mathbf{r}) = \sum_{k=1}^{n}\mathbf{r}_k - \sum_{k=1}^{n}\mathbf{r} = \sum_{k=1}^{n}\mathbf{r}_k - n\mathbf{r} = 0, \ \mathbf{r} = \frac{1}{n}\sum_{k=1}^{n}\mathbf{r}_k.$

77. **(a)** $\mathbf{r}_1 = -\mathbf{i} + 2\mathbf{j}, \ \mathbf{r}_2 = 2\mathbf{i} + 3\mathbf{j}, \ \mathbf{r}_3 = 5\mathbf{i} - 2\mathbf{j}, \ \mathbf{r}_4 = -\mathbf{j}, \ \mathbf{r} = \frac{1}{4}\sum_{k=1}^{4}\mathbf{r}_k = \frac{3}{2}\mathbf{i} + \frac{1}{2}\mathbf{j};$

centroid is at $(3/2, 1/2)$.

EXERCISE SET 14.3

1. **(a)** $(1)(6) + (2)(-8) = -10$

   (c) $(1)(8) + (-3)(-2) + (7)(-2) = 0$

   (b) $(-7)(0) + (-3)(1) = -3$

   (d) $(-3)(4) + (1)(2) + (2)(-5) = -20$

3. **(a)** $\mathbf{u} \cdot \mathbf{v} = -34 < 0,$ obtuse

   (c) $\mathbf{u} \cdot \mathbf{v} = -1 < 0,$ obtuse

   (b) $\mathbf{u} \cdot \mathbf{v} = 6 > 0,$ acute

   (d) $\mathbf{u} \cdot \mathbf{v} = 0,$ orthogonal

5. **(a)** $(14/13)\mathbf{i} + (21/13)\mathbf{j}$

   (c) $-(11/13)\mathbf{i} + \mathbf{j} + (55/13)\mathbf{k}$

   (b) $\langle 2, 6 \rangle$

   (d) $\langle -32/89, -12/89, 73/89 \rangle$

9. **(a)** $\langle 1, 2 \rangle \cdot (\langle 28, -14 \rangle + \langle 6, 0 \rangle) = \langle 1, 2 \rangle \cdot \langle 34, -14 \rangle = 6$

   (b) $\|6\mathbf{w}\| = 6\|\mathbf{w}\| = 36$ **(c)** $24\sqrt{5}$ **(d)** $24\sqrt{5}$

11. Let A, B, and C be the vertices $(-1, 0),$ $(2, -1),$ and $(1, 4)$ with corresponding interior angles $\alpha,$ $\beta,$ and $\gamma,$ then

$$\cos \alpha = \frac{\overrightarrow{AB} \cdot \overrightarrow{AC}}{\|\overrightarrow{AB}\| \, \|\overrightarrow{AC}\|} = \frac{\langle 3,-1\rangle \cdot \langle 2,4\rangle}{\sqrt{10}\sqrt{20}} = 1/(5\sqrt{2})$$

$$\cos \beta = \frac{\overrightarrow{BA} \cdot \overrightarrow{BC}}{\|\overrightarrow{BA}\| \, \|\overrightarrow{BC}\|} = \frac{\langle -3,1\rangle \cdot \langle -1,5\rangle}{\sqrt{10}\sqrt{26}} = 4/\sqrt{65}$$

$$\cos \gamma = \frac{\overrightarrow{CA} \cdot \overrightarrow{CB}}{\|\overrightarrow{CA}\| \, \|\overrightarrow{CB}\|} = \frac{\langle -2,-4\rangle \cdot \langle 1,-5\rangle}{\sqrt{20}\sqrt{26}} = 9/\sqrt{130}$$

13. $\overrightarrow{AB} = \langle 1,3,-2\rangle, \overrightarrow{BC} = \langle 4,-2,-1\rangle, \overrightarrow{AB} \cdot \overrightarrow{BC} = 0$ so $\overrightarrow{AB}$ and $\overrightarrow{BC}$ are orthogonal; it is a right triangle with the right angle at vertex B.

15. (a) $\mathbf{r} \cdot \mathbf{r}_0 = x_0 x + y_0 y = 0$ where $\mathbf{r}$ is perpendicular to $\mathbf{r}_0$; the line through the origin and perpendicular to $\mathbf{r}_0$.

(b) $(\mathbf{r} - \mathbf{r}_0) \cdot \mathbf{r}_0 = x_0(x - x_0) + y_0(y - y_0) = 0$ where the vector from (x_0, y_0) to (x, y) is perpendicular to $\mathbf{r}_0$; the line through (x_0, y_0) and perpendicular to $\mathbf{r}_0$.

(c) $\mathbf{r} \cdot (\mathbf{r} - \mathbf{r}_0) = x(x - x_0) + y(y - y_0) = (x - x_0/2)^2 + (y - y_0/2)^2 - x_0^2/4 - y_0^2/4 = 0$; circle with center at the midpoint of $\mathbf{r}_0$ and radius $\|\mathbf{r}_0\|/2$.

17. (a) $\mathbf{a} \cdot \mathbf{b} = 0, 4k + 3 = 0, k = -3/4$

(b) Use $\mathbf{a} \cdot \mathbf{b} = \|\mathbf{a}\| \, \|\mathbf{b}\| \cos \theta$ to get $4k + 3 = \sqrt{k^2 + 1}(5) \cos(\pi/4), 4k + 3 = 5\sqrt{k^2 + 1}/\sqrt{2}$
Square both sides and rearrange to get $7k^2 + 48k - 7 = 0, (7k - 1)(k + 7) = 0$ so $k = -7$ (invalid) or $k = 1/7$.

(c) Proceed as in (b) with $\theta = \pi/6$ to get $11k^2 - 96k + 39 = 0$ and use the quadratic formula to get $k = (48 \pm 25\sqrt{3})/11$.

(d) If $\mathbf{a}$ and $\mathbf{b}$ are parallel then $\theta = 0$ or π so $\mathbf{a} \cdot \mathbf{b} = \pm\|\mathbf{a}\| \, \|\mathbf{b}\|, 4k + 3 = \pm 5\sqrt{k^2 + 1}$, $9k^2 - 24k + 16 = 0, (3k - 4)^2 = 0, k = 4/3$.

19. $\cos^2 \alpha + \cos^2 \beta + \cos^2 \gamma = \dfrac{u_1^2}{\|\mathbf{u}\|^2} + \dfrac{u_2^2}{\|\mathbf{u}\|^2} + \dfrac{u_3^2}{\|\mathbf{u}\|^2} = (u_1^2 + u_2^2 + u_3^2)/\|\mathbf{u}\|^2 = \|\mathbf{u}\|^2/\|\mathbf{u}\|^2 = 1$

21. (a) $D = |3 - 8 + 7|/\sqrt{9 + 16} = 2/5$ \qquad (b) $D = |-6 + 5 - 1|/\sqrt{4 + 1} = 2/\sqrt{5}$

(c) $D = |4 + 6 - 8|/\sqrt{4 + 1} = 2/\sqrt{5}$

23. (a) $\overrightarrow{AP} = 2\mathbf{i} + 2\mathbf{j} + 3\mathbf{k}, \overrightarrow{AB} = -2\mathbf{i} + \mathbf{j} + 2\mathbf{k}, \|\text{proj}_{\overrightarrow{AB}} \overrightarrow{AP}\| = |\overrightarrow{AP} \cdot \overrightarrow{AB}|/\|\overrightarrow{AB}\| = 4/3$

(b) $\|\overrightarrow{AP}\| = \sqrt{17}, \sqrt{17 - 16/9} = \sqrt{137}/3$

25. Let P and Q be the points $(1,3)$ and $(4,7)$ then $\overrightarrow{PQ} = 3\mathbf{i} + 4\mathbf{j}$ so $W = \mathbf{F} \cdot \overrightarrow{PQ} = -12$ ft $\cdot$ lb.

27. The diagonals have lengths $\|\mathbf{u}+\mathbf{v}\|$ and $\|\mathbf{u}-\mathbf{v}\|$ but

$\|\mathbf{u}+\mathbf{v}\|^2 = (\mathbf{u}+\mathbf{v})\cdot(\mathbf{u}+\mathbf{v}) = \|\mathbf{u}\|^2 + 2\mathbf{u}\cdot\mathbf{v} + \|\mathbf{v}\|^2$, and

$\|\mathbf{u}-\mathbf{v}\|^2 = (\mathbf{u}-\mathbf{v})\cdot(\mathbf{u}+\mathbf{v}) = \|\mathbf{u}\|^2 - 2\mathbf{u}\cdot\mathbf{v} + \|\mathbf{v}\|^2$. If the parallelogram is a rectangle then $\mathbf{u}\cdot\mathbf{v} = 0$ so $\|\mathbf{u}+\mathbf{v}\|^2 = \|\mathbf{u}-\mathbf{v}\|^2$; the diagonals are equal. If the diagonals are equal, then $4\mathbf{u}\cdot\mathbf{v} = 0$, $\mathbf{u}\cdot\mathbf{v} = 0$ so $\mathbf{u}$ is perpendicular to $\mathbf{v}$ and hence the parallelogram is a rectangle.

29. $\|\mathbf{u}+\mathbf{v}\|^2 = (\mathbf{u}+\mathbf{v})\cdot(\mathbf{u}+\mathbf{v}) = \|\mathbf{u}\|^2 + 2\mathbf{u}\cdot\mathbf{v} + \|\mathbf{v}\|^2$ and

$\|\mathbf{u}-\mathbf{v}\|^2 = (\mathbf{u}-\mathbf{v})\cdot(\mathbf{u}-\mathbf{v}) = \|\mathbf{u}\|^2 - 2\mathbf{u}\cdot\mathbf{v} + \|\mathbf{v}\|^2$, subtract to get

$\|\mathbf{u}+\mathbf{v}\|^2 - \|\mathbf{u}-\mathbf{v}\|^2 = 4\mathbf{u}\cdot\mathbf{v}$, the result follows by dividing both sides by 4.

31. With the cube as shown in the diagram, and a the length of each edge,

$\mathbf{d}_1 = a\mathbf{i} + a\mathbf{j} + a\mathbf{k},\ \mathbf{d}_2 = a\mathbf{i} + a\mathbf{j} - a\mathbf{k},$

$\cos\theta = (\mathbf{d}_1\cdot\mathbf{d}_2)\,/\,(\|\mathbf{d}_1\|\,\|\mathbf{d}_2\|) = 1/3,$

$\theta \approx 71°$

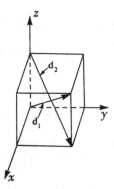

33. $\mathbf{v} = c_1\mathbf{v}_1 + c_2\mathbf{v}_2 + c_3\mathbf{v}_3$ so $\mathbf{v}\cdot\mathbf{v}_i = c_i\mathbf{v}_i\cdot\mathbf{v}_i$ because $\mathbf{v}_i\cdot\mathbf{v}_j = 0$ if $i \neq j$, thus $\mathbf{v}\cdot\mathbf{v}_i = c_i\|\mathbf{v}_i\|^2, c_i = \mathbf{v}\cdot\mathbf{v}_i/\|\mathbf{v}_i\|^2$ for $i = 1, 2, 3$.

35. If $\mathbf{v}$ is orthogonal to $\mathbf{w}_1$ and $\mathbf{w}_2$ then $\mathbf{v}\cdot\mathbf{w}_1 = \mathbf{v}\cdot\mathbf{w}_2 = 0$,

$\mathbf{v}\cdot(k_1\mathbf{w}_1 + k_2\mathbf{w}_2) = k_1\mathbf{v}\cdot\mathbf{w}_1 + k_2\mathbf{v}\cdot\mathbf{w}_2 = k_1(0) + k_2(0) = 0$ so $\mathbf{v}$ is orthogonal to $k_1\mathbf{w}_1 + k_2\mathbf{w}_2$ for all scalars k_1 and k_2.

EXERCISE SET 14.4

1. $\langle 7, 10, 9 \rangle$ **3.** $\langle -4, -6, -3 \rangle$

5. **(a)** $\mathbf{v}\times\mathbf{w} = \langle -23, 7, -1 \rangle, \mathbf{u}\times(\mathbf{v}\times\mathbf{w}) = \langle -20, -67, -9 \rangle$
   (b) $\mathbf{u}\times\mathbf{v} = \langle -10, -14, 2 \rangle, (\mathbf{u}\times\mathbf{v})\times\mathbf{w} = \langle -78, 52, -26 \rangle$
   (c) $\mathbf{v} - 2\mathbf{w} = \langle -2, -7, -3 \rangle, \mathbf{u}\times(\mathbf{v}-2\mathbf{w}) = \langle 24, 0, -16 \rangle$
   (d) $(\mathbf{u}\times\mathbf{v}) - 2\mathbf{w} = \langle -10, -14, 2 \rangle - \langle 2, 8, 10 \rangle = \langle -12, -22, -8 \rangle$
   (e) $(\mathbf{u}\times\mathbf{v})\times(\mathbf{v}\times\mathbf{w}) = \langle -10, -14, 2 \rangle \times \langle -23, 7, -1 \rangle = \langle 0, -56, -392 \rangle$
   (f) $(\mathbf{v}\times\mathbf{w})\times(\mathbf{u}\times\mathbf{v}) = \langle 0, 56, 392 \rangle$

9. A vector parallel to the yz-plane must be perpendicular to **i**;
    $\mathbf{i} \times (3\mathbf{i} - \mathbf{j} + 2\mathbf{k}) = -2\mathbf{j} - \mathbf{k}$, $\| -2\mathbf{j} - \mathbf{k} \| = \sqrt{5}$, the unit vectors are $\pm(2\mathbf{j} + \mathbf{k})/\sqrt{5}$.

11. $(\mathbf{u} + \mathbf{v}) \times (\mathbf{u} - \mathbf{v}) = \mathbf{u} \times \mathbf{u} - \mathbf{u} \times \mathbf{v} + \mathbf{v} \times \mathbf{u} - \mathbf{v} \times \mathbf{v} = 2\mathbf{v} \times \mathbf{u}$ because $\mathbf{u} \times \mathbf{u} = \mathbf{v} \times \mathbf{v} = 0$
    and $-\mathbf{u} \times \mathbf{v} = \mathbf{v} \times \mathbf{u}$.

13. (a) $A = \frac{1}{2} \| \overrightarrow{PQ} \times \overrightarrow{PR} \| = \frac{1}{2} \| \langle -1, -5, 2 \rangle \times \langle 2, 0, 3 \rangle \| = \frac{1}{2} \| \langle -15, 7, 10 \rangle \| = \sqrt{374}/2$

    (b) $A = \frac{1}{2} \| \overrightarrow{PQ} \times \overrightarrow{PR} \| = \frac{1}{2} \| \langle -1, 4, 8 \rangle \times \langle 5, 2, 12 \rangle \| = \frac{1}{2} \| \langle 32, 52, -22 \rangle \| = 9\sqrt{13}$

15. (a) $\overrightarrow{AB} = -\mathbf{i} + 2\mathbf{j} + 2\mathbf{k}$, $\overrightarrow{AC} = \mathbf{i} + \mathbf{j} - \mathbf{k}$, $\overrightarrow{AB} \times \overrightarrow{AC} = -4\mathbf{i} + \mathbf{j} - 3\mathbf{k}$, area $= \frac{1}{2} \| \overrightarrow{AB} \times \overrightarrow{AC} \| = \sqrt{26}/2$

    (b) area $= \frac{1}{2} h \| \overrightarrow{AB} \| = \frac{3}{2} h = \frac{1}{2}\sqrt{26}$, $h = \sqrt{26}/3$

17. (a) $\mathbf{u} = \overrightarrow{AP} = -4\mathbf{i} + 2\mathbf{k}$, $\mathbf{v} = \overrightarrow{AB} = -3\mathbf{i} + 2\mathbf{j} - 4\mathbf{k}$, $\mathbf{u} \times \mathbf{v} = -4\mathbf{i} - 22\mathbf{j} - 8\mathbf{k}$;
    distance $= \| \mathbf{u} \times \mathbf{v} \| / \| \mathbf{v} \| = 2\sqrt{141/29}$

    (b) $\mathbf{u} = \overrightarrow{AP} = 2\mathbf{i} + 2\mathbf{j} + 3\mathbf{k}$, $\mathbf{v} = \overrightarrow{AB} = -2\mathbf{i} + \mathbf{j} + 2\mathbf{k}$, $\mathbf{u} \times \mathbf{v} = \mathbf{i} - 10\mathbf{j} + 6\mathbf{k}$;
    distance $= \| \mathbf{u} \times \mathbf{v} \| / \| \mathbf{v} \| = \sqrt{137}/3$

19. ambiguous (needs parentheses)   **21.** 80   **23.** 1

25. (a) $V = |\mathbf{a} \cdot (\mathbf{b} \times \mathbf{c})| = |-16| = 16$ (b) $V = |\mathbf{a} \cdot (\mathbf{b} \times \mathbf{c})| = |45| = 45$

27. (a) $V = |\mathbf{a} \cdot (\mathbf{b} \times \mathbf{c})| = |-9| = 9$
    (b) $A = \| \mathbf{a} \times \mathbf{c} \| = \| 3\mathbf{i} - 8\mathbf{j} + 7\mathbf{k} \| = \sqrt{122}$
    (c) $\mathbf{b} \times \mathbf{c} = -3\mathbf{i} - \mathbf{j} + 2\mathbf{k}$ is perpendicular to the plane determined by **b** and **c**; let θ be the
        angle between **a** and $\mathbf{b} \times \mathbf{c}$ then

    $$\cos\theta = \frac{\mathbf{a} \cdot (\mathbf{b} \times \mathbf{c})}{\|\mathbf{a}\| \, \|\mathbf{b} \times \mathbf{c}\|} = \frac{-9}{\sqrt{14}\sqrt{14}} = -9/14$$

    so the acute angle ϕ that **a** makes with the plane determined by **b** and **c** is
    $\phi = \theta - \pi/2 = \sin^{-1}(9/14)$.

29. (a) $(\mathbf{u} + k\mathbf{v}) \times \mathbf{v} = \mathbf{u} \times \mathbf{v} + k\mathbf{v} \times \mathbf{v} = \mathbf{u} \times \mathbf{v} + k(0) = \mathbf{u} \times \mathbf{v}$
    (b) $\mathbf{u} \cdot \mathbf{v} \times \mathbf{z} = \mathbf{u} \cdot (\mathbf{v} \times \mathbf{z}) = \mathbf{z} \cdot (\mathbf{u} \times \mathbf{v})$ from formula (9)

    $$= (\mathbf{u} \times \mathbf{v}) \cdot \mathbf{z} = \mathbf{u} \times \mathbf{v} \cdot \mathbf{z}$$

31. Part (b) : let $\mathbf{u} = \langle u_1, u_2, u_3 \rangle$, $\mathbf{v} = \langle v_1, v_2, v_3 \rangle$, and $\mathbf{w} = \langle w_1, w_2, w_3 \rangle$; show that
    $\mathbf{u} \times (\mathbf{v} + \mathbf{w})$ and $(\mathbf{u} \times \mathbf{v}) + (\mathbf{u} \times \mathbf{w})$ are the same.

Part (c) : $(\mathbf{u}+\mathbf{v}) \times \mathbf{w} = -[\mathbf{w} \times (\mathbf{u}+\mathbf{v})]$ from part (a)

$$= -[(\mathbf{w} \times \mathbf{u}) + (\mathbf{w} \times \mathbf{v})] \text{ from part (b)}$$
$$= (\mathbf{u} \times \mathbf{w}) + (\mathbf{v} \times \mathbf{w}) \text{ from part (a)}$$

33.   Let $\mathbf{x} = x_1\mathbf{i} + x_2\mathbf{j} + x_3\mathbf{k}$ and $\mathbf{y} = y_1\mathbf{i} + y_2\mathbf{j} + y_3\mathbf{k}$. If $\mathbf{z} = \mathbf{i}$ then both $\mathbf{x} \times (\mathbf{y} \times \mathbf{i})$ and $(\mathbf{x} \cdot \mathbf{i})\mathbf{y} - (\mathbf{x} \cdot \mathbf{y})\mathbf{i}$ can be shown to be the same. The cases $\mathbf{z} = \mathbf{j}$ and $\mathbf{z} = \mathbf{k}$ are treated the same way. Finally, for $\mathbf{z} = z_1\mathbf{i} + z_2\mathbf{j} + z_3\mathbf{k}$,

$$\mathbf{x} \times (\mathbf{y} \times \mathbf{z}) = \mathbf{x} \times (z_1\mathbf{y} \times \mathbf{i} + z_2\mathbf{y} \times \mathbf{j} + z_3\mathbf{y} \times \mathbf{k})$$

$$= z_1[\mathbf{x} \times (\mathbf{y} \times \mathbf{i})] + z_2[\mathbf{x} \times (\mathbf{y} \times \mathbf{j})] + z_3[\mathbf{x} \times (\mathbf{y} \times \mathbf{k})]$$

$$= z_1[(\mathbf{x} \cdot \mathbf{i})\mathbf{y} - (\mathbf{x} \cdot \mathbf{y})\mathbf{i}] + z_2[(\mathbf{x} \cdot \mathbf{j})\mathbf{y} - (\mathbf{x} \cdot \mathbf{y})\mathbf{j}] + z_3[(\mathbf{x} \cdot \mathbf{k})\mathbf{y} - (\mathbf{x} \cdot \mathbf{y})\mathbf{k}]$$

$$= [\mathbf{x} \cdot (z_1\mathbf{i} + z_2\mathbf{j} + z_3\mathbf{k})]\,\mathbf{y} - (\mathbf{x} \cdot \mathbf{y})(z_1\mathbf{i} + z_2\mathbf{j} + z_3\mathbf{k}) = (\mathbf{x} \cdot \mathbf{z})\mathbf{y} - (\mathbf{x} \cdot \mathbf{y})\mathbf{z}$$

35.   If $\mathbf{a}, \mathbf{b}, \mathbf{c}$, and $\mathbf{d}$ lie in the same plane then $\mathbf{a} \times \mathbf{b}$ and $\mathbf{c} \times \mathbf{d}$ are parallel so $(\mathbf{a} \times \mathbf{b}) \times (\mathbf{c} \times \mathbf{d}) = 0$

37.   (a) $\overrightarrow{PQ} = \langle 3, -1, -3 \rangle, \overrightarrow{PR} = \langle 2, -2, 1 \rangle, \overrightarrow{PS} = \langle 4, -4, 3 \rangle.$

$$V = \frac{1}{6}|\overrightarrow{PQ} \cdot (\overrightarrow{PR} \times \overrightarrow{PS})| = \frac{1}{6}|-4| = 2/3.$$

     (b) $\overrightarrow{PQ} = \langle 1, 2, -1 \rangle, \overrightarrow{PR} = \langle 3, 4, 0 \rangle, \overrightarrow{PS} = \langle -1, -3, 4 \rangle.$

$$V = \frac{1}{6}|\overrightarrow{PQ} \cdot (\overrightarrow{PR} \times \overrightarrow{PS})| = \frac{1}{6}|-3| = 1/2.$$

EXERCISE SET 14.5

1. $\overrightarrow{P_1P_2} = \langle 2, 3 \rangle$ so $x = 3 + 2t$, $y = -2 + 3t$

3. $\overrightarrow{P_1P_2} = \langle 0, 2 \rangle$ so $x = 4$, $y = 1 + 2t$

5. $\overrightarrow{P_1P_2} = \langle -3, 6, 1 \rangle$ so $x = 5 - 3t$, $y = -2 + 6t$, $z = 1 + t$

7. $\overrightarrow{P_1P_2} = \langle -1, 6, 1 \rangle$ so $x = -t$, $y = 6t$, $z = t$

9.   Exercise 1 with $0 \le t \le 1$.   11.   Exercise 3 with $0 \le t \le 1$.

13.   Exercise 5 with $0 \le t \le 1$.   15.   Exercise 7 with $0 \le t \le 1$.

17. $x = -5 + 2t$, $y = 2 - 3t$

19. $2x + 2yy' = 0$, $y' = -x/y = -(3)/(-4) = 3/4$, $\mathbf{v} = 4\mathbf{i} + 3\mathbf{j}$; $x = 3 + 4t$, $y = -4 + 3t$

21. $x = -1 + 3t$, $y = 2 - 4t$, $z = 4 + t$

23. The line is parallel to the vector $\langle 2, -1, 2 \rangle$ so $x = -2 + 2t$, $y = -t$, $z = 5 + 2t$.

25. The line is parallel to the vector $\langle 1, 0, 0 \rangle$ so $x = 3 + t$, $y = 7$, $z = 0$.

27. $3 + 4t = 4t^2$, $4t^2 - 4t - 3 = 0$, $(2t + 1)(2t - 3) = 0$; $t = -1/2, 3/2$. If $t = -1/2$, then $x = -1$, $y = 1$; if $t = 3/2$, then $x = 3$, $y = 9$. The points of intersection are $(-1, 1)$ and $(3,9)$.

29. **(a)** $z = 0$ when $t = 3$ so the point is $(-2, 10, 0)$
 (b) $y = 0$ when $t = -2$ so the point is $(-2, 0, -5)$
 (c) x is always -2 so the line does not intersect the yz-plane

31. $(1 + t)^2 + (3 - t)^2 = 16$, $t^2 - 2t - 3 = 0$, $(t + 1)(t - 3) = 0$; $t = -1, 3$. The points of intersection are $(0, 4, -2)$ and $(4,0,6)$.

33. The line is parallel to the vector $\langle a, b, c \rangle$ so $x = x_1 + at$, $y = y_1 + bt$, $z = z_1 + ct$

35. The lines intersect if we can find values of t_1 and t_2 that satisfy the equations $2 + t_1 = 2 + t_2$, $2 + 3t_1 = 3 + 4t_2$, and $3 + t_1 = 4 + 2t_2$. Solutions of the first two of these equations are $t_1 = -1$, $t_2 = -1$ which also satisfy the third equation so the lines intersect at $(1, -1, 2)$.

37. The lines are parallel, respectively, to the vectors $\langle 7, 1, -3 \rangle$ and $\langle -1, 0, 2 \rangle$. These vectors are not parallel so the lines are not parallel. The system of equations $1 + 7t_1 = 4 - t_2$, $3 + t_1 = 6$, and $5 - 3t_1 = 7 + 2t_2$ has no solution so the lines do not intersect.

39. The points lie on the same line if $\overrightarrow{P_1 P_2}$ is parallel to $\overrightarrow{P_2 P_3}$.

 (a) $\overrightarrow{P_1 P_2} = \langle 3, -7, -7 \rangle$, $\overrightarrow{P_2 P_3} = \langle -9, -7, -3 \rangle$; these vectors are not parallel so the points do not lie on the same line.

 (b) $\overrightarrow{P_1 P_2} = \langle 2, -4, -4 \rangle$, $\overrightarrow{P_2 P_3} = \langle 1, -2, -2 \rangle$; $\overrightarrow{P_1 P_2} = 2 \ \overrightarrow{P_2 P_3}$ so the vectors are parallel and the points lie on the same line.

41. Let the desired point be $P(x_0, y_0, z_0)$, then $\overrightarrow{P_1 P} = (2/3) \ \overrightarrow{P_1 P_2}$,
 $\langle x_0 - 1, y_0 - 4, z_0 + 3 \rangle = (2/3) \langle 0, 1, 2 \rangle = \langle 0, 2/3, 4/3 \rangle$; equate corresponding components to get $x_0 = 1$, $y_0 = 14/3$, $z_0 = -5/3$.

43. Show that two different points on one line lie on the other line. For example, with $t = 0$ and 1 in the equations for the first line we find that $(3,1)$ and $(2,3)$ are on the line. These points are also on the second line for $t = 4/3$ and 1.

45. **(a)** $\langle x, y \rangle = \langle 2, -1 \rangle + t \langle -7, 4 \rangle$ **(b)** $\langle x, y \rangle = \langle 0, 3 \rangle + t \langle 4, 0 \rangle$

47. The line segment joining the points $(1,0)$ and $(-3,6)$.

49. $(3,0,1)$ is on the line $(t = 0)$ so $\mathbf{u} = -5\mathbf{i} + \mathbf{j}$, $\mathbf{v} = -\mathbf{i} + \mathbf{j} + 2\mathbf{k}$, $\mathbf{u} \times \mathbf{v} = 2\mathbf{i} + 10\mathbf{j} - 4\mathbf{k}$;
distance $= \|\mathbf{u} \times \mathbf{v}\|/\|\mathbf{v}\| = 2\sqrt{5}$.

51. The vectors $\mathbf{v}_1 = -\mathbf{i} + 2\mathbf{j} + \mathbf{k}$ and $\mathbf{v}_2 = 2\mathbf{i} - 4\mathbf{j} - 2\mathbf{k}$ are parallel to the lines, $\mathbf{v}_2 = -2\mathbf{v}_1$ so $\mathbf{v}_1$
and $\mathbf{v}_2$ are parallel. Let $t = 0$ to get the points $P(2,0,1)$ and $Q(1,3,5)$ on the first and second
lines, respectively. Let $\mathbf{u} = \overrightarrow{PQ} = -\mathbf{i} + 3\mathbf{j} + 4\mathbf{k}$, $\mathbf{v} = \frac{1}{2}\mathbf{v}_2 = \mathbf{i} - 2\mathbf{j} - \mathbf{k}$; $\mathbf{u} \times \mathbf{v} = 5\mathbf{i} + 3\mathbf{j} - \mathbf{k}$,
distance $= \|\mathbf{u} \times \mathbf{v}\|/\|\mathbf{v}\| = \sqrt{35/6}$.

53. **(a)** Let $t = 3$ and $t = -2$, respectively, in the equations for L_1 and L_2.

    (b) $\mathbf{u} = 2\mathbf{i} - \mathbf{j} - 2\mathbf{k}$ and $\mathbf{v} = \mathbf{i} + 3\mathbf{j} - \mathbf{k}$ are parallel to L_1 and L_2,
        $\cos\theta = \mathbf{u} \cdot \mathbf{v}/(\|\mathbf{u}\|\,\|\mathbf{v}\|) = 1/(3\sqrt{11})$, $\theta \approx 84°$.

    (c) $\mathbf{u} \times \mathbf{v} = 7\mathbf{i} + 7\mathbf{k}$ is perpendicular to both L_1 and L_2, and hence so is $\mathbf{i} + \mathbf{k}$, thus $x = 7 + t$,
        $y = -1$, $z = -2 + t$.

55. $(0,1,2)$ is on the given line $(t = 0)$ so $\mathbf{u} = \mathbf{j} - \mathbf{k}$ is a vector from this point to the point
$(0,2,1)$, $\mathbf{v} = 2\mathbf{i} - \mathbf{j} + \mathbf{k}$ is parallel to the given line. $\mathbf{u} \times \mathbf{v} = -2\mathbf{j} - 2\mathbf{k}$, and hence $\mathbf{w} = \mathbf{j} + \mathbf{k}$, is
perpendicular to both lines so $\mathbf{v} \times \mathbf{w} = -2\mathbf{i} - 2\mathbf{j} + 2\mathbf{k}$, and hence $\mathbf{i} + \mathbf{j} - \mathbf{k}$, is parallel to the
line we seek. Thus $x = t$, $y = 2 + t$, $z = 1 - t$ are parametric equations of the line.

57. **(a)** When $t = 0$ the particles are at $(4,1,2)$ and $(0,1,1)$ so the distance between them is
        $\sqrt{4^2 + 0^2 + 1^2} = \sqrt{17}$ cm.

    (b) At any time t the distance D between the particles is
        $D = \sqrt{[t - (4 - t)]^2 + [(1 + t) - (1 + 2t)]^2 + [(1 + 2t) - (2 + t)]^2} = \sqrt{6t^2 - 18t + 17}$,
        $dD/dt = (6t - 9)/\sqrt{6t^2 - 18t + 17} = 0$ when $t = 3/2$; the minimum
        distance is $\sqrt{6(3/2)^2 - 18(3/2) + 17} = \sqrt{14}/2$ cm.

EXERCISE SET 14.6

1. $(x - 2) + 4(y - 6) + 2(z - 1) = 0$, $x + 4y + 2z = 28$ **3.** $z = 0$

5. Denote the given points by P_1, P_2, and P_3, respectively, then $\overrightarrow{P_1P_2} \times \overrightarrow{P_1P_3}$ is a normal to the
plane.

    (a) $\overrightarrow{P_1P_2} \times \overrightarrow{P_1P_3} = \langle 2,1,2 \rangle \times \langle 3,-1,-2 \rangle = \langle 0,10,-5 \rangle$, for convenience choose $\langle 0,2,-1 \rangle$ which
        is also normal to the plane. Use any of the given points to get $2y - z = 1$

    (b) $\overrightarrow{P_1P_2} \times \overrightarrow{P_1P_3} = \langle -1,-1,-2 \rangle \times \langle -4,1,1 \rangle = \langle 1,9,-5 \rangle$, $x + 9y - 5z = 16$

7. **(a)** yes, because $\langle 2, -1, -4 \rangle$ and $\langle 3, 2, 1 \rangle$ are perpendicular
   (b) no, because $\langle 1, 2, 3 \rangle$ and $\langle 1, -1, 2 \rangle$ are not perpendicular

9. **(a)** yes, because $\langle 2, 1, -1 \rangle$ and $\langle 4, 2, -2 \rangle$ are parallel
   (b) no, because $\langle -1, 1, -3 \rangle$ and $\langle 2, 2, 0 \rangle$ are not parallel

11. **(a)** $\mathbf{n}_1 = \langle 1, 0, 0 \rangle, \mathbf{n}_2 = \langle 2, -1, 1 \rangle, \mathbf{n}_1 \cdot \mathbf{n}_2 = 2$ so

$$\cos \theta = \frac{\mathbf{n}_1 \cdot \mathbf{n}_2}{\|\mathbf{n}_1\| \, \|\mathbf{n}_2\|} = \frac{2}{\sqrt{1}\sqrt{6}} = 2/\sqrt{6}, \theta = \cos^{-1}(2/\sqrt{6}) \approx 35°$$

   (b) $\mathbf{n}_1 = \langle 1, 2, -2 \rangle, \mathbf{n}_2 = \langle 6, -3, 2 \rangle, \mathbf{n}_1 \cdot \mathbf{n}_2 = -4$ so

$$\cos \theta = \frac{(-\mathbf{n}_1) \cdot \mathbf{n}_2}{\| -\mathbf{n}_1\| \, \|\mathbf{n}_2\|} = \frac{4}{(3)(7)} = 4/21, \theta = \cos^{-1}(4/21) \approx 79°$$

13. **(a)** $z = 0$ **(b)** $y = 0$ **(c)** $x = 0$

15. $\langle 4, -2, 7 \rangle$ is normal to the desired plane and $(0,0,0)$ is a point on it; $4x - 2y + 7z = 0$

17. Find two points P_1 and P_2 on the line of intersection of the given planes and then find an equation of the plane that contains P_1, P_2, and the given point $P_0(-1, 4, 2)$. Let (x_0, y_0, z_0) be on the line of intersection of the given planes then $4x_0 - y_0 + z_0 - 2 = 0$ and $2x_0 + y_0 - 2z_0 - 3 = 0$, eliminate y_0 by addition of the equations to get $6x_0 - z_0 - 5 = 0$; if $x_0 = 0$ then $z_0 = -5$, if $x_0 = 1$ then $z_0 = 1$. Substitution of these values of x_0 and z_0 into either of the equations of the planes gives the corresponding values $y_0 = -7$ and $y_0 = 3$ so $P_1(0, -7, -5)$ and $P_2(1, 3, 1)$ are on the line of intersection of the planes. $\overrightarrow{P_0 P_1} \times \overrightarrow{P_0 P_2} = \langle 4, -13, 21 \rangle$ is normal to the desired plane whose equation is $4x - 13y + 21z = -14$.

19. The line is parallel to the line of intersection of the planes if it is parallel to both planes. Normals to the given planes are $\mathbf{n}_1 = \langle 1, -4, 2 \rangle$ and $\mathbf{n}_2 = \langle 2, 3, -1 \rangle$ so $\mathbf{n}_1 \times \mathbf{n}_2 = \langle -2, 5, 11 \rangle$ is parallel to the line of intersection of the planes and hence parallel to the desired line whose equations are $x = 5 - 2t, y = 5t, z = -2 + 11t$.

21. $\mathbf{n}_1 = \langle 2, 1, 1 \rangle$ and $\mathbf{n}_2 = \langle 1, 2, 1 \rangle$ are normals to the given planes, $\mathbf{n}_1 \times \mathbf{n}_2 = \langle -1, -1, 3 \rangle$ so $\langle 1, 1, -3 \rangle$ is normal to the desired plane whose equation is $x + y - 3z = 6$.

23. $\mathbf{v}_1 = \langle 1, 2, -1 \rangle$ and $\mathbf{v}_2 = \langle -1, -2, 1 \rangle$ are parallel, respectively, to the given lines and to each other so the lines are parallel. Let $t = 0$ to find the points $P_1(-2, 3, 4)$ and $P_2(3, 4, 0)$ that lie, respectively, on the given lines. $\mathbf{v}_1 \times \overrightarrow{P_1 P_2} = \langle -7, -1, -9 \rangle$ so $\langle 7, 1, 9 \rangle$ is normal to the desired plane whose equation is $7x + y + 9z = 25$.

25. The plane is the perpendicular bisector of the line segment that joins $P_1(2, -1, 1)$ and $P_2(3, 1, 5)$. The midpoint of the line segment is $(5/2, 0, 3)$ and $\overrightarrow{P_1 P_2} = \langle 1, 2, 4 \rangle$ is normal to the plane so an equation is $x + 2y + 4z = 29/2$.

27. $\mathbf{v} = \langle 0, 1, 1 \rangle$ is parallel to the line.

   (a) $\mathbf{n} = \langle 6, 4, -4 \rangle$ is normal to the plane, $\mathbf{v} \cdot \mathbf{n} = 0$ so the line is parallel to the plane because $\mathbf{v}$ and $\mathbf{n}$ are perpendicular. (0,0,0) lies on the line and the plane so the entire line must lie in the plane.

   (b) $\mathbf{n} = \langle 5, -3, 3 \rangle$ is normal to the plane, $\mathbf{v} \cdot \mathbf{n} = 0$ so the line is parallel to the plane. (0,0,0) is on the line, $(0, 0, 1/3)$ is on the plane. The line is below the plane because (0,0,0) is below $(0, 0, 1/3)$.

   (c) $\mathbf{n} = \langle 6, 2, -2 \rangle$, $\mathbf{v} \cdot \mathbf{n} = 0$ so the line is parallel to the plane. (0,0,0) is on the line, $(0, 0, -3/2)$ is on the plane. The line is above the plane because (0,0,0) is above $(0, 0, -3/2)$.

29. (a) $\mathbf{n}_1 = \langle -2, 3, 7 \rangle$ and $\mathbf{n}_2 = \langle 1, 2, -3 \rangle$ are normals to the planes, $\mathbf{n}_1 \times \mathbf{n}_2 = \langle -23, 1, -7 \rangle$ is parallel to the line of intersection. Let $z = 0$ in both equations and solve for x and y to get $x = -11/7$, $y = -12/7$ so $(-11/7, -12/7, 0)$ is on the line whose equations are $x = -11/7 - 23t$, $y = -12/7 + t$, $z = -7t$

   (b) Similar to part (a) with $\mathbf{n}_1 = \langle 3, -5, 2 \rangle$, $\mathbf{n}_2 = \langle 0, 0, 1 \rangle$, $\mathbf{n}_1 \times \mathbf{n}_2 = \langle -5, -3, 0 \rangle$. $z = 0$ so $3x - 5y = 0$, let $x = 0$ then $y = 0$ and (0,0,0) is on the line whose equations are $x = -5t$, $y = -3t$, $z = 0$.

31. $D = |2(1) - 2(-2) + (3) - 4|/\sqrt{4 + 4 + 1} = 5/3$

33. $D = |20(7) - 4(2) - 5(-1)|/\sqrt{400 + 16 + 25} = 137/21$

35. (0,0,0) is on the first plane so $D = |6(0) - 3(0) - 3(0) - 5|/\sqrt{36 + 9 + 9} = 5/\sqrt{54}$

37. (1,3,5) and (4,6,7) are on L_1 and L_2, respectively. $\mathbf{v}_1 = \langle 7, 1, -3 \rangle$ and $\mathbf{v}_2 = \langle -1, 0, 2 \rangle$ are parallel to L_1 and L_2, $\mathbf{v}_1 \times \mathbf{v}_2 = \langle 2, -11, 1 \rangle$ so the plane $2x - 11y + z + 51 = 0$ contains L_2, $D = |2(1) - 11(3) + (5) + 51|/\sqrt{4 + 121 + 1} = 25/\sqrt{126}$

39. (2,6,0) and (3,5,6) are on L_1 and L_2, respectively. $\mathbf{v}_1 = \langle 4, -4, 5 \rangle$ and $\mathbf{v}_2 = \langle 8, -3, 1 \rangle$ are parallel to L_1 and L_2, $\mathbf{v}_1 \times \mathbf{v}_2 = \langle 11, 36, 20 \rangle$ so $11x + 36y + 20z - 333 = 0$ contains L_2, $D = |11(2) + 36(6) + 20(0) - 333|/\sqrt{121 + 1296 + 400} = 95/\sqrt{1817}$.

41. $\mathbf{n}_1 = \langle a_1, b_1, c_1 \rangle$ and $\mathbf{n}_2 = \langle a_2, b_2, c_2 \rangle$ are normals to the planes, the planes are perpendicular if and only if their normals are perpendicular so $\mathbf{n}_1 \cdot \mathbf{n}_2 = 0$, $a_1 a_2 + b_1 b_2 + c_1 c_2 = 0$.

43. The distance between $(2, 1, -3)$ and the plane is $|2 - 3(1) + 2(-3) - 4|/\sqrt{1 + 9 + 4} = 11/\sqrt{14}$ which is the radius of the sphere; an equation is $(x - 2)^2 + (y - 1)^2 + (z + 3)^2 = 121/14$.

EXERCISE SET 14.7

1. (a) $4x^2 + y^2 = 4$; ellipse   (b) $y^2 + z^2 = 3$; circle   (c) $4x^2 + z^2 = 3$; ellipse

3. (a) $9x^2 - z^2 = 16$; hyperbola   (b) $y^2 + z^2 = 20$; circle   (c) $9x^2 - y^2 = 20$; hyperbola

5. (a) $z = 4y^2$; parabola   (b) $z = 9x^2 + 16$; parabola   (c) $9x^2 + 4y^2 = 4$; ellipse

7.

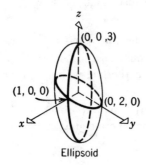

Ellipsoid

9.

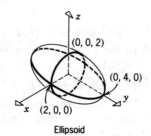

Ellipsoid

11.

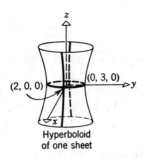

Hyperboloid
of one sheet

13.

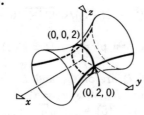

Hyperboloid
of one sheet

15.

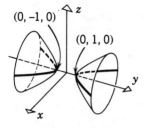

$(0, -1, 0)$

$(0, 1, 0)$

Hyperboloid
of two sheets

17.

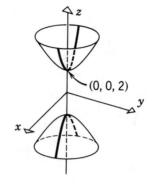

$(0, 0, 2)$

Hyperboloid
of two sheets

19.

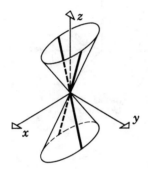

Elliptic cone

21.

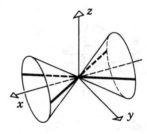

Circular cone

23.

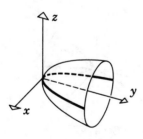

Circular paraboloid

25.

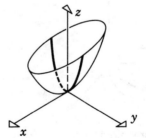

Elliptic paraboloid

27.

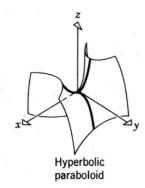

Hyperbolic
paraboloid

29. **(a)**

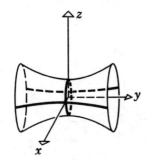

Hyperboloid
of one sheet

(b)

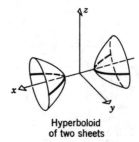

Hyperboloid
of two sheets

(c)

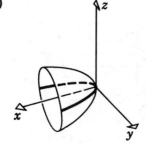

Paraboloid

(d)

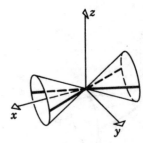

Cone

(e)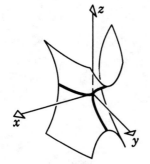

Hyperbolic
paraboloid

(f)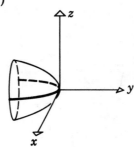

Paraboloid

31.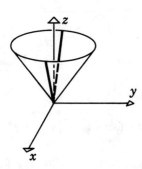

33.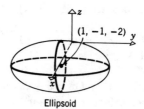

$(0, 1, 0)$

$(1, 0, 0)$

35.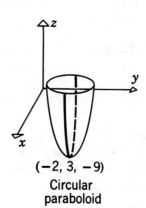

$(-2, 3, -9)$

Circular
paraboloid

37.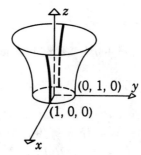

$(1, -1, -2)$

Ellipsoid

39.

$(0, -1, 5)$

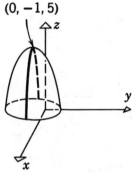

Circular
paraboloid

45. $x^2 + y^2 = 4 - x^2 - y^2, x^2 + y^2 = 2$; circle

47. $x^2 + y^2 = 2x, x^2 + y^2 - 2x = 0$; circle   **49.** $x^2 + y^2 = (y+1)^2, y = \frac{1}{2}x^2 - \frac{1}{2}$; parabola

51. $x^2 + y^2 + 4(\sqrt{x})^2 = 5, x^2 + y^2 + 4x = 5$ for $x \geq 0$; circular arc

53. **(a)** $z = k^2/9 + y^2/4, y^2 = 4(z - k^2/9)$ so $4p = 4$, $p = 1$; the focus is at $(k, 0, k^2/9 + 1)$ and the vertex is at $(k, 0, k^2/9)$.

   (b) $k = x^2/9 + y^2/4, x^2/(9k) + y^2/(4k) = 1$ so $a = 3\sqrt{k}, b = 2\sqrt{k}, c = \sqrt{5k}$; the foci are at $(\pm\sqrt{5k}, 0, k)$, the endpoints of the major axis are $(\pm 3\sqrt{k}, 0, k)$, and the endpoints of the minor axis are $(0, \pm 2\sqrt{k}, k)$.

55. $|z - (-1)| = \sqrt{x^2 + y^2 + (z-1)^2}, z^2 + 2z + 1 = x^2 + y^2 + z^2 - 2z + 1, z = (x^2 + y^2)/4$; paraboloid.

57. **(a)** $(3+t)^2 - (2+t)^2 = 5 + 2t$ and $(3+t)^2 - (2-t)^2 = 5 + 10t$ so both lines lie completely on the surface $z = x^2 - y^2$.

   (b) $(x_0 + t)^2 - (y_0 + at)^2 = x_0^2 - y_0^2 + 2(x_0 - ay_0)t + (1 - a^2)t^2 = z_0 + bt$ if $z_0 = x_0^2 - y_0^2$, $2(x_0 - ay_0) = b$, and $1 - a^2 = 0$. But $z_0 = x_0^2 - y_0^2$ because (x_0, y_0, z_0) is on the surface so it remains to find a and b so that $2(x_0 - ay_0) = b$ and $1 - a^2 = 0$. Solve the second equation to get $a = \pm 1$ so $b = 2(x_0 - y_0)$ and $b = 2(x_0 + y_0)$.

EXERCISE SET 14.8

1. **(a)** $(8, \pi/6, -4)$ **(b)** $(5\sqrt{2}, 3\pi/4, 6)$ **(c)** $(2, \pi/2, 0)$

   (d) $(8, 5\pi/3, 6)$ **(e)** $(2, 7\pi/4, 1)$ **(f)** $(0, 0, 1)$

3. $\rho = \sqrt{x^2 + y^2 + z^2}$, $\tan \theta = y/x$, $\cos \phi = z/\rho$

    (a) $\left(2\sqrt{2}, \pi/3, 3\pi/4\right)$ **(b)** $\left(2, 7\pi/4, \pi/4\right)$ **(c)** $\left(6, \pi/2, \pi/3\right)$

    (d) $\left(10, 5\pi/6, \pi/2\right)$ **(e)** $\left(8\sqrt{2}, \pi/4, \pi/6\right)$ **(f)** $\left(2\sqrt{2}, 5\pi/3, 3\pi/4\right)$

5. $\rho = \sqrt{r^2 + z^2}$, $\theta = \theta$, $\tan \phi = r/z$

    (a) $\left(2\sqrt{3}, \pi/6, \pi/6\right)$ **(b)** $\left(\sqrt{2}, \pi/4, 3\pi/4\right)$ **(c)** $\left(2, 3\pi/4, \pi/2\right)$

    (d) $\left(4\sqrt{3}, 1, 2\pi/3\right)$ **(e)** $\left(4\sqrt{2}, 5\pi/6, \pi/4\right)$ **(f)** $\left(2\sqrt{2}, 0, 3\pi/4\right)$

7.

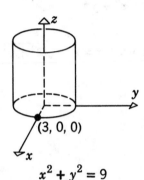

$(3, 0, 0)$

$x^2 + y^2 = 9$

9.

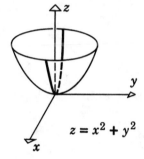

$z = x^2 + y^2$

11.

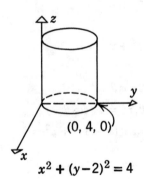

$(0, 4, 0)$

$x^2 + (y-2)^2 = 4$

13.

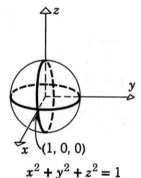

$(1, 0, 0)$

$x^2 + y^2 + z^2 = 1$

15.

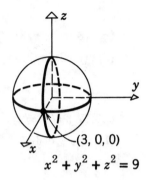

(3, 0, 0)

$$x^2 + y^2 + z^2 = 9$$

17.

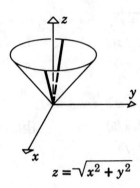

$$z = \sqrt{x^2 + y^2}$$

$$r/z = \tan \phi = 1, z = r = \sqrt{x^2 + y^2}$$

19.

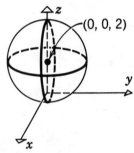

(0, 0, 2)

$$x^2 + y^2 + (z - 2)^2 = 4$$

$$\rho = 4 \cos \phi, \rho^2 = 4\rho \cos \phi$$
$$x^2 + y^2 + z^2 = 4z$$
$$x^2 + y^2 + (z - 2)^2 = 4$$

21.

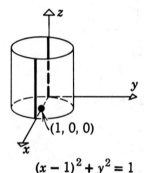

(1, 0, 0)

$$(x - 1)^2 + y^2 = 1$$

$$\rho \sin \phi = 2 \cos \theta$$
$$r = 2 \cos \theta, r^2 = 2r \cos \theta$$
$$x^2 + y^2 = 2x, (x - 1)^2 + y^2 = 1$$

23.   **(a)** $z = 3$ **(b)** $\rho \cos \phi = 3, \rho = 3 \sec \phi$

25.   **(a)** $z = 3r^2$ **(b)** $\rho \cos \phi = 3\rho^2 \sin^2 \phi, \rho = \dfrac{1}{3} \csc \phi \cot \phi$

27.   **(a)** $r = 2$ **(b)** $\rho \sin \phi = 2, \rho = 2 \csc \phi$

29.   **(a)** $r^2 + z^2 = 9$ **(b)** $\rho = 3$

31.   **(a)** $2r \cos \theta + 3r \sin \theta + 4z = 1$
        (b) $2\rho \sin \phi \cos \theta + 3\rho \sin \phi \sin \theta + 4\rho \cos \phi = 1$

33. **(a)** $r^2 \cos^2 \theta = 16 - z^2$

(b) $x^2 = 16 - z^2$, $x^2 + y^2 + z^2 = 16 + y^2$, $\rho^2 = 16 + \rho^2 \sin^2 \phi \sin^2 \theta$, $\rho^2 \left(1 - \sin^2 \phi \sin^2 \theta\right) = 16$

35. All points on or above the paraboloid $z = x^2 + y^2$, that are also on or below the plane $z = 4$

37. All points on or between concentric spheres of radii 1 and 3

39. $\theta = \pi/6$, $\phi = \pi/6$, spherical $(4000, \pi/6, \pi/6)$, rectangular $\left(1000\sqrt{3}, 1000, 2000\sqrt{3}\right)$

41.

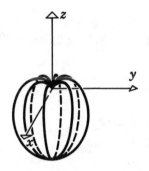

SUPPLEMENTARY EXERCISES CHAPTER 14

1. **(a)** $\overrightarrow{P_1 P_2} = (5 - 2)\mathbf{i} + (-1 - 3)\mathbf{j} = 3\mathbf{i} - 4\mathbf{j}$, $\| \overrightarrow{P_1 P_2} \| = 5$

 (b) $\overrightarrow{P_1 P_2} = (1 - 2)\mathbf{i} + (3 + 1)\mathbf{j} = -\mathbf{i} + 4\mathbf{j}$, $\| \overrightarrow{P_1 P_2} \| = \sqrt{17}$

3. $-(3\mathbf{i} - 4\mathbf{j}) = -3\mathbf{i} + 4\mathbf{j}$

5. $4\mathbf{i} + 3\mathbf{j}$ is the vector from $(1,2)$ to $(5,5)$ so the desired vector is $(3/5)(4\mathbf{i} + 3\mathbf{j})$

7. $12 \cos 120°\mathbf{i} + 12 \sin 120°\mathbf{j} = -6\mathbf{i} + 6\sqrt{3}\mathbf{j}$

9. $3\mathbf{u} - (\mathbf{i} + \mathbf{j}) = \mathbf{i} + \mathbf{u}$, $2\mathbf{u} = 2\mathbf{i} + \mathbf{j}$, $\mathbf{u} = \mathbf{i} + (1/2)\mathbf{j}$

11. The effect of $\mathbf{F}_1$ and $\mathbf{F}_2$ is the same as the effect of $\mathbf{F}_1 + \mathbf{F}_2 = -\mathbf{i} - 5\mathbf{j}$ acting at the point, to cancel the effect a force $\mathbf{F}_3 = -(\mathbf{F}_1 + \mathbf{F}_2) = \mathbf{i} + 5\mathbf{j}$ must be applied at the point.

13. **(a)** $\sqrt{6}$ **(b)** -3 **(c)** $\langle 5, -5, -5 \rangle$

 (d) $\langle -5, 5, 5 \rangle$ **(e)** $5\sqrt{3}/2$ **(f)** $\langle -1, 8, -9 \rangle$

15. **(a)** $2/3$ **(b)** $2/5$ **(c)** $\cos^{-1}(-2/15)$ **(d)** $3/5, -4/5, 0$

17. Both sides reduce to $2\mathbf{i} - 2\mathbf{j} + \mathbf{k}$ **19.** $\langle -3/\sqrt{2}, 0, 3/\sqrt{2} \rangle$

21. (a) $\text{proj}_{\mathbf{v}}\mathbf{u} = \dfrac{\mathbf{u} \cdot \mathbf{v}}{\|\mathbf{v}\|^2}\mathbf{v} = \dfrac{(-3)}{6}(\mathbf{i} + \mathbf{j} + 2\mathbf{k}) = -\dfrac{1}{2}\mathbf{i} - \dfrac{1}{2}\mathbf{j} - \mathbf{k}$

   (b) $\mathbf{u} - \text{proj}_{\mathbf{v}}\mathbf{u} = \dfrac{3}{2}\mathbf{i} + \dfrac{5}{2}\mathbf{j} - 2\mathbf{k}$

23. $\cos^2\alpha + \cos^2\beta + \cos^2\gamma = 1$, let $\alpha = 50°$, $\beta = 70°$,

   $\cos^2\gamma = 1 - \cos^2(50°) - \cos^2(70°) \approx 0.46985$, $\gamma \approx 62°$.

25. (a) $(\mathbf{u} + \mathbf{v}) \cdot (\mathbf{u} - \mathbf{v}) = 0$, $\mathbf{u} \cdot \mathbf{u} - \mathbf{u} \cdot \mathbf{v} + \mathbf{v} \cdot \mathbf{u} - \mathbf{v} \cdot \mathbf{v} = 0$, $\|\mathbf{u}\|^2 - \|\mathbf{v}\|^2 = 0$, $\|\mathbf{u}\| = \|\mathbf{v}\|$

   (b) $(\mathbf{a} \cdot \mathbf{b})^2 + \|\mathbf{a} \times \mathbf{b}\|^2 = (\|\mathbf{a}\| \|\mathbf{b}\| \cos\theta)^2 + (\|\mathbf{a}\| \|\mathbf{b}\| \sin\theta)^2$

   $$= \|\mathbf{a}\|^2\|\mathbf{b}\|^2(\cos^2\theta + \sin^2\theta) = \|\mathbf{a}\|^2\|\mathbf{b}\|^2$$

27. $\mathbf{a} \times \mathbf{b} = \langle 5, 7, -1 \rangle$ is orthogonal to both $\mathbf{a}$ and $\mathbf{b}$, $\|\mathbf{a} \times \mathbf{b}\| = 5\sqrt{3}$ so

   $\pm\langle 1/\sqrt{3}, 7/(5\sqrt{3}), -1/(5\sqrt{3}) \rangle$ are unit vectors orthogonal to both $\mathbf{a}$ and $\mathbf{b}$.

29. The plane contains $\overrightarrow{AB}$ and is parallel to $\mathbf{v}$ thus $\mathbf{v} \times \overrightarrow{AB}$ is normal to the plane,

   $\mathbf{v} \times \overrightarrow{AB} = \langle 5, -5, -5 \rangle$ so $\langle 1, -1, -1 \rangle$ is also a normal to the plane whose equation is

   $x - y - z = -4$.

31. $\overrightarrow{PQ} \times \overrightarrow{PR} = \langle 1, 2, -1 \rangle \times \langle 1, 0, 1 \rangle = \langle 2, -2, -2 \rangle$ is normal to the plane and hence so is $\langle 1, -1, -1 \rangle$, an equation of the plane is $x - y - z = -1$.

33. (a) Parametric equations of L are $x = 1 + 3t$, $y = 2 - t$, $z = 8 - 4t$. If Q is on L then for some t_0, $k = 1 + 3t_0$, $3 = 2 - t_0$, $\ell = 8 - 4t_0$. The second of these equations yields $t_0 = -1$ so $k = -2$, $\ell = 12$.

   (b) Use parametric equations in part (a) for L, solve the system $1 + 3t_1 = -8 - 3t_2$, $2 - t_1 = 5 + t_2$, $8 - 4t_1 = 0$ to get $t_1 = 2$, $t_2 = -5$ so L' intersects L at $(7, 0, 0)$.

   (c) An equation of the plane is $3x - 2y + 6z = 6$, use the parametric equations in part (a) to get $3(1 + 3t) - 2(2 - t) + 6(8 - 4t) = 6$, $t = 41/13$ so L intersects the plane at $(136/13, -15/13, -60/13)$.

35. (a) $\overrightarrow{P_1P_2} = \langle 2, 3, -3 \rangle$, use P_1 to get $x = 1 + 2t$, $y = -1 + 3t$, $z = 2 - 3t$.

   (b) $\overrightarrow{P_1P_2} = \langle 0, 5, -7 \rangle$, use P_1 to get $x = 1$, $y = -3 + 5t$, $z = 4 - 7t$.

37. (a) $\mathbf{n}_1 = \langle 2, 1, -1 \rangle$ and $\mathbf{n}_2 = \langle 1, 2, 1 \rangle$ are normals to the planes so $\mathbf{n}_1 \times \mathbf{n}_2 = \langle 3, -3, 3 \rangle$ is parallel to the line of intersection and hence so is $\langle 1, -1, 1 \rangle$. To find a point on the line of intersection, let $x = 0$ in the equations of the planes to get $y - z = 3$ and $2y + z = 3$ which yield $y = 2$, $z = -1$ so $(0, 2, -1)$ is on the line whose equations are $x = t$, $y = 2 - t$, $z = -1 + t$.

(b) $\mathbf{n}_1 \cdot \mathbf{n}_2 = 3 > 0$ so $\cos\theta = \dfrac{\mathbf{n}_1 \cdot \mathbf{n}_2}{\|\mathbf{n}_1\| \, \|\mathbf{n}_2\|} = \dfrac{3}{\sqrt{6}\sqrt{6}} = 1/2$, $\theta = 60°$.

39. **(a)** the region above the elliptic paraboloid $z = 4x^2 + 9y^2$

   (b) the point $(0,0,0)$

41. $z^2 = \dfrac{x^2}{(36/100)} + \dfrac{y^2}{(36/225)}$, elliptic cone

43. $x^2 + y^2/16 + z^2/25 = 1$, ellipsoid

45. $x^2/25 + y^2/4 - z^2/16 = -1$, hyperboloid of two sheets

47. $W = \mathbf{F} \cdot \overrightarrow{PQ} = (3\mathbf{i} - 4\mathbf{j} + \mathbf{k}) \cdot (\mathbf{i} - \mathbf{j} + 6\mathbf{k}) = 13$ ft $\cdot$ lb

49. **(a)** $(1,1,1)$ **(b)** $(\sqrt{3}, \pi/4, \tan^{-1}\sqrt{2})$

51. **(a)** $z = r^2(\cos^2\theta - \sin^2\theta)$, $z = x^2 - y^2$ **(b)** $(\rho\sin\phi\cos\theta)(\rho\cos\phi) = 1$, $xz = 1$

CHAPTER 15
Vector-Valued Functions

EXERCISE SET 15.1

1. $(-\infty, +\infty)$; $\mathbf{r}(\pi) = -\mathbf{i} - 3\pi\mathbf{j}$.

3. $[2, +\infty)$; $\mathbf{r}(3) = -\mathbf{i} - \ln 3\mathbf{j} + \mathbf{k}$.

5. $\mathbf{r} = 3\cos t\,\mathbf{i} + (t + \sin t)\mathbf{j}$

7. $\mathbf{r} = 2t\mathbf{i} + 2\sin 3t\mathbf{j} + 5\cos 3t\mathbf{k}$.

9. $x = 3t^2$, $y = -2$, $z = 0$.

11. $x = 2t - 1$, $y = -3\sqrt{t}$, $z = \sin 3t$.

13. The line in 2-space through the point $(2, 0)$ and parallel to the vector $-3\mathbf{i} - 4\mathbf{j}$.

15. The line in 3-space through the point $(0, -3, 1)$ and parallel to the vector $2\mathbf{i} + 3\mathbf{k}$.

17. An ellipse in the plane $z = -1$, center at $(0, 0, -1)$, major axis of length 6 parallel to x-axis, minor axis of length 4 parallel to y-axis.

19. The line is parallel to the vector $-2\mathbf{i} + 3\mathbf{j}$; the slope is $-3/2$.

21. $y = 0$ in the xz-plane so $1 - 2t = 0$, $t = 1/2$ thus $x = 2 + 1/2 = 5/2$ and $z = 3(1/2) = 3/2$; the coordinates are $(5/2, 0, 3/2)$.

23. $x = 2$

25. $(x - 1)^2 + (y - 3)^2 = 1$

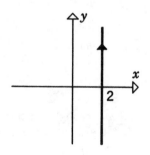

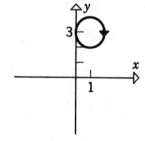

27. $x^2 - y^2 = 1$, $x \geq 1$

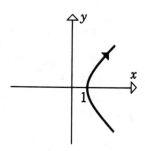

29.

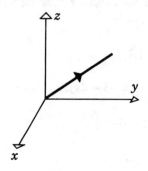

31.

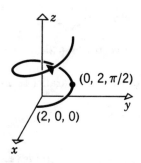

33.

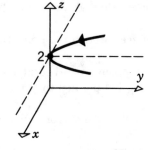

35. $x^2 + y^2 = (t\sin t)^2 + (t\cos t)^2 = t^2(\sin^2 t + \cos^2 t) = t^2 = z$

37. $x = \sin t$, $y = 2\cos t$, $z = \sqrt{3}\sin t$ so $x^2 + y^2 + z^2 = \sin^2 t + 4\cos^2 t + 3\sin^2 t = 4$ and $z = \sqrt{3}x$; it is the curve of intersection of the sphere $x^2 + y^2 + z^2 = 4$ and the plane $z = \sqrt{3}x$, which is a circle with center at $(0,0,0)$ and radius 2.

39. The helix makes one turn as t varies from 0 to 2π so $z = c(2\pi) = 3$, $c = 3/(2\pi)$.

41. $x^2 + y^2 = t^2\cos^2 t + t^2\sin^2 t = t^2$, $\sqrt{x^2 + y^2} = t = z$; a conical helix.

43. The plane is parallel to a line on the surface of the cone and does not go through the vertex so the curve of intersection is a parabola. Eliminate z to get $y + 2 = \sqrt{x^2 + y^2}$, $(y+2)^2 = x^2 + y^2$, $y = x^2/4 - 1$; let $x = t$, then $y = t^2/4 - 1$ and $z = t^2/4 + 1$.

45. Let $x = 3\cos t$ and $y = 3\sin t$, then $z = 9\cos^2 t$.

47. $x^2 + 4y^2 = 2x$, $(x-1)^2 + 4y^2 = 1$; let $x = 1 + \cos t$, $y = \dfrac{1}{2}\sin t$, $z = 2 + 2\cos t$.

EXERCISE SET 15.2

1. $9\mathbf{i} + 6\mathbf{j}$

3. $\mathbf{j}$

5. $2\mathbf{i} - 3\mathbf{j} + 4\mathbf{k}$

7. $\dfrac{\pi}{2}\mathbf{i} + \mathbf{k}$

9. $\displaystyle\lim_{t \to \pi/2} \mathbf{r}(t) = 3\mathbf{i} - \pi\mathbf{j} = \mathbf{r}(\pi/2)$

11. $\mathbf{r}'(t) = 5\mathbf{i} + (1 - 2t)\mathbf{j}$

13. $-\dfrac{1}{t^2}\mathbf{i} + \sec^2 t\,\mathbf{j} + 2e^{2t}\mathbf{k}$

15. $\mathbf{r}'(t) = \langle 1, 2t \rangle$,
$\mathbf{r}'(2) = \langle 1, 4 \rangle$,
$\mathbf{r}(2) = \langle 2, 4 \rangle$

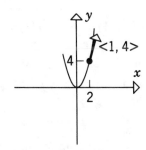

17. $\mathbf{r}'(t) = \langle -e^{-t}, 2e^{2t} \rangle$,
$\mathbf{r}'(\ln 2) = \langle -1/2, 8 \rangle$,
$\mathbf{r}(\ln 2) = \langle 1/2, 4 \rangle$

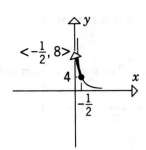

19. $\mathbf{r}'(t) = 2\cos t\,\mathbf{i} - 2\sin t\,\mathbf{k}$,
$\mathbf{r}'(\pi/2) = -2\mathbf{k}$,
$\mathbf{r}(\pi/2) = 2\mathbf{i} + \mathbf{j}$

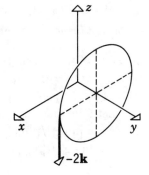

21. $r'(t) = j - 2tk$
$r'(1) = j - 2k,$
$r(1) = 3i + j + k$

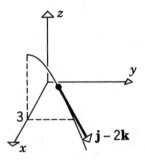

23. $r'(t) = 2ti - \dfrac{1}{t}j,\ r'(1) = 2i - j,\ r(1) = i + 2j;\ x = 1 + 2t,\ y = 2 - t,\ z = 0.$

25. $r'(t) = -2\pi \sin \pi t i + 2\pi \cos \pi t j + 3k,\ r'(1/3) = -\sqrt{3}\,\pi i + \pi j + 3k,$
$r(1/3) = i + \sqrt{3}j + k;\ x = 1 - \sqrt{3}\,\pi t,\ y = \sqrt{3} + \pi t,\ z = 1 + 3t.$

27. $r'(t) = 2i + \dfrac{3}{2\sqrt{3t + 4}}j,\ t = 0$ at P_0 so $r'(0) = 2i + \frac{3}{4}j,$
$r(0) = -i + 2j;\ r = (-i + 2j) + t(2i + \frac{3}{4}j).$

29. $r'(t) = 2ti + \dfrac{1}{(t + 1)^2}j - 2tk,\ t = -2$ at P_0 so $r'(-2) = -4i + j + 4k,$
$r(0) = 4i + j;\ r = (4i + j) + t(-4i + j + 4k).$

31. $r' = 3\cos t i + 2\sin t j + k,\ r'(\pi/2) = 2j + k,\ r(\pi/2) = 3i + \dfrac{\pi}{2}k;\ 2j + k$ is normal to the plane,
and $(3, 0, \pi/2)$ is on the plane so an equation of the plane is $2y + z = \pi/2.$

33. **(a)** $2t - t^2 - 3t = -2,\ t^2 + t - 2 = 0,\ (t + 2)(t - 1) = 0$ so $t = -2, 1.$ The points of intersection
are $(-2, 4, 6)$ and $(1, 1, -3).$

(b) $r' = i + 2tj - 3k;\ r'(-2) = i - 4j - 3k,\ r'(1) = i + 2j - 3k,$ and $n = 2i - j + k$ is normal
to the plane. Let θ be the acute angle, then
for $t = -2$: $\cos\theta = |n \cdot r'|/(\|n\|\ \|r'\|) = 3/\sqrt{156},\ \theta \approx 76°;$
for $t = 1$: $\cos\theta = |n \cdot r'|/(\|n\|\ \|r'\|) = 3/\sqrt{84},\ \theta \approx 71°.$

35. $r_1(1) = r_2(2) = i + j + 3k$ so the graphs intersect at P; $r_1'(t) = 2ti + j + 9t^2k$ and
$r_2'(t) = i + \frac{1}{2}tj - k$ so $r_1'(1) = 2i + j + 9k$ and $r_2'(2) = i + j - k$ are tangent to the graphs at P,
thus $\cos\theta = \dfrac{r_1'(1) \cdot r_2'(2)}{\|r_1'(1)\|\ \|r_2'(2)\|} = -\dfrac{6}{\sqrt{86}\sqrt{3}},\ \theta = \cos^{-1}(6/\sqrt{258}) \approx 68°.$

37. Eliminate the parameter to get $y = x^2$ for both graphs; $r_1'(t) = i + 2tj$ which is never 0, but
$r_2'(t) = 3t^2i + 6t^5j = 0$ when $t = 0.$

39. $\mathbf{r}' = 3t^2\mathbf{i} + (6t - 2)\mathbf{j} + 2t\mathbf{k}$; smooth

41. $\mathbf{r}' = (1 - t)e^{-t}\mathbf{i} + (2t - 2)\mathbf{j} - \pi\sin(\pi t)\mathbf{k}$; not smooth, $\mathbf{r}'(1) = 0$

43. (a) $7t^6$

    (b) $12(t\tan t + 1)\sec t - (\sin t)/t - (\cos t)\ln t$

45. $(d\mathbf{r}/dt)(dt/du) = (\mathbf{i} + 2t\mathbf{j})(4) = 4\mathbf{i} + 8t\mathbf{j} = 4\mathbf{i} + 8(4u + 1)\mathbf{j}$

47. $(d\mathbf{r}/dt)(dt/du) = (e^t\mathbf{i} - 4e^{-t}\mathbf{j})(2u) = 2ue^{u^2}\mathbf{i} - 8ue^{-u^2}\mathbf{j}$

49. $\dfrac{d}{dt}[\mathbf{r}(t) \times \mathbf{r}'(t)] = \mathbf{r}(t) \times \mathbf{r}''(t) + \mathbf{r}'(t) \times \mathbf{r}'(t) = \mathbf{r}(t) \times \mathbf{r}''(t) + 0 = \mathbf{r}(t) \times \mathbf{r}''(t)$

51. In Exercise 50, write each triple scalar product as a determinant.

53. $\dfrac{d}{dt}\left[\dfrac{1}{\|\mathbf{r}\|}\mathbf{r}\right] = \dfrac{1}{\|\mathbf{r}\|}\mathbf{r}' + \mathbf{r}\dfrac{d}{dt}[\|\mathbf{r}\|^{-1}] = \dfrac{1}{\|\mathbf{r}\|}\mathbf{r}' + \mathbf{r}(-\|\mathbf{r}\|^{-2})\dfrac{d}{dt}[\|\mathbf{r}\|] = \dfrac{1}{\|\mathbf{r}\|}\mathbf{r}' - \dfrac{\mathbf{r} \cdot \mathbf{r}'}{\|\mathbf{r}\|^3}\mathbf{r}$

55. Let $\mathbf{r}_1(t) = x_1(t)\mathbf{i} + y_1(t)\mathbf{j} + z_1(t)\mathbf{k}$ and $\mathbf{r}_2(t) = x_2(t)\mathbf{i} + y_2(t)\mathbf{j} + z_2(t)\mathbf{k}$, in both (6) and (7); show that the left and right members of the equalities are the same.

57. Let $\mathbf{r}(t) = x(t)\mathbf{i} + y(t)\mathbf{j}$ and use Theorem 15.2.2 and the chain rule.

59. Let $\mathbf{r}(t) = x(t)\mathbf{i} + y(t)\mathbf{j}$ and use (1) and the conditions for $\mathbf{r}$ to be continuous at t_0.

EXERCISE SET 15.3

1. $3t\mathbf{i} + 2t^2\mathbf{j} + \mathbf{C}$

3. $\left\langle \dfrac{1}{3}\sin 3t, \dfrac{1}{3}\cos 3t \right\rangle\Big]_0^{\pi/3} = \langle 0, -2/3 \rangle$

5. $\left(\dfrac{2}{3}t^{3/2}\mathbf{i} + 2t^{1/2}\mathbf{j}\right)\Big]_1^9 = \dfrac{52}{3}\mathbf{i} + 4\mathbf{j}$

7. $\langle (t - 1)e^t, t(\ln t - 1) \rangle + \mathbf{C}$

9. $(t^3/3)\mathbf{i} - t^2\mathbf{j} + \ln|t|\mathbf{k} + \mathbf{C}$

11. $\dfrac{1}{2}(e^2 - 1)\mathbf{i} + (1 - e^{-1})\mathbf{j} + \dfrac{1}{2}\mathbf{k}$

13. (a) $\mathbf{F} = 3t\mathbf{i} - 2\mathbf{j} - 3t^2\mathbf{k}$

    (b) $\displaystyle\int_0^2 \mathbf{F} \cdot (d\mathbf{r}/dt)\,dt = \int_0^2 (3t + 12t)\,dt = \int_0^2 15t\,dt = 30$

15. $\mathbf{r}(t) = \displaystyle\int \mathbf{r}'(t)dt = (\sin t)\mathbf{i} - (\cos t)\mathbf{j} + \mathbf{C}$,

$\mathbf{r}(0) = -\mathbf{j} + \mathbf{C} = \mathbf{i} - \mathbf{j}$ so $\mathbf{C} = \mathbf{i}$ and $\mathbf{r}(t) = (1 + \sin t)\mathbf{i} - (\cos t)\mathbf{j}$.

17. $\mathbf{r}'(t) = \displaystyle\int \mathbf{r}''(t)dt = 4t^3\mathbf{i} - 2t\mathbf{j} + \mathbf{C}_1$, $\mathbf{r}'(0) = \mathbf{C}_1 = 0$, $\mathbf{r}'(t) = 4t^3\mathbf{i} - 2t\mathbf{j}$

$\mathbf{r}(t) = \displaystyle\int \mathbf{r}'(t)dt = t^4\mathbf{i} - t^2\mathbf{j} + \mathbf{C}_2$, $\mathbf{r}(0) = \mathbf{C}_2 = 2\mathbf{i} - 4\mathbf{j}$, $\mathbf{r}(t) = (t^4 + 2)\mathbf{i} - (t^2 + 4)\mathbf{j}$

19. $\mathbf{r}(t) = \displaystyle\int \mathbf{r}'(t)dt = 2t\mathbf{i} + \frac{1}{2}\ln(t^2 + 1)\mathbf{j} + \frac{1}{2}t^2\mathbf{k} + \mathbf{C}$,

$\mathbf{r}(1) = 2\mathbf{i} + \frac{1}{2}\ln 2\mathbf{j} + \frac{1}{2}\mathbf{k} + \mathbf{C} = 0$ so $\mathbf{C} = -2\mathbf{i} - \frac{1}{2}\ln 2\mathbf{j} - \frac{1}{2}\mathbf{k}$ and

$\mathbf{r}(t) = 2(t - 1)\mathbf{i} + \frac{1}{2}\ln\frac{t^2 + 1}{2}\mathbf{j} + \frac{1}{2}(t^2 - 1)\mathbf{k}$.

21. $\mathbf{r}'(t) = 3\mathbf{i} - 2\mathbf{j} + \mathbf{k}$, $\|\mathbf{r}'(t)\| = \sqrt{14}$, $L = \displaystyle\int_3^4 \sqrt{14}\,dt = \sqrt{14}$

23. $\mathbf{r}'(t) = 3t^2\mathbf{i} + \mathbf{j} + \sqrt{6}\,t\mathbf{k}$, $\|\mathbf{r}'(t)\| = 3t^2 + 1$, $L = \displaystyle\int_1^3 (3t^2 + 1)dt = 28$

25. $\mathbf{r}'(t) = \langle e^t, -e^{-t}, \sqrt{2}\rangle$, $\|\mathbf{r}'(t)\| = e^t + e^{-t}$, $L = \displaystyle\int_0^1 (e^t + e^{-t})dt = e - e^{-1}$

27. $dx/dt = -a\sin t$, $dy/dt = a\cos t$, $dz/dt = c$,

$L = \displaystyle\int_0^{t_0} \sqrt{a^2\sin^2 t + a^2\cos^2 t + c^2}\,dt = \int_0^{t_0} \sqrt{a^2 + c^2}\,dt = t_0\sqrt{a^2 + c^2}$

29. $x = 3u - 2$, $y = 4u + 3$, $(dx/du)^2 + (dy/du)^2 = 25$,

$s = \displaystyle\int_0^t 5du = 5t$ so $t = s/5$, $x = (3/5)s - 2$, $y = (4/5)s + 3$.

31. $x = 3 + \cos u$, $y = 2 + \sin u$, $(dx/du)^2 + (dy/du)^2 = 1$,

$s = \displaystyle\int_0^t du = t$ so $t = s$, $x = 3 + \cos s$, $y = 2 + \sin s$ for $0 \le s \le 2\pi$.

33. $x = u^3/3$, $y = u^2/2$, $(dx/du)^2 + (dy/du)^2 = u^2(u^2 + 1)$,

$s = \displaystyle\int_0^t u(u^2 + 1)^{1/2}du = \frac{1}{3}[(t^2 + 1)^{3/2} - 1]$ so $t = [(3s + 1)^{2/3} - 1]^{1/2}$,

$x = \frac{1}{3}[(3s + 1)^{2/3} - 1]^{3/2}$, $y = \frac{1}{2}[(3s + 1)^{2/3} - 1]$ for $s \ge 0$.

35. $x = e^u \cos u$, $y = e^u \sin u$, $(dx/du)^2 + (dy/du)^2 = 2e^{2u}$, $s = \int_0^t \sqrt{2}\, e^u\, du = \sqrt{2}(e^t - 1)$ so
$t = \ln(s/\sqrt{2} + 1)$, $x = (s/\sqrt{2} + 1)\cos[\ln(s/\sqrt{2} + 1)]$, $y = (s/\sqrt{2} + 1)\sin[\ln(s/\sqrt{2} + 1)]$
for $0 \le s \le \sqrt{2}(e^{\pi/2} - 1)$.

37. $x = u \cos u$, $y = u \sin u$, $z = \dfrac{2}{3}\sqrt{2}\, u^{3/2}$,

$(dx/du)^2 + (dy/du)^2 + (dz/du)^2 = u^2 + 2u + 1 = (u + 1)^2$, $s = \int_0^t (u + 1)\,du = \dfrac{1}{2}t^2 + t$ so
$t = \sqrt{2s + 1} - 1$, $x = (\sqrt{2s + 1} - 1)\cos(\sqrt{2s + 1} - 1)$, $y = (\sqrt{2s + 1} - 1)\sin(\sqrt{2s + 1} - 1)$,
$z = \dfrac{2}{3}\sqrt{2}[\sqrt{2s + 1} - 1]^{3/2}$ for $s \ge 0$.

39. $x = a \cos u$, $y = a \sin u$, $z = cu$, $(dx/du)^2 + (dy/du)^2 + (dz/du)^2 = a^2 + c^2 = w^2$,
$s = \int_0^t w\, du = wt$ so $t = s/w$; $x = a\cos(s/w)$, $y = a\sin(s/w)$, $z = cs/w$ for $s \ge 0$.

41. **(a)** $x = OB + CP = a\cos\theta + a\theta\sin\theta$,
$y = AB - AC = a\sin\theta - a\theta\cos\theta$.

(b) $x = a(\cos u + u\sin u)$, $y = a(\sin u - u\cos u)$,
$dx/du = au\cos u$, $dy/du = au\sin u$,
$(dx/dy)^2 + (dy/du)^2 = a^2u^2$,
$s = \int_0^\theta au\, du = \dfrac{1}{2}a\theta^2$ so $\theta = \sqrt{2s/a}$,
$x = a(\cos\sqrt{2s/a} + \sqrt{2s/a}\sin\sqrt{2s/a})$,
$y = a(\sin\sqrt{2s/a} - \sqrt{2s/a}\cos\sqrt{2s/a})$ for $s \ge 0$.

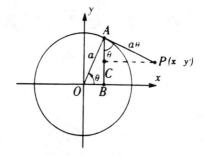

43. **(a)** $(dr/dt)^2 + r^2(d\theta/dt)^2 + (dz/dt)^2 = 9e^{4t}$, $L = \int_0^{\ln 2} 3e^{2t}dt = \dfrac{3}{2}e^{2t}\Big]_0^{\ln 2} = 9/2$.

(b) $(dr/dt)^2 + r^2(d\theta/dt)^2 + (dz/dt)^2 = 5t^2 + t^4 = t^2(5 + t^2)$,
$L = \int_1^2 t(5 + t^2)^{1/2}dt = 9 - 2\sqrt{6}$.

45. **(a)** $(d\rho/dt)^2 + \rho^2\sin^2\phi(d\theta/dt)^2 + \rho^2(d\phi/dt)^2 = 3e^{-2t}$, $L = \int_0^2 \sqrt{3}e^{-t}dt = \sqrt{3}(1 - e^{-2})$.

(b) $(d\rho/dt)^2 + \rho^2\sin^2\phi(d\theta/dt)^2 + \rho^2(d\phi/dt)^2 = 5$, $L = \int_1^5 \sqrt{5}dt = 4\sqrt{5}$.

47. Let $\mathbf{R}(t) = X(t)\mathbf{i} + Y(t)\mathbf{j}$ and $\mathbf{r}(t) = x(t)\mathbf{i} + y(t)\mathbf{j}$ where $X'(t) = x(t)$; $Y'(t) = y(t)$ and use Definition 15.3.1 and properties of indefinite and definite integrals.

EXERCISE SET 15.4

1. $\mathbf{r}'(t) = -5\sin t\mathbf{i} + 5\cos t\mathbf{j}$, $\|\mathbf{r}'(t)\| = 5$

$\mathbf{T}(t) = -\sin t\mathbf{i} + \cos t\mathbf{j}$, $\mathbf{T}'(t) = -\cos t\mathbf{i} - \sin t\mathbf{j}$;

$\mathbf{T}(\pi/3) = -\dfrac{\sqrt{3}}{2}\mathbf{i} + \dfrac{1}{2}\mathbf{j}$, $\mathbf{T}'(\pi/3) = -\dfrac{1}{2}\mathbf{i} - \dfrac{\sqrt{3}}{2}\mathbf{j}$, $\mathbf{N}(\pi/3) = -\dfrac{1}{2}\mathbf{i} - \dfrac{\sqrt{3}}{2}\mathbf{j}$.

3. $\mathbf{r}'(t) = 2t\mathbf{i} + \mathbf{j}$, $\|\mathbf{r}'(t)\| = \sqrt{4t^2 + 1}$, $\mathbf{T}(t) = (4t^2 + 1)^{-1/2}(2t\mathbf{i} + \mathbf{j})$,

$\mathbf{T}'(t) = (4t^2 + 1)^{-1/2}(2\mathbf{i}) - 4t(4t^2 + 1)^{-3/2}(2t\mathbf{i} + \mathbf{j})$;

$\mathbf{T}(1) = \dfrac{2}{\sqrt{5}}\mathbf{i} + \dfrac{1}{\sqrt{5}}\mathbf{j}$, $\mathbf{T}'(1) = \dfrac{2}{5\sqrt{5}}(\mathbf{i} - 2\mathbf{j})$, $\mathbf{N}(1) = \dfrac{1}{\sqrt{5}}\mathbf{i} - \dfrac{2}{\sqrt{5}}\mathbf{j}$.

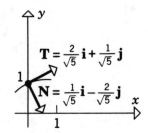

5. $\mathbf{r}'(t) = t\mathbf{i} + t^2\mathbf{j}$, $\mathbf{T}(t) = (t^2 + t^4)^{-1/2}(t\mathbf{i} + t^2\mathbf{j})$,

$\mathbf{T}'(t) = (t^2 + t^4)^{-1/2}(\mathbf{i} + 2t\mathbf{j}) - (t + 2t^3)(t^2 + t^4)^{-3/2}(t\mathbf{i} + t^2\mathbf{j})$;

$\mathbf{T}(1) = \dfrac{1}{\sqrt{2}}\mathbf{i} + \dfrac{1}{\sqrt{2}}\mathbf{j}$, $\mathbf{T}'(1) = \dfrac{1}{2\sqrt{2}}(-\mathbf{i} + \mathbf{j})$, $\mathbf{N}(1) = -\dfrac{1}{\sqrt{2}}\mathbf{i} + \dfrac{1}{\sqrt{2}}\mathbf{j}$.

7. $r'(t) = -4\sin t\mathbf{i} + 9\cos t\mathbf{j}$, $T(t) = (16\sin^2 t + 81\cos^2 t)^{-1/2}(-4\sin t\mathbf{i} + 9\cos t\mathbf{j})$,

$T'(t) = (16\sin^2 t + 81\cos^2 t)^{-1/2}(-4\cos t\mathbf{i} - 9\sin t\mathbf{j})$

$\qquad\qquad + 65\sin t\cos t(16\sin^2 t + 81\cos^2 t)^{-3/2}(-4\sin t\mathbf{i} + 9\cos t\mathbf{j})$;

$T(\pi/4) = -\dfrac{4}{\sqrt{97}}\mathbf{i} + \dfrac{9}{\sqrt{97}}\mathbf{j}$, $T'(\pi/4) = \dfrac{72}{97\sqrt{97}}(-9\mathbf{i} - 4\mathbf{j})$, $N(\pi/4) = -\dfrac{9}{\sqrt{97}}\mathbf{i} - \dfrac{4}{\sqrt{97}}\mathbf{j}$.

9. $r'(t) = -4\sin t\mathbf{i} + 4\cos t\mathbf{j} + \mathbf{k}$, $T(t) = \dfrac{1}{\sqrt{17}}(-4\sin t\mathbf{i} + 4\cos t\mathbf{j} + \mathbf{k})$,

$T'(t) = \dfrac{1}{\sqrt{17}}(-4\cos t\mathbf{i} - 4\sin t\mathbf{j})$, $T(\pi/2) = -\dfrac{4}{\sqrt{17}}\mathbf{i} + \dfrac{1}{\sqrt{17}}\mathbf{k}$

$T'(\pi/2) = -\dfrac{4}{\sqrt{17}}\mathbf{j}$, $N(\pi/2) = -\mathbf{j}$.

11. $r'(t) = \mathbf{i} + t\mathbf{j} + t^2\mathbf{k}$, $T(t) = (1 + t^2 + t^4)^{-1/2}(\mathbf{i} + t\mathbf{j} + t^2\mathbf{k})$,

$T'(t) = (1 + t^2 + t^4)^{-1/2}(\mathbf{j} + 2t\mathbf{k}) - (t + 2t^3)(1 + t^2 + t^4)^{-3/2}(\mathbf{i} + t\mathbf{j} + t^2\mathbf{k})$,

$T(0) = \mathbf{i}$, $T'(0) = \mathbf{j} = N(0)$.

13. $r'(t) = -3\sin t\mathbf{i} + 4\cos t\mathbf{j} + \mathbf{k}$, $T(t) = (9\sin^2 t + 16\cos^2 t + 1)^{-1/2}(-3\sin t\mathbf{i} + 4\cos t\mathbf{j} + \mathbf{k})$,

$T'(t) = (9\sin^2 t + 16\cos^2 t + 1)^{-1/2}(-3\cos t\mathbf{i} - 4\sin t\mathbf{j})$

$\qquad\qquad + 7\sin t\cos t(9\sin^2 t + 16\cos^2 +1)^{-3/2}(-3\sin t\mathbf{i} + 4\cos t\mathbf{j} + \mathbf{k})$,

$T(\pi/2) = -\dfrac{3}{\sqrt{10}}\mathbf{i} + \dfrac{1}{\sqrt{10}}\mathbf{k}$, $T'(\pi/2) = -\dfrac{4}{\sqrt{10}}\mathbf{j}$, $N(\pi/2) = -\mathbf{j}$.

15. $r'(t) = \mathbf{j} + 2t\mathbf{k}$, $T(t) = (1 + 4t^2)^{-1/2}(\mathbf{j} + 2t\mathbf{k})$,

$T'(t) = (1 + 4t^2)^{-1/2}(2\mathbf{k}) - 4t(1 + 4t^2)^{-3/2}(\mathbf{j} + 2t\mathbf{k})$,

$T(1) = \dfrac{1}{\sqrt{5}}\mathbf{j} + \dfrac{2}{\sqrt{5}}\mathbf{k}$, $T'(1) = \dfrac{2}{5\sqrt{5}}(-2\mathbf{j} + \mathbf{k})$, $N(1) = -\dfrac{2}{\sqrt{5}}\mathbf{j} + \dfrac{1}{\sqrt{5}}\mathbf{k}$.

17. $r(t) = a\cos t\mathbf{i} + a\sin t\mathbf{j} + ct\mathbf{k}$ where $a\cos t\mathbf{i} + a\sin t\mathbf{j}$ points from the z-axis to a point on the curve, but N is oppositely directed so N points directly toward the z-axis.

19. $\mathbf{r}'(t) = -5\sin t\mathbf{i} + 5\cos t\mathbf{j}$, $\mathbf{r}''(t) = -5\cos t\mathbf{i} - 5\sin t\mathbf{j}$,

$\mathbf{r}'(\pi/3) = -\dfrac{5\sqrt{3}}{2}\mathbf{i} + \dfrac{5}{2}\mathbf{j}$, $\mathbf{r}''(\pi/3) = -\dfrac{5}{2}\mathbf{i} - \dfrac{5\sqrt{3}}{2}\mathbf{j}$,

$\mathbf{u} = 25(-5/2\mathbf{i} - 5\sqrt{3}/2\mathbf{j}) - (0)\mathbf{r}' = \dfrac{125}{2}(-\mathbf{i} - \sqrt{3}\mathbf{j})$; $\mathbf{N} = -(1/2)\mathbf{i} - (\sqrt{3}/2)\mathbf{j}$.

21. $\mathbf{r}'(t) = 2t\mathbf{i} + \mathbf{j}$, $\mathbf{r}''(t) = 2\mathbf{i}$, $\mathbf{r}'(1) = 2\mathbf{i} + \mathbf{j}$, $\mathbf{r}''(1) = 2\mathbf{i}$,
$\mathbf{u} = 5(2\mathbf{i}) - (4)(2\mathbf{i} + \mathbf{j}) = 2(\mathbf{i} - 2\mathbf{j})$; $\mathbf{N} = (1/\sqrt{5})\mathbf{i} - (2/\sqrt{5})\mathbf{j}$.

23. $\mathbf{r}'(t) = -4\sin t\mathbf{i} + 4\cos t\mathbf{j} + \mathbf{k}$, $\mathbf{r}''(t) = -4\cos t\mathbf{i} - 4\sin t\mathbf{j}$,
$\mathbf{r}'(\pi/2) = -4\mathbf{i} + \mathbf{k}$, $\mathbf{r}''(\pi/2) = -4\mathbf{j}$, $\mathbf{u} = 17(-4\mathbf{j}) - (0)\mathbf{r}' = -68\mathbf{j}$; $\mathbf{N} = -\mathbf{j}$.

25. $\mathbf{r}'(t) = \mathbf{i} + t\mathbf{j} + t^2\mathbf{k}$, $\mathbf{r}''(t) = \mathbf{j} + 2t\mathbf{k}$, $\mathbf{r}'(0) = \mathbf{i}$, $\mathbf{r}''(0) = \mathbf{j}$, $\mathbf{u} = (1)(\mathbf{j}) - (0)\mathbf{r}' = \mathbf{j} = \mathbf{N}$.

27. $\mathbf{T} = \dfrac{3}{5}\cos t\mathbf{i} - \dfrac{3}{5}\sin t\mathbf{j} + \dfrac{4}{5}\mathbf{k}$, $\mathbf{N} = -\sin t\mathbf{i} - \cos t\mathbf{j}$, $\mathbf{B} = \mathbf{T} \times \mathbf{N} = \dfrac{4}{5}\cos t\mathbf{i} - \dfrac{4}{5}\sin t\mathbf{j} - \dfrac{3}{5}\mathbf{k}$

29. (a) $\mathbf{n}$ is perpendicular to S; if $\mathbf{r} = \mathbf{r}(t)$ lies in S, then $\mathbf{r} - \mathbf{r}_0$ is perpendicular to $\mathbf{n}$ so
$\mathbf{n} \cdot (\mathbf{r} - \mathbf{r}_0) = 0$. If $\mathbf{n} \cdot (\mathbf{r} - \mathbf{r}_0) = 0$ then either $\mathbf{r} = \mathbf{r}_0$ or $\mathbf{r} - \mathbf{r}_0$ is perpendicular to $\mathbf{n}$ so
$\mathbf{r} = \mathbf{r}(t)$ lies in S.

   (b) Differentiate $\mathbf{n} \cdot (\mathbf{r} - \mathbf{r}_0)$ twice with respect to t to get $\mathbf{n} \cdot \mathbf{r}' = 0$ and $\mathbf{n} \cdot \mathbf{r}'' = 0$ so both
$\mathbf{r}'$ and $\mathbf{r}''$ are perpendicular to $\mathbf{n}$ and thus in S, $\mathbf{T}$ is in the direction of $\mathbf{r}'$ so $\mathbf{T}$ is in S.
From Exercise 18, $\mathbf{N}$ is in the direction of $\mathbf{u}$, but $\mathbf{n} \cdot \mathbf{u} = \|\mathbf{r}'\|^2\mathbf{n} \cdot \mathbf{r}'' - (\mathbf{r}' \cdot \mathbf{r}'')\mathbf{n} \cdot \mathbf{r}' = 0$
so $\mathbf{u}$, and hence $\mathbf{N}$, are perpendicular to $\mathbf{n}$ and thus in S.

EXERCISE SET 15.5

1. $\mathbf{r}'(t) = 2t\mathbf{i} + 3t^2\mathbf{j}$, $\mathbf{r}''(t) = 2\mathbf{i} + 6t\mathbf{j}$, $\mathbf{r}'(1/2) = \mathbf{i} + 3/4\mathbf{j}$,
$\mathbf{r}''(1/2) = 2\mathbf{i} + 3\mathbf{j}$; $\kappa = \|3/2\mathbf{k}\|/\|\mathbf{i} + 3/4\mathbf{j}\|^3 = 96/125$.

3. $\mathbf{r}'(t) = 3e^{3t}\mathbf{i} - e^{-t}\mathbf{j}$, $\mathbf{r}''(t) = 9e^{3t}\mathbf{i} + e^{-t}\mathbf{j}$,
$\mathbf{r}'(0) = 3\mathbf{i} - \mathbf{j}$, $\mathbf{r}''(0) = 9\mathbf{i} + \mathbf{j}$; $\kappa = \|12\mathbf{k}\|/\|3\mathbf{i} - \mathbf{j}\|^3 = 6/(5\sqrt{10})$.

5. $\mathbf{r}'(t) = (\cos t - t\sin t)\mathbf{i} + (\sin t + t\cos t)\mathbf{j}$, $\mathbf{r}''(t) = -(2\sin t + t\cos t)\mathbf{i} + (2\cos t - t\sin t)\mathbf{j}$,
$\|\mathbf{r}'(t) \times \mathbf{r}''(t)\| = 2 + t^2$, $\|\mathbf{r}'(t)\| = (1 + t^2)^{1/2}$; $\kappa(1) = 3/(2\sqrt{2})$.

7. $\mathbf{r}'(t) = -4\sin t\mathbf{i} + 4\cos t\mathbf{j} + \mathbf{k}$, $\mathbf{r}''(t) = -4\cos t\mathbf{i} - 4\sin t\mathbf{j}$,
$\mathbf{r}'(\pi/2) = -4\mathbf{i} + \mathbf{k}$, $\mathbf{r}''(\pi/2) = -4\mathbf{j}$; $\kappa = \|4\mathbf{i} + 16\mathbf{k}\|/\| - 4\mathbf{i} + \mathbf{k}\|^3 = 4/17$.

9. $\mathbf{r}'(t) = \mathbf{i} + t\mathbf{j} + t^2\mathbf{k}$, $\mathbf{r}''(t) = \mathbf{j} + 2t\mathbf{k}$, $\mathbf{r}'(0) = \mathbf{i}$, $\mathbf{r}''(0) = \mathbf{j}$; $\kappa = \|\mathbf{k}\|/\|\mathbf{i}\|^3 = 1$.

11. $\mathbf{r}'(t) = -3\sin t\mathbf{i} + 4\cos t\mathbf{j} + \mathbf{k}$, $\mathbf{r}''(t) = -3\cos t\mathbf{i} - 4\sin t\mathbf{j}$,
    $\mathbf{r}'(\pi/2) = -3\mathbf{i} + \mathbf{k}$, $\mathbf{r}''(\pi/2) = -4\mathbf{j}$; $\kappa = \|4\mathbf{i} + 12\mathbf{k}\|/\|-3\mathbf{i} + \mathbf{k}\|^3 = 2/5$.

13. $\mathbf{r}'(t) = \mathbf{j} + 2t\mathbf{k}$, $\mathbf{r}''(t) = 2\mathbf{k}$, $\mathbf{r}'(1) = \mathbf{j} + 2\mathbf{k}$, $\mathbf{r}''(1) = 2\mathbf{k}$; $\kappa = \|2\mathbf{i}\|/\|\mathbf{j} + 2\mathbf{k}\|^3 = 2/(5\sqrt{5})$.

15. $\mathbf{r}'(x) = \mathbf{i} + (dy/dx)\mathbf{j}$, $\mathbf{r}''(x) = (d^2y/dx^2)\mathbf{j}$;
    $\kappa(x) = \|(d^2y/dx^2)\mathbf{k}\|/\|\mathbf{i} + (dy/dx)\mathbf{j}\|^3 = |d^2y/dx^2|/[1 + (dy/dx)^2]^{3/2}$.

17. $\kappa(x) = \dfrac{|\sin x|}{(1 + \cos^2 x)^{3/2}}$, $\kappa(\pi/2) = 1$ 19. $\kappa(x) = \dfrac{2|x|^3}{(x^4 + 1)^{3/2}}$, $\kappa(1) = 1/\sqrt{2}$

21. $\kappa(x) = \dfrac{2\sec^2 x|\tan x|}{(1 + \sec^4 x)^{3/2}}$, $\kappa(\pi/4) = 4/(5\sqrt{5})$

23. $\kappa(x) = \dfrac{\sec^2 x}{(1 + \tan^2 x)^{3/2}} = \cos x$; $\kappa(x)$ is maximum for $x = 0$.

25. $x'(t) = 2t$, $y'(t) = 3t^2$, $x''(t) = 2$, $y''(t) = 6t$,
    $x'(1/2) = 1$, $y'(1/2) = 3/4$, $x''(1/2) = 2$, $y''(1/2) = 3$; $\kappa = 96/125$.

27. $x'(t) = 3e^{3t}$, $y'(t) = -e^{-t}$, $x''(t) = 9e^{3t}$, $y''(t) = e^{-t}$,
    $x'(0) = 3$, $y'(0) = -1$, $x''(0) = 9$, $y''(0) = 1$; $\kappa = 6/(5\sqrt{10})$.

29. $x'(t) = \cos t - t\sin t$, $y'(t) = \sin t + t\cos t$,
    $x''(t) = -2\sin t - t\cos t$, $y''(t) = 2\cos t - t\sin t$,
    $x'y'' - y'x'' = 2 + t^2$, $x'^2 + y'^2 = 1 + t^2$; $\kappa(1) = 3/(2\sqrt{2})$.

31. $x'(t) = -a\sin t$, $y'(t) = b\cos t$, $x''(t) = -a\cos t$, $y''(t) = -b\sin t$;
    $\kappa(0) = a/b^2$, $\kappa(\pi/2) = b/a^2$.

33. $\kappa(\theta) = 1$, $\kappa(\pi/6) = 1$

35. $\kappa(\theta) = \dfrac{3}{2\sqrt{2}a(1 + \cos\theta)^{1/2}}$, $\kappa(\pi/2) = \dfrac{3}{2\sqrt{2}a}$

37. $x'(t) = a(1 - \cos t)$, $y'(t) = a \sin t$,
$x''(t) = a \sin t$, $y''(t) = a \cos t$;

$$\kappa(t) = \frac{a^2(1 - \cos t)}{[2a^2(1 - \cos t)]^{3/2}}$$
$$= \frac{1}{2\sqrt{2}a(1 - \cos t)^{1/2}},$$

but $1 - \cos t = 2 \sin^2(t/2)$ so

$$\kappa(t) = \frac{1}{4a \sin(t/2)} = \frac{1}{4a} \csc(t/2).$$

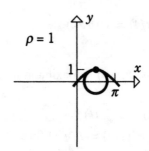

39. $\kappa(x) = \frac{|\sin x|}{(1 + \cos^2 x)^{3/2}}$,

$\kappa(\pi/2) = 1$, $\rho = 1/\kappa = 1$

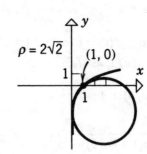

41. $\kappa(x) = (1 + x^2)^{-3/2}$,
$\kappa(-1) = 2^{-3/2}, \rho = 2^{3/2} = 2\sqrt{2}$

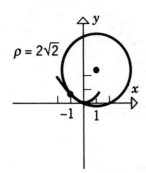

43. $\kappa(x) = \frac{|x|}{(x^2 + 1)^{3/2}}$,

$\kappa(1) = 2^{-3/2}, \rho = 2^{3/2} = 2\sqrt{2}$

45. $\kappa(t) = \dfrac{|\cos t - 1|}{(2 - 2\cos t)^{3/2}}$,

$\kappa(\pi) = 1/4, \rho = 4$

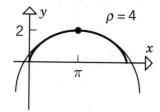

47. $\kappa(t) = \dfrac{2}{(4\sin^2 t + \cos^2 t)^{3/2}}$,

$\rho(t) = \dfrac{1}{2}(4\sin^2 t + \cos^2 t)^{3/2}$,

$\rho(0) = 1/2, \; \rho(\pi/2) = 4$

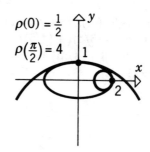

49. Let $y = t$, then $x = \dfrac{t^2}{4p}$ and $\kappa(t) = \dfrac{1/|2p|}{[t^2/(4p^2) + 1]^{3/2}}$;

$t = 0$ when $(x, y) = (0, 0)$ so $\kappa(0) = 1/|2p|, \; \rho = 2|p|$.

51. Let $x = 3\cos t, \; y = 2\sin t$ for $0 \le t < 2\pi, \; \kappa(t) = \dfrac{6}{(9\sin^2 t + 4\cos^2 t)^{3/2}}$ so

$\rho(t) = \dfrac{1}{6}(9\sin^2 t + 4\cos^2 t)^{3/2} = \dfrac{1}{6}(5\sin^2 t + 4)^{3/2}$ which, by inspection, is minimum when

$t = 0$ or π. The radius of curvature is minimum at $(3, 0)$ and $(-3, 0)$.

53. $\mathbf{r}'(t) = -\sin t\,\mathbf{i} + \cos t\,\mathbf{j} - \sin t\,\mathbf{k}, \; \mathbf{r}''(t) = -\cos t\,\mathbf{i} - \sin t\,\mathbf{j} - \cos t\,\mathbf{k}$,

$\|\mathbf{r}'(t) \times \mathbf{r}''(t)\| = \| - \mathbf{i} + \mathbf{k}\| = \sqrt{2}, \; \|\mathbf{r}'(t)\| = (1 + \sin^2 t)^{1/2}; \; \kappa(t) = \sqrt{2}/(1 + \sin^2 t)^{3/2}$,

$\rho(t) = (1 + \sin^2 t)^{3/2}/\sqrt{2}$. The minimum value of ρ is $1/\sqrt{2}$; the maximum value is 2.

55. From Exercise 32: $dr/d\theta = ae^{a\theta} = ar, \; d^2r/d\theta^2 = a^2 e^{a\theta} = a^2 r; \; \kappa = 1/[\sqrt{1 + a^2}\, r]$.

57. $|d\phi/ds| = \kappa = \|d\mathbf{T}/ds\| = 0.05$ radians/cm $= (9/\pi)^\circ/$cm $\approx 2.86^\circ/$cm.

59. $\kappa = 0$ along $y = 0$; along $y = x^2, \; \kappa(x) = 2/(1 + 4x^2)^{3/2}, \; \kappa(0) = 2$. Along $y = x^3$,

$\kappa(x) = 6|x|/(1 + 9x^4)^{3/2}, \; \kappa(0) = 0$.

61. $\kappa = 1/r$ along the circle; along $y = ax^2, \; \kappa(x) = 2a/(1 + 4a^2 x^2)^{3/2}, \; \kappa(0) = 2a$ so $2a = 1/r$,

$a = 1/(2r)$.

63. $\mathbf{r}'(t) = (1/t)\mathbf{i} + 2\mathbf{j} + 2t\mathbf{k}$

    (a) $\|\mathbf{r}'(t)\| = \sqrt{1/t^2 + 4 + 4t^2} = \sqrt{(2t + 1/t)^2} = 2t + 1/t$

    (b) $\dfrac{ds}{dt} = 2t + 1/t$ **(c)** $\displaystyle\int_1^3 (2t + 1/t)\,dt = 8 + \ln 3$

65. **(a)** $\mathbf{r}(s) = (3s/5 + 1)\mathbf{i} + (4s/5 - 2)\mathbf{j},$
            $\mathbf{T} = d\mathbf{r}/ds = (3/5)\mathbf{i} + (4/5)\mathbf{j}$

     (b) $\mathbf{T}(5) = (3/5)\mathbf{i} + (4/5)\mathbf{j}$
          $\mathbf{r}(5) = 4\mathbf{i} + 2\mathbf{j}$

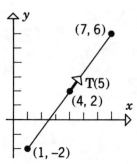

67. **(a)** $\kappa = \dfrac{d\phi}{ds} = as,\ \phi = \dfrac{1}{2}as^2 + C,$ but $\phi = 0$ when $s = 0$ so $C = 0,\ \phi = \dfrac{1}{2}as^2.$

     (b) $\mathbf{T} = \dfrac{d\mathbf{r}}{ds} = \dfrac{dx}{ds}\mathbf{i} + \dfrac{dy}{ds}\mathbf{j}$ so $\dfrac{dx}{ds} = \cos\left(\tfrac{1}{2}as^2\right)$ and $\dfrac{dy}{ds} = \sin\left(\tfrac{1}{2}as^2\right),\ x = \displaystyle\int_0^s \cos\left(\tfrac{1}{2}au^2\right)du$ and
     $y = \displaystyle\int_0^s \left(\sin\tfrac{1}{2}au^2\right)du.$

69. $\dfrac{d\mathbf{N}}{ds} = \mathbf{B} \times \dfrac{d\mathbf{T}}{ds} + \dfrac{d\mathbf{B}}{ds} \times \mathbf{T} = \mathbf{B} \times (\kappa\mathbf{N}) + (-\tau\mathbf{N}) \times \mathbf{T} = \kappa\mathbf{B} \times \mathbf{N} - \tau\mathbf{N} \times \mathbf{T},$ but $\mathbf{B} \times \mathbf{N} = -\mathbf{T}$
    and $\mathbf{N} \times \mathbf{T} = -\mathbf{B}$ so $\dfrac{d\mathbf{N}}{ds} = -\kappa\mathbf{T} + \tau\mathbf{B}.$

71. $\mathbf{r} = a\cos(s/w)\mathbf{i} + a\sin(s/w)\mathbf{j} + (cs/w)\mathbf{k},\ \mathbf{r}' = -(a/w)\sin(s/w)\mathbf{i} + (a/w)\cos(s/w)\mathbf{j} + (c/w)\mathbf{k},$
    $\mathbf{r}'' = -(a/w^2)\cos(s/w)\mathbf{i} - (a/w^2)\sin(s/w)\mathbf{j},\ \mathbf{r}''' = (a/w^3)\sin(s/w)\mathbf{i} - (a/w^3)\cos(s/w)\mathbf{j},$
    $\mathbf{r}' \times \mathbf{r}'' = (ac/w^3)\sin(s/w)\mathbf{i} - (ac/w^3)\cos(s/w)\mathbf{j} + (a^2/w^3)\mathbf{k},\ (\mathbf{r}' \times \mathbf{r}'') \cdot \mathbf{r}''' = a^2c/w^6,$
    $\|\mathbf{r}''(s)\| = a/w^2,$ so $\tau = c/w^2$ and $\mathbf{B} = (c/w)\sin(s/w)\mathbf{i} - (c/w)\cos(s/w)\mathbf{j} + (a/w)\mathbf{k}.$

73. $\mathbf{r}' = 2\mathbf{i} + 2t\mathbf{j} + t^2\mathbf{k},\ \mathbf{r}'' = 2\mathbf{j} + 2t\mathbf{k},\ \mathbf{r}''' = 2\mathbf{k},\ \mathbf{r}' \times \mathbf{r}'' = 2t^2\mathbf{i} - 4t\mathbf{j} + 4\mathbf{k},\ \|\mathbf{r}' \times \mathbf{r}''\| = 2(t^2 + 2),$
    $\tau = 8/[2(t^2 + 2)]^2 = 2/(t^2 + 2)^2.$

75. $\mathbf{r}' = e^t\mathbf{i} - e^{-t}\mathbf{j} + \sqrt{2}\mathbf{k},\ \mathbf{r}'' = e^t\mathbf{i} + e^{-t}\mathbf{j},\ \mathbf{r}''' = e^t\mathbf{i} - e^{-t}\mathbf{j},\ \mathbf{r}' \times \mathbf{r}'' = -\sqrt{2}e^{-t}\mathbf{i} + \sqrt{2}e^t\mathbf{j} + 2\mathbf{k},$
    $\|\mathbf{r}' \times \mathbf{r}''\| = \sqrt{2}(e^t + e^{-t}),\ \tau = (-2\sqrt{2})/[2(e^t + e^{-t})^2] = -\sqrt{2}/(e^t + e^{-t})^2.$

EXERCISE SET 15.6

1. $\mathbf{v}(t) = -3\sin t\mathbf{i} + 3\cos t\mathbf{j}$
    $\mathbf{a}(t) = -3\cos t\mathbf{i} - 3\sin t\mathbf{j}$
    $\|\mathbf{v}(t)\| = \sqrt{9\sin^2 t + 9\cos^2 t} = 3$
    $\mathbf{r}(\pi/3) = (3/2)\mathbf{i} + (3\sqrt{3}/2)\mathbf{j}$
    $\mathbf{v}(\pi/3) = -(3\sqrt{3}/2)\mathbf{i} + (3/2)\mathbf{j}$
    $\mathbf{a}(\pi/3) = -(3/2)\mathbf{i} - (3\sqrt{3}/2)\mathbf{j}$

    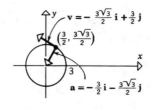

3. $\mathbf{v}(t) = e^t\mathbf{i} - e^{-t}\mathbf{j}$
    $\mathbf{a}(t) = e^t\mathbf{i} + e^{-t}\mathbf{j}$
    $\|\mathbf{v}(t)\| = \sqrt{e^{2t} + e^{-2t}}$
    $\mathbf{r}(0) = \mathbf{i} + \mathbf{j}$
    $\mathbf{v}(0) = \mathbf{i} - \mathbf{j}$
    $\mathbf{a}(0) = \mathbf{i} + \mathbf{j}$

    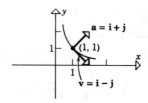

5. $\mathbf{v}(t) = \sinh t\mathbf{i} + \cosh t\mathbf{j}$
    $\mathbf{a}(t) = \cosh t\mathbf{i} + \sinh t\mathbf{j}$
    $\|\mathbf{v}(t)\| = \sqrt{\sinh^2 t + \cosh^2 t}$
    $\mathbf{r}(\ln 2) = (5/4)\mathbf{i} + (3/4)\mathbf{j}$
    $\mathbf{v}(\ln 2) = (3/4)\mathbf{i} + (5/4)\mathbf{j}$
    $\mathbf{a}(\ln 2) = (5/4)\mathbf{i} + (3/4)\mathbf{j}$

    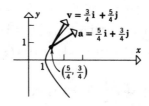

7. $\mathbf{v} = \mathbf{i} + t\mathbf{j} + t^2\mathbf{k}$, $\mathbf{a} = \mathbf{j} + 2t\mathbf{k}$; at $t = 1$, $\mathbf{v} = \mathbf{i} + \mathbf{j} + \mathbf{k}$, $\|\mathbf{v}\| = \sqrt{3}$, $\mathbf{a} = \mathbf{j} + 2\mathbf{k}$

9. $\mathbf{v} = -2\sin t\mathbf{i} + 2\cos t\mathbf{j} + \mathbf{k}$, $\mathbf{a} = -2\cos t\mathbf{i} - 2\sin t\mathbf{j}$;
    at $t = \pi/4$, $\mathbf{v} = -\sqrt{2}\mathbf{i} + \sqrt{2}\mathbf{j} + \mathbf{k}$, $\|\mathbf{v}\| = \sqrt{5}$, $\mathbf{a} = -\sqrt{2}\mathbf{i} - \sqrt{2}\mathbf{j}$

11. $\mathbf{v} = e^t(\cos t + \sin t)\mathbf{i} + e^t(\cos t - \sin t)\mathbf{j} + \mathbf{k}$, $\mathbf{a} = 2e^t\cos t\mathbf{i} - 2e^t\sin t\mathbf{j}$; at $t = \pi/2$,
    $\mathbf{v} = e^{\pi/2}\mathbf{i} - e^{\pi/2}\mathbf{j} + \mathbf{k}$, $\|\mathbf{v}\| = (1 + 2e^\pi)^{1/2}$, $\mathbf{a} = -2e^{\pi/2}\mathbf{j}$

13. $\mathbf{v} = (6/\sqrt{t})\mathbf{i} + (3/2)t^{1/2}\mathbf{j}$, $\|\mathbf{v}\| = \sqrt{36/t + 9t/4}$, $d\|\mathbf{v}\|/dt = (-36/t^2 + 9/4)/(2\sqrt{36/t + 9t/4}) = 0$ if $t = 4$ which yields a minimum by the first derivative test. The minimum speed is $3\sqrt{2}$ when $\mathbf{r} = 24\mathbf{i} + 8\mathbf{j}$.

15. $\mathbf{v} = 3\cos 3t\mathbf{i} + 6\sin 3t\mathbf{j}$, $\|\mathbf{v}\| = \sqrt{9\cos^2 3t + 36\sin^2 3t} = 3\sqrt{1 + 3\sin^2 3t}$; the maximum speed is 6, the minimum speed is 3.

17. $\mathbf{v} = 3t^2\mathbf{i} + 2t\mathbf{j}$, $\mathbf{a} = 6t\mathbf{i} + 2\mathbf{j}$; $\mathbf{v} = 3\mathbf{i} + 2\mathbf{j}$ and $\mathbf{a} = 6\mathbf{i} + 2\mathbf{j}$ when $t = 1$ so $\cos\theta = (\mathbf{v} \cdot \mathbf{a})/(\|\mathbf{v}\| \|\mathbf{a}\|) = 11/\sqrt{130}$, $\theta \approx 15°$.

19. $\mathbf{v} = (2t - 5)\mathbf{i} + 2\mathbf{j} + 6t\mathbf{k}$, $\mathbf{a} = 2\mathbf{i} + 6\mathbf{k}$;
    $\mathbf{v} \cdot \mathbf{a} = 40t - 10 = 0$ if $t = 1/4$, $\mathbf{r} = -19/16\mathbf{i} + 3/2\mathbf{j} + 3/16\mathbf{k}$.

21. In both cases, the equation of the path in rectangular coordinates is $x^2 + y^2 = 4$, the particles move counterclockwise around this circle; $\mathbf{v}_1 = -6\sin 3t\mathbf{i} + 6\cos 3t\mathbf{j}$ and $\mathbf{v}_2 = -4t\sin(t^2)\mathbf{i} + 4t\cos(t^2)\mathbf{j}$ so $\|\mathbf{v}_1\| = 6$ and $\|\mathbf{v}_2\| = 4t$.

23. $\mathbf{v}(t) = \int(-32\mathbf{j})dt = -32t\mathbf{j} + \mathbf{C}_1$, $\mathbf{v}(0) = \mathbf{C}_1 = 0$ so $\mathbf{v}(t) = -32t\mathbf{j}$;

    $\mathbf{r}(t) = \int(-32t\mathbf{j})dt = -16t^2\mathbf{j} + \mathbf{C}_2$, $\mathbf{r}(0) = \mathbf{C}_2 = 0$ so $\mathbf{r}(t) = -16t^2\mathbf{j}$

25. $\mathbf{v}(t) = -\sin t\mathbf{i} + \cos t\mathbf{j} + \mathbf{C}_1$, $\mathbf{v}(0) = \mathbf{j} + \mathbf{C}_1 = \mathbf{i}$, $\mathbf{C}_1 = \mathbf{i} - \mathbf{j}$, $\mathbf{v}(t) = (1 - \sin t)\mathbf{i} + (\cos t - 1)\mathbf{j}$;
    $\mathbf{r}(t) = (t + \cos t)\mathbf{i} + (\sin t - t)\mathbf{j} + \mathbf{C}_2$, $\mathbf{r}(0) = \mathbf{i} + \mathbf{C}_2 = \mathbf{j}$,
    $\mathbf{C}_2 = -\mathbf{i} + \mathbf{j}$ so $\mathbf{r}(t) = (t + \cos t - 1)\mathbf{i} + (\sin t - t + 1)\mathbf{j}$

27. $\mathbf{v}(t) = \int(\mathbf{i} + t\mathbf{k})dt = t\mathbf{i} + \frac{1}{2}t^2\mathbf{k} + \mathbf{C}_1$, $\mathbf{v}(0) = \mathbf{C}_1 = 0$ so

    $\mathbf{v}(t) = t\mathbf{i} + \frac{1}{2}t^2\mathbf{k}$; $\mathbf{r}(t) = \int\left(t\mathbf{i} + \frac{1}{2}t^2\mathbf{k}\right)dt = \frac{1}{2}t^2\mathbf{i} + \frac{1}{6}t^3\mathbf{k} + \mathbf{C}_2$,

    $\mathbf{r}(0) = \mathbf{C}_2 = \mathbf{j}$ so $\mathbf{r}(t) = \frac{1}{2}t^2\mathbf{i} + \mathbf{j} + \frac{1}{6}t^3\mathbf{k}$.

29. $\mathbf{v}(t) = -\cos t\mathbf{i} + \sin t\mathbf{j} + e^t\mathbf{k} + \mathbf{C}_1$, $\mathbf{v}(0) = -\mathbf{i} + \mathbf{k} + \mathbf{C}_1 = \mathbf{k}$ so
    $\mathbf{C}_1 = \mathbf{i}$, $\mathbf{v}(t) = (1 - \cos t)\mathbf{i} + \sin t\mathbf{j} + e^t\mathbf{k}$; $\mathbf{r}(t) = (t - \sin t)\mathbf{i} - \cos t\mathbf{j} + e^t\mathbf{k} + \mathbf{C}_2$,
    $\mathbf{r}(0) = -\mathbf{j} + \mathbf{k} + \mathbf{C}_2 = -\mathbf{i} + \mathbf{k}$ so $\mathbf{C}_2 = -\mathbf{i} + \mathbf{j}$, $\mathbf{r}(t) = (t - \sin t - 1)\mathbf{i} + (1 - \cos t)\mathbf{j} + e^t\mathbf{k}$.

31. $\mathbf{a} = \mathbf{r}''(t) = 0$, $\mathbf{r}'(t) = \langle b_1, b_2, b_3\rangle$, $\mathbf{r}(t) = \langle b_1t + c_1, b_2t + c_2, b_3t + c_3\rangle$ where $b_1, b_2, b_3, c_1, c_2,$ c_3 are constants so $x = b_1t + c_1$, $y = b_2t + c_2$, $z = b_3t + c_3$ which is a line.

33. $\Delta\mathbf{r} = \mathbf{r}(3) - \mathbf{r}(1) = 8\mathbf{i} + 26/3\mathbf{j}$; $\mathbf{v} = 2t\mathbf{i} + t^2\mathbf{j}$, $L = \int_1^3 t\sqrt{4 + t^2}\,dt = (13\sqrt{13} - 5\sqrt{5})/3$.

35. $\Delta\mathbf{r} = \mathbf{r}(2\pi) - \mathbf{r}(0) = \mathbf{0}$; $\mathbf{v} = 6\cos 3t\mathbf{i} - 6\sin 3t\mathbf{j}$, $L = \int_0^{2\pi} 6\,dt = 12\pi$.

37. $\Delta\mathbf{r} = \mathbf{r}(\ln 3) - \mathbf{r}(0) = 2\mathbf{i} - 2/3\mathbf{j} + \sqrt{2}\ln 3\mathbf{k}$; $\mathbf{v} = e^t\mathbf{i} - e^{-t}\mathbf{j} + \sqrt{2}\mathbf{k}$, $L = \int_0^{\ln 3}(e^t + e^{-t})\,dt = 8/3$.

39. $\mathbf{v} = -2\sin t\mathbf{i} + 2\cos t\mathbf{j}$, $\mathbf{a} = -2\cos t\mathbf{i} - 2\sin t\mathbf{j}$; when $t = \pi/3$, $\mathbf{v} = -\sqrt{3}\mathbf{i} + \mathbf{j}$, $\mathbf{a} = -\mathbf{i} - \sqrt{3}\mathbf{j}$, $\|\mathbf{v}\| = 2$, $\mathbf{v}\cdot\mathbf{a} = 0$, $\mathbf{v}\times\mathbf{a} = 4\mathbf{k}$ so $a_T = 0$, $a_N = 2$.

41. $\mathbf{v} = -e^{-t}\mathbf{i} + e^t\mathbf{j}$, $\mathbf{a} = e^{-t}\mathbf{i} + e^t\mathbf{j}$; when $t = 0$, $\mathbf{v} = -\mathbf{i} + \mathbf{j}$, $\mathbf{a} = \mathbf{i} + \mathbf{j}$, $\|\mathbf{v}\| = \sqrt{2}$, $\mathbf{v}\cdot\mathbf{a} = 0$, $\mathbf{v}\times\mathbf{a} = -2\mathbf{k}$ so $a_T = 0$, $a_N = \sqrt{2}$.

43. $\mathbf{v} = (3t^2 - 2)\mathbf{i} + 2t\mathbf{j}$, $\mathbf{a} = 6t\mathbf{i} + 2\mathbf{j}$; when $t = 1$, $\mathbf{v} = \mathbf{i} + 2\mathbf{j}$, $\mathbf{a} = 6\mathbf{i} + 2\mathbf{j}$, $\|\mathbf{v}\| = \sqrt{5}$, $\mathbf{v}\cdot\mathbf{a} = 10$, $\mathbf{v}\times\mathbf{a} = -10\mathbf{k}$ so $a_T = 2\sqrt{5}$, $a_N = 2\sqrt{5}$.

45. $\mathbf{v} = \mathbf{i} + 2t\mathbf{j} + 3t^2\mathbf{k}$, $\mathbf{a} = 2\mathbf{j} + 6t\mathbf{k}$; when $t = 1$, $\mathbf{v} = \mathbf{i} + 2\mathbf{j} + 3\mathbf{k}$, $\mathbf{a} = 2\mathbf{j} + 6\mathbf{k}$, $\|\mathbf{v}\| = \sqrt{14}$, $\mathbf{v}\cdot\mathbf{a} = 22$, $\mathbf{v}\times\mathbf{a} = 6\mathbf{i} - 6\mathbf{j} - 2\mathbf{k}$ so $a_T = 22/\sqrt{14}$, $a_N = \sqrt{76}/\sqrt{14} = \sqrt{38/7}$.

47. $\mathbf{v} = 3\cos t\mathbf{i} - 2\sin t\mathbf{j} - 2\cos 2t\mathbf{k}$, $\mathbf{a} = -3\sin t\mathbf{i} - 2\cos t\mathbf{j} + 4\sin 2t\mathbf{k}$; when $t = \pi/2$, $\mathbf{v} = -2\mathbf{j} + 2\mathbf{k}$, $\mathbf{a} = -3\mathbf{i}$, $\|\mathbf{v}\| = 2\sqrt{2}$, $\mathbf{v}\cdot\mathbf{a} = 0$, $\mathbf{v}\times\mathbf{a} = -6\mathbf{j} - 6\mathbf{k}$ so $a_T = 0$, $a_N = 3$.

49. $\|\mathbf{v}\| = 4$, $\mathbf{v}\cdot\mathbf{a} = -12$, $\mathbf{v}\times\mathbf{a} = 8\mathbf{k}$ so $a_T = -3$, $a_N = 2$, $\mathbf{T} = -\mathbf{j}$, $\mathbf{N} = (\mathbf{a} - a_T\mathbf{T})/a_N = \mathbf{i}$.

51. $\|\mathbf{v}\| = 3$, $\mathbf{v}\cdot\mathbf{a} = 4$, $\mathbf{v}\times\mathbf{a} = 4\mathbf{i} - 3\mathbf{j} - 2\mathbf{k}$ so $a_T = 4/3$, $a_N = \sqrt{29}/3$, $\mathbf{T} = (1/3)(2\mathbf{i} + 2\mathbf{j} + \mathbf{k})$, $\mathbf{N} = (\mathbf{a} - a_T\mathbf{T})/a_N = (\mathbf{i} - 8\mathbf{j} + 14\mathbf{k})/(3\sqrt{29})$.

53. $a_T = \dfrac{d^2s}{dt^2} = \dfrac{d}{dt}\sqrt{3t^2 + 4} = 3t/\sqrt{3t^2 + 4}$ so when $t = 2$, $a_T = 3/2$.

55. $a_T = \dfrac{d^2s}{dt^2} = \dfrac{d}{dt}\sqrt{(4t - 1)^2 + \cos^2\pi t} = [4(t - 1) - \pi\cos\pi t\sin\pi t]/\sqrt{(4t - 1)^2 + \cos^2\pi t}$ so when $t = 1/4$, $a_T = -\pi/\sqrt{2}$.

57. $\|\mathbf{v}\| = 4$, $\mathbf{v}\times\mathbf{a} = 8\mathbf{k}$ so $\kappa = \|\mathbf{v}\times\mathbf{a}\|/\|\mathbf{v}\|^3 = 1/8$.

59. $\|\mathbf{v}\| = 3$, $\mathbf{v}\times\mathbf{a} = 4\mathbf{i} - 3\mathbf{j} - 2\mathbf{k}$ so $\kappa = \|\mathbf{v}\times\mathbf{a}\|/\|\mathbf{v}\|^3 = \sqrt{29}/27$.

61. $a_N = \kappa(ds/dt)^2 = (1/\rho)(ds/dt)^2 = (1/1)(3\times 10^5)^2 = 9\times 10^{10}$ kilometers/sec^2

63. $a_N = \kappa(ds/dt)^2 = [2/(1 + 4x^2)^{3/2}](3)^2 = 18/(1 + 4x^2)^{3/2}$.

65. **(a)** $v_0 = 320$, $\alpha = 60°$, $s_0 = 0$ so $x = 160t$, $y = 160\sqrt{3}t - 16t^2$.

(b) $dy/dt = 160\sqrt{3} - 32t$, $dy/dt = 0$ when $t = 5\sqrt{3}$ so
$y_{max} = 160\sqrt{3}(5\sqrt{3}) - 16(5\sqrt{3})^2 = 1200$ ft.

(c) $y = 16t(10\sqrt{3} - t)$, $y = 0$ when $t = 0$ or $10\sqrt{3}$ so $x_{max} = 160(10\sqrt{3}) = 1600\sqrt{3}$ ft.

(d) $\mathbf{v}(t) = 160\mathbf{i} + (160\sqrt{3} - 32t)\mathbf{j}$, $\mathbf{v}(10\sqrt{3}) = 160(\mathbf{i} - \sqrt{3}\mathbf{j})$, $\|\mathbf{v}(10\sqrt{3})\| = 320$ ft/sec.

67. $v_0 = 80$, $\alpha = -60°$, $s_0 = 168$ so $x = 40t$, $y = 168 - 40\sqrt{3}\,t - 16t^2$; $y = 0$ when
$t = -7\sqrt{3}/2$ (invalid) or $t = \sqrt{3}$ so $x(\sqrt{3}) = 40\sqrt{3}$ ft.

69. $\alpha = 30°$, $s_0 = 0$ so $x = \sqrt{3}v_0t/2$, $y = v_0t/2 - 16t^2$; $dy/dt = v_0/2 - 32t$, $dy/dt = 0$ when
$t = v_0/64$ so $y_{max} = v_0^2/256 = 2500$, $v_0 = 800$ ft/sec.

71. $v_0 = 800$, $s_0 = 0$ so $x = (800\cos\alpha)t$, $y = (800\sin\alpha)t - 16t^2 = 16t(50\sin\alpha - t)$; $y = 0$ when
$t = 0$ or $50\sin\alpha$ so $x_{max} = 40,000\sin\alpha\cos\alpha = 20,000\sin 2\alpha = 10,000$, $2\alpha = 30°$ or $150°$,
$\alpha = 15°$ or $75°$.

73. $s_0 = 0$ so $x = (v_0\cos\alpha)t$, $y = (v_0\sin\alpha)t - gt^2/2$

(a) $dy/dt = v_0\sin\alpha - gt$ so $dy/dt = 0$ when $t = (v_0\sin\alpha)/g$, $y_{max} = (v_0\sin\alpha)^2/(2g)$

(b) $y = 0$ when $t = 0$ or $(2v_0\sin\alpha)/g$, so $x = R = (2v_0^2\sin\alpha\cos\alpha)/g = (v_0^2\sin 2\alpha)/g$ when
$t = (2v_0\sin\alpha)/g$; R is maximum when $2\alpha = 90°$, $\alpha = 45°$, and the maximum value of R
is v_0^2/g.

75. $v_0 = 80$, $\alpha = 30°$, $s_0 = 5$ so $x = 40\sqrt{3}t$, $y = 5 + 40t - 16t^2$

(a) $y = 0$ when $t = (-40 \pm \sqrt{(40)^2 - 4(-16)(5)})/(-32) = (5 \pm \sqrt{30})/4$, reject $(5 - \sqrt{30})/4$
to get $t = (5 + \sqrt{30})/4 \approx 2.62$ sec.

(b) $x \approx 40\sqrt{3}(2.62) \approx 181.5$ ft.

SUPPLEMENTARY EXERCISES CHAPTER 15

1.   **(a)** $\mathbf{v} = \frac{1}{2}(t+4)^{-1/2}\mathbf{i} + 2\mathbf{j}$, $\mathbf{a} = -\frac{1}{4}(t+4)^{-3/2}\mathbf{i}$

     (b) $\mathbf{r}'(-3) = (1/2)\mathbf{i} + 2\mathbf{j}$, $\mathbf{r}''(-3) = -(1/4)\mathbf{i}$
           $\mathbf{r}'(0) = (1/4)\mathbf{i} + 2\mathbf{j}$, $\mathbf{r}''(0) = -(1/32)\mathbf{i}$

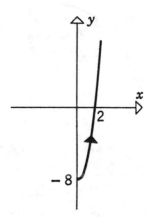

$y = 2x^2 - 8$, $x \geq 0$

3.   **(a)** $\mathbf{v} = \langle 6t^2, 3t^2 \rangle$, $\mathbf{a} = \langle 12t, 6t \rangle$

     (b) $\mathbf{r}'(0) = \langle 0, 0 \rangle$, $\mathbf{r}''(0) = \langle 0, 0 \rangle$
           $\mathbf{r}'(-1/2) = \langle 3/2, 3/4 \rangle$, $\mathbf{r}''(-1/2) = \langle -6, -3 \rangle$

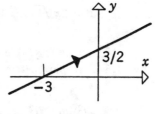

$x - 2y = -3$

5.   **(a)** $(k\mathbf{i} + m\mathbf{j})t + \mathbf{C}$

     (b) $\langle e^{2t}/2, 2e^t \rangle \big]_0^{\ln 3} = \langle 9/2, 6 \rangle - \langle 1/2, 2 \rangle = \langle 4, 4 \rangle$

     (c) $\displaystyle\int_0^2 dt = 2$ **(d)** $\sqrt{t^2 + 3}\,\mathbf{i} + \ln(\sin t)\mathbf{j} + \mathbf{C}$

7.   **(a)** $ds/dt = \|\mathbf{r}'(t)\| = \|\langle (t^2 - 1)/t^2, 2/t \rangle\| = 1 + 1/t^2$

     (b) $s = \displaystyle\int_1^t (1 + 1/u^2)\,du = t - 1/t$, $t^2 - st - 1 = 0$, $t = (s \pm \sqrt{s^2 + 4})/2$, but $t \geq 0$
           so $t = (s + \sqrt{s^2 + 4})/2$ and $x = \sqrt{s^2 + 4}$, $y = 2\ln[(s + \sqrt{s^2 + 4})/2]$.

9.

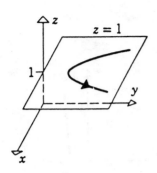

11. $\mathbf{r}(t) = a\sin t\mathbf{i} + a\cos t\mathbf{j} + a\ln(\cos t)\mathbf{k}$, $\mathbf{v} = a\cos t\mathbf{i} - a\sin t\mathbf{j} - a\tan t\mathbf{k}$

$\|\mathbf{v}\| = a(\cos^2 t + \sin^2 t + \tan^2 t)^{1/2} = a(1 + \tan^2 t)^{1/2} = a\sec t$

$\mathbf{a} = -a\sin t\mathbf{i} - a\cos t\mathbf{j} - a\sec^2 t\mathbf{k}$, $\mathbf{T} = \mathbf{v}/\|\mathbf{v}\| = \cos^2 t\mathbf{i} - \sin t\cos t\mathbf{j} - \sin t\mathbf{k}$

$d\mathbf{T}/dt = -2\sin t\cos t\mathbf{i} - (\cos^2 t - \sin^2 t)\mathbf{j} - \cos t\mathbf{k}$; at $t = 0$, $\mathbf{v} = a\mathbf{i}$, $\|\mathbf{v}\| = a$, $\mathbf{a} = -a(\mathbf{j} + \mathbf{k})$,

$\mathbf{T} = \mathbf{i}$, $\mathbf{N} = (d\mathbf{T}/dt)/\|d\mathbf{T}/dt\| = (-\mathbf{j} - \mathbf{k})/\sqrt{2}$, $\kappa = \|\mathbf{v} \times \mathbf{a}\|/\|\mathbf{v}\|^3 = \|a^2\mathbf{j} - a^2\mathbf{k}\|/a^3 = \sqrt{2}/a$

13. $(dx/dt)^2 + (dy/dt)^2 + (dz/dt)^2 = 4 + 144\cos^2 3t + 144\sin^2 3t = 148$,

$$L = \int_0^{2\pi} \sqrt{148}\, dt = 2\pi\sqrt{148} = 4\pi\sqrt{37}.$$

15. $\mathbf{r}'(t) = \langle -e^{-t}, 2e^{2t}, 3t^2\rangle$, $\mathbf{r}'(0) = \langle -1, 2, 0\rangle$ is parallel to the tangent line to the curve at the tip of $\mathbf{r}(0) = \langle 1, 1, 1\rangle$ so parametric equations of the tangent line are $x = 1 - t$, $y = 1 + 2t$, $z = 1$.

17. **(a)** $\displaystyle\int_0^3 \langle 2t, 3, -t^2\rangle dt = \langle t^2, 3t, -t^3/3\rangle\Big]_0^3 = \langle 9, 9, -9\rangle$

(b) $\mathbf{u} \times \mathbf{v} = \langle 3t + t^4, -2t^2, 2t^3\rangle$, $d(\mathbf{u} \times \mathbf{v})/dt = \langle 3 + 4t^3, -4t, 6t^2\rangle$.

19. **(a)** $\mathbf{r}'(t) = 2t\mathbf{i} - (1/t^2)\mathbf{j}$, $t = 1$ at P_0 so $\mathbf{T} = \mathbf{r}'(1)/\|\mathbf{r}'(1)\| = (2\mathbf{i} - \mathbf{j})/\sqrt{5}$

(b) $\kappa(t) = \dfrac{6t^4}{(4t^6 + 1)^{3/2}}$ so $\kappa(1) = \dfrac{6}{5^{3/2}}$

21. **(a)** Let $y = t$, then $\mathbf{r}(t) = (t - 1)^2\mathbf{i} + t\mathbf{j}$,

$\mathbf{r}'(t) = 2(t - 1)\mathbf{i} + \mathbf{j}$, $t = 1$ at P_0 so $\mathbf{T} = \mathbf{r}'(1)/\|\mathbf{r}'(1)\| = \mathbf{j}$

(b) $\kappa(t) = 2/[4(t - 1)^2 + 1]^{3/2}$, $\kappa(1) = 2$

23. $x = t + t^3$, $y = t + t^2$, $\kappa(t) = \dfrac{|2 - 6t - 6t^2|}{[(1 + 3t^2)^2 + (1 + 2t)^2]^{3/2}}$, $t = 1$ at P_0, $\kappa(1) = 2/25$

25. Let $y = t$, then $x = \ln(\sec t)$, $\kappa(t) = |\cos t|$, $t = 0$ at P_0, $\kappa(0) = 1$

27. $\kappa(x) = \dfrac{2}{[1+4(x-1)^2]^{3/2}}$, $\kappa(1) = 2$, $\rho = 1/2$. The parabola opens upward and has its vertex at $(1,0)$ so the center of curvature is at $(1, 1/2)$ and the oscillating circle is $(x-1)^2 + (y-1/2)^2 = 1/4$. $y' = 0$ and $y'' = 2$ at $(1,0)$ for both the parabola and the circle.

29. $d\mathbf{r}/du = (d\mathbf{r}/dt)(dt/du) = (1/u)\langle e^t, 4e^{2t}\rangle = \langle 1, 4u\rangle$

31. $\mathbf{v} = \langle t\sin t, t\cos t\rangle$, $\mathbf{a} = \langle \sin t + t\cos t, \cos t - t\sin t\rangle$, $\|\mathbf{v}\| = t$, $\mathbf{v}\cdot\mathbf{a} = t$, $\mathbf{v}\times\mathbf{a} = -t^2\mathbf{k}$ so $a_T = 1$, $a_N = t$.

33. (a) $\mathbf{v} = \langle -e^{-t}, e^t\rangle$, $\mathbf{a} = \langle e^{-t}, e^t\rangle$; $t = 0$ at P_0 so $\mathbf{v} = \langle -1, 1\rangle$, $\mathbf{a} = \langle 1, 1\rangle$, $ds/dt = \|\mathbf{v}\| = \sqrt{2}$
   (b) $\mathbf{v}\times\mathbf{a} = -2\mathbf{k}$, $\kappa = \|\mathbf{v}\times\mathbf{a}\|/\|\mathbf{v}\|^3 = 1/\sqrt{2}$
   (c) $\mathbf{v}\cdot\mathbf{a} = 0$, $a_T = (\mathbf{v}\cdot\mathbf{a})/\|\mathbf{v}\| = 0$, $a_N = \|\mathbf{v}\times\mathbf{a}\|/\|\mathbf{v}\| = \sqrt{2}$
   (d) The trajectory is the branch of the hyperbola $y = 1/x$ in the first quadrant, traced so that y increases with t.
   (e) The radius is $1/\kappa = \sqrt{2}$. If the center is (h, k), then $(x-h)^2 + (y-k)^2 = 2$ is an equation of the circle. The circle must be tangent to the curve at P_0 so $(1-h)^2 + (1-k)^2 = 2$ and, equating slopes, $-(1-h)/(1-k) = -1$, $1-h = 1-k$ thus $(1-h)^2 = 1$, $(1-h) = \pm 1$, $h = 0$ (reject, the center must be to the right of $x = 1$) or $h = 2$, $k = h = 2$. The center is at $(2, 2)$.

35. $\mathbf{r}(0) = 0$, $\mathbf{v}(0) = \mathbf{i} + 2\mathbf{j}$. Use $\mathbf{F} = m\mathbf{a}$ with $m = 1$ to get
   $$\mathbf{a} = \sin t\,\mathbf{i} + 4e^{2t}\mathbf{j}, \ \mathbf{v}(t) = \int \mathbf{a}\,dt = -\cos t\,\mathbf{i} + 2e^{2t}\mathbf{j} + \mathbf{C}_1,$$
   $$\mathbf{v}(0) = -\mathbf{i} + 2\mathbf{j} + \mathbf{C}_1 = \mathbf{i} + 2\mathbf{j}, \ \mathbf{C}_1 = 2\mathbf{i} \text{ so}, \ \mathbf{v}(t) = (2 - \cos t)\mathbf{i} + 2e^{2t}\mathbf{j},$$
   $$\mathbf{r}(t) = \int \mathbf{v}\,dt = (2t - \sin t)\mathbf{i} + e^{2t}\mathbf{j} + \mathbf{C}_2, \ \mathbf{r}(0) = \mathbf{j} + \mathbf{C}_2 = 0 \text{ so } \mathbf{C}_2 = -\mathbf{j},$$
   $$\mathbf{r}(t) = (2t - \sin t)\mathbf{i} + (e^{2t} - 1)\mathbf{j}$$

37. $dx/dt = 4$, by the chain rule $dy/dt = (2 - 2x)(dx/dt) = 8(1 - x)$,
   $ds/dt = [(dx/dt)^2 + (dy/dt)^2]^{1/2} = 4[1 + 4(1 - x)^2]^{1/2}$,
   $d^2s/dt^2 = 2[1 + 4(1 - x)^2]^{-1/2}[8(1 - x)](-dx/dt) = -64(1 - x)[1 + 4(1 - x)^2]^{-1/2}$
   so $a_T = -64(1 - x)/\sqrt{1 + 4(1 - x)^2}$; $d^2x/dt^2 = 0$, $d^2y/dt^2 = -8dx/dt = -32$,
   $\|\mathbf{a}\|^2 = (d^2x/dt^2)^2 + (d^2y/dt^2)^2 = 0 + (-32)^2 = 1024$.
   (a) $a_T = 0$, $a_N^2 = \|\mathbf{a}\|^2 - a_T^2 = 1024$, $a_N = 32$
   (b) $a_T = -64/\sqrt{5}$, $a_N^2 = 1024 - (64/\sqrt{5})^2 = 1024/5$, $a_N = 32/\sqrt{5}$

39. In one revolution the weight travels a distance that is equal to the circumference of a circle of radius 2 m so $ds/dt = 2\pi(2) = 4\pi$ m/sec, $a_T = d^2s/dt^2 = 0$,
   $$a_N = \kappa(ds/dt)^2 = (1/2)(4\pi)^2 = 8\pi^2 \text{ m/sec}^2.$$

41.   **(a)** $\mathbf{r}'(t) = e^t \langle 2\cos 2t, -2\sin 2t \rangle + e^t \langle \sin 2t, \cos 2t \rangle = e^t \langle 2\cos 2t + \sin 2t, \cos 2t - 2\sin 2t \rangle$,

$ds/dt = \|\mathbf{r}'(t)\| = e^t[(2\cos 2t + \sin 2t)^2 + (\cos 2t - 2\sin 2t)^2]^{1/2} = \sqrt{5}\, e^t$

(b) $L = \displaystyle\int_0^{\ln 3} (ds/dt)\,dt = \int_0^{\ln 3} \sqrt{5}\, e^t\, dt = 2\sqrt{5}.$

CHAPTER 16
Partial Derivatives

EXERCISE SET 16.1

1. (a) $f(2,1) = (2)^2(1) + 1 = 5$ (b) $f(1,2) = (1)^2(2) + 1 = 3$
   (c) $f(0,0) = (0)^2(0) + 1 = 1$ (d) $f(1,-3) = (1)^2(-3) + 1 = -2$
   (e) $f(3a,a) = (3a)^2(a) + 1 = 9a^3 + 1$
   (f) $f(ab, a-b) = (ab)^2(a-b) + 1 = a^3b^2 - a^2b^3 + 1$

3. (a) $f(x+y, x-y) = (x+y)(x-y) + 3 = x^2 - y^2 + 3$
   (b) $f\left(xy, 3x^2y^3\right) = (xy)\left(3x^2y^3\right) + 3 = 3x^3y^4 + 3$

5. $F(g(x), h(y)) = F\left(x^3, 3y+1\right) = x^3 e^{x^3(3y+1)}$

7. (a) $t^2 + 3t^{10}$ (b) 0 (c) 3076

9. (a) 19 (b) -9 (c) 3
   (d) $a^6 + 3$ (e) $-t^8 + 3$ (f) $(a+b)(a-b)^2 b^3 + 3$

11. $F\left(x^2, y+1, z^2\right) = (y+1)e^{x^2(y+1)z^2}$

13. (a) t^{14} (b) 0 (c) $16,384$

15.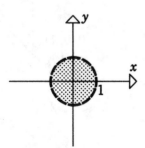

17.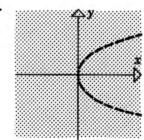

19.

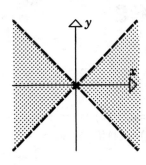

21. all points above or on the line $y = -2$

23. all points above the line $y = 2x$

25. all points not on the plane $x + y + z = 0$

27. all points inside the cylinder $x^2 + y^2 = 1$

29.

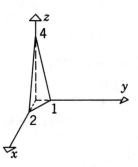

31.

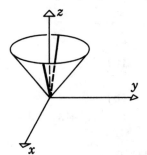

33.

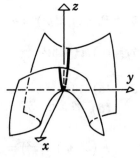

35.

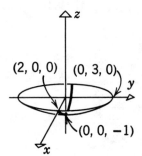

37.

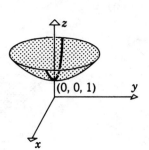

39.

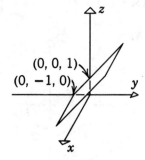

41.

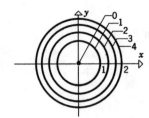

43.

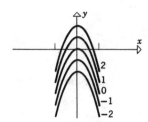

45.

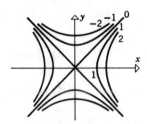

47.

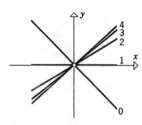

49.

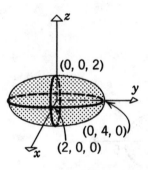

51.

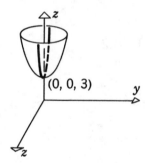

53. concentric spheres, common center at $(2,0,0)$

55. concentric cylinders, common axis the y-axis

57. **(a)** $f(-1,1) = 0;\ x^2 - 2x^3 + 3xy = 0$ **(b)** $f(0,0) = 0;\ x^2 - 2x^3 + 3xy = 0$
     (c) $f(2,-1) = -18;\ x^2 - 2x^3 + 3xy = -18$

59. **(a)** $f(1,-2,0) = 5;\ x^2 + y^2 - z = 5$ **(b)** $f(1,0,3) = -2;\ x^2 + y^2 - z = -2$
     (c) $f(0,0,0) = 0;\ x^2 + y^2 - z = 0$

61. $V = 8/\sqrt{16 + x^2 + y^2}$

$\sqrt{16 + x^2 + y^2} = 8/V$

$x^2 + y^2 = 64/V^2 - 16,$

the equipotential curves are circles.

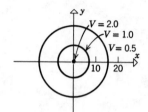

63. **(a)** A **(b)** B

65. **(a)** decrease   **(b)** increase   **(c)** increase   **(d)** decrease

67. **(a)** open   **(b)** neither   **(c)** closed   **(d)** closed

69. **(a)** bounded   **(b)** unbounded   **(c)** unbounded   **(d)** unbounded

EXERCISE SET 16.2

1.

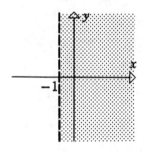

3.

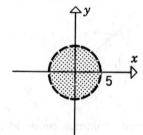

5.

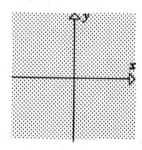

7.

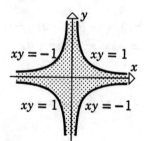

9. all of 3-space

11. all points not on the cylinder $x^2 + z^2 = 1$

13. 35   **15.** −8   **17.** 0

19. along $y = 0$: $\lim\limits_{x \to 0} \dfrac{x}{x^2} = \lim\limits_{x \to 0} \dfrac{1}{x}$ does not exist because $\left|\dfrac{1}{x}\right| \to +\infty$ as $x \to 0$ so the original limit does not exist.

21. Let $z = x^2 + y^2$, then $\lim\limits_{(x,y) \to (0,0)} \dfrac{\sin\left(x^2 + y^2\right)}{x^2 + y^2} = \lim\limits_{z \to 0^+} \dfrac{\sin z}{z} = 1$

23. $\lim\limits_{(x,y) \to (0,0)} \dfrac{\left(x^2 + y^2\right)\left(x^2 - y^2\right)}{x^2 + y^2} = \lim\limits_{(x,y) \to (0,0)} \left(x^2 - y^2\right) = 0$

25. along $y = 0$: $\lim\limits_{x \to 0} \dfrac{0}{3x^2} = \lim\limits_{x \to 0} 0 = 0$; along $y = x$: $\lim\limits_{x \to 0} \dfrac{x^2}{5x^2} = \lim\limits_{x \to 0} 1/5 = 1/5$
so the limit does not exist.

27. Let $z = x^2 + y^2$, then $\lim\limits_{(x,y) \to (0,0)} e^{-1/\left(x^2 + y^2\right)} = \lim\limits_{z \to 0^+} e^{-1/z} = 0$

29. Use polar coordinates: $y = r \sin \theta$ and $x^2 + y^2 = r^2$ so
$y \ln\left(x^2 + y^2\right) = r \sin \theta \ln r^2 = 2r \sin \theta \ln r$. But $|\sin \theta| \le 1$ so $\left|y \ln(x^2 + y^2)\right| \le |2r \ln r|$;
$\lim\limits_{r \to 0^+} 2r \ln r = 0$ thus $\lim\limits_{(x,y) \to (0,0)} y \ln\left(x^2 + y^2\right) = 0$

31. 8/3

33. Let $t = \sqrt{x^2 + y^2 + z^2}$, then $\lim\limits_{(x,y) \to (0,0,0)} \dfrac{\sin\left(x^2 + y^2 + z^2\right)}{\sqrt{x^2 + y^2 + z^2}} = \lim\limits_{t \to 0^+} \dfrac{\sin\left(t^2\right)}{t} = 0$

35. along the z-axis: $\lim\limits_{z \to 0} \left(0/z^2\right) = \lim\limits_{z \to 0} 0 = 0$;

along the line $x = t$, $y = t$, $z = t$: $\lim\limits_{t \to 0} \dfrac{t^2}{3t^2} = \lim\limits_{t \to 0} 1/3 = 1/3$ so the limit does not exist.

37. (a) $\lim\limits_{x \to 0} \dfrac{mx^3}{x^4 + m^2 x^2} = \lim\limits_{x \to 0} \dfrac{mx}{x^2 + m^2} = 0$ **(b)** $\lim\limits_{x \to 0} \dfrac{x^4}{2x^4} = \lim\limits_{x \to 0} 1/2 = 1/2$

39. (a) $\lim\limits_{t \to 0} \dfrac{abct^3}{a^2 t^2 + b^4 t^4 + c^4 t^4} = \lim\limits_{t \to 0} \dfrac{abct}{a^2 + b^4 t^2 + c^4 t^2} = 0$

(b) $\lim\limits_{t \to 0} \dfrac{t^4}{t^4 + t^4 + t^4} = \lim\limits_{t \to 0} 1/3 = 1/3$

41. $-\pi/2$ because $\dfrac{x^2 - 1}{x^2 + (y - 1)^2} \to -\infty$ as $(x, y) \to (0, 1)$

43. No, because $\displaystyle\lim_{(x,y)\to(0,0)} \frac{x^2}{x^2+y^2}$ does not exist.

Along $x = 0 : \displaystyle\lim_{y\to 0}\left(0/y^2\right) = \lim_{y\to 0} 0 = 0$; along $y = 0 : \displaystyle\lim_{x\to 0}\left(x^2/x^2\right) = \lim_{x\to 0} 1 = 1$.

45. $x^2 + y^2 = r^2$, $|x^2 + y^2 - 0| = |r^2| = r^2 < \epsilon$ if $r < \sqrt{\epsilon}$; choose $\delta = \sqrt{\epsilon}$.

47. $x^2 + y^2 + z^2 = \rho^2$, $|x^2 + y^2 + z^2 - 0| = |\rho^2| = \rho^2 < \epsilon$ if $\rho < \sqrt{\epsilon}$; choose $\delta = \sqrt{\epsilon}$.

EXERCISE SET 16.3

1. $\partial z/\partial x = 9x^2 y^2$, $\partial z/\partial y = 6x^3 y$ **3.** $\partial z/\partial x = 8xy^3 e^{x^2 y^3}$, $\partial z/\partial y = 12x^2 y^2 e^{x^2 y^3}$

5. $\partial z/\partial x = x^3/(y^{3/5} + x) + 3x^2 \ln(1 + xy^{-3/5})$, $\partial z/\partial y = -(3/5)x^4/(y^{8/5} + xy)$

7. $f_x(x,y) = (3/2)x^2 y\left(5x^2 - 7\right)\left(3x^5 y - 7x^3 y\right)^{-1/2}$

$f_y(x,y) = (1/2)x^3\left(3x^2 - 7\right)\left(3x^5 y - 7x^3 y\right)^{-1/2}$

9. $f_x(x,y) = \dfrac{y^{-1/2}}{y^2 + x^2}$, $f_y(x,y) = -\dfrac{xy^{-3/2}}{y^2 + x^2} - \dfrac{3}{2}y^{-5/2}\tan^{-1}(x/y)$

11. $f_x(x,y) = -(4/3)y^2 \sec^2 x\left(y^2 \tan x\right)^{-7/3}$, $f_y(x,y) = -(8/3)y \tan x\left(y^2 \tan x\right)^{-7/3}$

13. (a) $f_x(x,y) = -2x$, $f_x(3,1) = -6$ (b) $f_y(x,y) = -21y^2$, $f_y(3,1) = -21$

15. (a) $\partial z/\partial x = x(x^2 + 4y^2)^{-1/2}$, $\partial z/\partial x\,|_{(1,2)} = 1/\sqrt{17}$

(b) $\partial z/\partial y = 4y(x^2 + 4y^2)^{-1/2}$, $\partial z/\partial y\,|_{(1,2)} = 8/\sqrt{17}$

17. $\dfrac{3}{2}\left(x^2 + y^2 + z^2\right)^{1/2}\left(2x + 2z\dfrac{\partial z}{\partial x}\right) = 0$, $\partial z/\partial x = -x/z$; similarly, $\partial z/\partial y = -y/z$

19. $2x + z\left(xy\dfrac{\partial z}{\partial x} + yz\right)\cos xyz + \dfrac{\partial z}{\partial x}\sin xyz = 0$, $\dfrac{\partial z}{\partial x} = -\dfrac{2x + yz^2 \cos xyz}{xyz\cos xyz + \sin xyz}$;

$z\left(xy\dfrac{\partial z}{\partial y} + xz\right)\cos xyz + \dfrac{\partial z}{\partial y}\sin xyz = 0$, $\dfrac{\partial z}{\partial y} = -\dfrac{xz^2 \cos xyz}{xyz\cos xyz + \sin xyz}$

21. $f_{xx} = 8$, $f_{yy} = -96xy^2 + 140y^3$, $f_{xy} = f_{yx} = -32y^3$

23. $f_{xx} = e^x \cos y$, $f_{yy} = -e^x \cos y$, $f_{xy} = f_{yx} = -e^x \sin y$

25. $f_{xx} = -16/(4x - 5y)^2$, $f_{yy} = -25/(4x - 5y)^2$, $f_{xy} = f_{yx} = 20/(4x - 5y)^2$

27. (a) $30xy^4 - 4$ (b) $60x^2y^3$ (c) $60x^3y^2$

29. (a) $f_{xyy} = -30ye^{-5x}$, $f_{xyy}(0,1) = -30$ (b) $f_{xxx} = -125y^3e^{-5x}$, $f_{xxx}(0,1) = -125$
(c) $f_{yyxx} = 150ye^{-5x}$, $f_{yyxx}(0,1) = 150$

31. (a) $\dfrac{\partial^3 f}{\partial x^3}$ (b) $\dfrac{\partial^3 f}{\partial y^2 \partial x}$ (c) $\dfrac{\partial^4 f}{\partial x^2 \partial y^2}$ (d) $\dfrac{\partial^4 f}{\partial y^3 \partial x}$

33. $\partial w/\partial x = 2xy^4z^3 + y$, $\partial w/\partial y = 4x^2y^3z^3 + x$, $\partial w/\partial z = 3x^2y^4z^2 + 2z$

35. $\partial w/\partial x = 2x/\left(y^2 + z^2\right)$, $\partial w/\partial y = -2y\left(x^2 + z^2\right)/\left(y^2 + z^2\right)^2$, $\partial w/\partial z = 2z\left(y^2 - x^2\right)/\left(y^2 + z^2\right)^2$

37. $\partial w/\partial x = x/\sqrt{x^2 + y^2 + z^2}$, $\partial w/\partial y = y/\sqrt{x^2 + y^2 + z^2}$, $\partial w/\partial z = z/\sqrt{x^2 + y^2 + z^2}$

39. $f_x = -y^2z^3/\left(1 + x^2y^4z^6\right)$, $f_y = -2xyz^3/\left(1 + x^2y^4z^6\right)$, $f_z = -3xy^2z^2/\left(1 + x^2y^4z^6\right)$

41. $f_x = 4xyz \cos h\sqrt{z} \sin h\left(x^2yz\right) \cos h\left(x^2yz\right)$, $f_y = 2x^2z \cos h\sqrt{z} \sin h\left(x^2yz\right) \cos h\left(x^2yz\right)$,
$f_z = 2x^2y \cos h\sqrt{z} \sin h\left(x^2yz\right) \cos h\left(x^2yz\right) + (1/2)z^{-1/2} \sin h\sqrt{z} \sin h^2\left(x^2yz\right)$

43. (a) -80 (b) 40 (c) -60

45. (a) $2/\sqrt{7}$ (b) $4/\sqrt{7}$ (c) $1/\sqrt{7}$

47. $(3/2)\left(x^2 + y^2 + z^2 + w^2\right)^{1/2}\left(2x + 2w\dfrac{\partial w}{\partial x}\right) = 0$, $\partial w/\partial x = -x/w$; similarly, $\partial w/\partial y = -y/w$
and $\partial w/\partial z = -z/w$

49. $\dfrac{\partial w}{\partial x} = -\dfrac{yzw \cos xyz}{2w + \sin xyz}$, $\dfrac{\partial w}{\partial y} = -\dfrac{xzw \cos xyz}{2w + \sin xyz}$, $\dfrac{\partial w}{\partial z} = -\dfrac{xyw \cos xyz}{2w + \sin xyz}$

51. (a) $f_{xy} = 15x^2y^4z^7 + 2y$ (b) $f_{yz} = 35x^3y^4z^6 + 3y^2$
(c) $f_{xz} = 21x^2y^5z^6$ (d) $f_{zz} = 42x^3y^5z^5$
(e) $f_{zyy} = 140x^3y^3z^6 + 6y$ (f) $f_{xxy} = 30xy^4z^7$
(g) $f_{zyx} = 105x^2y^4z^6$ (h) $f_{xxyz} = 210xy^4z^6$

53. (a) $\partial^2 f/\partial x^2 = e^x \sin y - e^y \cos x = -\partial^2 f/\partial y^2$

(b) $\partial^2 f/\partial x^2 = 2(y^2 - x^2)/(x^2 + y^2)^2 = -\partial^2 f/\partial y^2$

(c) $\partial^2 f/\partial x^2 = 4xy/(x^2 + y^2)^2 = -\partial^2 f/\partial y^2$

55. $\partial z/\partial y = 6y$, $\partial z/\partial y|_{(2,1)} = 6$

57. **(a)** $\partial z/\partial y = 8y$, $\partial z/\partial y|_{(-1,1)} = 8$ **(b)** $\partial z/\partial x = 2x$, $\partial z/\partial x|_{(-1,1)} = -2$

59. **(a)** $\partial V/\partial r = 2\pi rh$ **(b)** $\partial V/\partial h = \pi r^2$

(c) $\partial V/\partial r|_{r=6,\ h=4} = 48\pi$ **(d)** $\partial V/\partial h|_{r=8,\ h=10} = 64\pi$

61. **(a)** $P = 10T/V$, $\partial P/\partial T = 10/V$, $\partial P/\partial T|_{T=80,\ V=50} = 1/5$

(b) $V = 10T/P, \partial V/\partial P = -10T/P^2$, if $V = 50$ and $T = 80$ then
$P = 10(80)/(50) = 16, \partial V/\partial P|_{T=80,\ P=16} = -25/8$

63. $\left(1 + \dfrac{\partial z}{\partial x}\right)\cos(x+z) + \cos(x-y) = 0$, $\dfrac{\partial z}{\partial x} = -1 - \dfrac{\cos(x-y)}{\cos(x+z)}$; $\dfrac{\partial z}{\partial y}\cos(x+z) - \cos(x-y) = 0$,

$\dfrac{\partial z}{\partial y} = \dfrac{\cos(x-y)}{\cos(x+z)}$; $\dfrac{\partial^2 z}{\partial x \partial y} = \dfrac{-\cos(x+z)\sin(x-y) + \cos(x-y)\sin(x+z)(\partial z/\partial x)}{\cos^2(x+z)}$,

substitute for $\partial z/\partial x$ and simplify to get

$\dfrac{\partial^2 z}{\partial x \partial y} = -\dfrac{\cos^2(x+z)\sin(x-y) + \cos^2(x-y)\sin(x+z) + \cos(x-y)\cos(x+z)\sin(x+z)}{\cos^3(x+z)}$

65. **(a)** $\partial T/\partial x = 3x^2 + 1$, $\partial T/\partial x|_{(1,2)} = 4$ **(b)** $\partial T/\partial y = 4y$, $\partial T/\partial y|_{(1,2)} = 8$

67. $\partial u/\partial x = \partial v/\partial y$ and $\partial u/\partial y = -\partial v/\partial x$ so $\partial^2 u/\partial x^2 = \partial^2 v/\partial x \partial y$, and $\partial^2 u/\partial y^2 = -\partial^2 v/\partial y \partial x$,
$\partial^2 u/\partial x^2 + \partial^2 u/\partial y^2 = \partial^2 v/\partial x \partial y - \partial^2 v/\partial y \partial x$, if $\partial^2 v/\partial x \partial y = \partial^2 v/\partial y \partial x$ then

$\partial^2 u/\partial x^2 + \partial^2 u/\partial y^2 = 0$ thus u satisfies Laplace's equation. The proof that v satisfies Laplace's equation is similar.

69. **(a)** Both are positive; $\partial T/\partial x$ has the largest absolute value.

(b) All are negative.

71. $f_x(x, y) = \dfrac{2}{3}(x^2 + y^2)^{-1/3}(2x) = \dfrac{4x}{(x^2 + y^2)^{1/3}}$, $(x, y) \neq (0, 0)$;

$f_x(0, 0) = \dfrac{d}{dx}[f(x, 0)]\Big|_{x=0} = \dfrac{d}{dx}[x^{4/3}]\Big|_{x=0} = \dfrac{4}{3}x^{1/3}\Big|_{x=0} = 0.$

EXERCISE SET 16.4

1. $\Delta f = f(1.1, 3.2) - f(1, 3) = (1.1)^2(3.2) - (1)^2(3) = 0.872$

3. $\Delta f = f(3, 1) - f(-1, 2) = 3/1 - (-1)/2 = 7/2$

5. $f_x(x, y) = y,\ f_y(x, y) = x,$
   $\Delta f = (x + \Delta x)(y + \Delta y) - xy = y\Delta x + x\Delta y + \Delta x\Delta y$
   $\quad = f_x(x, y)\Delta x + f_y(x, y)\Delta y + (0)\Delta x + (\Delta x)\Delta y$ where $\epsilon_1 = 0$ and $\epsilon_2 = \Delta x$.

7. $f_x(x, y) = 2xy,\ f_y(x, y) = x^2,$
   $\Delta f = (x + \Delta x)^2(y + \Delta y) - x^2 y = 2xy\Delta x + x^2\Delta y + y(\Delta x)^2 + (\Delta x)^2\Delta y + 2x\Delta x\Delta y$
   $\quad = f_x(x, y)\Delta x + f_y(x, y)\Delta y + (y\Delta x + \Delta x\Delta y)\Delta x + (2x\Delta x)\Delta y$
   where $\epsilon_1 = y\Delta x + \Delta x\Delta y$ and $\epsilon_2 = 2x\Delta x$.

9. **(a)** $\displaystyle\lim_{(x,y)\to(0,0)} f(x, y) = 0 = f(0, 0)$

   (b) $\displaystyle\lim_{h\to 0} \frac{f(0 + h, 0) - f(0, 0)}{h} = \lim_{h\to 0} \frac{f(h, 0)}{h} = \lim_{h\to 0} \frac{|h|}{h}$, which does not exist because
   $\displaystyle\lim_{h\to 0^+} |h|/h = 1$ and $\displaystyle\lim_{h\to 0^-} |h|/h = -1$.

11. $f_x(0, 0) = \displaystyle\lim_{h\to 0} \frac{f(h, 0) - f(0, 0)}{h} = \lim_{h\to 0} \frac{0 - 0}{h} = \lim_{h\to 0} 0 = 0$

    $f_y(0, 0) = \displaystyle\lim_{h\to 0} \frac{f(0, h) - f(0, 0)}{h} = \lim_{h\to 0} \frac{0 - 0}{h} = 0$; along $y = 0$, $\displaystyle\lim_{x\to 0} \frac{0}{x^2} = \lim_{x\to 0} 0 = 0$

    along $y = x$, $\displaystyle\lim_{x\to 0} \frac{x^2}{2x^2} = \lim_{x\to 0} 1/2 = 1/2$ so $\displaystyle\lim_{(x,y)\to(0,0)} f(x, y)$ does not exist

13. $f_{xy} = f_{yx} = 12x^2 + 6x$

15. $f_{xy} = f_{yx} = -6xy^2 \sin(x^2 + y^3)$

17. **(a)** $4:\ f_{xxx}, f_{xxy} = f_{xyx} = f_{yxx}, f_{xyy} = f_{yxy} = f_{yyx}, f_{yyy}$

    (b) $5:\ f_{xxxx}, f_{xxxy} = f_{xxyx} = f_{xyxx} = f_{yxxx},$
    $\qquad f_{xxyy} = f_{xyxy} = f_{xyyx} = f_{yxyx} = f_{yyxx} = f_{yxxy},$
    $\qquad f_{xyyy} = f_{yxyy} = f_{yyxy} = f_{yyyx}, f_{yyyy}$

19. $42t^{13}$

21. $3t^{-2}\sin(1/t)$

23. $-\dfrac{10}{3}t^{7/3}e^{1-t^{10/3}}$

25. $\partial z/\partial u = 24u^2v^2 - 16uv^3 - 2v + 3$, $\partial z/\partial v = 16u^3v - 24u^2v^2 - 2u - 3$

27. $\partial z/\partial u = -\dfrac{2\sin u}{3\sin v}$, $\partial z/\partial v = -\dfrac{2\cos u\cos v}{3\sin^2 v}$

29. $\partial z/\partial u = e^u$, $\partial z/\partial v = 0$

31. $\partial z/\partial u = 2e^{2u}/(1+e^{4u})$, $\partial z/\partial v = 0$

33. $\partial T/\partial r = 3r^2\sin\theta\cos^2\theta - 4r^3\sin^3\theta\cos\theta$

$\partial T/\partial\theta = -2r^3\sin^2\theta\cos\theta + r^4\sin^4\theta + r^3\cos^3\theta - 3r^4\sin^2\theta\cos^2\theta$

35. $\partial t/\partial x = \left(x^2 + y^2\right)/\left(4x^2y^3\right)$, $\partial t/\partial y = \left(y^2 - 3x^2\right)/\left(4xy^4\right)$

37. $-\pi$ **39.** $\sqrt{3}e^{\sqrt{3}}$, $\left(2 - 4\sqrt{3}\right)e^{\sqrt{3}}$

41. $F(x,y) = x^3 - 3xy^2 + y^3 - 5$, $\dfrac{dy}{dx} = -\dfrac{3x^2 - 3y^2}{-6xy + 3y^2} = \dfrac{x^2 - y^2}{2xy - y^2}$

43. $F(x,y) = x - (xy)^{1/2} + 3y - 4$, $\dfrac{dy}{dx} = -\dfrac{1 - (1/2)(xy)^{-1/2}y}{-(1/2)(xy)^{-1/2}x + 3} = \dfrac{2\sqrt{xy} - y}{x - 6\sqrt{xy}}$

45. $D = \left(x^2 + y^2\right)^{1/2}$ where x and y are the distances of cars A and B, respectively, from the intersection and D is the distance between them.

$dD/dt = \left[x/\left(x^2 + y^2\right)^{1/2}\right](dx/dt) + \left[y/\left(x^2 + y^2\right)^{1/2}\right](dy/dt)$, $dx/dt = -25$ and $dy/dt = -30$ when $x = 0.3$ and $y = 0.4$ so $dD/dt = (0.3/0.5)(-25) + (0.4/0.5)(-30) = -39$ mph.

47. $A = \dfrac{1}{2}ab\sin\theta$ but $\theta = \pi/6$ when $a = 4$ and $b = 3$ so $A = \dfrac{1}{2}(4)(3)\sin(\pi/6) = 3$.

Solve $\dfrac{1}{2}ab\sin\theta = 3$ for θ to get $\theta = \sin^{-1}\left(\dfrac{6}{ab}\right)$, $0 \le \theta \le \pi/2$.

$\dfrac{d\theta}{dt} = \dfrac{\partial\theta}{\partial a}\dfrac{da}{dt} + \dfrac{\partial\theta}{\partial b}\dfrac{db}{dt} = \dfrac{1}{\sqrt{1 - \dfrac{36}{a^2b^2}}}\left(-\dfrac{6}{a^2b}\right)\dfrac{da}{dt} + \dfrac{1}{\sqrt{1 - \dfrac{36}{a^2b^2}}}\left(-\dfrac{6}{ab^2}\right)\dfrac{db}{dt}$

$= -\dfrac{6}{\sqrt{a^2b^2 - 36}}\left(\dfrac{1}{a}\dfrac{da}{dt} + \dfrac{1}{b}\dfrac{db}{dt}\right)$, $\dfrac{da}{dt} = 1$ and $\dfrac{db}{dt} = 1$

when $a = 4$ and $b = 3$ so $\dfrac{d\theta}{dt} = -\dfrac{6}{\sqrt{144 - 36}}\left(\dfrac{1}{4} + \dfrac{1}{3}\right) = -\dfrac{7}{12\sqrt{3}} = -\dfrac{7}{36}\sqrt{3}$ radians/sec

49. $\dfrac{dT}{dt} = \dfrac{\partial T}{\partial x}\dfrac{dx}{dt} + \dfrac{\partial T}{\partial y}\dfrac{dy}{dt} = \dfrac{y^2}{x}\dfrac{dx}{dt} + 2y\ln x\dfrac{dy}{dt}$, $dx/dt = 1$ and $dy/dt = -4$ at $(3,2)$ so

$dT/dt = (4/3)(1) + (4\ln 3)(-4) = 4/3 - 16\ln 3\,°\text{C/sec}$.

51. $z = f(u)$, $u = x^2 - y^2$; $\partial z/\partial x = (dz/du)(\partial u/\partial x) = 2x\,dz/du$,

$\partial z/\partial y = (dz/du)(\partial u/\partial y) = -2y\,dz/du$, $y\partial z/\partial x + x\partial z/\partial y = 2xy\,dz/du - 2xy\,dz/du = 0$.

53. **(a)** $\dfrac{\partial z}{\partial x} = \dfrac{dz}{du}\dfrac{\partial u}{\partial x}$, $\dfrac{\partial^2 z}{\partial x^2} = \dfrac{dz}{du}\dfrac{\partial^2 u}{\partial x^2} + \dfrac{\partial}{\partial x}\left(\dfrac{dz}{du}\right)\dfrac{\partial u}{\partial x} = \dfrac{dz}{du}\dfrac{\partial^2 u}{\partial x^2} + \dfrac{d^2 z}{du^2}\left(\dfrac{\partial u}{\partial x}\right)^2$; proceed in a similar fashion for $\partial^2 z/\partial y^2$.

(b) $\dfrac{\partial^2 z}{\partial y\partial x} = \dfrac{dz}{du}\dfrac{\partial^2 u}{\partial y\partial x} + \dfrac{\partial}{\partial y}\left(\dfrac{dz}{du}\right)\dfrac{\partial u}{\partial x} = \dfrac{dz}{du}\dfrac{\partial^2 u}{\partial y\partial x} + \dfrac{d^2 z}{du^2}\dfrac{\partial u}{\partial x}\dfrac{\partial u}{\partial y}$.

55. **(a)** $\dfrac{\partial^2 z}{\partial x^2} + \dfrac{\partial^2 z}{\partial y^2} = \dfrac{dz}{dr}\dfrac{\partial^2 r}{\partial x^2} + \dfrac{d^2 z}{dr^2}\left(\dfrac{\partial r}{\partial x}\right)^2 + \dfrac{dz}{dr}\dfrac{\partial^2 r}{\partial y^2} + \dfrac{d^2 z}{dr^2}\left(\dfrac{\partial r}{\partial y}\right)^2$

$= \dfrac{dz}{dr}\dfrac{y^2}{r^3} + \dfrac{d^2 z}{dr^2}\dfrac{x^2}{r^2} + \dfrac{dz}{dr}\dfrac{x^2}{r^3} + \dfrac{d^2 z}{dr^2}\dfrac{y^2}{r^2} = \dfrac{d^2 z}{dr^2} + \dfrac{1}{r}\dfrac{dz}{dr} = 0$ so $r\dfrac{d^2 z}{dr^2} + \dfrac{dz}{dr} = 0$.

(b) $r\dfrac{d^2 z}{dr^2} + \dfrac{dz}{dr} = \dfrac{d}{dr}\left(r\dfrac{dz}{dr}\right) = 0$, $r\dfrac{dz}{dr} = C_1$, $\dfrac{dz}{dr} = C_1/r$, $z = C_1\ln r + C_2$.

57. $z = f(u) + g(v)$ where $u = y + cx$ and $v = y - cx$, $\dfrac{\partial z}{\partial x} = \dfrac{\partial z}{\partial u}\dfrac{\partial u}{\partial x} + \dfrac{\partial z}{\partial v}\dfrac{\partial v}{\partial x} = cf'(u) - cg'(v)$

and $\dfrac{\partial^2 z}{\partial x^2} = \dfrac{\partial}{\partial u}\left(\dfrac{\partial z}{\partial x}\right)\dfrac{\partial u}{\partial x} + \dfrac{\partial}{\partial v}\left(\dfrac{\partial z}{\partial x}\right)\dfrac{\partial v}{\partial x} = c^2 f''(u) + c^2 g''(v)$,

similarly we find that $\dfrac{\partial^2 z}{\partial y^2} = f''(u) + g''(v)$ so $\dfrac{\partial^2 z}{\partial x^2} = c^2\dfrac{\partial^2 z}{\partial y^2}$.

59. The result follows by a direct application of the chain rule.

61. **(a)** $1 = -r\sin\theta\dfrac{\partial\theta}{\partial x} + \cos\theta\dfrac{\partial r}{\partial x}$ and $0 = r\cos\theta\dfrac{\partial\theta}{\partial x} + \sin\theta\dfrac{\partial r}{\partial x}$; solve for $\partial r/\partial x$ and $\partial\theta/\partial x$.

(b) $0 = -r\sin\theta\dfrac{\partial\theta}{\partial y} + \cos\theta\dfrac{\partial r}{\partial y}$ and $1 = r\cos\theta\dfrac{\partial\theta}{\partial y} + \sin\theta\dfrac{\partial r}{\partial y}$; solve for $\partial r/\partial y$ and $\partial\theta/\partial y$.

63. Square and add the results of parts (a) and (b).

65. **(a)** $f(tx,ty) = 3t^2 x^2 + t^2 y^2 = t^2 f(x,y)$; $n = 2$.

(b) $f(tx,ty) = \sqrt{t^2 x^2 + t^2 y^2} = tf(x,y)$; $n = 1$.

(c) $f(tx,ty) = t^3 x^2 y - 2t^3 y^3 = t^3 f(x,y)$; $n = 3$.

(d) $f(tx,ty) = 5/\left(t^2 x^2 + 2t^2 y^2\right)^2 = t^{-4} f(x,y)$; $n = -4$.

67. **(a)** $\dfrac{\partial w}{\partial x} = \dfrac{\partial f}{\partial x} + \dfrac{\partial f}{\partial y}\dfrac{\partial y}{\partial x}$ **(b)** $\dfrac{\partial w}{\partial z} = \dfrac{\partial f}{\partial y}\dfrac{\partial y}{\delta z}$

69. Represent the line segment C that joins A and B by $x = x_0 + (x_1 - x_0)t$, $y = y_0 + (y_1 - y_0)t$ for $0 \le t \le 1$. $f(x,y) = F(t)$ for (x,y) on C, moreover $f(x_1, y_1) - f(x_0, y_0) = F(1) - F(0)$. Apply the Mean Value Theorem to $F(t)$ on the interval $[0,1]$ to get $[F(1) - F(0)]/(1 - 0) = F'(t^*)$, $F(1) - F(0) = F'(t^*)$ for some t^* in $(0,1)$ so $f(x_1, y_1) - f(x_0, y_0) = F'(t^*)$. By the chain rule, $F'(t) = f_x(x,y)(dx/dt) + f_y(x,y)(dy/dt) = f_x(x,y)(x_1 - x_0) + f_y(x,y)(y_1 - y_0)$. Let (x^*, y^*) be the point on C for $t = t^*$ then
$$f(x_1, y_1) - f(x_0, y_0) = F'(t^*) = f_x(x^*, y^*)(x_1 - x_0) + f_y(x^*, y^*)(y_1 - y_0).$$

EXERCISE SET 16.5

1. At P, $\partial z/\partial x = 48$ and $\partial z/\partial y = -14$, tangent plane $48x - 14y - z = 64$, normal line $x = 1 + 48t$, $y = -2 - 14t$, $z = 12 - t$.

3. At P, $\partial z/\partial x = 1$ and $\partial z/\partial y = -1$, tangent plane $x - y - z = 0$, normal line $x = 1 + t$, $y = -t$, $z = 1 - t$.

5. At P, $\partial z/\partial x = 0$ and $\partial z/\partial y = 3$, tangent plane $3y - z = -1$, normal line $x = \pi/6$, $y = 3t$, $z = 1 - t$.

7. By implicit differentiation $\partial z/\partial x = -x/z$, $\partial z/\partial y = -y/z$ so at P, $\partial z/\partial x = 3/4$ and $\partial z/\partial y = 0$, tangent plane $3x - 4z = -25$, normal line $x = -3 + 3t/4$, $y = 0$, $z = 4 - t$.

9. $dz = 7dx - 2dy$

11. $dz = \left[y/\left(1 + x^2 y^2\right)\right] dx + \left[x/\left(1 + x^2 y^2\right)\right] dy$

13. $df = (2x + 2y - 4)dx + 2x dy$; $x = 1$, $y = 2$, $dx = 0.01$, $dy = 0.04$ so $df = 0.10$

15. $df = -x^{-2} dx - y^{-2} dy$; $x = -1$, $y = -2$, $dx = -0.02$, $dy = -0.04$ so $df = 0.03$

17. The tangent plane is horizontal if the normal $\partial z/\partial x \mathbf{i} + \partial z/\partial y \mathbf{j} - \mathbf{k}$ is parallel to $\mathbf{k}$ which occurs when $\partial z/\partial x = \partial z/\partial y = 0$.

(a) $\partial z/\partial x = 3x^2 y^2$, $\partial z/\partial y = 2x^3 y$; $3x^2 y^2 = 0$ and $2x^3 y = 0$ for all (x,y) on the x-axis or y-axis, and $z = 0$ for these points, the tangent plane is horizontal at all points on the x-axis or y-axis.

(b) $\partial z/\partial x = 2x - y - 2$, $\partial z/\partial y = -x + 2y + 4$; solve the system $2x - y - 2 = 0$, $-x + 2y + 4 = 0$, to get $x = 0$, $y = -2$. $z = -4$ at $(0, -2)$, the tangent plane is horizontal at $(0, -2, -4)$.

19. $\partial z/\partial x = -6x$, $\partial z/\partial y = -4y$ so $-6x_0\mathbf{i}-4y_0\mathbf{j} - \mathbf{k}$ is normal to the surface at a point (x_0, y_0, z_0) on the surface. This normal must be parallel to the given line and hence to the vector

     $-3\mathbf{i} + 8\mathbf{j} - \mathbf{k}$ which is parallel to the line so $-6x_0 = -3$, $x_0 = 1/2$ and $-4y_0 = 8$, $y_0 = -2$. $z = -3/4$ at $(1/2, -2)$. The point on the surface is $(1/2, -2, -3/4)$.

21. $(2, 2, 2\sqrt{2})$ satisfies both equations. $\mathbf{n}_1 = -\left(1/\sqrt{2}\right)\mathbf{i} - \left(1/\sqrt{2}\right)\mathbf{j} - \mathbf{k}$ and $\mathbf{n}_2 = \left(1/\sqrt{2}\right)\mathbf{i} + \left(1/\sqrt{2}\right)\mathbf{j} - \mathbf{k}$ are normal, respectively, to each of the surfaces at $(2, 2, 2\sqrt{2})$. $\mathbf{n}_1 \cdot \mathbf{n}_2 = 0$ so the normals are perpendicular and hence so are the tangent planes.

23. $z = \sqrt{x^2 + y^2}$, $dz = x\left(x^2 + y^2\right)^{-1/2} dx + y\left(x^2 + y^2\right)^{-1/2} dy$; $x = 3$, $y = 4$, $dx = 0.2$, $dy = -0.04$ so $dz = 0.088$ cm.

25. $A = xy$, $dA = y\,dx + x\,dy$, $dA/A = dx/x + dy/y$, $|dx/x| \le 0.03$ and $|dy/y| \le 0.05$, $|dA/A| \le |dx/x| + |dy/y| \le 0.08 = 8\%$

27. $z = \sqrt{x^2 + y^2}$, $dz = \dfrac{x}{\sqrt{x^2 + y^2}}dx + \dfrac{y}{\sqrt{x^2 + y^2}}dy$,

     $\dfrac{dz}{z} = \dfrac{x}{x^2 + y^2}dx + \dfrac{y}{x^2 + y^2}dy = \dfrac{x^2}{x^2 + y^2}\left(\dfrac{dx}{x}\right) + \dfrac{y^2}{x^2 + y^2}\left(\dfrac{dy}{y}\right)$,

     $\left|\dfrac{dz}{z}\right| \le \dfrac{x^2}{x^2 + y^2}\left|\dfrac{dx}{x}\right| + \dfrac{y^2}{x^2 + y^2}\left|\dfrac{dy}{y}\right|$, if $\left|\dfrac{dx}{x}\right| \le r/100$ and $\left|\dfrac{dy}{y}\right| \le r/100$ then

     $\left|\dfrac{dz}{z}\right| \le \dfrac{x^2}{x^2 + y^2}(r/100) + \dfrac{y^2}{x^2 + y^2}(r/100) = \dfrac{r}{100}$ so the percentage error in z is at most $r\%$.

29. $dR = \dfrac{R_2^2}{(R_1 + R_2)^2}dR_1 + \dfrac{R_1^2}{(R_1 + R_2)^2}dR_2$, $\dfrac{dR}{R} = \dfrac{R_2}{R_1 + R_2}\left(\dfrac{dR_1}{R_1}\right) + \dfrac{R_1}{R_1 + R_2}\left(\dfrac{dR_2}{R_2}\right)$,

     $\left|\dfrac{dR}{R}\right| \le \dfrac{R_2}{R_1 + R_2}\left|\dfrac{dR_1}{R_1}\right| + \dfrac{R_1}{R_1 + R_2}\left|\dfrac{dR_2}{R_2}\right|$; if $R_1 = 200$, $R_2 = 400$, $|dR_1/R_1| \le 0.02$, and $|dR_2/R_2| \le 0.02$ then $|dR/R| \le (400/600)(0.02) + (200/600)(0.02) = 0.02 = 2\%$.

31. $d\theta = \dfrac{1}{\sqrt{c^2 - a^2}}da - \dfrac{a}{c\sqrt{c^2 - a^2}}dc$; if $a = 3$, $c = 5$, $|da| \le 0.01$, and $|dc| \le 0.01$ then

     $|d\theta| \le (1/4)(0.01) + (3/20)(0.01) = 0.004$ radians.

33. $dT = \dfrac{\pi}{g\sqrt{L/g}}dL - \dfrac{\pi L}{g^2\sqrt{L/g}}dg$, $\dfrac{dT}{T} = \dfrac{1}{2}\dfrac{dL}{L} - \dfrac{1}{2}\dfrac{dg}{g}$; $|dL/L| \le 0.005$ and $|dg/g| \le 0.001$ so $|dT/T| \le (1/2)(0.005) + (1/2)(0.001) = 0.003 = 0.3\%$

35.   **(a)** $z = xy$, $dz = y\,dx + x\,dy$, $dz/z = dx/x + dy/y$; $(r + s)\%$.

(b) $z = x/y$, $dz = dx/y - xdy/y^2$, $dz/z = dx/x - dy/y$; $(r + s)\%$.

(c) $z = x^2y^3$, $dz = 2xy^3dx + 3x^2y^2dy$, $dz/z = 2dx/x + 3dy/y$; $(2r + 3s)\%$.

(d) $z = x^3y^{1/2}$, $dz = 3x^2y^{1/2}dx + x^3dy/(2y^{1/2})$, $dz/z = 3dx/x + (1/2)dy/y$; $(3r + s/2)\%$.

37. **(a)** $2t + 7 = (-1 + t)^2 + (2 + t)^2$, $t^2 = 1$, $t = \pm 1$ so the points of intersection are $(-2, 1, 5)$ and $(0, 3, 9)$.

(b) $\partial z/\partial x = 2x$, $\partial z/\partial y = 2y$ so at $(-2, 1, 5)$ the vector $\mathbf{n} = -4\mathbf{i} + 2\mathbf{j} - \mathbf{k}$ is normal to the surface. $\mathbf{v} = \mathbf{i} + \mathbf{j} + 2\mathbf{k}$ is parallel to the line; $\mathbf{n} \cdot \mathbf{v} = -4$ so the cosine of the acute angle is $[\mathbf{n} \cdot (-\mathbf{v})]/(\|\mathbf{n}\| \| - \mathbf{v}\|) = 4/(\sqrt{21}\sqrt{6}) = 4/(3\sqrt{14})$. Similarly, at $(0, 3, 9)$ the vector $\mathbf{n} = 6\mathbf{j} - \mathbf{k}$ is normal to the surface, $\mathbf{n} \cdot \mathbf{v} = 4$ so the cosine of the acute angle is $4/(\sqrt{37}\sqrt{6}) = 4/\sqrt{222}$.

39. Use implicit differentiation to get $\partial z/\partial x = -c^2x/(a^2z)$, $\partial z/\partial y = -c^2y/(b^2z)$. At (x_0, y_0, z_0), $z_0 \neq 0$, a normal to the surface is $-[c^2x_0/(a^2z_0)]\mathbf{i} - [c^2y_0/(b^2z_0)]\mathbf{j} - \mathbf{k}$ so the tangent

plane is $-\dfrac{c^2x_0}{a^2z_0}x - \dfrac{c^2y_0}{b^2z_0}y - z = -\dfrac{c^2x_0^2}{a^2z_0} - \dfrac{c^2y_0^2}{b^2z_0} - z_0$, $\dfrac{x_0x}{a^2} + \dfrac{y_0y}{b^2} - \dfrac{z_0z}{c^2} = \dfrac{x_0^2}{a^2} + \dfrac{y_0^2}{b^2} + \dfrac{z_0^2}{c^2} = 1$

41. $\mathbf{n}_1 = f_x(x_0, y_0)\mathbf{i} + f_y(x_0, y_0)\mathbf{j} - \mathbf{k}$ and $\mathbf{n}_2 = g_x(x_0, y_0)\mathbf{i} + g_y(x_0, y_0)\mathbf{j} - \mathbf{k}$ are normal, respectively, to $z = f(x, y)$ and $z = g(x, y)$ at P; $\mathbf{n}_1$ and $\mathbf{n}_2$ are perpendicular if and only if $\mathbf{n}_1 \cdot \mathbf{n}_2 = 0$, $f_x(x_0, y_0)g_x(x_0, y_0) + f_y(x_0, y_0)g_y(x_0, y_0) + 1 = 0$, $f_x(x_0, y_0)g_x(x_0, y_0) + f_y(x_0, y_0)g_y(x_0, y_0) = -1$.

EXERCISE SET 16.6

1. $\nabla z = 4\mathbf{i} - 8\mathbf{j}$

3. $\nabla z = \dfrac{x}{x^2 + y^2}\mathbf{i} + \dfrac{y}{x^2 + y^2}\mathbf{j}$

5. $\nabla f(x, y) = 3(2x + y)(x^2 + xy)^2\mathbf{i} + 3x(x^2 + xy)^2\mathbf{j}$, $\nabla f(-1, -1) = -36\mathbf{i} - 12\mathbf{j}$

7. $\nabla f(x, y) = [y/(x + y)]\mathbf{i} + [y/(x + y) + \ln(x + y)]\mathbf{j}$, $\nabla f(-3, 4) = 4\mathbf{i} + 4\mathbf{j}$

9. $\nabla f(x, y) = (3y/2)(1 + xy)^{1/2}\mathbf{i} + (3x/2)(1 + xy)^{1/2}\mathbf{j}$, $\nabla f(3, 1) = 3\mathbf{i} + 9\mathbf{j}$,
$D_{\mathbf{u}}f = \nabla f \cdot \mathbf{u} = 12/\sqrt{2} = 6\sqrt{2}$

11. $\nabla f(x, y) = [2x/(1 + x^2 + y)]\mathbf{i} + [1/(1 + x^2 + y)]\mathbf{j}$, $\nabla f(0, 0) = \mathbf{j}$, $D_{\mathbf{u}}f = -3/\sqrt{10}$

13. $\nabla f(x, y) = 12x^2y^2\mathbf{i} + 8x^3y\mathbf{j}$, $\nabla f(2, 1) = 48\mathbf{i} + 64\mathbf{j}$, $\mathbf{u} = (4/5)\mathbf{i} - (3/5)\mathbf{j}$, $D_{\mathbf{u}}f = \nabla f \cdot \mathbf{u} = 0$

15. $\nabla f(x, y) = (y^2/x)\mathbf{i} + 2y\ln x\mathbf{j}$, $\nabla f(1, 4) = 16\mathbf{i}$, $\mathbf{u} = (-\mathbf{i} + \mathbf{j})/\sqrt{2}$, $D_{\mathbf{u}}f = -8\sqrt{2}$

17. $\nabla f(x,y) = -\left[y/\left(x^2+y^2\right)\right]\mathbf{i} + \left[x/\left(x^2+y^2\right)\right]\mathbf{j}$, $\nabla f(-2,2) = -(\mathbf{i}+\mathbf{j})/4$, $\mathbf{u} = -(\mathbf{i}+\mathbf{j})/\sqrt{2}$,
$D_{\mathbf{u}}f = \sqrt{2}/4$

19. $\nabla f(x,y) = (y/2)(xy)^{-1/2}\mathbf{i} + (x/2)(xy)^{-1/2}\mathbf{j}$, $\nabla f(1,4) = \mathbf{i} + (1/4)\mathbf{j}$,
$\mathbf{u} = \cos\theta\mathbf{i} + \sin\theta\mathbf{j} = (1/2)\mathbf{i} + (\sqrt{3}/2)\mathbf{j}$, $D_{\mathbf{u}}f = 1/2 + \sqrt{3}/8$

21. $\nabla f(x,y) = 2\sec^2(2x+y)\mathbf{i} + \sec^2(2x+y)\mathbf{j}$, $\nabla f(\pi/6,\pi/3) = 8\mathbf{i}+4\mathbf{j}$, $\mathbf{u} = (\mathbf{i}-\mathbf{j})/\sqrt{2}$, $D_{\mathbf{u}}f = 2\sqrt{2}$

23. $f(1,2) = 3$, level curve
$4x - 2y + 3 = 3$,
$4x - 2y = 0$.
$\nabla f(x,y) = 4\mathbf{i} - 2\mathbf{j}$
$\nabla f(1,2) = 4\mathbf{i} - 2\mathbf{j}$

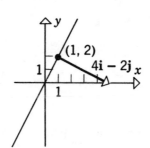

25. $f(-2,0) = 4$, level curve
$x^2 + 4y^2 = 4$, $x^2/4 + y^2 = 1$.
$\nabla f(x,y) = 2x\mathbf{i} + 8y\mathbf{j}$
$\nabla f(-2,0) = -4\mathbf{i}$

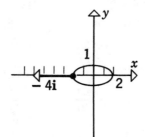

27. $\nabla f(x,y) = 12x^2y^2\mathbf{i} + 8x^3y\mathbf{j}$, $\nabla f(-1,1) = 12\mathbf{i} - 8\mathbf{j}$, $\mathbf{u} = (3\mathbf{i} - 2\mathbf{j})/\sqrt{13}$, $\|\nabla f(-1,1)\| = 4\sqrt{13}$

29. $\nabla f(x,y) = x\left(x^2+y^2\right)^{-1/2}\mathbf{i} + y\left(x^2+y^2\right)^{-1/2}\mathbf{j}$, $\nabla f(4,-3) = (4\mathbf{i} - 3\mathbf{j})/5$, $\mathbf{u} = (4\mathbf{i} - 3\mathbf{j})/5$,
$\|\nabla f(4,-3)\| = 1$

31. $\nabla f(x,y) = -2x\mathbf{i} - 2y\mathbf{j}$, $\nabla f(-1,-3) = 2\mathbf{i} + 6\mathbf{j}$, $\mathbf{u} = -(\mathbf{i} + 3\mathbf{j})/\sqrt{10}$, $-\|\nabla f(-1,-3)\| = -2\sqrt{10}$

33. $\nabla f(x,y) = -3\sin(3x - y)\mathbf{i} + \sin(3x - y)\mathbf{j}$, $\nabla f(\pi/6,\pi/4) = (-3\mathbf{i} + \mathbf{j})/\sqrt{2}$, $\mathbf{u} = (3\mathbf{i} - \mathbf{j})/\sqrt{10}$,
$-\|\nabla f(\pi/6,\pi/4)\| = -\sqrt{5}$

35. $\nabla f(x,y) = y(x + y)^{-2}\mathbf{i} - x(x + y)^{-2}\mathbf{j}$, $\nabla f(1,0) = -\mathbf{j}$, $\overrightarrow{PQ} = -2\mathbf{i} - \mathbf{j}$, $\mathbf{u} = (-2\mathbf{i} - \mathbf{j})/\sqrt{5}$,
$D_{\mathbf{u}}f = 1/\sqrt{5}$

37. $\nabla f(x,y) = \dfrac{ye^y}{2\sqrt{xy}}\mathbf{i} + \left(\sqrt{xy}e^y + \dfrac{xe^y}{2\sqrt{xy}}\right)\mathbf{j}$, $\nabla f(1,1) = (e/2)(\mathbf{i} + 3\mathbf{j})$, $\mathbf{u} = -\mathbf{j}$, $D_{\mathbf{u}}f = -3e/2$

39. $\nabla f(x,y) = 8xy\mathbf{i} + 4x^2\mathbf{j}$, $\nabla f(1,-2) = -16\mathbf{i} + 4\mathbf{j}$ is perpendicular to the level curve through P so $\mathbf{u} = \pm(-4\mathbf{i} + \mathbf{j})/\sqrt{17}$.

41. Solve the system $(3/5)f_x(1,2) - (4/5)f_y(1,2) = -5$, $(4/5)f_x(1,2) + (3/5)f_y(1,2) = 10$ for $f_x(1,2)$ and $f_y(1,2)$ to get $f_x(1,2) = 5$, $f_y(1,2) = 10$. For (c), $\nabla f(1,2) = 5\mathbf{i} + 10\mathbf{j}$, $\mathbf{u} = (-\mathbf{i} - 2\mathbf{j})/\sqrt{5}$, $D_{\mathbf{u}}f = -5\sqrt{5}$.

43. $\nabla f(4,-5) = 2\mathbf{i} - \mathbf{j}$, $\mathbf{u} = (5\mathbf{i} + 2\mathbf{j})/\sqrt{29}$, $D_{\mathbf{u}}f = 8/\sqrt{29}$

45. $\nabla z = 6x\mathbf{i} - 2y\mathbf{j}$, $\|\nabla z\| = \sqrt{36x^2 + 4y^2} = 6$ if $36x^2 + 4y^2 = 36$; all points on the ellipse $9x^2 + y^2 = 9$.

47. $\mathbf{r} = t\mathbf{i} - t^2\mathbf{j}$, $d\mathbf{r}/dt = \mathbf{i} - 2t\mathbf{j} = \mathbf{i} - 4\mathbf{j}$ at the point $(2,-4)$, $\mathbf{u} = (\mathbf{i} - 4\mathbf{j})/\sqrt{17}$; $dz/ds = D_{\mathbf{u}}z = \nabla z \cdot \mathbf{u} = 36/\sqrt{17}$.

49. (a) $\nabla V(x,y) = -2e^{-2x}\cos 2y\mathbf{i} - 2e^{-2x}\sin 2y\mathbf{j}$, $\mathbf{E} = -\nabla V(\pi/4, 0) = 2e^{-\pi/2}\mathbf{i}$
   (b) $V(x,y)$ decreases most rapidly in the direction of $-\nabla V(x,y)$ which is $\mathbf{E}$.

51. (a) $\nabla r = \dfrac{x}{\sqrt{x^2 + y^2}}\mathbf{i} + \dfrac{y}{\sqrt{x^2 + y^2}}\mathbf{j} = \mathbf{r}/r$

   (b) $\nabla f(r) = \dfrac{\partial f(r)}{\partial x}\mathbf{i} + \dfrac{\partial f(r)}{\partial y}\mathbf{j} = f'(r)\dfrac{\partial r}{\partial x}\mathbf{i} + f'(r)\dfrac{\partial r}{\partial y}\mathbf{j} = f'(r)\nabla r$

53. $\mathbf{u}_r = \cos\theta\mathbf{i} + \sin\theta\mathbf{j}$, $\mathbf{u}_\theta = -\sin\theta\mathbf{i} + \cos\theta\mathbf{j}$,

$$\nabla z = \frac{\partial z}{\partial x}\mathbf{i} + \frac{\partial z}{\partial y}\mathbf{j} = \left(\frac{\partial z}{\partial r}\cos\theta - \frac{1}{r}\frac{\partial z}{\partial\theta}\sin\theta\right)\mathbf{i} + \left(\frac{\partial z}{\partial r}\sin\theta + \frac{1}{r}\frac{\partial z}{\partial x}\cos\theta\right)\mathbf{j}$$

$$= \frac{\partial z}{\partial r}(\cos\theta\mathbf{i} + \sin\theta\mathbf{j}) + \frac{1}{r}\frac{\partial z}{\partial\theta}(-\sin\theta\mathbf{i} + \cos\theta\mathbf{j}) = \frac{\partial z}{\partial r}\mathbf{u}_r + \frac{1}{r}\frac{\partial z}{\partial\theta}\mathbf{u}_\theta$$

55. $dz/dt = (\partial z/\partial x)(dx/dt) + (\partial z/\partial y)(dy/dt)$
$$= (\partial z/\partial x\mathbf{i} + \partial z/\partial y\mathbf{j}) \cdot (dx/dt\mathbf{i} + dy/dt\mathbf{j}) = \nabla z \cdot \mathbf{r}'(t)$$

57. Let $\mathbf{u}_1$ and $\mathbf{u}_2$ be nonparallel unit vectors for which the directional derivative is zero. Let $\mathbf{u}$ be any other unit vector, then $\mathbf{u} = c_1\mathbf{u}_1 + c_2\mathbf{u}_2$ for some choice of scalars c_1 and c_2,
$$D_{\mathbf{u}}f(x,y) = \nabla f(x,y) \cdot \mathbf{u} = c_1\nabla f(x,y) \cdot \mathbf{u}_1 + c_2\nabla f(x,y) \cdot \mathbf{u}_2$$
$$= c_1 D_{\mathbf{u}_1}f(x,y) + c_2 D_{\mathbf{u}_2}f(x,y) = 0.$$

EXERCISE SET 16.7

1. $165t^{32}$ **3.** $-2t \cos\left(t^2\right)$ **5.** 3264

7. $\nabla f(x,y,z) = 20x^4y^2z^3\mathbf{i} + 8x^5yz^3\mathbf{j} + 12x^5y^2z^2\mathbf{k}$, $\nabla f(2,-1,1) = 320\mathbf{i} - 256\mathbf{j} + 384\mathbf{k}$, $D_{\mathbf{u}}f = -320$

9. $\nabla f(x,y,z) = \dfrac{2x}{x^2 + 2y^2 + 3z^2}\mathbf{i} + \dfrac{4y}{x^2 + 2y^2 + 3z^2}\mathbf{j} + \dfrac{6z}{x^2 + 2y^2 + 3z^2}\mathbf{k}$,
$\nabla f(-1,2,4) = (-2/57)\mathbf{i} + (8/57)\mathbf{j} + (24/57)\mathbf{k}$, $D_{\mathbf{u}}f = -314/741$

11. $\nabla f(x,y,z) = \left(3x^2z - 2xy\right)\mathbf{i} - x^2\mathbf{j} + \left(x^3 + 2z\right)\mathbf{k}$, $\nabla f(2,-1,1) = 16\mathbf{i} - 4\mathbf{j} + 10\mathbf{k}$,
$\mathbf{u} = (3\mathbf{i} - \mathbf{j} + 2\mathbf{k})/\sqrt{14}$, $D_{\mathbf{u}}f = 72/\sqrt{14}$

13. $\nabla f(x,y,z) = -\dfrac{1}{z+y}\mathbf{i} - \dfrac{z-x}{(z+y)^2}\mathbf{j} + \dfrac{y+x}{(z+y)^2}\mathbf{k}$, $\nabla f(1,0,-3) = (1/3)\mathbf{i} + (4/9)\mathbf{j} + (1/9)\mathbf{k}$,
$\mathbf{u} = (-6\mathbf{i} + 3\mathbf{j} - 2\mathbf{k})/7$, $D_{\mathbf{u}}f = -8/63$

15. $\nabla f(1,1,-1) = 3\mathbf{i} - 3\mathbf{j}$, $\mathbf{u} = (\mathbf{i} - \mathbf{j})/\sqrt{2}$, $\|\nabla f(1,1,-1)\| = 3\sqrt{2}$

17. $\nabla f(1,2,-2) = (-\mathbf{i} + \mathbf{j})/2$, $\mathbf{u} = (-\mathbf{i} + \mathbf{j})/\sqrt{2}$, $\|\nabla f(1,2,-2)\| = 1/\sqrt{2}$

19. $\nabla f(5,7,6) = -\mathbf{i} + 11\mathbf{j} - 12\mathbf{k}$, $\mathbf{u} = (\mathbf{i} - 11\mathbf{j} + 12\mathbf{k})/\sqrt{266}$, $-\|\nabla f(5,7,6)\| = -\sqrt{266}$

21. $\nabla f(2,1,-1) = -\mathbf{i} + \mathbf{j} - \mathbf{k}$. $\overrightarrow{PQ} = -3\mathbf{i} + \mathbf{j} + \mathbf{k}$, $\mathbf{u} = (-3\mathbf{i} + \mathbf{j} + \mathbf{k})/\sqrt{11}$, $D_{\mathbf{u}}f = 3/\sqrt{11}$

23. Let $\mathbf{u}$ be the unit vector in the direction of $\mathbf{a}$, then
$D_{\mathbf{u}}f(3,-2,1) = \nabla f(3,-2,1) \cdot \mathbf{u} = \|\nabla f(3,-2,1)\| \cos\theta = 5\cos\theta = -5$, $\cos\theta = -1$, $\theta = \pi$ so
$\nabla f(3,-2,1)$ is oppositely directed to $\mathbf{u}$; $\nabla f(3,-2,1) = -5\mathbf{u} = -10/3\mathbf{i} + 5/3\mathbf{j} + 10/3\mathbf{k}$.

25. $f(x,y,z) = x^2 + y^2 + z^2$, $\nabla f(-3,2,-6) = -2(3\mathbf{i} - 2\mathbf{j} + 6\mathbf{k})$; tangent plane $3x - 2y + 6z = -49$;
normal line $x = -3 + 3t$, $y = 2 - 2t$, $z = -6 + 6t$

27. $f(x,y,z) = \sqrt{\dfrac{z+x}{y-1}} - z^2$, $\nabla f(3,5,1) = (\mathbf{i} - \mathbf{j} - 15\mathbf{k})/8$; tangent plane $x - y - 15z = -17$;
normal line $x = 3 + t$, $y = 5 - t$, $z = 1 - 15t$

29. $f(x,y,z) = x^2 + y^2 + z^2$, if (x_0, y_0, z_0) is on the sphere then $\nabla f(x_0, y_0, z_0) = 2(x_0\mathbf{i} + y_0\mathbf{j} + z_0\mathbf{k})$
is normal to the sphere at (x_0, y_0, z_0), the normal line is $x = x_0 + x_0t$, $y = y_0 + y_0t$, $z = z_0 + z_0t$
which passes through the origin when $t = -1$.

31. $f(x,y,z) = x^2 + y^2 - z^2$, if (x_0, y_0, z_0) is on the surface then $\nabla f(x_0, y_0, z_0) = 2(x_0\mathbf{i} + y_0\mathbf{j} - z_0\mathbf{k})$
is normal there and hence so is $\mathbf{n}_1 = x_0\mathbf{i} + y_0\mathbf{j} - z_0\mathbf{k}$; $\mathbf{n}_1$ must be parallel to $\overrightarrow{PQ} = 3\mathbf{i} + 2\mathbf{j} - 2\mathbf{k}$

so $n_1 = c \ \overrightarrow{PQ}$ for some constant c. Equate components to get $x_0 = 3c$, $y_0 = 2c$ and $z_0 = 2c$ which when substituted into the equation of the surface yields $9c^2 + 4c^2 - 4c^2 = 1$, $c^2 = 1/9$, $c = \pm 1/3$ so the points are $(1, 2/3, 2/3)$ and $(-1, -2/3, -2/3)$.

33. $dw = 8dx - 3dy + 4dz$

35. $dw = \dfrac{yz}{1 + x^2 y^2 z^2} dx + \dfrac{xz}{1 + x^2 y^2 z^2} dy + \dfrac{xy}{1 + x^2 y^2 z^2} dz$

37. $df = 2y^2 z^3 dx + 4xyz^3 dy + 6xy^2 z^2 dz$
    $= 2(-1)^2 (2)^3 (-0.01) + 4(1)(-1)(2)^3 (-0.02) + 6(1)(-1)^2 (2)^2 (0.02) = 0.96$

39. $V = \ell h$, $dV = wh d\ell + \ell h dw + \ell w dh$,
    $|dV| \leq wh |d\ell| + \ell h |dw| + \ell w |dh| \leq (4)(5)(0.05) + (3)(5)(0.05) + (3)(4)(0.05) = 2.35$ cm^3

41. $dA = \dfrac{1}{2} b \sin \theta \, da + \dfrac{1}{2}$ a $\sin \theta \, db + \dfrac{1}{2} ab \cos \theta \, d\theta$,

    $|dA| \leq \dfrac{1}{2} b \sin \theta |da| + \dfrac{1}{2} a \sin \theta |db| + \dfrac{1}{2} ab |\cos \theta| |d\theta|$

    $\leq \dfrac{1}{2} (50)(1/2)(1/2) + \dfrac{1}{2} (40)(1/2)(1/4) + \dfrac{1}{2} (40)(50) \left(\sqrt{3}/2 \right) (\pi/90)$

    $= 35/4 + 50\pi \sqrt{3}/9 \approx 39$ ft^2

43. $dw = y^2 z^3 dx + 2xyz^3 dy + 3xy^2 z^2 dz$,
    $|dw/w| \leq |dx/x| + 2|dy/y| + 3|dz/z| \leq (0.01) + 2(0.02) + 3(0.03) = 0.14 = 14\%$

45. $f(x, y, z) = x^2 + y^2 + z^2 - a^2$, $g(x, y, z) = z^2 - x^2 - y^2$. If (x_0, y_0, z_0) is a point of intersection then $f_x g_x + f_y g_y + f_z g_z = (2x_0)(-2x_0) + (2y_0)(-2y_0) + (2z_0)(2z_0) = 4 \left(z_0^2 - x_0^2 - y_0^2 \right)$, but $z_0^2 - x_0^2 - y_0^2 = 0$ because (x_0, y_0, z_0) is on $z^2 = x^2 + y^2$.

EXERCISE SET 16.8

1. $\partial f / \partial v = 8vw^3 x^4 y^5$, $\partial f / \partial w = 12v^2 w^2 x^4 y^5$, $\partial f / \partial x = 16v^2 w^3 x^3 y^5$, $\partial f / \partial y = 20v^2 w^3 x^4 y^4$

3. $\partial f / \partial v_1 = 2v_1 / \left(v_3^2 + v_4^2 \right)$, $\partial f / \partial v_2 = -2v_2 / \left(v_3^2 + v_4^2 \right)$, $\partial f / \partial v_3 = -2v_3 \left(v_1^2 - v_2^2 \right) / \left(v_3^2 + v_4^2 \right)^2$,
   $\partial f / \partial v_4 = -2v_4 \left(v_1^2 - v_2^2 \right) / \left(v_3^2 + v_4^2 \right)^2$

5. $128, -512, 32, 64/3$

7. $210t^{29}$

9. $\partial z/\partial r = (dz/dx)(\partial x/\partial r) = 2r\cos^2\theta/\left(r^2\cos^2\theta + 1\right)$,

      $\partial z/\partial\theta = (dz/dx)(\partial x/\partial\theta) = -2r^2\sin\theta\cos\theta/\left(r^2\cos^2\theta + 1\right)$

11. $\partial w/\partial\rho = 2\rho\left(4\sin^2\phi + \cos^2\phi\right)$, $\partial w/\partial\phi = 6\rho^2\sin\phi\cos\phi$, $\partial w/\partial\theta = 0$

13. $\dfrac{dw}{dy} = \dfrac{\partial w}{\partial x}\dfrac{dx}{dy} + \dfrac{\partial w}{\partial y} + \dfrac{\partial w}{\partial z}\dfrac{dz}{dy} = (-2\sin 2y\cos 2y + y + 1/2)/\sqrt{\cos^2 2y + y^2 + y}$

15.   Let a, b, and c be the lengths of the sides opposite angles A, B, and C, respectively. By the law of cosines $a = \left(b^2 + c^2 - 2bc\cos A\right)^{1/2}$ so

      $\dfrac{da}{dt} = \dfrac{b - c\cos A}{a}\dfrac{db}{dt} + \dfrac{c - b\cos A}{a}\dfrac{dc}{dt} + \dfrac{bc\sin A}{a}\dfrac{dA}{dt} = \dfrac{10 - 10}{10\sqrt{3}}(4) + \dfrac{20 - 5}{10\sqrt{3}}(2) + \dfrac{100\sqrt{3}}{10\sqrt{3}}(\pi/60)$

      $= \sqrt{3} + \pi/6$ cm/sec, increasing

17. $2x + 4x\partial z/\partial x + 4z + 2z\partial z/\partial x - 3y\partial z/\partial x = 0$, $\partial z/\partial x = -(2x + 4z)/(4x + 2z - 3y)$;

      $4x\partial z/\partial y + 2z\partial z/\partial y - 3y\partial z/\partial y - 3z = 0$, $\partial z/\partial y = 3z/(4x + 2z - 3y)$

19. $f_{ww} = 0$, $f_{xx} = -2wxyz/\left(x^2 + y^2\right)^2$, $f_{yy} = 2wxyz/\left(x^2 + y^2\right)^2$, $f_{zz} = 0$

21.   Let $z = f(u)$ where $u = x^2 + y^2$ then $\partial z/\partial x = (dz/du)(\partial u/\partial x) = 2x\,dz/du$,

      $\partial z/\partial y = (dz/du)(\partial u/\partial y) = 2y\,dz/du$ so $y\,\partial z/\partial x - x\partial z/\partial y = 2xy\,dz/du - 2xy\,dz/du = 0$

23.   Let $w = f(r, s, t)$ where $r = x - y$, $s = y - z$, $t = z - x$;

      $\partial w/\partial x = (\partial w/\partial r)(\partial r/\partial x) + (\partial w/\partial t)(\partial t/\partial x) = \partial w/\partial r - \partial w/\partial t$, similarily

      $\partial w/\partial y = -\partial w/\partial r + \partial w/\partial s$ and $\partial w/\partial z = -\partial w/\partial s + \partial w/\partial t$ so $\partial w/\partial x + \partial w/\partial y + \partial w/\partial z = 0$

25. $\dfrac{\partial F}{\partial x} + \dfrac{\partial F}{\partial z}\dfrac{\partial z}{\partial x} = 0$ so $\dfrac{\partial z}{\partial x} = -\dfrac{\partial F/\partial x}{\partial F/\partial z}$, $\dfrac{\partial F}{\partial y} + \dfrac{\partial F}{\partial z}\dfrac{\partial z}{\partial y} = 0$ so $\dfrac{\partial z}{\partial y} = -\dfrac{\partial F/\partial y}{\partial F/\partial z}$.

27. $ye^x - 5\sin 3z - 3z = 0$; $\dfrac{\partial z}{\partial x} = -\dfrac{ye^x}{-15\cos 3z - 3} = \dfrac{ye^x}{15\cos 3z + 3}$, $\dfrac{\partial z}{\partial y} = \dfrac{e^x}{15\cos 3z + 3}$.

29. $\nabla f(u, v, w) = \dfrac{\partial f}{\partial x}\mathbf{i} + \dfrac{\partial f}{\partial y}\mathbf{j} + \dfrac{\partial f}{\partial z}\mathbf{k}$

      $= \left(\dfrac{\partial f}{\partial u}\dfrac{\partial u}{\partial x} + \dfrac{\partial f}{\partial v}\dfrac{\partial v}{\partial x} + \dfrac{\partial f}{\partial w}\dfrac{\partial w}{\partial x}\right)\mathbf{i} + \left(\dfrac{\partial f}{\partial u}\dfrac{\partial u}{\partial y} + \dfrac{\partial f}{\partial v}\dfrac{\partial v}{\partial y} + \dfrac{\partial f}{\partial w}\dfrac{\partial w}{\partial y}\right)\mathbf{j}$

         $+ \left(\dfrac{\partial f}{\partial u}\dfrac{\partial u}{\partial z} + \dfrac{\partial f}{\partial v}\dfrac{\partial v}{\partial z} + \dfrac{\partial f}{\partial w}\dfrac{\partial w}{\partial z}\right)\mathbf{k} = \dfrac{\partial f}{\partial u}\nabla u + \dfrac{\partial f}{\partial v}\nabla v + \dfrac{\partial f}{\partial w}\nabla w$

31. $w_r = e^r/(e^r + e^s + e^t + e^u)$, $w_{rs} = -e^r e^s/(e^r + e^s + e^t + e^u)^2$,

  $w_{rst} = 2e^r e^s e^t/(e^r + e^s + e^t + e^u)^3$,

  $w_{rstu} = -6e^r e^s e^t e^u/(e^r + e^s + e^t + e^u)^4 = -6e^{r+s+t+u}/e^{4w} = -6e^{r+s+t+u-4w}$

33.   **(a)** $dw/dt = \displaystyle\sum_{i=1}^{4}(\partial w/\partial x_i)(dx_i/dt)$

  (b) $\partial w/\partial v_j = \displaystyle\sum_{i=1}^{4}(\partial w/\partial x_i)(\partial x_i/\partial v_j)$ for $j = 1, 2, 3$

35. $dF/dx = (\partial F/\partial u)(du/dx) + (\partial F/\partial v)(dv/dx)$

  $= -f(u)a'(x) + f(v)b'(x) = f(b(x))b'(x) - f(a(x))a'(x)$

EXERCISE SET 16.9

1. $f_x = 6x + 2y = 0$, $f_y = 2x + 2y = 0$; critical point $(0,0)$; $D > 0$ and $f_{xx} > 0$ at $(0,0)$, relative minimum.

3. $f_x = y + 2 = 0$, $f_y = 2y + x + 3 = 0$; critical point $(1, -2)$; $D < 0$ at $(1, -2)$, saddle point.

5. $f_x = 2x + y - 3 = 0$, $f_y = x + 2y = 0$; critical point $(2, -1)$; $D > 0$ and $f_{xx} > 0$ at $(2, -1)$, relative minimum.

7. $f_x = 2x - 2xy = 0$, $f_y = 4y - x^2 = 0$; critical points $(0,0)$ and $(\pm 2, 1)$; $D > 0$ and $f_{xx} > 0$ at $(0,0)$, relative minimum; $D < 0$ at $(\pm 2, 1)$, saddle points.

9. $f_x = 2x - 2/(x^2 y) = 0$, $f_y = 2y - 2/(xy^2) = 0$; critical points $(-1, -1)$ and $(1,1)$; $D > 0$ and $f_{xx} > 0$ at $(-1, -1)$ and $(1,1)$, relative minima.

11. $f_x = 2x = 0$, $f_y = 1 - e^y = 0$; critical point $(0,0)$; $D < 0$ at $(0,0)$, saddle point.

13. $f_x = e^x \sin y = 0$, $f_y = e^x \cos y = 0$, $\sin y = \cos y = 0$ is impossible, no critical points.

15. $f_x = 2y^2 - 2xy + 4y = 0$, $f_y = 4xy - x^2 + 4x = 0$; $2y(y - x + 2) = 0$ and $x(4y - x + 4) = 0$, critical points $(0,0)$, $(0, -2)$, $(4,0)$, and $(4/3, -2/3)$; $D < 0$ at $(0,0)$, $(0, -2)$, and $(4,0)$, saddle points; $D > 0$ and $f_{xx} > 0$ at $(4/3, -2/3)$, relative minimum.

17. $f_x = -2(x + 1)e^{-(x^2+y^2+2x)} = 0$, $f_y = -2ye^{-(x^2+y^2+2x)} = 0$; critical point $(-1,0)$; $D > 0$ and $f_{xx} < 0$ at $(-1,0)$, relative maximum.

19. $f_x = \cos x = 0$, $f_y = \cos y = 0$; critical point $(\pi/2, \pi/2)$; $D > 0$ and $f_{xx} < 0$ at $(\pi/2, \pi/2)$, relative maximum.

21. (a) critical point $(0,0)$; $D = 0$

(b) $f(0,0) = 0$, $x^4 + y^4 \geq 0$ so $f(x,y) \geq f(0,0)$, relative minimum.

23. (a) $f_x = 3e^y - 3x^2 = 3\left(e^y - x^2\right) = 0$, $f_y = 3xe^y - 3e^{3y} = 3e^y\left(x - e^{2y}\right) = 0$, $e^y = x^2$ and $e^{2y} = x$, $x^4 = x$, $x\left(x^3 - 1\right) = 0$ so $x = 0, 1$; critical point $(1,0)$; $D > 0$ and $f_{xx} < 0$ at $(1,0)$, relative maximum.

(b) $\displaystyle\lim_{x \to -\infty} f(x,0) = \lim_{x \to -\infty}\left(3x - x^3 - 1\right) = +\infty$ so no absolute maximum.

25. $f_x = y - 1 = 0$, $f_y = x - 3 = 0$; critical point $(3,1)$.

$\qquad$ Along $y = 0$: $u(x) = -x$; no critical points,

$\qquad$ along $x = 0$: $v(y) = -3y$; no critical points,

along $y = -\dfrac{4}{5}x + 4$: $w(x) = -\dfrac{4}{5}x^2 + \dfrac{27}{5}x - 12$; critical point $(27/8, 13/10)$.

(x,y)	$(3,1)$	$(0,0)$	$(5,0)$	$(0,4)$	$(27/8, 13/10)$
$f(x,y)$	-3	0	-5	-12	$-231/80$

Absolute maxium value is 0, absolute minimum value is -12.

27. $f_x = 2x - 2 = 0$, $f_y = -6y + 6 = 0$; critical point $(1,1)$.

Along $y = 0$: $u_1(x) = x^2 - 2x$; critical point $(1,0)$,

along $y = 2$: $u_2(x) = x^2 - 2x$; critical point $(1,2)$

along $x = 0$: $v_1(y) = -3y^2 + 6y$; critical point $(0,1)$,

along $x = 2$: $v_2(y) = -3y^2 + 6y$; critical point $(2,1)$

(x,y)	$(1,1)$	$(1,0)$	$(1,2)$	$(0,1)$	$(2,1)$	$(0,0)$	$(0,2)$	$(2,0)$	$(2,2)$
$f(x,y)$	2	-1	-1	3	3	0	0	0	0

Absolute maxium value is 3, absolute minimum value is -1.

29. $f_x = 2x - 1 = 0$, $f_y = 4y = 0$; critical point $(1/2, 0)$.

Along $x^2 + y^2 = 4$: $y^2 = 4 - x^2$, $u(x) = 8 - x - x^2$ for $-2 \leq x \leq 2$; critical points $(-1/2, \pm\sqrt{15}/2)$.

(x,y)	$(1/2, 0)$	$(-1/2, \sqrt{15}/2)$	$(-1/2, -\sqrt{15}/2)$	$(-2,0)$	$(2,0)$
$f(x,y)$	$-1/4$	$33/4$	$33/4$	6	2

Absolute maximum value is $33/4$, absolute minimum value is $-1/4$.

31. $f(x, y) = (y - x)^2$; if (x_0, y_0) is on the line $y = x$ then $f(x_0, y_0) = 0$, but $(y - x)^2 \geq 0$ so $f(x, y) \geq f(x_0, y_0)$ for all points (x, y) thus f has an absolute minimum at (x_0, y_0).

33. Minimize $S = x^2 + y^2 + z^2$ subject to $x + y + z = 27$, $x > 0$, $y > 0$, $z > 0$. $z = 27 - x - y$ so $S = x^2 + y^2 + (27 - x - y)^2$, $S_x = 4x + 2y - 54 = 0$, $S_y = 2x + 4y - 54 = 0$; critical point $(9,9)$; $S_{xx}S_{yy} - S_{xy}^2 > 0$ and $S_{xx} > 0$ at $(9,9)$, relative minimum. $z = 9$ when $x = y = 9$, the sum of the squares is minimum for the numbers 9,9,9.

35. Minimize $w = D^2 = x^2 + y^2 + z^2$ subject to $x^2 - yz = 5$. $x^2 = 5 + yz$ so $w = 5 + yz + y^2 + z^2$, $w_y = z + 2y = 0$, $w_z = y + 2z = 0$; critical point when $y = z = 0$; $w_{yy}w_{zz} - w_{yz}^2 > 0$ and $w_{yy} > 0$ when $y = z = 0$, relative minimum. $x^2 = 5$, $x = \pm\sqrt{5}$ when $y = z = 0$. The points $(\pm\sqrt{5}, 0, 0)$ are closest to the origin.

37. Maximize $V = xyz$ subject to $x + y + z = 1$, $x > 0$, $y > 0$, $z > 0$. $z = 1 - x - y$ so $V = xy - x^2y - xy^2$, $V_x = y(1 - 2x - y) = 0$, $V_y = x(1 - x - 2y) = 0$, $1 - 2x - y = 0$ and $1 - x - 2y = 0$; critical point $(1/3, 1/3)$; $V_{xx}V_{yy} - V_{xy}^2 > 0$ and $V_{xx} < 0$ at $(1/3, 1/3)$, relative maximum. The maximum volume is $V = (1/3)(1/3)(1/3) = 1/27$.

39. Let x, y, and z be, respectively, the length, width, and height of the box. Minimize $C = 10(2xy) + 5(2xz + 2yz) = 10(2xy + xz + yz)$ subject to $xyz = 16$. $z = 16/(xy)$ so $C = 20(xy + 8/y + 8/x)$, $C_x = 20(y - 8/x^2) = 0$, $C_y = 20(x - 8/y^2) = 0$; critical point $(2,2)$; $C_{xx}C_{yy} - C_{xy}^2 > 0$ and $C_{xx} > 0$ at $(2,2)$, relative minimum. $z = 4$ when $x = y = 2$. The cost of materials is minimum if the length and width are 2 ft and the height is 4 ft.

41. Minimize $w = D^2 = (x + 1)^2 + (y - 3)^2 + (z - 2)^2$ subject to $x - 2y + z = 4$. $z = 4 - x + 2y$ so $w = (x + 1)^2 + (y - 3)^2 + (2 - x + 2y)^2$, $w_x = 2(2x - 2y - 1) = 0$, $w_y = 2(-2x + 5y + 1) = 0$; critical point $(1/2, 0)$; $w_{xx}w_{yy} - w_{xy}^2 > 0$ and $w_{xx} > 0$ at $(1/2, 0)$, relative minimum. $w = 27/2$, $D = \sqrt{w} = 3\sqrt{6}/2$ is the minimum distance.

43. Minimize $S = xy + 2xz + 2yz$ subject to $xyz = V$, $x > 0$, $y > 0$, $z > 0$ where x, y, and z are, respectively, the length, width, and height of the box. $z = V/(xy)$ so $S = xy + 2V/y + 2V/x$, $S_x = y - 2V/x^2 = 0$, $S_y = x - 2V/y^2 = 0$; critical point $(\sqrt[3]{2V}, \sqrt[3]{2V})$; $S_{xx}S_{yy} - S_{xy}^2 > 0$ and $S_{xx} > 0$ at this point so there is a relative minimum there. The length and width are each $\sqrt[3]{2V}$, the height is $z = \sqrt[3]{2V}/2$.

45. (a) $\dfrac{\partial f}{\partial a} = \Sigma 2(ax_k + b - y_k)x_k = 2(a\Sigma x_k^2 + b\Sigma x_k - \Sigma x_k y_k) = 0$ if

$(\Sigma x_k^2)a + (\Sigma x_k)b = \Sigma x_k y_k$, $\dfrac{\partial f}{\partial b} = \Sigma 2(ax_k + b - y_k) = 2(a\Sigma x_k + bn - \Sigma y_k) = 0$ if $(\Sigma x_k)a + nb = \Sigma y_k$.

47. $\Sigma x_k = 10$, $\Sigma y_k = 8.2$, $\Sigma x_k^2 = 30$, $\Sigma x_k y_k = 23$, $n = 4$; $a = 0.5$, $b = 0.8$, $y = 0.5x + 0.8$.

49. For example, for $0 \le x \le 1$ let $z = \begin{cases} y & \text{if } 0 < y < 1 \\ 1/2 & \text{if } y = 0 \text{ or } y = 1 \end{cases}$; let $z = y$ for $-\infty < x < +\infty$, $y > 0$.

EXERCISE SET 16.10

1. $y = 8x\lambda$, $x = 16y\lambda$; $y/(8x) = x/(16y)$, $x^2 = 2y^2$ so $4\left(2y^2\right) + 8y^2 = 16$, $y^2 = 1$, $y = \pm 1$. Test $(\pm\sqrt{2}, -1)$ and $(\pm\sqrt{2}, 1)$. $f\left(-\sqrt{2}, -1\right) = f\left(\sqrt{2}, 1\right) = \sqrt{2}$, $f\left(-\sqrt{2}, 1\right) = f\left(\sqrt{2}, -1\right) = -\sqrt{2}$. Maximum $\sqrt{2}$ at $\left(-\sqrt{2}, -1\right)$ and $\left(\sqrt{2}, 1\right)$, minimum $-\sqrt{2}$ at $\left(-\sqrt{2}, 1\right)$ and $\left(\sqrt{2}, -1\right)$.

3. $12x^2 = 4x\lambda$, $2y = 2y\lambda$. If $y \ne 0$ then $\lambda = 1$ and $12x^2 = 4x$, $12x(x - 1/3) = 0$, $x = 0$ or $x = 1/3$ so from $2x^2 + y^2 = 1$ we find that $y = \pm 1$ when $x = 0$, $y = \pm\sqrt{7}/3$ when $x = 1/3$. If $y = 0$ then $2x^2 + (0)^2 = 1$, $x = \pm 1/\sqrt{2}$. Test $(0, \pm 1)$, $(1/3, \pm\sqrt{7}/3)$, and $(\pm 1/\sqrt{2}, 0)$. $f(0, \pm 1) = 1$, $f\left(1/3, \pm\sqrt{7}/3\right) = 25/27$, $f\left(1/\sqrt{2}, 0\right) = \sqrt{2}$, $f\left(-1/\sqrt{2}, 0\right) = -\sqrt{2}$. Maximum $\sqrt{2}$ at $\left(1/\sqrt{2}, 0\right)$, minimum $-\sqrt{2}$ at $\left(-1/\sqrt{2}, 0\right)$.

5. $2 = 2x\lambda$, $1 = 2y\lambda$, $-2 = 2z\lambda$; $1/x = 1/(2y) = -1/z$ thus $x = 2y$, $z = -2y$ so $(2y)^2 + y^2 + (-2y)^2 = 4$, $y^2 = 4/9$, $y = \pm 2/3$. Test $(-4/3, -2/3, 4/3)$ and $(4/3, 2/3, -4/3)$. $f(-4/3, -2/3, 4/3) = -6$, $f(4/3, 2/3, -4/3) = 6$. Maximum 6 at $(4/3, 2/3, -4/3)$, minimum -6 at $(-4/3, -2/3, 4/3)$.

7. $yz = 2x\lambda$, $xz = 2y\lambda$, $xy = 2z\lambda$; $yz/(2x) = xz/(2y) = xy/(2z)$ thus $y^2 = x^2$, $z^2 = x^2$ so $x^2 + x^2 + x^2 = 1$, $x = \pm 1/\sqrt{3}$. Test the eight possibilities with $x = \pm 1/\sqrt{3}$, $y = \pm 1/\sqrt{3}$, and $z = \pm 1/\sqrt{3}$ to find the maximum is $1/\left(3\sqrt{3}\right)$ at $(1/\sqrt{3}, 1/\sqrt{3}, 1/\sqrt{3})$, $(1/\sqrt{3}, -1/\sqrt{3}, -1/\sqrt{3})$, $(-1/\sqrt{3}, 1/\sqrt{3}, -1/\sqrt{3})$, and $(-1/\sqrt{3}, -1/\sqrt{3}, 1/\sqrt{3})$; the minimum is $-1/\left(3\sqrt{3}\right)$ at $(1/\sqrt{3}, 1/\sqrt{3}, -1/\sqrt{3})$, $(1/\sqrt{3}, -1/\sqrt{3}, 1/\sqrt{3})$, $(-1/\sqrt{3}, 1/\sqrt{3}, 1/\sqrt{3})$, and $(-1/\sqrt{3}, -1/\sqrt{3}, -1/\sqrt{3})$.

9. $f(x, y) = (x - 4)^2 + (y - 2)^2$, $g(x, y) = y - 2x$; $2(x - 4) = -2\lambda$, $2(y - 2) = \lambda$; $x - 4 = -2(y - 2)$, $x = -2y + 8$ so $y = 2(-2y + 8) + 3$, $y = 19/5$. The point is $(2/5, 19/5)$.

11. $f(x, y, z) = (x - 1)^2 + (y + 1)^2 + (z - 1)^2$; $2(x - 1) = 4\lambda$, $2(y + 1) = 3\lambda$, $2(z - 1) = \lambda$; $x = 4z - 3$, $y = 3z - 4$ so $4(4z - 3) + 3(3z - 4) + z = 2$, $z = 1$. The point is $(1, -1, 1)$.

13. $f(x, y) = \sin x \sin y$; $x + y = \pi/2$; $\cos x \sin y = \lambda$, $\sin x \cos y = \lambda$; $\cos x \sin y = \sin x \cos y$ but $\cos x \ne 0$ and $\cos y \ne 0$ thus $\tan y = \tan x$, $y = x$ so $x + x = \pi/2$, $x = \pi/4$. The maximum value of $f(x, y)$ is $f(\pi/4, \pi/4) = 1/2$.

15. $f(x,y) = (x-1)^2 + (y-2)^2$; $2(x-1) = 2x\lambda$, $2(y-2) = 2y\lambda$; $(x-1)/x = (y-2)/y$, $y = 2x$
so $x^2 + (2x)^2 = 45$, $x = \pm 3$. $f(-3,-6) = 80$ and $f(3,6) = 20$ so $(3,6)$ is closest and $(-3,-6)$
is farthest.

17. $f(x,y,z) = x^2 + y^2 + z^2$, $x+y+z = 27$; $2x = \lambda$, $2y = \lambda$, $2z = \lambda$, $y = x$, $z = x$ so $x+x+x = 27$,
$x = 9$. The numbers are 9,9,9.

19. $f(x,y,z) = x^2 + y^2 + z^2$; $2x = 2x\lambda$, $2y = -z\lambda$, $2z = -y\lambda$. If $x \neq 0$ then $\lambda = 1$ thus $2y = -z$
and $2z = -y$ so $y = z = 0$; use $x^2 - yz = 5$ to get $x = \pm\sqrt{5}$. If $x = 0$ then $(0)^2 - yz = 5$,
$y = -5/z$ and also (from $2y = -z\lambda$, $2z = -y\lambda$)$y^2 = z^2$ so $(-5/z)^2 = z^2$, $z^4 = 25$, $z = \pm\sqrt{5}$.
$f\left(\pm\sqrt{5},0,0\right) = 5$ and $f\left(0,\sqrt{5},-\sqrt{5}\right) = f\left(0,-\sqrt{5},\sqrt{5}\right) = 10$ so $\left(\pm\sqrt{5},0,0\right)$ are closest to the
origin.

21. $f(x,y,z) = 20xy + 10xz + 10yz$, $xyz = 16$; $20y + 10z = yz\lambda$, $20x + 10z = xz\lambda$, $10x + 10y = xy\lambda$;
$(20y + 10z)/(yz) = (20x + 10z)/(xz) = (10x + 10y)/(xy)$; $y = x$, $z = 2x$ so $x(x)(2x) = 16$,
$x = 2$. The length and width are each 2 ft and the height is 4 ft.

23. $f(a,b,\alpha) = ab\sin\alpha$, $2a + 2b = \ell$; $b\sin\alpha = 2\lambda$, $a\sin\alpha = 2\lambda$, $ab\cos\alpha = 0$; $\cos\alpha = 0$, $\alpha = \pi/2$
and $a = b$ so $2b + 2b = \ell$, $b = \ell/4$.

SUPPLEMENTARY EXERCISES CHAPTER 16

1. (a) 1   **(b)** xy **(c)** $e^{r+s} \ln rs$

3. (a) upper half of the elliptic cone $z^2 = x^2 + 4y^2$
     (b) the plane with x, y, and z intercepts of 1, a, and b.

5. $1/(x \sin yz) - 3y(\csc yz \cot yz) \ln xy$

7. $\pi/2, 0, 1, -\pi^2/4$ **9.**   2

11. $(2x - 2y + 4r)/\left(x^2 + y^2 + 2z\right) = \dfrac{2}{r+s}$

13. $\partial^2 w/\partial x^2 = \partial^2 w/\partial y^2 = -(x-y)^{-2} - \cos(x+y)$

15. $f_{xyzz} = f_{zzxy} = 0$

17. (a) $dP/dt = (\partial P/\partial T)(dT/dt) = (10/V)(dT/dt) = (10/2.5)(3) = 12$ newtons/m²/min
     (b) $dP/dt = (\partial P/\partial V)(dV/dt) = -\left(10T/V^2\right)(dV/dt)$
     $= -(500/6.25)(-3) = 240$ newtons/m²/min

19. **(a)** $\displaystyle\lim_{(x,y)\to(0,0)} \frac{(x^2-y^2)(x^2+y^2)}{x^2+y^2} = \lim_{(x,y)\to(0,0)} (x^2-y^2) = 0$

(b) continuous at $(0,0)$ because $\displaystyle\lim_{(x,y)\to(0,0)} f(x,y) = f(0,0)$

21. **(a)** $-(6x-5y+y\sec^2 xy)/\left(-5x+x\sec^2 xy\right)$ **(b)** $-[\ln y + \cos(x-y)]/[x/y - \cos(x-y)]$

23. $dV/dt = (\partial V/\partial E)(dE/dt) + (\partial V/\partial r)(dr/dt) = \dfrac{R}{r+R}\dfrac{dE}{dt} - \dfrac{RE}{(r+R)^2}\dfrac{dr}{dt}$

25. **(a)** $\nabla f(3,1) = \langle 6, 45\rangle$

(b) $\overrightarrow{P_0 P_1} = \langle 1,-4\rangle$, $\mathbf{u} = \langle 1,-4\rangle/\sqrt{17}$, $D_{\mathbf{u}}f = -174/\sqrt{17}$

27. **(a)** $\nabla f(3,2,6) = \langle 1/3, 1/2, 1/6\rangle$ **(b)** $D_{\mathbf{u}}f = \sqrt{3}/9$

29. **(a)** $\nabla f(1,-1,2) = \langle 1,3,0\rangle$

(b) $\overrightarrow{P_0 P_1} = \langle 10,11,-2\rangle$, $\mathbf{u} = \langle 10,11,-2\rangle/15$, $D_{\mathbf{u}}f = 43/15$

31. **(a)** $\nabla f(1,-1) = 2(-\mathbf{i}+\mathbf{j})$, $D_{\mathbf{u}}f = 0$ if $\mathbf{u}$ is normal to ∇f so $\mathbf{u} = \pm(\mathbf{i}+\mathbf{j})/\sqrt{2}$

(b) $\nabla f(-2,0) = \mathbf{i} - 2\mathbf{j}$, $\mathbf{u} = \pm(2\mathbf{i}+\mathbf{j})/\sqrt{5}$

33. $\nabla f(1,2) = a\mathbf{i}+b\mathbf{j}$; $D_{\mathbf{u}}f = 2\sqrt{2}$ when $\mathbf{u} = (\mathbf{i}+\mathbf{j})/\sqrt{2}$ and $D_{\mathbf{u}}f = -3$ when $\mathbf{u} = -\mathbf{j}$ so $(a+b)/\sqrt{2} = 2\sqrt{2}$ and $-b = -3$, $a = 1$, $b = 3$. If $\mathbf{u} = -(\mathbf{i}+2\mathbf{j})/\sqrt{5}$ then $D_{\mathbf{u}}f = -7/\sqrt{5}$.

35. **(a)** $\langle f_x(4,-3), f_y(4,-3), -1\rangle = \langle 8/5, -6/5, -1\rangle$, let $\mathbf{N} = \langle 8,-6,-5\rangle$

(b) $8x - 6y - 5z = 0$

37. Let $f(x,y,z) = z + xy$; $\nabla f(x_0,y_0,z_0) = \langle y_0, x_0, 1\rangle$ is normal to the surface at $P_0\,(x_0,y_0,z_0)$. The normal line passes through the origin when $\langle x_0,y_0,z_0\rangle$ and $\langle y_0,x_0,1\rangle$ are parallel so $\langle x_0,y_0,z_0\rangle = k\,\langle y_0,x_0,1\rangle = \langle ky_0, kx_0, k\rangle$ for some value of k. Equate the third component of these vectors to find that $k = z_0$ so $x_0 = y_0 z_0$ and $y_0 = x_0 z_0$, eliminate y_0 to get $x_0 = x_0 z_0^2$, $x_0\left(1 - z_0^2\right) = 0$, $x_0 = 0$ or $z_0 = \pm 1$. If $x_0 = 0$ then $y_0 = (0)z_0 = 0$ and, from the equation of the surface, $z_0 = 2 - (0)(0) = 2$ so $(0,0,2)$ is one of the points. If $z_0 = 1$ then $y_0 = x_0$ so $1 = 2 - x_0^2$, $x_0^2 = 1$, $x_0 = \pm 1$ so $(1,1,1)$ and $(-1,-1,1)$ are also points where the normal line passes through the origin. If $z_0 = -1$ then $y_0 = -x_0$ so $-1 = 2 + x_0^2$, $x_0^2 = -3$ which has no real solution.

39. $\langle 18x_0, 8y_0, -1\rangle$ is normal to the surface at a point (x_0,y_0,z_0). $\overrightarrow{PQ} = \langle -6,-4,-1\rangle$ so the normal line is parallel to $\overrightarrow{PQ}$ if $\langle 18x_0, 8y_0, -1\rangle = k\langle -6,-4,-1\rangle$ for some value of k. By inspection $k = 1$ so $18x_0 = -6$ and $8y_0 = -4$, $x_0 = -1/3$ and $y_0 = -1/2$ thus $z_0 = 2$. The only point is $(-1/3, -1/2, 2)$.

41. $dV = (2/3)xhdx + (1/3)x^2dh = (2/3)(1)(2)(-0.1) + (1/3)(1)^2(0.2) = -0.2/3 \approx -0.067 \ \text{m}^3$
$\Delta V = (1/3)(0.9)^2(2.2) - (1/3)(1)^2(2) = -0.218/3 \approx -0.073 \ \text{m}^3$

43. $f_x = 2x + 3y - 6 = 0$, $f_y = 3x + 6y + 3 = 0$; critical point $(15, -8)$; $f_{xx}f_{yy} - f_{xy}^2 > 0$ and $f_{xx} > 0$ at $(15, -8)$, relative minimum.

45. $f_x = 3x^2 - 3y = 0$, $f_y = -3x + y = 0$; critical points $(0,0)$ and $(3,9)$; $D < 0$ at $(0,0)$, saddle point; $D > 0$ and $f_{xx} > 0$ at $(3,9)$, relative minimum.

47.   **(a)** Let (x, y, z) be a point on the portion of the ellipsoid that is in the first octant then $V = (2x)(2y)(2z) = 8xyz$. For convenience introduce the new variables $u = x/a$, $v = y/b$, and $w = z/c$ so $V = (8abc)uvw$ where $u^2 + v^2 + w^2 = 1$. Also for convenience we will maximize $S = u^2v^2w^2$ instead of V. $w^2 = 1 - u^2 - v^2$ so $S = u^2v^2 - u^4v^2 - u^2v^4$, $S_u = 2uv^2(1 - 2u^2 - v^2) = 0$, $S_v = 2vu^2 (1 - u^2 - 2v^2) = 0$; critical point $(1/\sqrt{3}, 1/\sqrt{3})$; $S_{uu}S_{vv} - S_{uv}^2 > 0$ and $S_{uu} < 0$ at this point so a relative maximum occurs there. If $u = v = 1/\sqrt{3}$ then $w = 1/\sqrt{3}$ so $x = a/\sqrt{3}$, $y = b/\sqrt{3}$, and $z = c/\sqrt{3}$. The dimensions of the box are $2a/\sqrt{3}$, $2b/\sqrt{3}$, and $2c/\sqrt{3}$.

  (b) $f(x, y, z) = 8xyz$, $(x/a)^2 + (y/b)^2 + (z/c)^2 = 1$; $8yz = (2x/a^2)\lambda$, $8xz = (2y/b^2)\lambda$, $8xy = (2z/c^2)\lambda$; $4a^2yz/x = 4b^2xz/y = 4c^2xy/z$, $y^2/b^2 = x^2/a^2$ and $z^2/c^2 = x^2/a^2$ so $3(x^2/a^2) = 1$, $x = a/\sqrt{3}$ and therefore $y = b/\sqrt{3}$ and $z = c/\sqrt{3}$. The dimensions agree with those in part (a).

49. $f(I_1, I_2, I_3) = I_1^2 R_1 + I_2^2 R_2 + I_3^2 R_3$, $I_1 + I_2 + I_3 = I$. $2I_1 R_1 = \lambda$, $2I_2 R_2 = \lambda$, $2I_3 R_3 = \lambda$; $2I_1 R_1 = 2I_2 R_2 = 2I_3 R_3$, $I_1/I_2 = R_2/R_1 = R_1^{-1}/R_2^{-1}$ and $I_2/I_3 = R_2^{-1}/R_3^{-1}$ so $I_1 : I_2 : I_3 = R_1^{-1} : R_2^{-1} : R_3^{-1}$.

CHAPTER 17
Multiple Integrals

EXERCISE SET 17.1

1. $\displaystyle\int_0^1 \int_0^2 (x+3)dy\,dx = \int_0^1 (2x+6)dx = 7$

3. $\displaystyle\int_2^4 \int_0^1 x^2 y\,dx\,dy = \int_2^4 \frac{1}{3}y\,dy = 2$

5. $\displaystyle\int_0^{\ln 3} \int_0^{\ln 2} e^{x+y}\,dy\,dx = \int_0^{\ln 3} e^x dx = 2$

7. $\displaystyle\int_0^3 \int_0^1 x(x^2+y)^{1/2}dx\,dy = \int_0^3 \frac{1}{3}[(1+y)^{3/2} - y^{3/2}]dy = 2(31 - 9\sqrt{3})/15$

9. $\displaystyle\int_{-1}^0 \int_2^5 dx\,dy = \int_{-1}^0 3\,dy = 3$

11. $\displaystyle\int_0^1 \int_0^1 \frac{x}{(xy+1)^2}dy\,dx = \int_0^1 \left(1 - \frac{1}{x+1}\right)dx = 1 - \ln 2$

13. $\displaystyle\int_0^{\ln 2} \int_0^1 xy\,e^{y^2 x}dy\,dx = \int_0^{\ln 2} \frac{1}{2}(e^x - 1)dx = (1 - \ln 2)/2$

15. $\displaystyle\int_{-1}^1 \int_{-2}^2 4xy^3 dy\,dx = \int_{-1}^1 0\,dx = 0$

17. $\displaystyle\int_0^1 \int_2^3 x\sqrt{1-x^2}\,dy\,dx = \int_0^1 x(1-x^2)^{1/2}dx = 1/3$

19. $\displaystyle\int_{-\pi/4}^{\pi/4} \int_0^{\pi/4} \cos(x+y)dy\,dx = \int_{-\pi/4}^{\pi/4} [\sin(x + \pi/4) - \sin x]dx = 1$

21.

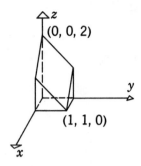

23.

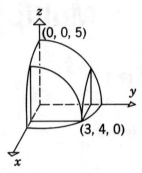

25.

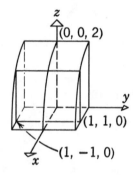

27. $V = \int_1^3 \int_0^2 (3x^3 + 3x^2 y)dy\, dx = \int_1^3 (6x^3 + 6x^2)dx = 172$

29. $V = \int_0^2 \int_0^3 x^2 dy\, dx = \int_0^2 3x^2 dx = 8$

31. $\displaystyle\iint\limits_R f(x,y)dA = \int_a^b \left[\int_c^d g(x)h(y)dy \right] dx = \int_a^b g(x) \left[\int_c^d h(y)dy \right] dx$

$$= \left[\int_a^b g(x)dx \right] \left[\int_c^d h(y)dy \right]$$

EXERCISE SET 17.2

1. $\int_0^1 \int_{x^2}^x xy^2 dy\, dx = \int_0^1 \frac{1}{3}(x^4 - x^7)dx = 1/40$

3. $\displaystyle\int_0^3\int_0^{\sqrt{9-y^2}} y\, dx\, dy = \int_0^3 y\sqrt{9-y^2}\, dy = 9$

5. $\displaystyle\int_{\sqrt{\pi}}^{\sqrt{2\pi}}\int_0^{x^3}\sin(y/x)dy\, dx = \int_{\sqrt{\pi}}^{\sqrt{2\pi}}[-x\cos(x^2)+x]dx = \pi/2$

7. $\displaystyle\int_{\pi/2}^{\pi}\int_0^{x^2}\frac{1}{x}\cos(y/x)dy\, dx = \int_{\pi/2}^{\pi}\sin x\, dx = 1$

9. $\displaystyle\int_0^a\int_0^{\sqrt{a^2-x^2}} (x+y)dy\, dx = \int_0^a [x\sqrt{a^2-x^2}+(a^2-x^2)/2]dx = 2a^3/3$

11. $\displaystyle\int_0^1\int_0^x y\sqrt{x^2-y^2}\, dy\, dx = \int_0^1 \frac{1}{3}x^3 dx = 1/12$

13. $\displaystyle\int_0^2\int_0^{x^2} 6xy\, dy\, dx = \int_0^2 3x^5 dx = 32$

15. $\displaystyle\int_1^2\int_{\pi/2}^{2\pi/x} x\cos xy\, dy\, dx = -\int_1^2 \sin(\pi x/2)dx = -2/\pi$

17. $\displaystyle\int_4^8\int_{16/x}^x x^2 dy\, dx = \int_4^8 (x^3-16x)dx = 576$

19. $\displaystyle\int_0^4\int_0^{\sqrt{y}} x(1+y^2)^{-1/2}dx\, dy = \int_0^4 \frac{1}{2}y(1+y^2)^{-1/2}dy = (\sqrt{17}-1)/2$

21. $\displaystyle\int_{-1}^1\int_{-\sqrt{1-x^2}}^{\sqrt{1-x^2}} (3x-2y)dy\, dx = \int_{-1}^1 6x\sqrt{1-x^2}\, dx = 0$

23. $\displaystyle\int_0^1\int_x^1 \frac{1}{1+x^2}dy\, dx = \int_0^1 \left[\frac{1}{1+x^2}-\frac{x}{1+x^2}\right]dx = \frac{\pi}{4}-\frac{1}{2}\ln 2$

25. $\displaystyle\int_0^2\int_{y^2}^{6-y} xy\, dx\, dy = \int_0^2 \frac{1}{2}(36y-12y^2+y^3-y^5)dy = 50/3$

27. $\displaystyle\int_{-1}^0\int_x^{x^3} (x-1)dy\, dx + \int_0^1\int_{x^3}^x (x-1)dy\, dx$

$$= \int_{-1}^0 (x^4-x^3-x^2+x)dx + \int_0^1(-x^4+x^3+x^2-x)dx = -1/2$$

29. $A = \int_0^5 \int_0^{5-x} dy\,dx = \int_0^5 (5-x)dx = 25/2$

31. $A = \int_0^{\pi/4} \int_{\sin x}^{\cos x} dy\,dx = \int_0^{\pi/4} (\cos x - \sin x)dx = \sqrt{2} - 1$

33. $A = \int_{-3}^3 \int_{1-y^2/9}^{9-y^2} dx\,dy = \int_{-3}^3 8(1 - y^2/9)dy = 32$

35. $V = \int_0^{5/2} \int_0^{5-2x} (5 - 2x - y)dy\,dx = \int_0^{5/2} \frac{1}{2}(5 - 2x)^2 dx = 125/12$

37. $V = \int_{-1}^1 \int_0^{1-x^2} (x + 2y + 2)dy\,dx = \int_{-1}^1 (x^4 - x^3 - 4x^2 + x + 3)dx = 56/15$

39. $V = \int_0^3 \int_0^2 (9x^2 + y^2)dy\,dx = \int_0^3 (18x^2 + 8/3)dx = 170$

41. $V = \int_{-3/2}^{3/2} \int_{-\sqrt{9-4x^2}}^{\sqrt{9-4x^2}} (y + 3)dy\,dx = \int_{-3/2}^{3/2} 6\sqrt{9 - 4x^2}\,dx = 27\pi/2$

43. $V = \int_0^4 \int_0^{2-x/2} (x/4 + 2y)dy\,dx = \int_0^4 (x^2/8 - 3x/2 + 4)dx = 20/3$

45. $V = 4\int_0^1 \int_0^{\sqrt{1-x^2}} (1 - x^2 - y^2)dy\,dx = \frac{8}{3}\int_0^1 (1 - x^2)^{3/2}dx = \pi/2$

47. $V = 8\int_0^5 \int_0^{\sqrt{25-x^2}} \sqrt{25 - x^2}\,dy\,dx = 8\int_0^5 (25 - x^2)dx = 2000/3$

49. $\int_0^{\sqrt{2}} \int_{y^2}^2 f(x,y)dx\,dy$

51. $\int_1^{e^2} \int_{\ln x}^2 f(x,y)dy\,dx$

53. $\int_{-1}^1 \int_{-2\sqrt{1-y^2}}^{2\sqrt{1-y^2}} f(x,y)dx\,dy$

55. $\int_0^{\pi/2} \int_0^{\sin x} f(x,y)dy\,dx$

57. $\int_0^4 \int_0^{y/4} e^{-y^2} dx\,dy = \int_0^4 \frac{1}{4}ye^{-y^2}dy = (1 - e^{-16})/8$

59. $\displaystyle\int_0^2\int_0^{x^2} e^{x^3}\,dy\,dx = \int_0^2 x^2 e^{x^3}\,dx = (e^8-1)/3$

61. $\displaystyle\int_0^{\pi/2}\int_0^{\cos y} x\,dx\,dy = \frac{1}{2}\int_0^{\pi/2}\cos^2 y\,dy = \pi/8$

63. $\displaystyle\int_0^2\int_0^{y^2}\sin(y^3)\,dx\,dy = \int_0^2 y^2\sin(y^3)\,dy = (1-\cos 8)/3$

65. (a) $\displaystyle\int_{-2}^{-1}\int_0^2 xy^2\,dy\,dx + \int_{-1}^1\int_1^2 xy^2\,dy\,dx + \int_1^2\int_0^2 xy^2\,dy\,dx$

$$= \int_{-2}^{-1}\frac{8}{3}x\,dx + \int_{-1}^1\frac{7}{3}x\,dx + \int_1^2\frac{8}{3}x\,dx = 0$$

(b) $\displaystyle\int_1^4\int_{-\sqrt{y}}^{2-y} xy^2\,dx\,dy + \int_{1/2}^1\int_{-\sqrt{y}}^{\sqrt{y}} xy^2\,dx\,dy$

$$= \int_1^4\frac{1}{2}(4y^2 - 5y^3 + y^4)\,dy + \int_{1/2}^1 (0)\,dy = -603/40$$

67. $\displaystyle 2\int_{-1}^1\int_0^{\sqrt{1-x^2}}\,dy\,dx + \int_{-1}^1\int_0^{\sqrt{1-x^2}} x\sqrt{9-y^2}\,dy\,dx = 2\cdot\frac{1}{2}\pi(1)^2 + 0 = \pi$ because the first integral gives the area of a semicircle of radius 1, and with $f(x,y) = x\sqrt{9-y^2}$, $f(-x,y) = -f(x,y)$ so the second integral evaluates to 0.

69. $f(x,y) = x^3 y$, $f(-x,y) = -f(x,y)$ so $\displaystyle\iint_R x^3 y\,dA = 0$

EXERCISE SET 17.3

1. $\displaystyle\int_0^{\pi/2}\int_0^{\sin\theta} r\cos\theta\,dr\,d\theta = \int_0^{\pi/2}\frac{1}{2}\sin^2\theta\cos\theta\,d\theta = 1/6$

3. $\displaystyle\int_{-\pi/2}^{\pi/2}\int_0^{a\sin\theta} r^2\,dr\,d\theta = \int_{-\pi/2}^{\pi/2}\frac{a^3}{3}\sin^3\theta\,d\theta = 0$

5. $\displaystyle\int_0^\pi\int_0^{1-\sin\theta} r^2\cos\theta\,dr\,d\theta = \int_0^\pi\frac{1}{3}(1-\sin\theta)^3\cos\theta\,d\theta = 0$

7. $A = \int_0^{2\pi} \int_0^{1-\cos\theta} r\,dr\,d\theta = \int_0^{2\pi} \frac{1}{2}(1-\cos\theta)^2\,d\theta = 3\pi/2$

9. $A = \int_{\pi/4}^{\pi/2} \int_{\sin 2\theta}^{1} r\,dr\,d\theta = \int_{\pi/4}^{\pi/2} \frac{1}{2}(1-\sin^2 2\theta)d\theta = \pi/16$

11. $A = 2\int_{\pi/6}^{\pi/2} \int_2^{4\sin\theta} r\,dr\,d\theta = \int_{\pi/6}^{\pi/2}(16\sin^2\theta - 4)d\theta = 4\pi/3 + 2\sqrt{3}$

13. $V = 8\int_0^{\pi/2} \int_0^1 r\sqrt{9-r^2}\,dr\,d\theta = \frac{8}{3}(27-16\sqrt{2})\int_0^{\pi/2} d\theta = 4(27-16\sqrt{2})\pi/3$

15. $V = 2\int_0^{\pi/2} \int_0^{2\sin\theta} r^2 dr\,d\theta = \frac{16}{3}\int_0^{\pi/2} \sin^3\theta\,d\theta = 32/9$

17. $V = 2\int_0^{\pi/2} \int_0^{\cos\theta}(1-r^2)r\,dr\,d\theta = \frac{1}{2}\int_0^{\pi/2}(1-\sin^4\theta)d\theta = 5\pi/32$

19. $\int_0^{2\pi} \int_0^1 e^{-r^2}r\,dr\,d\theta = \frac{1}{2}(1-e^{-1})\int_0^{2\pi} d\theta = (1-e^{-1})\pi$

21. $\int_0^{\pi/4} \int_0^2 \frac{1}{1+r^2}r\,dr\,d\theta = \frac{1}{2}\ln 5 \int_0^{\pi/4} d\theta = \frac{\pi}{8}\ln 5$

23. $\int_0^{\pi/2} \int_0^1 r^3 dr\,d\theta = \frac{1}{4}\int_0^{\pi/2} d\theta = \pi/8$

25. $\int_0^{\pi/2} \int_0^{2\cos\theta} r^2 dr\,d\theta = \frac{8}{3}\int_0^{\pi/2}\cos^3\theta\,d\theta = 16/9$

27. $\int_0^{\pi/2} \int_0^a \frac{r}{(1+r^2)^{3/2}}dr\,d\theta = \frac{\pi}{2}(1-1/\sqrt{1+a^2})$

29. $\int_0^{\pi/4} \int_0^2 \frac{r}{\sqrt{1+r^2}}dr\,d\theta = \frac{\pi}{4}(\sqrt{5}-1)$

31. $V = 2\int_0^{\pi/2} \int_0^{a\sin\theta} \frac{c}{a}(a^2-r^2)^{1/2}r\,dr\,d\theta = \frac{2}{3}a^2c\int_0^{\pi/2}(1-\cos^3\theta)d\theta = (3\pi-4)a^2c/9$

33. $A = \displaystyle\int_{\pi/6}^{\pi/4} \int_{\sqrt{8\cos 2\theta}}^{4\sin\theta} r\,dr\,d\theta + \int_{\pi/4}^{\pi/2} \int_0^{4\sin\theta} r\,dr\,d\theta$

$= \displaystyle\int_{\pi/6}^{\pi/4} (8\sin^2\theta - 4\cos 2\theta)d\theta + \int_{\pi/4}^{\pi/2} 8\sin^2\theta\,d\theta = 4\pi/3 + 2\sqrt{3} - 2$

35. **(a)** $I^2 = \left[\displaystyle\int_0^{+\infty} e^{-x^2}dx\right]\left[\int_0^{+\infty} e^{-y^2}dy\right] = \int_0^{+\infty}\left[\int_0^{+\infty} e^{-x^2}dx\right]e^{-y^2}dy$

$= \displaystyle\int_0^{+\infty}\int_0^{+\infty} e^{-x^2}e^{-y^2}dx\,dy = \int_0^{+\infty}\int_0^{+\infty} e^{-(x^2+y^2)}dx\,dy$

(b) $I^2 = \displaystyle\int_0^{\pi/2}\int_0^{+\infty} e^{-r^2}r\,dr\,d\theta = \frac{1}{2}\int_0^{\pi/2} d\theta = \pi/4$

(c) $I = \sqrt{\pi}/2$

EXERCISE SET 17.4

1. $z = \sqrt{9-y^2}$, $z_x = 0$, $z_y = -y/\sqrt{9-y^2}$, $z_x^2 + z_y^2 + 1 = 9/(9-y^2)$,

$S = \displaystyle\int_0^2 \int_{-3}^3 \frac{3}{\sqrt{9-y^2}}\,dy\,dx = \int_0^2 3\pi\,dx = 6\pi$

3. $z^2 = 4x^2 + 4y^2$, $2zz_x = 8x$ so $z_x = 4x/z$, similarly $z_y = 4y/z$ thus

$z_x^2 + z_y^2 + 1 = (16x^2 + 16y^2)/z^2 + 1 = 5$, $S = \displaystyle\int_0^1 \int_{x^2}^x \sqrt{5}\,dy\,dx = \sqrt{5}\int_0^1 (x - x^2)dx = \sqrt{5}/6$

5. $z_x = -2x$, $z_y = -2y$, $z_x^2 + z_y^2 + 1 = 4x^2 + 4y^2 + 1$,

$S = \displaystyle\iint_R \sqrt{4x^2 + 4y^2 + 1}\,dA = \int_0^{2\pi} \int_0^1 r\sqrt{4r^2 + 1}\,dr\,d\theta$

$= \displaystyle\frac{1}{12}(5\sqrt{5} - 1)\int_0^{2\pi} d\theta = (5\sqrt{5} - 1)\pi/6$

7. $z_x = y$, $z_y = x$, $z_x^2 + z_y^2 + 1 = x^2 + y^2 + 1$,

$S = \displaystyle\iint_R \sqrt{x^2 + y^2 + 1}\,dA = \int_0^{\pi/6} \int_0^3 r\sqrt{r^2 + 1}\,dr\,d\theta$

$= \displaystyle\frac{1}{3}(10\sqrt{10} - 1)\int_0^{\pi/6} d\theta = (10\sqrt{10} - 1)\pi/18$

9. On the sphere, $z_x = -x/z$ and $z_y = -y/z$ so

$z_x^2 + z_y^2 + 1 = (x^2 + y^2 + z^2)/z^2 = 16/(16 - x^2 - y^2)$; the planes $z = 1$ and $z = 2$ intersect the sphere along the circles $x^2 + y^2 = 15$ and $x^2 + y^2 = 12$;

$$S = \iint_R \frac{4}{\sqrt{16 - x^2 - y^2}} dA = \int_0^{2\pi} \int_{\sqrt{12}}^{\sqrt{15}} \frac{4r}{\sqrt{16 - r^2}} dr\, d\theta = 4 \int_0^{2\pi} d\theta = 8\pi$$

11. On both upper and lower halves of the sphere, $z_x = -x/z$ and $z_y = -y/z$ so

$z_x^2 + z_y^2 + 1 = (x^2 + y^2 + z^2)/z^2 = a^2/(a^2 - x^2 - y^2)$,

$$S = (2)(2) \int_0^{\pi/2} \int_0^{a \sin\theta} \frac{ar}{\sqrt{a^2 - r^2}} dr\, d\theta = 4a^2 \int_0^{\pi/2} (1 - \cos\theta) d\theta = 2(\pi - 2)a^2$$

13. $z_x = -x/z$ and $z_y = 0$ on $x^2 + z^2 = 16$ so $z_x^2 + z_y^2 + 1 = (x^2 + z^2)/z^2 = 16/(16 - x^2)$,

$$S = 8 \int_0^4 \int_0^{\sqrt{16-x^2}} \frac{4}{\sqrt{16 - x^2}} dy\, dx = 32 \int_0^4 dx = 128$$

15. $z_x = \dfrac{h}{a} \dfrac{x}{\sqrt{x^2 + y^2}}$, $z_y = \dfrac{h}{a} \dfrac{y}{\sqrt{x^2 + y^2}}$, $z_x^2 + z_y^2 + 1 = \dfrac{h^2 x^2 + h^2 y^2}{a^2(x^2 + y^2)} + 1 = (a^2 + h^2)/a^2$,

$$S = \int_0^{2\pi} \int_0^a \frac{\sqrt{a^2 + h^2}}{a} r\, dr\, d\theta = \frac{1}{2} a\sqrt{a^2 + h^2} \int_0^{2\pi} d\theta = \pi a \sqrt{a^2 + h^2}$$

EXERCISE SET 17.5

1. $\displaystyle\int_{-1}^1 \int_0^2 \int_0^1 (x^2 + y^2 + z^2) dx\, dy\, dz = \int_{-1}^1 \int_0^2 (1/3 + y^2 + z^2) dy\, dz = \int_{-1}^1 (10/3 + 2z^2) dz = 8$

3. $\displaystyle\int_0^2 \int_{-1}^{y^2} \int_1^z yz\, dx\, dz\, dy = \int_0^2 \int_{-1}^{y^2} (yz^2 - yz) dz\, dy = \int_0^2 \left(\frac{1}{3}y^7 - \frac{1}{2}y^5 + \frac{5}{6}y\right) dy = 7$

5. $\displaystyle\int_0^3 \int_0^{\sqrt{9-z^2}} \int_0^x xy\, dy\, dx\, dz = \int_0^3 \int_0^{\sqrt{9-z^2}} \frac{1}{2}x^3 dx\, dz = \int_0^3 \frac{1}{8}(81 - 18z^2 + z^4) dz = 81/5$

7. $\displaystyle\int_0^2 \int_0^{\sqrt{4-x^2}} \int_{-5+x^2+y^2}^{3-x^2-y^2} x\, dz\, dy\, dx = \int_0^2 \int_0^{\sqrt{4-x^2}} [2x(4 - x^2) - 2xy^2] dy\, dx$

$$= \int_0^2 \frac{4}{3}x(4 - x^2)^{3/2} dx = 128/15$$

9. $\displaystyle\int_0^\pi \int_0^1 \int_0^{\pi/6} xy\sin yz\,dz\,dy\,dx = \int_0^\pi \int_0^1 x[1-\cos(\pi y/6)]dy\,dx$

$$= \int_0^\pi (1-3/\pi)x\,dx = \pi(\pi-3)/2$$

11. $\displaystyle\int_0^{\sqrt 2} \int_0^x \int_0^{2-x^2} xyz\,dz\,dy\,dx = \int_0^{\sqrt 2}\int_0^x \frac{1}{2}xy(2-x^2)^2 dy\,dx = \int_0^{\sqrt 2}\frac14 x^3(2-x^2)^2 dx = 1/6$

13. $\displaystyle V = \int_0^4 \int_0^{(4-x)/2}\int_0^{(12-3x-6y)/4} dz\,dy\,dx$

$$= \int_0^4 \int_0^{(4-x)/2} \frac14(12-3x-6y)dy\,dx = \int_0^4 \frac{3}{16}(4-x)^2 dx = 4$$

15. $\displaystyle V = 2\int_0^2 \int_{x^2}^4 \int_0^{4-y} dz\,dy\,dx = 2\int_0^2\int_{x^2}^4(4-y)dy\,dx = 2\int_0^2\left(8-4x^2+\frac12 x^4\right)dx = 256/15$

17. $\displaystyle V = 2\int_{-3}^3 \int_0^{\sqrt{9-x^2}/3}\int_0^{x+3} dz\,dy\,dx = 2\int_{-3}^3\int_0^{\sqrt{9-x^2}/3}(x+3)dy\,dx$

$$= 2\int_{-3}^3 \frac13(x+3)\sqrt{9-x^2}\,dx = 9\pi$$

19. The projection of the curve of intersection onto the xy-plane is $x^2+y^2=1$,

$$V = 4\int_0^1\int_0^{\sqrt{1-x^2}}\int_{4x^2+y^2}^{4-3y^2} dz\,dy\,dx = 16\int_0^1\int_0^{\sqrt{1-x^2}}(1-x^2-y^2)dy\,dx$$

$$= \frac{32}{3}\int_0^1(1-x^2)^{3/2}dx = 2\pi$$

21. The projection of the curve of intersection onto the xy-plane is $x^2+y^2=a^2$,

$$V = 4\int_0^a\int_0^{\sqrt{a^2-x^2}}\int_{(x^2+y^2)/a}^{\sqrt{2a^2-x^2-y^2}} dz\,dy\,dx$$

$$= 4\int_0^a\int_0^{\sqrt{a^2-x^2}}\left[\sqrt{2a^2-x^2-y^2}-\frac1a(x^2+y^2)\right]dy\,dx$$

$$= 4\int_0^{\pi/2}\int_0^a\left(r\sqrt{2a^2-r^2}-\frac1a r^3\right)dr\,d\theta = 4\int_0^{\pi/2}\frac{1}{12}(8\sqrt2-7)a^3 d\theta = (8\sqrt2-7)\pi a^3/6$$

23. (a)

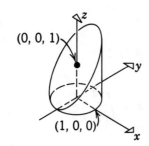

(b)

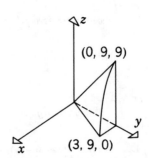

25. (a) $\displaystyle\int_0^a \int_0^{b(1-x/a)} \int_0^{c(1-x/a-y/b)} dz\,dy\,dx, \quad \int_0^b \int_0^{a(1-y/b)} \int_0^{c(1-x/a-y/b)} dz\,dx\,dy,$

$\displaystyle\int_0^c \int_0^{a(1-z/c)} \int_0^{b(1-x/a-z/c)} dy\,dx\,dz, \quad \int_0^a \int_0^{c(1-x/a)} \int_0^{b(1-x/a-z/c)} dy\,dz\,dx,$

$\displaystyle\int_0^c \int_0^{b(1-z/c)} \int_0^{a(1-y/b-z/c)} dx\,dy\,dz, \quad \int_0^b \int_0^{c(1-y/b)} \int_0^{a(1-y/b-z/c)} dx\,dz\,dy$

(b) Use the first integral in part (a) to get

$$\int_0^a \int_0^{b(1-x/a)} c\left(1 - \frac{x}{a} - \frac{y}{b}\right) dy\,dx = \int_0^a \frac{1}{2}bc\left(1 - \frac{x}{a}\right)^2 dx = \frac{1}{6}abc$$

27. $\displaystyle V = \int_0^a \int_0^x \int_0^{x-y} dz\,dy\,dx = \int_0^a \int_0^x (x-y)dy\,dx = \int_0^a \frac{1}{2}x^2 dx = a^3/6$

29. (a) $\displaystyle\left[\int_{-1}^1 x\,dx\right]\left[\int_0^1 y^2 dy\right]\left[\int_0^{\pi/2} \sin z\,dz\right] = (0)(1/3)(1) = 0$

(b) $\displaystyle\left[\int_0^1 e^{2x} dx\right]\left[\int_0^{\ln 3} e^y dy\right]\left[\int_0^{\ln 2} e^{-z} dz\right] = [(e^2-1)/2](2)(1/2) = (e^2-1)/2$

31. (a) 10 **(b)** 0

33. The region is symmetric with respect to the xz-plane, and for $f(x,y,z) = x^2 y^3 e^z$, $f(x,-y,z) = -f(x,y,z)$ so the integral evaluates to 0.

EXERCISE SET 17.6

1. Introduce an x-axis with origin at m_1, let a be the unknown coordinate of the fulcrum then the total moment about the fulcrum is $5(0 - a) + 10(5 - a) + 20(15 - a) = 350 - 35a$ so for equilibrium $350 - 35a = 0$, $a = 10$. The fulcrum should be placed 10 ft to the right of m_1.

3. $M = \int_0^1 \int_0^{\sqrt{x}} (x + y) dy\, dx = 13/20$, $M_x = \int_0^1 \int_0^{\sqrt{x}} (x + y)y\, dy\, dx = 3/10$,

   $M_y = \int_0^1 \int_0^{\sqrt{x}} (x + y)x\, dy\, dx = 19/42$, $\bar{x} = M_y/M = 190/273$, $\bar{y} = M_x/M = 6/13$;

   the mass is $13/30$ and the center of gravity is at $(190/273, 6/13)$.

5. $M = \int_0^{\pi/2} \int_0^a r^3 \sin\theta \cos\theta\, dr\, d\theta = a^4/8$, $\bar{x} = \bar{y}$ from the symmetry of the density and the

   region, $M_y = \int_0^{\pi/2} \int_0^a r^4 \sin\theta \cos^2\theta\, dr\, d\theta = a^5/15$, $\bar{x} = 8a/15$; mass $a^4/8$, center of gravity $(8a/15, 8a/15)$.

7. $A = 1/2$, $\iint_R x\, dA = \int_0^1 \int_0^x x\, dy\, dx = 1/3$, $\iint_R y\, dA = \int_0^1 \int_0^x y\, dy\, dx = 1/6$;

   centroid $(2/3, 1/3)$

9. $A = \int_{-2}^1 \int_x^{2-x^2} dy\, dx = 9/2$, $\iint_R x\, dA = \int_{-2}^1 \int_x^{2-x^2} x\, dy\, dx = -9/4$,

   $\iint_R y\, dA = \int_{-2}^1 \int_x^{2-x^2} y\, dy\, dx = 9/5$; centroid $(-1/2, 2/5)$

11. $\bar{x} = 0$ from the symmetry of the region,

    $A = \frac{1}{2}\pi(b^2 - a^2)$, $\iint_R y\, dA = \int_0^\pi \int_a^b r^2 \sin\theta\, dr\, d\theta = \frac{2}{3}(b^3 - a^3)$;

    centroid $\bar{x} = 0$, $\bar{y} = \dfrac{4(b^3 - a^3)}{3\pi(b^2 - a^2)}$.

13. $\bar{x} = 0$ from the symmetry of the region, $A = 16$,

    $\iint_R y\, dA = \int_{-4}^4 \int_{|x|}^4 y\, dy\, dx = 128/3$; centroid $(0, 8/3)$

15. $M = \int_0^a \int_0^a \int_0^a (a - x)dz\, dy\, dx = a^4/2$, $\bar{y} = \bar{z} = a/2$ from the symmetry of density and

region, $\bar{x} = \dfrac{1}{M} \int_0^a \int_0^a \int_0^a x(a - x)dz\, dy\, dx = (2/a^4)(a^5/6) = a/3$; mass $a^4/2$, center of

gravity $(a/3, a/2, a/2)$

17. $M = \int_{-1}^1 \int_0^1 \int_0^{1-y^2} yz\, dz\, dy\, dx = 1/6$, $\bar{x} = 0$ by the symmetry of density and region,

$\bar{y} = \dfrac{1}{M} \iiint_G y^2 z\, dV = (6)(8/105) = 16/35$, $\bar{z} = \dfrac{1}{M} \iiint_G yz^2 dV = (6)(1/12) = 1/2$;

mass $1/6$, center of gravity $(0, 16/35, 1/2)$

19. $\bar{x} = \bar{y} = \bar{z}$ from the symmetry of the region, $V = 1/6$,

$\bar{x} = \dfrac{1}{V} \int_0^1 \int_0^{1-x} \int_0^{1-x-y} x\, dz\, dy\, dx = (6)(1/24) = 1/4$; centroid $(1/4, 1/4, 1/4)$

21. $\bar{x} = 1/2$ and $\bar{y} = 0$ from the symmetry of the region,

$V = \int_0^1 \int_{-1}^1 \int_{y^2}^1 dz\, dy\, dx = 4/3$, $\bar{z} = \dfrac{1}{V} \iiint_G z\, dV = (3/4)(4/5) = 3/5$;

centroid $(1/2, 0, 3/5)$

23. $\bar{x} = \bar{y} = \bar{z}$ from the symmetry of the region, $V = \pi a^3/6$,

$\bar{x} = \dfrac{1}{V} \int_0^a \int_0^{\sqrt{a^2-x^2}} \int_0^{\sqrt{a^2-x^2-y^2}} x\, dz\, dy\, dx = \dfrac{1}{V} \int_0^a \int_0^{\sqrt{a^2-x^2}} x\sqrt{a^2 - x^2 - y^2}\, dy\, dx$

$= \dfrac{1}{V} \int_0^{\pi/2} \int_0^a r^2\sqrt{a^2 - r^2} \cos\theta\, dr\, d\theta = \dfrac{6}{\pi a^3}(\pi a^4/16) = 3a/8$; centroid $(3a/8, 3a/8, 3a/8)$

25. $M = \iint_R k\sqrt{x^2 + y^2}\, dA = \int_0^{2\pi} \int_0^a kr^2 dr\, d\theta = \dfrac{2}{3}\pi k a^3$

27. Let $x = r\cos\theta$, $y = r\sin\theta$, and $dA = r\, dr\, d\theta$ in formulas (15a) and (15b).

29. $\bar{x} = \bar{y}$ from the symmetry of the region, $A = \int_0^{\pi/2} \int_0^{\sin 2\theta} r\, dr\, d\theta = \pi/8$,

$\bar{x} = \dfrac{1}{A} \int_0^{\pi/2} \int_0^{\sin 2\theta} r^2 \cos\theta\, dr\, d\theta = (8/\pi)(16/105) = \dfrac{128}{105\pi}$; centroid $\left(\dfrac{128}{105\pi}, \dfrac{128}{105\pi} \right)$

31. $\bar{x} = 0$ from the symmetry of the region, $\pi a^2/2$ is the area of the semicircle, $2\pi\bar{y}$ is the distance traveled by the centroid to generate the sphere so $4\pi a^3/3 = (\pi a^2/2)(2\pi\bar{y})$, $\bar{y} = 4a/(3\pi)$

33. $\bar{x} = k$ so $V = (\pi ab)(2\pi k) = 2\pi^2 abk$

35.   The region generates a cone of volume $\dfrac{1}{3}\pi ab^2$ when it is revolved about the x-axis, the area of the region is $\dfrac{1}{2}ab$ so $\dfrac{1}{3}\pi ab^2 = \left(\dfrac{1}{2}ab\right)(2\pi\bar{y})$, $\bar{y} = b/3$. A cone of volume $\dfrac{1}{3}\pi a^2 b$ is generated when the region is revolved about the y-axis so $\dfrac{1}{3}\pi a^2 b = \left(\dfrac{1}{2}ab\right)(2\pi\bar{x})$, $\bar{x} = a/3$. The centroid is $(a/3, b/3)$.

37. $I_z = \displaystyle\int_{-a/2}^{a/2} \int_{-b/2}^{b/2} (x^2 + y^2)\delta\, dy\, dx = 4\delta \int_0^{a/2} \int_0^{b/2} (x^2 + y^2)dy\, dx$

      $= \delta ab(a^2 + b^2)/12; M = \delta ab$ so $I_z = M(a^2 + b^2)/12$.

39. $I_y = \displaystyle\int_0^{2\pi} \int_0^R r^3 \cos^2\theta\, \delta\, dr\, d\theta = \delta\pi R^4/4; M = \delta\pi R^2$ so $I_y = MR^2/4$.

41. $I_z = \displaystyle\int_0^\pi \int_0^{2R\sin\theta} r^3\delta\, dr\, d\theta = 3\delta\pi R^4/2; M = \delta\pi R^2$ so $I_z = 3MR^2/2$.

43. $I_z = \displaystyle\int_0^a \int_0^a \int_0^a (x^2 + y^2)\delta\, dz\, dy\, dx = 2\delta a^5/3; M = \delta a^3$ so $I_z = 2Ma^2/3$.

EXERCISE SET 17.7

1. $\displaystyle\int_0^{2\pi} \int_0^1 \int_0^{\sqrt{1-r^2}} zr\, dz\, dr\, d\theta = \int_0^{2\pi} \int_0^1 \frac{1}{2}(1 - r^2)r\, dr\, d\theta = \int_0^{2\pi} \frac{1}{8}d\theta = \pi/4$

3. $\displaystyle\int_0^{\pi/2} \int_0^{\pi/2} \int_0^1 \rho^3 \sin\phi\cos\phi\, d\rho\, d\phi\, d\theta = \int_0^{\pi/2} \int_0^{\pi/2} \frac{1}{4}\sin\phi\cos\phi\, d\phi\, d\theta = \int_0^{\pi/2} \frac{1}{8}d\theta = \pi/16$

5. $V = \displaystyle\int_0^{2\pi} \int_0^3 \int_{r^2}^9 r\, dz\, dr\, d\theta = \int_0^{2\pi} \int_0^3 r(9 - r^2)dr\, d\theta = \int_0^{2\pi} \frac{81}{4}d\theta = 81\pi/2$

7. $r^2 + z^2 = 20$ intersects $z = r^2$ in a circle of radius 2,

      $V = \displaystyle\int_0^{2\pi} \int_0^2 \int_{r^2}^{\sqrt{20-r^2}} r\, dz\, dr\, d\theta = \int_0^{2\pi} \int_0^2 (r\sqrt{20 - r^2} - r^3)dr\, d\theta$

                                  $= \dfrac{4}{3}(10\sqrt{5} - 19)\displaystyle\int_0^{2\pi} d\theta = 8(10\sqrt{5} - 19)\pi/3$

9. $x^2 + y^2 = 4x$ becomes $r = 4\cos\theta$ in cylindrical coordinates,

$$V = \int_0^{\pi/2} \int_0^{4\cos\theta} \int_0^{\sqrt{16-r^2}} r\,dz\,dr\,d\theta = \int_0^{\pi/2} \int_0^{4\cos\theta} r\sqrt{16-r^2}\,dr\,d\theta$$

$$= \int_0^{\pi/2} \frac{64}{3}(1-\sin^3\theta)\,d\theta = 32(3\pi-4)/9$$

11. $V = \int_0^{\pi/2} \int_{\pi/6}^{\pi/3} \int_0^2 \rho^2 \sin\phi\,d\rho\,d\phi\,d\theta = \int_0^{\pi/2} \int_{\pi/6}^{\pi/3} \frac{8}{3}\sin\phi\,d\phi\,d\theta$

$$= \frac{4}{3}(\sqrt{3}-1)\int_0^{\pi/2} d\theta = 2(\sqrt{3}-1)\pi/3$$

13. $V = \int_0^{2\pi} \int_{\pi/4}^{\pi/2} \int_0^3 \rho^2 \sin\phi\,d\rho\,d\phi\,d\theta = \int_0^{2\pi} \int_{\pi/4}^{\pi/2} 9\sin\phi\,d\phi\,d\theta = \frac{9\sqrt{2}}{2}\int_0^{2\pi} d\theta = 9\sqrt{2}\pi$

15. **(a)** $V = 2\int_0^{2\pi} \int_0^a \int_0^{\sqrt{a^2-r^2}} r\,dz\,dr\,d\theta = 4\pi a^3/3$

(b) $V = \int_0^{2\pi} \int_0^\pi \int_0^a \rho^2 \sin\phi\,d\rho\,d\phi\,d\theta = 4\pi a^3/3$

17. $M = \int_0^{\pi/2} \int_0^{2\cos\theta} \int_0^{4-r^2} zr\,dz\,dr\,d\theta = \int_0^{\pi/2} \int_0^{2\cos\theta} \frac{1}{2}r(4-r^2)^2\,dr\,d\theta$

$$= \frac{16}{3}\int_0^{\pi/2} (1-\sin^6\theta)\,d\theta = (16/3)(11\pi/32) = 11\pi/6$$

19. $M = \int_0^{2\pi} \int_0^{\pi/4} \int_0^1 \rho^3 \sin\phi\,d\rho\,d\phi\,d\theta = \int_0^{2\pi} \int_0^{\pi/4} \frac{1}{4}\sin\phi\,d\phi\,d\theta$

$$= \frac{1}{8}(2-\sqrt{2})\int_0^{2\pi} d\theta = (2-\sqrt{2})\pi/4$$

21. $M = \int_0^{2\pi} \int_0^\pi \int_0^a k\rho^3 \sin\phi\,d\rho\,d\phi\,d\theta = \int_0^{2\pi} \int_0^\pi \frac{1}{4}ka^4 \sin\phi\,d\phi\,d\theta = \frac{1}{2}ka^4 \int_0^{2\pi} d\theta = \pi ka^4$

23. $\bar{x} = \bar{y} = 0$ from the symmetry of the region,

$$V = \int_0^{2\pi} \int_0^1 \int_{r^2}^{\sqrt{2-r^2}} r\,dz\,dr\,d\theta = \int_0^{2\pi} \int_0^1 (r\sqrt{2-r^2} - r^3)\,dr\,d\theta = (8\sqrt{2}-7)\pi/6,$$

$$\bar{z} = \frac{1}{V} \int_0^{2\pi} \int_0^1 \int_{r^2}^{\sqrt{2-r^2}} zr\,dz\,dr\,d\theta = \frac{6}{(8\sqrt{2}-7)\pi}(7\pi/12) = 7/(16\sqrt{2}-14);$$

centroid $\left(0,0,\dfrac{7}{16\sqrt{2}-14}\right)$

25. $\bar{x} = \bar{y} = \bar{z}$ from the symmetry of the region, $V = \pi a^3/6$,

$$\bar{z} = \frac{1}{V} \int_0^{\pi/2} \int_0^{\pi/2} \int_0^a \rho^3 \cos\phi \sin\phi\,d\rho\,d\phi\,d\theta = \frac{6}{\pi a^3}(\pi a^4/16) = 3a/8;$$

centroid $(3a/8, 3a/8, 3a/8)$

27. $\bar{x} = \bar{z} = 0$ from the symmetry of the region, $V = 54\pi/3 - 16\pi/3 = 38\pi/3$,

$$\bar{y} = \frac{1}{V} \int_0^\pi \int_0^\pi \int_2^3 \rho^3 \sin^2\phi \sin\theta\,d\rho\,d\phi\,d\theta = \frac{1}{V} \int_0^\pi \int_0^\pi \frac{65}{4} \sin^2\phi \sin\theta\,d\phi\,d\theta$$

$$= \frac{1}{V} \int_0^\pi \frac{65\pi}{8} \sin\theta\,d\theta = \frac{3}{38\pi}(65\pi/4) = 195/152; \text{ centroid } (0, 195/152, 0)$$

29. $\displaystyle\int_0^\pi \int_0^{\pi/2} \int_0^1 e^{-\rho^3}\rho^2 \sin\phi\,d\rho\,d\phi\,d\theta = \frac{1}{3}(1-e^{-1}) \int_0^\pi \int_0^{\pi/2} \sin\phi\,d\phi\,d\theta = (1-e^{-1})\pi/3$

31. $\displaystyle\int_0^{2\pi} \int_0^\pi \int_0^3 \rho^3 \sin\phi\,d\rho\,d\phi\,d\theta = 81\pi$

33. $\bar{x} = \bar{y} = 0$ from the symmetry of density and region,

$$M = \int_0^{2\pi} \int_0^{\pi/2} \int_0^a k\rho^3 \sin\phi\,d\rho\,d\phi\,d\theta = \pi k a^4/2,$$

$$\bar{z} = \frac{1}{M} \int_0^{2\pi} \int_0^{\pi/2} \int_0^a k\rho^4 \sin\phi \cos\phi\,d\rho\,d\phi\,d\theta = \frac{2}{\pi k a^4}(\pi k a^5/5) = 2a/5;$$

center of gravity $(0, 0, 2a/5)$

35. $\bar{x} = \bar{y} = 0$ from the symmetry of density and region,

$$M = \int_0^{2\pi} \int_0^1 \int_0^{1-r^2} (r^2 + z^2)r\,dz\,dr\,d\theta = \pi/4,$$

$$\bar{z} = \frac{1}{M} \int_0^{2\pi} \int_0^1 \int_0^{1-r^2} z(r^2 + z^2)r\,dz\,dr\,d\theta = (4/\pi)(11\pi/120) = 11/30;$$

center of gravity $(0, 0, 11/30)$

37. In spherical coordinates the spheres are $\rho = 3$ and $\rho = 4\cos\phi$, respectively. They intersect when $\phi = \cos^{-1}(3/4)$ so

$$V = \int_0^{2\pi} \int_0^{\cos^{-1}(3/4)} \int_0^3 \rho^2 \sin\phi \, d\rho \, d\phi \, d\theta + \int_0^{2\pi} \int_{\cos^{-1}(3/4)}^{\pi/2} \int_0^{4\cos\phi} \rho^2 \sin\phi \, d\rho \, d\phi \, d\theta$$

$$= \int_0^{2\pi} \int_0^{\cos^{-1}(3/4)} 9 \sin\phi \, d\phi \, d\theta + \int_0^{2\pi} \int_{\cos^{-1}(3/4)}^{\pi/2} \frac{64}{3} \sin\phi \cos^3\phi \, d\phi \, d\theta$$

$$= \frac{9}{4} \int_0^{2\pi} d\theta + \frac{27}{16} \int_0^{2\pi} d\theta = 63\pi/8$$

39. **(a)** The sphere and cone intersect in a circle of radius $\rho_0 \sin\phi_0$,

$$V = \int_{\theta_1}^{\theta_2} \int_0^{\rho_0 \sin\phi_0} \int_{r\cot\phi_0}^{\sqrt{\rho_0^2 - r^2}} r \, dz \, dr \, d\theta = \int_{\theta_1}^{\theta_2} \int_0^{\rho_0 \sin\phi_0} \left(r\sqrt{\rho_0^2 - r^2} - r^2 \cot\phi_0 \right) dr \, d\theta$$

$$= \int_{\theta_1}^{\theta_2} \frac{1}{3}\rho_0^3 (1 - \cos^3\phi_0 - \sin^3\phi_0 \cot\phi_0) d\theta = \frac{1}{3}\rho_0^3 (1 - \cos^3\phi_0 - \sin^2\phi_0 \cos\phi_0)(\theta_2 - \theta_1)$$

$$= \frac{1}{3}\rho_0^3 (1 - \cos\phi_0)(\theta_2 - \theta_1).$$

(b) From part (a), the volume of the solid bounded by $\theta = \theta_1$, $\theta = \theta_2$, $\phi = \phi_1$, $\phi = \phi_2$, and $\rho = \rho_0$ is

$$\frac{1}{3}\rho_0^3 (1 - \cos\phi_2)(\theta_2 - \theta_1) - \frac{1}{3}\rho_0^3 (1 - \cos\phi_1)(\theta_2 - \theta_1) = \frac{1}{3}\rho_0^3 (\cos\phi_1 - \cos\phi_2)(\theta_2 - \theta_1)$$

so the volume of the spherical wedge between $\rho = \rho_1$ and $\rho = \rho_2$ is

$$\Delta V = \frac{1}{3}\rho_2^3 (\cos\phi_1 - \cos\phi_2)(\theta_2 - \theta_1) - \frac{1}{3}\rho_1^3 (\cos\phi_1 - \cos\phi_2)(\theta_2 - \theta_1)$$

$$= \frac{1}{3}(\rho_2^3 - \rho_1^3)(\cos\phi_1 - \cos\phi_2)(\theta_2 - \theta_1)$$

(c) $\dfrac{d}{d\phi}\cos\phi = -\sin\phi$ so from the Mean-Value Theorem $\cos\phi_2 - \cos\phi_1 = -\sin\phi^*(\phi_2 - \phi_1)$ where ϕ^* is between ϕ_1 and ϕ_2. Similarly $\dfrac{d}{d\rho}\rho^3 = 3\rho^2$ so $\rho_2^3 - \rho_1^3 = 3\rho^{*2}(\rho_2 - \rho_1)$ where ρ^* is between ρ_1 and ρ_2. Thus $\cos\phi_1 - \cos\phi_2 = \sin\phi^*\Delta\phi$ and $\rho_2^3 - \rho_1^3 = 3\rho^{*2}\Delta\rho$ so $\Delta V = \rho^{*2}\sin\phi^*\Delta\rho\Delta\phi\Delta\theta.$

41. $I_y = \displaystyle\int_0^{2\pi} \int_0^R \int_0^h (r^2\cos^2\theta + z^2)\delta r \, dz \, dr \, d\theta = \delta \int_0^{2\pi} \int_0^R \left(hr^3\cos^2\theta + \frac{1}{3}h^3 r \right) dr \, d\theta$

$$= \delta \int_0^{2\pi} \left(\frac{1}{4}R^4 h \cos^2\theta + \frac{1}{6}R^2 h^3 \right) d\theta = \delta \left(\frac{\pi}{4}R^4 h + \frac{\pi}{3}R^2 h^3 \right);$$

$M = \delta\pi R^2 h$ so $I_y = M(R^2/4 + h^2/3).$

43. $I_z = \int_0^{2\pi} \int_0^\pi \int_0^R (\rho^2 \sin^2 \phi)\delta \, \rho^2 \sin \phi \, d\rho \, d\phi \, d\theta$

$= \delta \int_0^{2\pi} \int_0^\pi \int_0^R \rho^4 \sin^3 \phi \, d\rho \, d\phi \, d\theta = \frac{8}{15}\delta\pi R^5; M = \frac{4}{3}\delta\pi R^3 \text{ so } I_z = \frac{2}{5}MR^2.$

45. **(a)** $\rho = \sqrt{r^2 + z^2}, \cos\phi = z/\rho = z/\sqrt{r^2 + z^2},$

$F_z = \int_0^{2\pi} \int_0^R \int_a^{a+h} \frac{k\delta zr}{(r^2 + z^2)^{3/2}} dz \, dr \, d\theta$

$= k\delta \int_0^{2\pi} \int_0^R \left[\frac{r}{\sqrt{r^2 + a^2}} - \frac{r}{\sqrt{r^2 + (a+h)^2}} \right] dr \, d\theta$

$= 2\pi k\delta(\sqrt{R^2 + a^2} - \sqrt{R^2 + (a+h)^2} + h).$

(b) The components of the force in the x and y directions are zero because of the symmetry of the solid with respect to the yz and xz planes.

47. In spherical coordinates the plane $z = h$ is $\rho \cos\phi = h$ so $\rho = h/\cos\phi,$

$F_z = \int_0^{2\pi} \int_0^{\tan^{-1}(R/h)} \int_0^{h/\cos\phi} \frac{k\delta \cos\phi}{\rho^2} \rho^2 \sin\phi \, d\rho \, d\phi \, d\theta$

$= k\delta \int_0^{2\pi} \int_0^{\tan^{-1}(R/h)} \int_0^{h/\cos\phi} \sin\phi \cos\phi \, d\rho \, d\phi \, d\theta$

$= 2\pi k\delta h \left[1 - \cos\left(\tan^{-1} \frac{R}{h} \right) \right] = 2\pi k\delta h(1 - h/\sqrt{R^2 + h^2}).$

SUPPLEMENTARY EXERCISES CHAPTER 17

1. $\int_{1/2}^1 \int_0^{2x} \cos(\pi x^2) dy \, dx = \int_{1/2}^1 2x \cos(\pi x^2) dx = -1/(\sqrt{2}\pi)$

3. $\int_{-1}^0 \int_0^{y^2} \int_{xy}^1 2y \, dz \, dx \, dy = \int_{-1}^0 \int_0^{y^2} (2y - 2xy^2) dx \, dy = \int_{-1}^0 (2y^3 - y^6) dy = -9/14$

5. $\int_0^1 \int_{2y}^2 e^x e^y \, dx \, dy$

7. $A = \int_0^1 \int_0^{2x^3} dy \, dx + \int_1^2 \int_0^{4-2x} dy \, dx = 1/2 + 1 = 3/2$

9. **(a)**
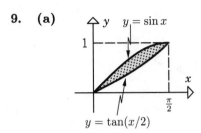

    (b)
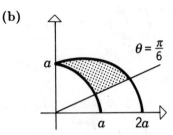

11. $\displaystyle\int_0^8 \int_{-\sqrt[3]{y}}^{\sqrt[3]{y}} x^2 \sin(y^2)dx\,dy = \int_0^8 \frac{2}{3}y\sin(y^2)dy = (1 - \cos 64)/3$

13. $\displaystyle V = \int_0^2 \int_0^{(6-3x)/2} (6 - 3x - 2y)dy\,dx = \int_0^2 \frac{1}{4}(6-3x)^2 dx = 6$

15. $\displaystyle V = \int_0^{2\pi} \int_{\sqrt{2}}^2 \frac{1}{r^3}dr\,d\theta = \int_0^{2\pi} \frac{1}{8}d\theta = \pi/4$

17. $\displaystyle\int_0^{2a} \int_0^{\sqrt{2ay-y^2}} \frac{2xy}{x^2+y^2}dx\,dy = \int_0^{2a} (y\ln 2a - y\ln y)dy = a^2$

19. $\displaystyle A = 6\int_0^{\pi/6} \int_0^{\cos 3\theta} r\,dr\,d\theta = 3\int_0^{\pi/6} \cos^2 3\theta\,d\theta = \pi/4$

21. $z_x = 6x$, $z_y = 6y$, $z_x^2 + z_y^2 + 1 = 36(x^2 + y^2) + 1$;
    $$S = \int_0^{2\pi} \int_0^1 r\sqrt{36r^2 + 1}\,dr\,d\theta = (37\sqrt{37} - 1)\pi/54$$

23. $z_x = x/z$, $z_y = y/z$, $z_x^2 + z_y^2 + 1 = 2$; $\displaystyle S = \int_0^{2\pi} \int_1^4 \sqrt{2}\,r\,dr\,d\theta = 15\sqrt{2}\pi$

25. $\displaystyle\int_0^{2\pi} \int_0^4 \int_0^{4-r\sin\theta} r^2\,dz\,dr\,d\theta = \int_0^{2\pi} \int_0^4 (4r^2 - r^3\sin\theta)dr\,d\theta$
    $$= \int_0^{2\pi} \frac{64}{3}(4 - 3\sin\theta)d\theta = 512\pi/3$$

27. **(a)** G is the region in the first octant bounded by the coordinate planes and the plane
    $z = 1 - x - 2y$; the integral is $\displaystyle\int_0^{1/2} \int_0^{1-2y} \int_0^{1-2y-z} z\,dx\,dz\,dy$

(b) G is the region in the first octant bounded by the planes $x = 0$, $z = 0$, $z = 4 - y$, and the parabolic cylinder $y = x^2$; the integral is $\int_0^4 \int_0^{4-y} \int_0^{\sqrt{y}} 3\, dx\, dz\, dy$

29. (a) $\int_0^{2\pi} \int_0^{\pi/3} \int_0^a \rho^4 \sin^3 \phi\, d\rho\, d\phi\, d\theta$ **(b)** $\int_0^{2\pi} \int_0^{\sqrt{3}a/2} \int_{r/\sqrt{3}}^{\sqrt{a^2-r^2}} r^3 dz\, dr\, d\theta$

(c) $\int_{-\sqrt{3}a/2}^{\sqrt{3}a/2} \int_{-\sqrt{3a^2/4-x^2}}^{\sqrt{3a^2/4-x^2}} \int_{\sqrt{x^2+y^2}/\sqrt{3}}^{\sqrt{a^2-x^2-y^2}} (x^2 + y^2)dz\, dy\, dx$

31. $V = \int_0^{2\pi} \int_0^{\pi} \int_0^{a(1+\cos \phi)} \rho^2 \sin \phi\, d\rho\, d\phi\, d\theta = \int_0^{2\pi} \int_0^{\pi} \frac{1}{3}a^3(1 + \cos \phi)^3 \sin \phi\, d\phi\, d\theta$

$$= \frac{4}{3}a^3 \int_0^{2\pi} d\theta = 8\pi a^3/3$$

33. $V = \int_0^{2\pi} \int_0^{a/\sqrt{3}} \int_{\sqrt{3}r}^a r\, dz\, dr\, d\theta = \int_0^{2\pi} \int_0^{a/\sqrt{3}} (ar - \sqrt{3}r^2)dr\, d\theta = \frac{1}{18}a^3 \int_0^{2\pi} d\theta = \pi a^3/9$

35. $\bar{x} = 0$ from the symmetry of the region,

$A = \int_0^{2\pi} \int_0^{a(1+\sin \theta)} r\, dr\, d\theta = 3\pi a^2/2,$

$\bar{y} = \frac{1}{A} \int_0^{2\pi} \int_0^{a(1+\sin \theta)} r^2 \sin \theta\, dr\, d\theta = \frac{1}{A} \int_0^{2\pi} \frac{1}{3}a^3(1 + \sin \theta)^3 \sin \theta\, d\theta$ but

$(1 + \sin \theta)^3 \sin \theta = \sin \theta + 3\sin^2 \theta + 3\sin^3 \theta + \sin^4 \theta$ and $\int_0^{2\pi} \sin \theta\, d\theta = \int_0^{2\pi} 3\sin^3 \theta\, d\theta = 0$

so $\bar{y} = \frac{1}{A} \int_0^{2\pi} \frac{1}{3}a^3(3\sin^2 \theta + \sin^4 \theta)d\theta = \frac{2}{3\pi a^2}(5\pi a^3/4) = 5a/6$; centroid $(0, 5a/6)$

37. $\bar{x} = 0$ from the symmetry of density and region, $M = 2\int_0^a \int_0^{b(1-x/a)} kx\, dy\, dx = ka^2b/3,$

$\bar{y} = \frac{2}{M} \int_0^a \int_0^{b(1-x/a)} k\, xy\, dy\, dx = \frac{6}{ka^2b}(ka^2b^2/24) = b/4$; center of gravity $(0, b/4)$

39. $M = \int_0^a \int_0^{b(1-x/a)} \int_0^{c(1-x/a-y/b)} kz\, dz\, dy\, dx = \frac{1}{2}kc^2 \int_0^a \int_0^{b(1-x/a)} (1 - x/a - y/b)^2 dy\, dx$

$$= \frac{1}{6}kbc^2 \int_0^a (1 - x/a)^3 dx = \frac{1}{24}kabc^2$$

41. $\bar{x} = 0$ from the symmetry of the region, $V = 2 \int_0^2 \int_{x^2}^4 \int_0^{4-y} dz\,dy\,dx = 256/15,$

$$\bar{y} = \frac{2}{V} \int_0^2 \int_{x^2}^4 \int_0^{4-y} y\,dz\,dy\,dx = (15/128)(512/35) = 12/7,$$

$$\bar{z} = \frac{2}{V} \int_0^2 \int_{x^2}^4 \int_0^{4-y} z\,dz\,dy\,dx = (15/128)(1024/105) = 8/7; \text{ centroid } (0, 12/7, 8/7)$$

43. $\bar{x} = \bar{y} = 0$ from the symmetry of the region, $V = \pi R^2 h/3,$

$$\bar{z} = \frac{1}{V} \int_0^{2\pi} \int_0^R \int_0^{h(1-r/R)} zr\,dz\,dr\,d\theta = \frac{3}{\pi R^2 h}(\pi R^2 h^2/12) = h/4; \text{ centroid } (0, 0, h/4)$$

CHAPTER 18
Topics In Vector Calculus

EXERCISE SET 18.1

1. **(a)** $\int_0^1 (2t + t^2)dt = 4/3$ **(b)** $\int_0^1 (t^2 - t^2)2t\, dt = 0$

   (c) $\int_C (2x + y)dx + \int_C (x^2 - y)dy = 4/3 + 0 = 4/3$

3. $\int_0^2 (t^2/2 - t^3)dt = -8/3$

5. $\int_0^{\pi/4} (8\cos^2 t - 16\sin^2 t - 20\sin t \cos t)dt = 1 - \pi$

7. $C : x = (3-t)^2/3,\ y = 3 - t,\ 0 \le t \le 3;\ \int_0^3 \frac{1}{3}(3-t)^2 dt = 3$

9. $C : x = \cos t,\ y = \sin t,\ 0 \le t \le \pi/2;\ \int_0^{\pi/2} (-\sin t - \cos^2 t)dt = -1 - \pi/4$

11. $\int_0^\pi (0)dt = 0$ 13. $\int_0^1 e^{-t}dt = 1 - e^{-1}$

15. $\int_0^1 (-3)e^{3t}dt = 1 - e^3$ 17. $\int_0^{\pi/2} (7\sin^2 t \cos t + 3\sin t \cos t)dt = 23/6$

19. $C : x = t^2,\ y = t,\ 0 \le t \le 1;\ W = \int_0^1 3t^4 dt = 3/5$

21. **(a)** $C_1 : (1,1)$ to $(2,2);\ x = 1 + t, y = 1 + t, 0 \le t \le 1$
    $C_2 : (2,2)$ to $(4,2);\ x = 2 + 2t, y = 2, 0 \le t \le 1$
    $$\int_0^1 \frac{5/2}{(1+t)^2}dt + \int_0^1 \frac{1/2}{1 + (1+t)^2}dt = 5/4 - \pi/8 + \frac{1}{2}\tan^{-1} 2$$
    (b) $C_1 : (0,3)$ to $(6,3);\ x = 6t, y = 3, 0 \le t \le 1$
    $C_2 : (6,3)$ to $(6,0);\ x = 6, y = 3 - 3t, 0 \le t \le 1$
    $$\int_0^1 \frac{6}{36t^2 + 9}dt + \int_0^1 \frac{-12}{36 + 9(1-t)^2}dt = \frac{1}{3}\tan^{-1} 2 - \frac{2}{3}\tan^{-1}(1/2)$$

417

(c) $C : x = 4\cos t, y = 4\sin t, 0 \le t \le \pi/2$

$$\int_0^{\pi/2} \left(-\frac{1}{4}\sin t + \cos t \right) dt = 3/4$$

23. $C_1 : (0,0,0)$ to $(1,3,1)$; $x = t$, $y = 3t$, $z = t$, $0 \le t \le 1$

$C_2 : (1,3,1)$ to $(2,-1,4)$; $x = 1+t$, $y = 3 - 4t$, $z = 1 + 3t$, $0 \le t \le 1$

$$W = \int_0^1 (4t + 8t^2)dt + \int_0^1 (-11 - 17t - 11t^2)dt = -37/2$$

25. $W = \int_0^1 \lambda[(1 - \lambda)t + (3\lambda - 1)t^2 - (1 + 2\lambda)t^3]dt = -\lambda/12$, $W = 1$ when $\lambda = -12$

27. $C : x = a\cos t$, $y = a\sin t$, $0 \le t \le \pi$;

$$\mathbf{F} = \frac{k(x\mathbf{i} + y\mathbf{j})}{(x^2 + y^2)^{3/2}} = \frac{k}{a^2}(\cos t\,\mathbf{i} + \sin t\,\mathbf{j}), \; d\mathbf{r}/dt = a(-\sin t\,\mathbf{i} + \cos t\,\mathbf{j}), \; W = \int_0^\pi (0)dt = 0$$

EXERCISE SET 18.2

1. $\partial x/\partial y = 0 = \partial y/\partial x$, conservative so $\partial\phi/\partial x = x$ and $\partial\phi/\partial y = y$, $\phi = x^2/2 + k(y)$, $k'(y) = y$, $k(y) = y^2/2 + K$, $\phi = x^2/2 + y^2/2 + K$

3. $\partial(x^2y)/\partial y = x^2$ and $\partial(5xy^2)/\partial x = 5y^2$, not conservative

5. $\partial(\cos y + y\cos x)/\partial y = -\sin y + \cos x = \partial(\sin x - x\sin y)/\partial x$, conservative so
   $\partial\phi/\partial x = \cos y + y\cos x$ and $\partial\phi/\partial y = \sin x - x\sin y$, $\phi = x\cos y + y\sin x + k(y)$,
   $-x\sin y + \sin x + k'(y) = \sin x - x\sin y$, $k'(y) = 0$, $k(y) = K$, $\phi = x\cos y + y\sin x + K$

7. $\partial(y^2)/\partial y = 2y = \partial(2xy)/\partial x$, independent of path

   (a) $\partial\phi/\partial x = y^2$ and $\partial\phi/\partial y = 2xy$, $\phi = xy^2 + k(y)$, $2xy + k'(y) = 2xy$, $k'(y) = 0$, $k(y) = K$,
   $\phi = xy^2 + K$. Let $K = 0$ to get $\phi(1,3) - \phi(-1,2) = 9 - (-4) = 13$

   (b) $C : x = -1 + 2t$, $y = 2 + t$, $0 \le t \le 1$; $\int_0^1 (4 + 14t + 6t^2)dt = 13$

9. $\partial(3y)/\partial y = 3 = \partial(3x)/\partial x$, $\phi = 3xy$, $\phi(4,0) - \phi(1,2) = -6$

11. $\partial(2xe^y)/\partial y = 2xe^y = \partial(x^2e^y)/\partial x$, $\phi = x^2e^y$, $\phi(3,2) - \phi(0,0) = 9e^2$

13. $\partial(2xy^3)/\partial y = 6xy^2 = \partial(3x^2y^2)/\partial x$, $\phi = x^2y^3$, $\phi(-1,0) - \phi(2,-2) = 32$

15. $\phi = x^2y^2/2$, $W = \phi(0,0) - \phi(1,1) = -1/2$

17. $\phi = \tan^{-1}(x/y)$, $W = \phi(3,3) - \phi(0,2) = \pi/4$

19. $\partial(e^y + ye^x)/\partial y = e^y + e^x = \partial(xe^y + e^x)/\partial x$ so **F** is conservative, $\phi = xe^y + ye^x$ so the work done in parts (a), (b), and (c) is $\phi(-a,0) - \phi(a,0) = -2a$. From Theorem 18.2.2 the work done in part (d) is 0.

21. $\mathbf{F} = \dfrac{kx}{(x^2 + y^2)^{3/2}}\mathbf{i} + \dfrac{ky}{(x^2 + y^2)^{3/2}}\mathbf{j}$, $\dfrac{\partial}{\partial y}\dfrac{kx}{(x^2 + y^2)^{3/2}} = -\dfrac{3kxy}{(x^2 + y^2)^{5/2}} = \dfrac{\partial}{\partial x}\dfrac{ky}{(x^2 + y^2)^{3/2}}$

so **F** is conservative

23. **(a)** If **F** is conservative then $\mathbf{F} = (\partial\phi/\partial x)\mathbf{i} + (\partial\phi/\partial y)\mathbf{j}$ for some function ϕ so $f(x,y) = \partial\phi/\partial x$ and $g(x,y) = \partial\phi/\partial y$. But $d\phi = (\partial\phi/\partial x)dx + (\partial\phi/\partial y)dy$ so $d\phi = f(x,y)dx + g(x,y)dy$. If ϕ is such that $d\phi = f(x,y)dx + g(x,y)dy$ then $f(x,y) = \partial\phi/\partial x$ and $g(x,y) = \partial\phi/\partial y$ so $\mathbf{F} = \nabla\phi$ and hence **F** is conservative.

(b) $\displaystyle\int_{(x_0,y_0)}^{(x_1,y_1)} d\phi = \int_{(x_0,y_0)}^{(x_1,y_1)} f(x,y)dx + g(x,y)dy = \phi(x_1,y_1) - \phi(x_0,y_0)$

EXERCISE SET 18.3

1. $\displaystyle\iint_R (2x - 2y)dA = \int_0^1 \int_0^1 (2x - 2y)dy\,dx = 0$

3. $\displaystyle\int_{-2}^4 \int_1^2 (2y - 3x)dy\,dx = 0$

5. $\displaystyle\int_0^{\pi/2} \int_0^{\pi/2} (-y\cos x + x\sin y)dy\,dx = 0$

7. $\displaystyle\iint_R [1 - (-1)]dA = 2\iint_R dA = 8\pi$

9. $\displaystyle\iint_R \left(-\dfrac{y}{1+y} - \dfrac{1}{1+y}\right)dA = -\iint_R dA = -4$

11. $\displaystyle\iint_R \left(-\dfrac{y^2}{1+y^2} - \dfrac{1}{1+y^2}\right)dA = -\iint_R dA = -1$

13. $\displaystyle\int_0^1 \int_{x^2}^{\sqrt{x}} (y^2 - x^2)dy\,dx = 0$

15. (a) $\displaystyle\int_C x\,dy = \int_0^{2\pi} ab\cos^2 t\,dt = \pi ab$ (b) $\displaystyle\int_C -y\,dx = \int_0^{2\pi} ab\sin^2 t\,dt = \pi ab$

17. $C_1 : (0,0)$ to $(a,0); x = at,\quad y = 0,\quad 0 \le t \le 1$
$C_2 : (a,0)$ to $(0,b); x = a - at, y = bt,\quad 0 \le t \le 1$
$C_3 : (0,b)$ to $(0,0); x = 0,\quad y = b - bt, 0 \le t \le 1$

$$A = \int_C x\,dy = \int_0^1 (0)dt + \int_0^1 ab(1-t)dt + \int_0^1 (0)dt = \frac{1}{2}ab$$

19. $\displaystyle W = \iint_R y\,dA = \int_0^\pi \int_0^5 r^2 \sin\theta\,dr\,d\theta = 250/3$

21. (a) $\displaystyle \bar{x} = \frac{1}{A}\iint_R x\,dA$, but $\displaystyle \int_C \frac{1}{2}x^2\,dy = \iint_R x\,dA$ from Green's Theorem so

$$\bar{x} = \frac{1}{A}\int_C \frac{1}{2}x^2\,dy = \frac{1}{2A}\int_C x^2\,dy. \text{ Similarly, } \bar{y} = -\frac{1}{2A}\int_C y^2\,dx.$$

(b) $\bar{x} = 0$ from the symmetry of the region,
$C_1 : (a,0)$ to $(-a,0)$ along $y = \sqrt{a^2 - x^2};\ x = a\cos t,\ y = a\sin t,\ 0 \le t \le \pi$
$C_2 : (-a,0)$ to $(a,0); x = -a + 2at,\ y = 0, 0 \le t \le 1$

$$A = \pi a^2/2, \quad \bar{y} = -\frac{1}{2A}\left[\int_0^\pi -a^3 \sin^3 t\,dt + \int_0^1 (0)dt\right]$$

$$= -\frac{1}{\pi a^2}\left(-\frac{4a^3}{3}\right) = \frac{4a}{3\pi}; \text{ centroid } \left(0, \frac{4a}{3\pi}\right)$$

23. (a) $C : x = a + (c-a)t, y = b + (d-b)t, 0 \le t \le 1$

$$\int_C x\,dy - y\,dx = \int_0^1 (ad - bc)dt = ad - bc$$

(b) Let C_1, C_2, and C_3 be the line segments from (x_1, y_1) to (x_2, y_2), (x_2, y_2) to (x_3, y_3), and (x_3, y_3) to (x_1, y_1), then from (4c)

$$A = \sum_{i=1}^3 \frac{1}{2}\int_{C_i} x\,dy - y\,dx \text{ and from the result of part (a)}$$

$$A = \frac{1}{2}[(x_1 y_2 - x_2 y_1) + (x_2 y_3 - x_3 y_2) + (x_3 y_1 - x_1 y_3)]$$

(c) $A = \dfrac{1}{2}[(x_1 y_2 - x_2 y_1) + (x_2 y_3 - x_3 y_2) + \cdots + (x_n y_1 - x_1 y_n)]$

(d) $A = \dfrac{1}{2}[(0-0) + (6+8) + (0+2) + (0-0)] = 8$

25. **(a)** $\displaystyle\int_0^{2\pi} (\sin^2 t + \cos^2 t)dt = \int_0^{2\pi} dt = 2\pi$

(b) $\partial g/\partial x = \dfrac{y^2 - x^2}{(x^2 + y^2)^2} = \partial f/\partial y$

(c) f and g do not have continuous first partial derivatives at the point $(0,0)$ in R so Green's Theorem is not applicable.

EXERCISE SET 18.4

1. R is the circular region enclosed by $x^2 + y^2 = 1$;

$$\iint_\sigma \delta_0 dS = \delta_0 \iint_R \sqrt{4x^2 + 4y^2 + 1}\, dA = \delta_0 \int_0^{2\pi} \int_0^1 \sqrt{4r^2 + 1}\, r\, dr\, d\theta$$

$$= \frac{1}{12}(5\sqrt{5} - 1)\delta_0 \int_0^{2\pi} d\theta = \frac{\pi}{6}(5\sqrt{5} - 1)\delta_0.$$

3. $z = \sqrt{4 - x^2}$, $\dfrac{\partial z}{\partial x} = -\dfrac{x}{\sqrt{4 - x^2}}$, $\dfrac{\partial z}{\partial y} = 0$;

$$\iint_\sigma \delta_0 dS = \delta_0 \iint_R \sqrt{\frac{x^2}{4 - x^2} + 1}\, dA = 2\delta_0 \int_0^4 \int_0^1 \frac{1}{\sqrt{4 - x^2}} dx\, dy = \frac{4}{3}\pi\delta_0.$$

5. R is the annular region between $x^2 + y^2 = 1$ and $x^2 + y^2 = 4$;

$$\iint_\sigma z^2 dS = \iint_R (x^2 + y^2)\sqrt{\frac{x^2}{x^2 + y^2} + \frac{y^2}{x^2 + y^2} + 1}\, dA$$

$$= \sqrt{2} \iint_R (x^2 + y^2)dA = \sqrt{2} \int_0^{2\pi} \int_1^2 r^3 dr\, d\theta = \frac{15}{2}\pi\sqrt{2}.$$

7. $z = \sqrt{1 - x^2}$, R is the rectangular region enclosed by $x = -1$, $x = 1$, $y = 0$ and $y = 1$; $\partial z/\partial x = -x/\sqrt{1 - x^2}$ which does not exist along the boundaries $x = \pm 1$. We avoid them by using $x = \pm x_0$ as boundaries where x_0 is slightly smaller than 1 and then let x_0 approach 1. By the symmetry of the surface and the integrand we can integrate over the region R' enclosed by $x = 0$, $x = x_0$, $y = 0$ and $y = 1$ and double the result so

$$\iint_\sigma x^2 y\, dS = \lim_{x_0 \to 1^-} 2 \iint_{R'} x^2 y \sqrt{\frac{x^2}{1 - x^2} + 1}\, dA = \lim_{x_0 \to 1^-} 2 \int_0^{x_0} \int_0^1 \frac{x^2 y}{\sqrt{1 - x^2}} dy\, dx$$

$$= \lim_{x_0 \to 1^-} \int_0^{x_0} \frac{x^2}{\sqrt{1 - x^2}} dx; \text{ use } x = \sin\theta \text{ to get}$$

$$\iint_{\sigma} x^2 y \, dS = \lim_{\theta_0 \to \pi/2-} \int_0^{\theta_0} \sin^2\theta \, d\theta \quad (\theta_0 = \sin^{-1} x_0)$$

$$= \lim_{\theta_0 \to \pi/2-} \left(\frac{1}{2}\theta_0 - \frac{1}{4}\sin 2\theta_0\right) = \frac{\pi}{4}.$$

9. If we use the projection of σ onto the xz-plane then $y = 1 - x$ and R is the rectangular region in the xz-plane enclosed by $x = 0$, $x = 1$, $z = 0$ and $z = 1$;

$$\iint_{\sigma}(x + y + z)dS = \iint_R (1 + z)\sqrt{2} \, dA = \sqrt{2}\int_0^1\int_0^1 (1 + z)dz \, dx = \frac{3\sqrt{2}}{2}.$$

11. There are six surfaces:

$\sigma_1 : z = 0; \ 0 \le x \le 1, \ 0 \le y \le 1$ (project onto xy-plane),

$\sigma_2 : x = 0; \ 0 \le y \le 1, \ 0 \le z \le 1$ (project onto yz-plane),

$\sigma_3 : y = 0; \ 0 \le x \le 1, \ 0 \le z \le 1$ (project onto xz-plane),

$\sigma_4 : z = 1; \ 0 \le x \le 1, \ 0 \le y \le 1$ (project onto xy-plane),

$\sigma_5 : x = 1; \ 0 \le y \le 1, \ 0 \le z \le 1$ (project onto yz-plane),

$\sigma_6 : y = 1; \ 0 \le x \le 1, \ 0 \le z \le 1$ (project onto xz-plane), so

$$\iint_{\sigma_1}(x + y + z)dS = \int_0^1\int_0^1 (x + y)dx \, dy = 1, \quad \iint_{\sigma_2}(x + y + z)dS = \int_0^1\int_0^1 (y + z)dy \, dz = 1,$$

$$\iint_{\sigma_3}(x + y + z)dS = \int_0^1\int_0^1 (x + z)dx \, dz = 1,$$

$$\iint_{\sigma_4}(x + y + z)dS = \int_0^1\int_0^1 (x + y + 1)dx \, dy = 2,$$

$$\iint_{\sigma_5}(x + y + z)dS = \int_0^1\int_0^1 (1 + y + z)dy \, dz = 2,$$

$$\iint_{\sigma_6}(x + y + z)dS = \int_0^1\int_0^1 (x + 1 + z)dx \, dz = 2,$$

thus, $\displaystyle\iint_{\sigma}(x + y + z)dS = 1 + 1 + 1 + 2 + 2 + 2 = 9.$

13. R is the circular region enclosed by $x^2 + y^2 = 1$;

$$\iint_\sigma \sqrt{x^2 + y^2 + z^2}\, dS = \iint_R \sqrt{2(x^2 + y^2)}\sqrt{\frac{x^2}{x^2 + y^2} + \frac{y^2}{x^2 + y^2} + 1}\, dA$$

$$= \lim_{r_0 \to 0^+} 2\iint_{R'} \sqrt{x^2 + y^2}\, dA$$

where R' is the annular region enclosed by $x^2 + y^2 = 1$ and $x^2 + y^2 = r_0^2$ with r_0 slightly larger

than 0 because $\sqrt{\dfrac{x^2}{x^2 + y^2} + \dfrac{y^2}{x^2 + y^2} + 1}$ is not defined for $x^2 + y^2 = 0$, so

$$\iint_\sigma \sqrt{x^2 + y^2 + z^2}\, dS = \lim_{r_0 \to 0^+} 2\int_0^{2\pi}\int_{r_0}^1 r^2\, dr\, d\theta = \lim_{r_0 \to 0^+} \frac{4\pi}{3}(1 - r_0^3) = \frac{4\pi}{3}.$$

15. Use $z = \sqrt{a^2 - x^2 - y^2}$ and let R be the circular region enclosed by $x^2 + y^2 = a^2$, $x = 0$, and $y = 0$ then by symmetry of the surface and the integrand

$$\iint_\sigma (x^2 + y^2)\, dS = 2\iint_R (x^2 + y^2)\sqrt{\frac{x^2}{a^2 - x^2 - y^2} + \frac{y^2}{a^2 - x^2 - y^2} + 1}\, dA$$

$$= 2a\iint_R \frac{x^2 + y^2}{\sqrt{a^2 - x^2 - y^2}}\, dA,$$

but the integrand is not defined along the boundary $x^2 + y^2 = a^2$ so

$$\iint_\sigma (x^2 + y^2)\, dS = \lim_{r_0 \to a^-} 2a\int_0^{2\pi}\int_0^{r_0} \frac{r^3}{\sqrt{a^2 - r^2}}\, dr\, d\theta$$

$$= \lim_{r_0 \to a^-} 4\pi a\left[\frac{2}{3}a^3 + \frac{1}{3}(a^2 - r_0^2)^{3/2} - a^2\sqrt{a^2 - r_0^2}\right] = \frac{8\pi}{3}a^4.$$

17. $z = 4 - y^2$, R is the rectangular region enclosed by $x = 0$, $x = 3$, $y = 0$ and $y = 3$;

$$\iint_\sigma y\, dS = \iint_R y\sqrt{4y^2 + 1}\, dA = \int_0^3\int_0^3 y\sqrt{4y^2 + 1}\, dy\, dx = \frac{1}{4}(37\sqrt{37} - 1).$$

19. (a) $\dfrac{\sqrt{29}}{16}\int_0^6\int_0^{(12-2x)/3} xy(12 - 2x - 3y)\, dy\, dx$

(b) $\dfrac{\sqrt{29}}{4}\int_0^3\int_0^{(12-4z)/3} yz(12 - 3y - 4z)\, dy\, dz$

(c) $\dfrac{\sqrt{29}}{9}\int_0^3\int_0^{6-2z} xz(12 - 2x - 4z)\, dx\, dz$

21. $\displaystyle\int_0^4 \int_1^2 y^3 z\sqrt{4y^2+1}\,dy\,dz;\ \frac{1}{2}\int_0^4 \int_1^4 xz\sqrt{1+4x}\,dx\,dz$

23. $\displaystyle M = \iint_\sigma \delta(x,y,z)dS = \iint_\sigma \delta_0 dS = \delta_0 \iint_\sigma dS = \delta_0 S$

EXERCISE SET 18.5

1. From the orientation of the plane surface we see that upward, right, and forward unit normals should be used; in each case the result is $\dfrac{2}{\sqrt{29}}\mathbf{i} + \dfrac{3}{\sqrt{29}}\mathbf{j} + \dfrac{4}{\sqrt{29}}\mathbf{k}$.

3. (a) Use an upward unit normal to get $\mathbf{n} = -\dfrac{2}{\sqrt{21}}\mathbf{i} - \dfrac{4}{\sqrt{21}}\mathbf{j} + \dfrac{1}{\sqrt{21}}\mathbf{k}$.

    (b) Use an upward unit normal to get $\mathbf{n} = \dfrac{3}{5\sqrt{2}}\mathbf{i} - \dfrac{4}{5\sqrt{2}}\mathbf{j} + \dfrac{1}{\sqrt{2}}\mathbf{k}$.

    (c) Use a downward unit normal with $z = -\sqrt{25 - x^2}$ to get $\mathbf{n} = \dfrac{3}{5}\mathbf{i} - \dfrac{4}{5}\mathbf{k}$.

5. R is the circular region enclosed by $x^2 + y^2 = 1$;

$$\iint_\sigma \mathbf{F} \cdot \mathbf{n}\,dS = \iint_R (2x^2 + 2y^2 + 2z)dA = 2\iint_R dA = (2)(\text{area of } R) = 2\pi.$$

7. R is the circular region enclosed by $x^2 + y^2 = 1$;

$$\iint_\sigma \mathbf{F} \cdot \mathbf{n}\,dS \lim_{r_0 \to 1^-} \iint_{R_{r_0}} z^2 dA = \lim_{r_0 \to 1^-} \iint_{R_{r_0}} (1 - x^2 - y^2)dA$$

where R_{r_0} is the region enclosed by $x^2 + y^2 = r_0^2$ with r_0 slightly less than 1 because $-\dfrac{\partial z}{\partial x}\mathbf{i} - \dfrac{\partial z}{\partial y}\mathbf{j} + \mathbf{k}$ is not defined along the boundary $x^2 + y^2 = 1$, so

$$\iint_\sigma \mathbf{F} \cdot \mathbf{n}\,dS \lim_{r_0 \to 1^-} \int_0^{2\pi} \int_0^{r_0} (1 - r^2)r\,dr\,d\theta = \lim_{r_0 \to 1^-} 2\pi\left(\frac{1}{2}r_0^2 - \frac{1}{4}r_0^4\right) = \frac{\pi}{2}.$$

9. R is the circular region enclosed by $x^2 + y^2 = 9$;

$$\iint_{\sigma} \mathbf{F} \cdot \mathbf{n}\, dS = \iint_{R} \left(\frac{x^2}{\sqrt{9 - x^2 - y^2}} + \frac{y^2}{\sqrt{9 - x^2 - y^2}} + z \right) dA$$

$$= 9 \iint_{R} \frac{1}{\sqrt{9 - x^2 - y^2}} dA = \lim_{r_0 \to 3^-} 9 \int_0^{2\pi} \int_0^{r_0} \frac{r}{\sqrt{9 - r^2}} dr\, d\theta$$

$$= \lim_{r_0 \to 3^-} 18\pi (3 - \sqrt{9 - r_0^2}) = 54\pi.$$

11. R is the annular region enclosed by $x^2 + y^2 = 1$ and $x^2 + y^2 = 4$;

$$\iint_{\sigma} \mathbf{F} \cdot \mathbf{n}\, dS = \iint_{R} \left(-\frac{x^2}{\sqrt{x^2 + y^2}} - \frac{y^2}{\sqrt{x^2 + y^2}} + 2z \right) dA$$

$$= \iint_{R} \sqrt{x^2 + y^2}\, dA = \int_0^{2\pi} \int_1^2 r^2 dr\, d\theta = \frac{14\pi}{3}.$$

13. R is the circular region enclosed by $x^2 + y^2 - y = 0$; $\iint_{\sigma} \mathbf{F} \cdot \mathbf{n}\, dS = \iint_{R} (-x) dA$, but in polar

coordinates the boundary $x^2 + y^2 - y = 0$ is $r = \sin\theta$, so

$$\iint_{\sigma} \mathbf{F} \cdot \mathbf{n}\, dS = -\int_0^{\pi} \int_0^{\sin\theta} r^2 \cos\theta\, dr\, d\theta = 0.$$

15. Divide the surface into two parts σ_1 and σ_2 corresponding to $z = \sqrt{a^2 - x^2 - y^2}$ and
$z = -\sqrt{a^2 - x^2 - y^2}$, respectively. For each part R is the circular region enclosed by
$x^2 + y^2 = a^2$; σ_1 is oriented by upward normals and σ_2 by downward normals so

$$\iint_{\sigma_1} \mathbf{F} \cdot \mathbf{n}\, dS = \iint_{R} \left(\frac{x^2}{\sqrt{a^2 - x^2 - y^2}} + \frac{y^2}{\sqrt{a^2 - x^2 - y^2}} + z \right) dA$$

$$= a^2 \iint_{R} \frac{1}{\sqrt{a^2 - x^2 - y^2}} dA = \lim_{r_0 \to a^-} a^2 \int_0^{2\pi} \int_0^{r_0} \frac{r}{\sqrt{a^2 - r^2}} dr\, d\theta$$

$$= \lim_{r_0 \to a^-} 2\pi a^2 (a - \sqrt{a^2 - r_0^2}) = 2\pi a^3;$$

similarly $\iint_{\sigma_2} \mathbf{F} \cdot \mathbf{n}\, dS = a^2 \iint_{R} \frac{1}{\sqrt{a^2 - x^2 - y^2}} dA = 2\pi a^3$ so

$$\iint_{\sigma} \mathbf{F} \cdot \mathbf{n}\, dS = 2\pi a^3 + 2\pi a^3 = 4\pi a^3.$$

17. Replace $\mathbf{n}$ by $-\mathbf{n}$ to reverse the orientation of σ, so $\displaystyle\iint_\sigma \mathbf{F} \cdot (-\mathbf{n})dS = -\iint_\sigma \mathbf{F} \cdot \mathbf{n}\,dS$.

19. (a) $\displaystyle\iint_R \mathbf{F} \cdot \left(\mathbf{i} - \frac{\partial x}{\partial y}\mathbf{j} - \frac{\partial x}{\partial z}\mathbf{k}\right) dA$, σ oriented by forward normals, and

   $\displaystyle\iint_R \mathbf{F} \cdot \left(-\mathbf{i} + \frac{\partial x}{\partial y}\mathbf{j} + \frac{\partial x}{\partial z}\mathbf{k}\right) dA$, σ oriented by backward normals where R is the projection of σ onto the yz-plane.

   (b) $\displaystyle\iint_R \mathbf{F} \cdot \left(-\frac{\partial y}{\partial x}\mathbf{i} + \mathbf{j} - \frac{\partial y}{\partial z}\mathbf{k}\right) dA$, σ oriented by right normals, and

   $\displaystyle\iint_R \mathbf{F} \cdot \left(\frac{\partial y}{\partial x}\mathbf{i} - \mathbf{j} + \frac{\partial y}{\partial z}\mathbf{k}\right) dA$, σ oriented by left normals, where R is the projection of σ
   onto the xz-plane.

21. R is the circular region in the xz-plane enclosed by $x^2 + z^2 = 1$;

   $$\iint_\sigma \mathbf{F} \cdot \mathbf{n}\,dS = \iint_R \left(\frac{x^2}{\sqrt{1 - x^2 - z^2}} + y + \frac{z^2}{\sqrt{1 - x^2 - z^2}}\right) dA = \iint_R \frac{1}{\sqrt{1 - x^2 - z^2}}dA,$$

   use polar coordinates in the xz-plane with $x = r\cos\theta$ and $z = r\sin\theta$ to get

   $$\iint_\sigma \mathbf{F} \cdot \mathbf{n}\,dS = \lim_{r_0 \to 1^-} \int_0^{2\pi} \int_0^{r_0} \frac{r}{\sqrt{1 - r^2}}dr\,d\theta = \lim_{r_0 \to 1^-} 2\pi(1 - \sqrt{1 - r_0^2}) = 2\pi$$

EXERCISE SET 18.6

1. $\operatorname{div} \mathbf{F} = z^3 + 8y^3x^2 + 10zy$ 3. $\operatorname{div} \mathbf{F} = ye^{xy} + \sin y + 2\sin z \cos z$

5. $\operatorname{div} \mathbf{F} = \dfrac{1}{x} + xze^{xyz} + \dfrac{x}{x^2 + z^2}$

7. G is the cube; $\displaystyle\iiint_G \operatorname{div} \mathbf{F}\,dV = 8\iiint_G dV = (8)(\text{volume of cube}) = (8)(1) = 8$.

9. G is the cylindrical solid;

   $$\iiint_G \operatorname{div} \mathbf{F}\,dV = 3\iiint_G dV = (3)(\text{volume of cylinder}) = (3)[\pi a^2(1)] = 3\pi a^2.$$

11. G is the cylindrical solid;

$$\iiint_G \operatorname{div} \mathbf{F} \, dV = 3 \iiint_G (x^2 + y^2 + z^2) dV = 3 \int_0^{2\pi} \int_0^2 \int_0^3 (r^2 + z^2) r \, dz \, dr \, d\theta = 180\pi.$$

13. G is the tetrahedron; $\displaystyle \iiint_G \operatorname{div} \mathbf{F} \, dV = \iiint_G x \, dV = \int_0^1 \int_0^{1-x} \int_0^{1-x-y} x \, dz \, dy \, dx = \frac{1}{24}.$

15. G is the conical solid;

$$\iiint_G \operatorname{div} \mathbf{F} \, dV = 2 \iiint_G (x + y + z) dV = 2 \int_0^{2\pi} \int_0^1 \int_r^1 (r\cos\theta + r\sin\theta + z) r \, dz \, dr \, d\theta = \frac{\pi}{2}.$$

17. G is the solid bounded by $z = 4 - x^2$, $y + z = 5$, and the coordinate planes;

$$\iiint_G \operatorname{div} \mathbf{F} \, dV = 4 \iiint_G x^2 dV = 4 \int_{-2}^2 \int_0^{4-x^2} \int_0^{5-z} x^2 dy \, dz \, dx = \frac{4608}{35}.$$

19. Refer to Exercise 15 in 18.5 where it is shown that $\displaystyle \iint_\sigma \mathbf{F} \cdot \mathbf{n} \, dS = 4\pi a^3$;

$$\iiint_G \operatorname{div} \mathbf{F} \, dV = 3 \iiint_G dV = (3)(\text{volume of sphere}) = (3)\left(\frac{4}{3}\pi a^3\right) = 4\pi a^3.$$

21. (a) Let $\mathbf{F} = f_1\mathbf{i} + g_1\mathbf{j} + h_1\mathbf{k}$ and $\mathbf{G} = f_2\mathbf{i} + g_2\mathbf{j} + h_2\mathbf{k}$ then
$\mathbf{F} + \mathbf{G} = (f_1 + f_2)\mathbf{i} + (g_1 + g_2)\mathbf{j} + (h_1 + h_2)\mathbf{k}$ and

$$\operatorname{div}(\mathbf{F} + \mathbf{G}) = \frac{\partial}{\partial x}(f_1 + f_2) + \frac{\partial}{\partial y}(g_1 + g_2) + \frac{\partial}{\partial z}(h_1 + h_2)$$

$$= \frac{\partial f_1}{\partial x} + \frac{\partial f_2}{\partial x} + \frac{\partial g_1}{\partial y} + \frac{\partial g_2}{\partial y} + \frac{\partial h_1}{\partial z} + \frac{\partial h_2}{\partial z}$$

$$= \left(\frac{\partial f_1}{\partial x} + \frac{\partial g_1}{\partial y} + \frac{\partial h_1}{\partial z}\right) + \left(\frac{\partial f_2}{\partial x} + \frac{\partial g_2}{\partial y} + \frac{\partial h_2}{\partial z}\right) = \operatorname{div} \mathbf{F} + \operatorname{div} \mathbf{G}$$

(b) Let $\mathbf{F} = f_1\mathbf{i} + g_1\mathbf{j} + h_1\mathbf{k}$ then $f\mathbf{F} = ff_1\mathbf{i} + fg_1\mathbf{j} + fh_1\mathbf{k}$ and

$$\operatorname{div}(f\mathbf{F}) = \frac{\partial}{\partial x}(ff_1) + \frac{\partial}{\partial y}(fg_1) + \frac{\partial}{\partial z}(fh_1)$$

$$= f\frac{\partial f_1}{\partial x} + \frac{\partial f}{\partial x}f_1 + f\frac{\partial g_1}{\partial y} + \frac{\partial f}{\partial y}g_1 + f\frac{\partial h_1}{\partial z} + \frac{\partial f}{\partial z}h_1$$

$$= f\left(\frac{\partial f_1}{\partial x} + \frac{\partial g_1}{\partial y} + \frac{\partial h_1}{\partial z}\right) + \left(\frac{\partial f}{\partial x}f_1 + \frac{\partial f}{\partial y}g_1 + \frac{\partial f}{\partial z}h_1\right) = f \operatorname{div} \mathbf{F} + (\nabla f) \cdot \mathbf{F}.$$

23. Let σ be the surface of the cylindrical solid G bounded by $x^2 + y^2 = a^2$, $z = 0$ and $z = h$. Divide σ into four parts;

$\sigma_1 : z = 0$ with $x^2 + y^2 \leq a^2$ and $\mathbf{n} = -\mathbf{k}$, $\sigma_2 : z = h$ with $x^2 + y^2 \leq a^2$ and $\mathbf{n} = \mathbf{k}$,

$\sigma_3 : y = \sqrt{a^2 - x^2}$ and $\sigma_4 : y = -\sqrt{a^2 - x^2}$, both with $0 \leq z \leq h$.

$$\iint_{\sigma_1} \mathbf{F} \cdot \mathbf{n}\, dS = \iint_{\sigma_1} 0\, dS = 0, \quad \iint_{\sigma_2} \mathbf{F} \cdot \mathbf{n}\, dS = h \iint_{\sigma_2} dS = \pi a^2 h,$$

$$\iint_{\sigma_3} \mathbf{F} \cdot \mathbf{n}\, dS = \iint_{\sigma_4} \mathbf{F} \cdot \mathbf{n}\, dS = a^2 \iint_R \frac{1}{\sqrt{a^2 - x^2}} dA$$

$$= \lim_{x_0 \to a^-} a^2 \int_0^h \int_{-x_0}^{x_0} \frac{1}{\sqrt{a^2 - x^2}} dx\, dz = \lim_{x_0 \to a^-} \left(2a^2 h \sin^{-1} \frac{x_0}{a}\right) = \pi a^2 h \text{ so}$$

$$\iint_\sigma \mathbf{F} \cdot \mathbf{n}\, dS = 0 + \pi a^2 h + \pi a^2 h + \pi a^2 h = 3\pi a^2 h \text{ and vol}(G) = \frac{1}{3}(3\pi a^2 h) = \pi a^2 h.$$

EXERCISE SET 18.7

1. curl $\mathbf{F} = 0$ **3.** curl $\mathbf{F} = -xe^{xy}\mathbf{k}$

5. curl $\mathbf{F} = -xye^{xyz}\mathbf{i} + \dfrac{z}{x^2 + z^2}\mathbf{j} + yze^{xyz}\mathbf{k}$

7. curl $\mathbf{F} = 2\mathbf{i} + 3\mathbf{j} + 4\mathbf{k}$;

$$\iint_\sigma (\text{curl } \mathbf{F}) \cdot \mathbf{n}\, dS = \iint_R (4x + 6y + 4) dA = \int_0^{2\pi} \int_0^2 (4r \cos\theta + 6r \sin\theta + 4)r\, dr\, d\theta = 16\pi.$$

9. curl $\mathbf{F} = x\mathbf{k}$, take σ as part of the plane $z = y$ oriented with upward normals, R is the circular region in the xy-plane enclosed by $x^2 + y^2 - y = 0$;

$$\iint_\sigma (\text{curl } \mathbf{F}) \cdot \mathbf{n}\, dS = \iint_R x\, dA = \int_0^\pi \int_0^{\sin\theta} r^2 \cos\theta\, dr\, d\theta = 0.$$

11. curl $\mathbf{F} = \mathbf{i} + \mathbf{j} + \mathbf{k}$, take σ as part of the plane $z = 0$ with $x^2 + y^2 \leq a^2$ and $\mathbf{n} = \mathbf{k}$;
$$\iint_\sigma (\text{curl } \mathbf{F}) \cdot \mathbf{n}\, dS = \iint_\sigma dS = \text{area of circle} = \pi a^2.$$

13. If σ is oriented with upward normals then C consists of three parts parametrized as

$C_1 : \mathbf{r}(t) = (1 - t)\mathbf{i} + t\mathbf{j}$ for $0 \leq t \leq 1$, $C_2 : \mathbf{r}(t) = (1 - t)\mathbf{j} + t\mathbf{k}$ for $0 \leq t \leq 1$,

$C_3 : \mathbf{r}(t) = t\mathbf{i} + (1 - t)\mathbf{k}$ for $0 \leq t \leq 1$.

$$\int_{C_1} \mathbf{F} \cdot d\mathbf{r} = \int_{C_2} \mathbf{F} \cdot d\mathbf{r} = \int_{C_3} \mathbf{F} \cdot d\mathbf{r} = \int_0^1 (3t-1)dt = \frac{1}{2} \text{ so}$$

$\int_C \mathbf{F} \cdot d\mathbf{r} = \frac{1}{2} + \frac{1}{2} + \frac{1}{2} = \frac{3}{2}$. curl $\mathbf{F} = \mathbf{i} + \mathbf{j} + \mathbf{k}$, $z = 1 - x - y$, R is the triangular region in the xy-plane enclosed by $x + y = 1$, $x = 0$, and $y = 0$;

$$\iint_\sigma (\text{curl } \mathbf{F}) \cdot \mathbf{n} \, dS = 3 \iint_R dA = (3)(\text{area of } R) = (3)\left[\frac{1}{2}(1)(1)\right] = \frac{3}{2}.$$

15. If σ is oriented with upward normals then C can be parametrized as $\mathbf{r}(t) = a \cos t\,\mathbf{i} + a \sin t\,\mathbf{j}$ for $0 \le t \le 2\pi$.

$$\int_C \mathbf{F} \cdot d\mathbf{r} = \int_0^{2\pi} 0 \, dt = 0; \text{ curl } \mathbf{F} = 0 \text{ so } \iint_\sigma (\text{curl } \mathbf{F}) \cdot \mathbf{n} \, dS = \iint_\sigma 0 \, dS = 0.$$

17. Take σ as part of the plane $z = 0$ for $x^2 + y^2 \le 1$ with $\mathbf{n} = \mathbf{k}$; $\mathbf{F} = z^2\mathbf{i} + 2x\mathbf{j} - y^3\mathbf{k}$,

curl $\mathbf{F} = -3y^2\mathbf{i} + 2z\mathbf{j} + 2\mathbf{k}$,

$$\iint_\sigma (\text{curl } \mathbf{F}) \cdot \mathbf{n} \, dS = 2 \iint_\sigma dS = (2)(\text{area of circle}) = (2)[\pi(1)^2] = 2\pi.$$

19. Let $\mathbf{F} = f\mathbf{i} + g\mathbf{j} + h\mathbf{k}$ then

$$\text{curl } \mathbf{F} = \left(\frac{\partial h}{\partial y} - \frac{\partial g}{\partial z}\right)\mathbf{i} + \left(\frac{\partial f}{\partial z} - \frac{\partial h}{\partial x}\right)\mathbf{j} + \left(\frac{\partial g}{\partial x} - \frac{\partial f}{\partial y}\right)\mathbf{k},$$

$$\text{div}(\text{curl } \mathbf{F}) = \frac{\partial^2 h}{\partial x \partial y} - \frac{\partial^2 g}{\partial x \partial z} + \frac{\partial^2 f}{\partial y \partial z} - \frac{\partial^2 h}{\partial y \partial x} + \frac{\partial^2 g}{\partial z \partial x} - \frac{\partial^2 f}{\partial z \partial y}$$

$$= \left(\frac{\partial^2 f}{\partial y \partial z} - \frac{\partial^2 f}{\partial z \partial y}\right) + \left(\frac{\partial^2 g}{\partial z \partial x} - \frac{\partial^2 g}{\partial x \partial z}\right) + \left(\frac{\partial^2 h}{\partial x \partial y} - \frac{\partial^2 h}{\partial y \partial x}\right)$$

but $\dfrac{\partial^2 f}{\partial y \partial z} = \dfrac{\partial^2 f}{\partial z \partial y}$, $\dfrac{\partial^2 g}{\partial z \partial x} = \dfrac{\partial^2 g}{\partial x \partial z}$, and $\dfrac{\partial^2 h}{\partial x \partial y} = \dfrac{\partial^2 h}{\partial y \partial x}$

because of the continuity assumptions so div(curl $\mathbf{F}$) = 0.

21. (a) $\nabla f = \dfrac{\partial f}{\partial x}\mathbf{i} + \dfrac{\partial f}{\partial y}\mathbf{j} + \dfrac{\partial f}{\partial z}\mathbf{k}$,

$$\text{curl } (\nabla f) = \left(\frac{\partial^2 f}{\partial y \partial z} - \frac{\partial^2 f}{\partial z \partial y}\right)\mathbf{i} + \left(\frac{\partial^2 f}{\partial z \partial x} - \frac{\partial^2 f}{\partial x \partial z}\right)\mathbf{j} + \left(\frac{\partial^2 f}{\partial x \partial y} - \frac{\partial^2 f}{\partial y \partial x}\right)\mathbf{k}$$

but the continuity conditions imply equality of mixed second partial derivatives so curl $(\nabla f) = 0$.

(b) curl $(\nabla f + \text{curl } \mathbf{F}) = $ curl $(\nabla f) + $ curl(curl $\mathbf{F}$) $= $ curl(curl $\mathbf{F}$) because curl $(\nabla f) = 0$ from part (a).

23. Let σ_1 and σ_2 be the upper and lower hemispheres oriented, respectively, by upward and downward normals so

$$\iint\limits_{\sigma} (\text{curl } \mathbf{F}) \cdot \mathbf{n} \, dS = \iint\limits_{\sigma_1} (\text{curl } \mathbf{F}) \cdot \mathbf{n} \, dS + \iint\limits_{\sigma_2} (\text{curl } \mathbf{F}) \cdot \mathbf{n} \, dS.$$

Let C be the boundary shared by σ_1 and σ_2 and suppose that C is positively oriented with respect to σ_1 and hence negatively oriented with respect to σ_2. Then by Stokes' Theorem

$$\iint\limits_{\sigma_1} (\text{curl } \mathbf{F}) \cdot \mathbf{n} \, dS = \int_C \mathbf{F} \cdot d\mathbf{r} \quad \text{and} \quad \iint\limits_{\sigma_2} (\text{curl } \mathbf{F}) \cdot \mathbf{n} \, dS = - \int_C \mathbf{F} \cdot d\mathbf{r} \quad \text{so}$$

$$\iint\limits_{\sigma} (\text{curl } \mathbf{F}) \cdot \mathbf{n} \, dS = \int_C \mathbf{F} \cdot d\mathbf{r} - \int_C \mathbf{F} \cdot d\mathbf{r} = 0.$$

EXERCISE SET 18.8

1. $\displaystyle \iint\limits_{\sigma} \mathbf{F} \cdot \mathbf{n} \, dS = 2 \iint\limits_{R} (x^2 + y^2 + z) dA = 2 \iint\limits_{R} dA = (2)(\text{area of } R) = 2\pi$

3. $\displaystyle \iint\limits_{\sigma} \mathbf{F} \cdot \mathbf{n} \, dS = \iint\limits_{R} \left(x + y + \frac{9 + xy}{\sqrt{9 - x^2 - y^2}} \right) dA$

$$= \lim_{r_0 \to 3^-} \int_0^{2\pi} \int_0^{r_0} \left(r\cos\theta + r\sin\theta + \frac{9 + r^2 \sin\theta\cos\theta}{\sqrt{9 - r^2}} \right) r \, dr \, d\theta$$

$$= \lim_{r_0 \to 3^-} \int_0^{2\pi} \int_0^{r_0} \left(r^2 \cos\theta + r^2 \sin\theta + \frac{9r}{\sqrt{9 - r^2}} + \frac{r^3}{\sqrt{9 - r^2}} \sin\theta\cos\theta \right) dr \, d\theta$$

$$= \lim_{r_0 \to 3^-} 18\pi(3 - \sqrt{9 - r_0^2}) = 54\pi.$$

5. $\displaystyle \iint\limits_{\sigma} \mathbf{F} \cdot \mathbf{n} \, dS = \iiint\limits_{G} \text{div } \mathbf{F} \, dV = 3 \iiint\limits_{G} dV = (3)(\text{volume of sphere}) = 4\pi a^3$

7. $\displaystyle \iint\limits_{\sigma} \mathbf{F} \cdot \mathbf{n} \, dS = \iiint\limits_{G} \text{div } \mathbf{F} \, dV = 3 \iiint\limits_{G} (x^2 + y^2 + z^2) dV$

$$= 3 \int_0^{2\pi} \int_0^2 \int_0^3 (r^2 + z^2) r \, dz \, dr \, d\theta = 180\pi$$

9. div $\mathbf{F} = 0$; no sources or sinks.

11. div $\mathbf{F} = 3x^2 + 3y^2 + 3z^2$; sources at all points except the origin, no sinks.

13. (a) Take σ as the part of the plane $2x+y+2z = 2$ in the first octant, oriented with downward normals; curl $\mathbf{F} = -x\mathbf{i} + (y-1)\mathbf{j} - \mathbf{k}$,

$$\int_C \mathbf{F} \cdot \mathbf{T}\, ds = \iint_\sigma (\text{curl } \mathbf{F}) \cdot \mathbf{n}\, dS$$

$$= \iint_R \left(x - \frac{1}{2}y + \frac{3}{2}\right) dA = \int_0^1 \int_0^{2-2x} \left(x - \frac{1}{2}y + \frac{3}{2}\right) dy\, dx = \frac{3}{2}.$$

(b) At the origin curl $\mathbf{F} = -\mathbf{j} - \mathbf{k}$ and with $\mathbf{n} = \mathbf{k}$, curl $\mathbf{F}(0,0,0) \cdot \mathbf{n} = (-\mathbf{j} - \mathbf{k}) \cdot \mathbf{k} = -1$.

(c) The rotation of $\mathbf{F}$ has its maximum value at the origin about the unit vector in the same direction as curl $\mathbf{F}(0,0,0)$ so $\mathbf{n} = -\frac{1}{\sqrt{2}}\mathbf{j} - \frac{1}{\sqrt{2}}\mathbf{k}$.

SUPPLEMENTARY EXERCISES CHAPTER 18

1. $x = t$, $y = \sin t$, $0 \le t \le \pi$; $\displaystyle\int_0^\pi (2\sin t + 3\cos t)dt = 4$

3. $x = t$, $y = \ln t$, $1 \le t \le 3$; $\displaystyle\int_1^3 (2t - 1)dt = 6$

5. $x = t$, $y = 2t$, $z = 3t$, $0 \le t \le 1$; $\displaystyle\int_0^1 4t\, dt = 2$

7. $\partial(y\sin xy)/\partial y = xy\cos xy + \sin xy$, $\partial(-x\cos xy)/\partial x = xy\sin xy - \cos xy$, not conservative

9. $\partial(3x^2 - y^2/x^2)/\partial y = -2y/x^2 = \partial(2y/x + 4y)/\partial x$, conservative so $\partial\phi/\partial x = 3x^2 - y^2/x^2$ and $\partial\phi/\partial y = 2y/x+4y$, $\phi = x^3 + y^2/x + k(y)$, $2y/x + k'(y) = 2y/x+4y$, $k'(y) = 4y$, $k(y) = 2y^2 + K$, $\phi = x^3 + y^2/x + 2y^2 + K$.

11. $\partial(\cos 2y - 3x^2y^2)/\partial y = -2\sin 2y - 6x^2y$ and $\partial(\cos 2y - 2x\sin 2y - 2x^3y)/\partial x = -2\sin 2y - 6x^2y$ so it is independent of path. The line segment from $(1, \pi/4)$ to $(2, \pi/4)$ is $x = 1+t$, $y = \pi/4$, $0 \le t \le 1$; the line integral along this path is

$$-\int_0^1 3(\pi/4)^2(1 + t)^3 dt = -7\pi^2/16.$$

13. $\partial(1/y)/\partial y = -1/y^2 = \partial(-x/y^2)/\partial x$, independent of path; $\phi = x/y$, $\phi(2, 1) - \phi(1, 2) = 2 - 1/2 = 3/2$

15. $\displaystyle\int_0^2 \int_0^{2x} (y^2 - 2x)\,dy\,dx = 0$

17. $\displaystyle\int_{-2}^2 \int_0^{4-y^2} 3\,dx\,dy = 32$

19. $\displaystyle\iint\limits_R (-3x^2 - 3y^2)\,dA = -3 \int_0^\pi \int_1^2 r^3\,dr\,d\theta = -45\pi/4$

21. $C : x = t,\ y = t^2,\ 0 \le t \le 1;\ W = \displaystyle\int_C \mathbf{F} \cdot d\mathbf{r} = \int_0^1 (2t + 3t^2 + 2t^4)\,dt = 12/5$

23. $C : x = t,\ y = 2t,\ z = 3t,\ 0 \le t \le 1;\ W = \displaystyle\int_C \mathbf{F} \cdot d\mathbf{r} = \int_0^1 4t\,dt = 2$

25. By symmetry $\bar{x} = \bar{y} = 0$.

$$\iint\limits_\sigma dS = \iint\limits_R \sqrt{x^2 + y^2 + 1}\,dA = \int_0^{2\pi} \int_0^{\sqrt{8}} \sqrt{r^2 + 1}\,r\,dr\,d\theta = \frac{52\pi}{3},$$

$$\iint\limits_\sigma z\,dS = \iint\limits_R z\sqrt{x^2 + y^2 + 1}\,dA = \frac{1}{2} \iint\limits_R (x^2 + y^2)\sqrt{x^2 + y^2 + 1}\,dA$$

$$= \frac{1}{2} \int_0^{2\pi} \int_0^{\sqrt{8}} r^3\sqrt{r^2 + 1}\,dr\,d\theta = \frac{596\pi}{15}$$

so $\bar{z} = \dfrac{596\pi/15}{52\pi/3} = \dfrac{149}{65}$. The centroid is $(\bar{x}, \bar{y}, \bar{z}) = (0, 0, 149/65)$.

27. $\nabla f = \dfrac{x}{x^2 + y^2 + z^2}\mathbf{i} + \dfrac{y}{x^2 + y^2 + z^2}\mathbf{j} + \dfrac{z}{x^2 + y^2 + z^2}\mathbf{k}$; on σ, $\nabla f = x\mathbf{i} + y\mathbf{j} + z\mathbf{k}$ because $x^2 + y^2 + z^2 = 1$. $D_\mathbf{n} f = \nabla f \cdot \mathbf{n}$ so

$$\iint\limits_\sigma D_\mathbf{n} f\,dS = \iint\limits_\sigma \nabla f \cdot \mathbf{n}\,dS = \iint\limits_R \frac{1}{\sqrt{1 - x^2 - y^2}}\,dA$$

$$= \lim_{r_0 \to 1^-} \int_0^{\pi/2} \int_0^{r_0} \frac{r}{\sqrt{1 - r^2}}\,dr\,d\theta = \lim_{r_0 \to 1^-} \frac{\pi}{2}\left(1 - \sqrt{1 - r_0^2}\right) = \frac{\pi}{2}.$$

29. $D_\mathbf{n}\phi = \nabla\phi \cdot \mathbf{n}$ so $\displaystyle\iint\limits_\sigma D_\mathbf{n}\phi\,dS = \iint\limits_\sigma \nabla\phi \cdot \mathbf{n}\,dS = \iiint\limits_G \text{div}\,(\nabla\phi)\,dV$ by the Divergence

Theorem. $\nabla\phi = \dfrac{\partial\phi}{\partial x}\mathbf{i} + \dfrac{\partial\phi}{\partial y}\mathbf{j} + \dfrac{\partial\phi}{\partial z}\mathbf{k}$ so div $(\nabla\phi) = \dfrac{\partial^2\phi}{\partial x^2} + \dfrac{\partial^2\phi}{\partial y^2} + \dfrac{\partial^2\phi}{\partial z^2}$ and

$$\iint\limits_\sigma D_\mathbf{n}\phi\,dS = \iiint\limits_G \left(\frac{\partial^2\phi}{\partial x^2} + \frac{\partial^2\phi}{\partial y^2} + \frac{\partial^2\phi}{\partial z^2}\right) dV.$$

CHAPTER 19
Second-Order Differential Equations

EXERCISE SET 19.1

1. **(a)** $y = e^{2x}$, $y' = 2e^{2x}$, $y'' = 4e^{2x}$; $y'' - y' - 2y = 0$
   $y = e^{-x}$, $y' = -e^{-x}$, $y'' = e^{-x}$; $y'' - y' - 2y = 0$.

   (b) $y = c_1 e^{2x} + c_2 e^{-x}$, $y' = 2c_1 e^{2x} - c_2 e^{-x}$, $y'' = 4c_1 e^{2x} + c_2 e^{-x}$; $y'' - y' - 2y = 0$

3. $m^2 + 3m - 4 = 0$, $(m-1)(m+4) = 0$; $m = 1, -4$ so $y = c_1 e^x + c_2 e^{-4x}$.

5. $m^2 - 2m + 1 = 0$, $(m-1)^2 = 0$; $m = 1$, so $y = c_1 e^x + c_2 x e^x$.

7. $m^2 + 5 = 0$, $m = \pm\sqrt{5}\,i$ so $y = c_1 \cos\sqrt{5}\,x + c_2 \sin\sqrt{5}\,x$.

9. $m^2 - m = 0$, $m(m-1) = 0$; $m = 0, 1$ so $y = c_1 + c_2 e^x$.

11. $m^2 + 4m + 4 = 0$, $(m+2)^2 = 0$; $m = -2$ so $y = c_1 e^{-2t} + c_2 t e^{-2t}$.

13. $m^2 - 4m + 13 = 0$, $m = 2 \pm 3i$ so $y = e^{2x}(c_1 \cos 3x + c_2 \sin 3x)$.

15. $8m^2 - 2m - 1 = 0$, $(4m+1)(2m-1) = 0$; $m = -1/4, 1/2$ so $y = c_1 e^{-x/4} + c_2 e^{x/2}$.

17. $m^2 + 2m - 3 = 0$, $(m+3)(m-1) = 0$; $m = -3, 1$ so $y = c_1 e^{-3x} + c_2 e^x$ and $y' = -3c_1 e^{-3x} + c_2 e^x$.
    Solve the system $c_1 + c_2 = 1$, $-3c_1 + c_2 = 5$ to get $c_1 = -1$, $c_2 = 2$ so $y = -e^{-3x} + 2e^x$.

19. $m^2 - 6m + 9 = 0$, $(m-3)^2 = 0$; $m = 3$ so $y = (c_1 + c_2 x)e^{3x}$ and $y' = (3c_1 + c_2 + 3c_2 x)e^{3x}$.
    Solve the system $c_1 = 2$, $3c_1 + c_2 = 1$ to get $c_1 = 2$, $c_2 = -5$ so $y = (2 - 5x)e^{3x}$.

21. $m^2 + 4m + 5 = 0$, $m = -2 \pm i$ so $y = e^{-2x}(c_1 \cos x + c_2 \sin x)$,
    $y' = e^{-2x}[(c_2 - 2c_1)\cos x - (c_1 + 2c_2)\sin x]$. Solve the system $c_1 = -3$, $c_2 - 2c_1 = 0$
    to get $c_1 = -3$, $c_2 = -6$ so $y = -e^{-2x}(3\cos x + 6\sin x)$.

23. **(a)** $m = 5, -2$ so $(m-5)(m+2) = 0$, $m^2 - 3m - 10 = 0$; $y'' - 3y' - 10y = 0$.
    (b) $m = 4, 4$ so $(m-4)^2 = 0$, $m^2 - 8m + 16 = 0$; $y'' - 8y' + 16y = 0$.
    (c) $m = -1 \pm 4i$ so $(m+1-4i)(m+1+4i) = 0$, $m^2 + 2m + 17 = 0$; $y'' + 2y' + 17y = 0$.

25. $m^2 + km + k = 0$, $m = \left(-k \pm \sqrt{k^2 - 4k}\right)/2$
    (a) $k^2 - 4k > 0$, $k(k-4) > 0$; $k < 0$ or $k > 4$

(b) $k^2 - 4k = 0; k = 0, 4$ **(c)** $k^2 - 4k < 0, k(k-4) < 0; 0 < k < 4$

27. **(a)** $\dfrac{d^2y}{dz^2} + 2\dfrac{dy}{dz} + 2y = 0$, $m^2 + 2m + 2 = 0$; $m = -1 \pm i$ so

$y = e^{-z}(c_1 \cos z + c_2 \sin z) = \dfrac{1}{x}[c_1 \cos(\ln x) + c_2 \sin(\ln x)]$.

(b) $\dfrac{d^2y}{dz^2} - 2\dfrac{dy}{dz} - 2y = 0$, $m^2 - 2m - 2 = 0$; $m = 1 \pm \sqrt{3}$ so

$y = c_1 e^{(1+\sqrt{3})z} + c_2 e^{(1-\sqrt{3})z} = c_1 x^{1+\sqrt{3}} + c_2 x^{1-\sqrt{3}}$

29. **(a)** $W(x) = \begin{vmatrix} e^{m_1 x} & e^{m_2 x} \\ m_1 e^{m_1 x} & m_2 e^{m_2 x} \end{vmatrix} = m_2 e^{(m_1+m_2)x} - m_1 e^{(m_1+m_2)x}$

$= (m_2 - m_1)e^{(m_1+m_2)x} \neq 0$ if $m_1 \neq m_2$.

(b) $W(x) = \begin{vmatrix} e^{mx} & x e^{mx} \\ m e^{mx} & (mx+1)e^{mx} \end{vmatrix} = e^{2mx} \neq 0$.

31. **(a)** The general solution is $c_1 e^{\mu x} + c_2 e^{mx}$; let $c_1 = 1/(\mu - m)$, $c_2 = -1/(\mu - m)$.

(b) $\displaystyle\lim_{\mu \to m} \dfrac{e^{\mu x} - e^{mx}}{\mu - m} = \lim_{\mu \to m} x e^{\mu x} = x e^{mx}$.

EXERCISE SET 19.2

1. $m^2 + 6m + 5 = 0$, $(m+1)(m+5) = 0$; $m = -1, -5$ so $y_c = c_1 e^{-x} + c_2 e^{-5x}$. Let
$y_p = Ae^{3x}$, then $y_p' = 3Ae^{3x}$, $y_p'' = 9Ae^{3x}$, $(9A + 18A + 5A)e^{3x} = 32Ae^{3x} = 2e^{3x}$,
$A = 1/16$; $y = c_1 e^{-x} + c_2 e^{-5x} + \frac{1}{16}e^{3x}$.

3. $m^2 - 9m + 20 = 0$, $(m-4)(m-5) = 0$; $m = 4, 5$ so $y_c = c_1 e^{4x} + c_2 e^{5x}$. Let $y_p = Axe^{5x}$,
then $y_p' = (5Ax + A)e^{5x}$, $y_p'' = (25Ax + 10A)e^{5x}$,
$(25Ax + 10A - 45Ax - 9A + 20Ax)e^{5x} = Ae^{5x} = -3e^{5x}$, $A = -3$; $y = c_1 e^{4x} + c_2 e^{5x} - 3xe^{5x}$.

5. $m^2 + 2m + 1 = 0$, $(m+1)^2 = 0$; $m = -1$ so $y_c = (c_1 + c_2 x)e^{-x}$. Let $y_p = Ax^2 e^{-x}$,
then $y_p' = (-Ax^2 + 2Ax)e^{-x}$, $y_p'' = (Ax^2 - 4Ax + 2A)e^{-x}$,
$(Ax^2 - 4Ax + 2A - 2Ax^2 + 4Ax + Ax^2)e^{-x} = 2Ae^{-x} = e^{-x}$, $A = 1/2$; $y = (c_1 + c_2 x)e^{-x} + \frac{1}{2}x^2 e^{-x}$.

7. $m^2 + m - 12 = 0$, $(m-3)(m+4) = 0$; $m = 3, -4$ so $y_c = c_1 e^{3x} + c_2 e^{-4x}$. Let
    $y_p = A_0 + A_1 x + A_2 x^2$, then $y_p' = A_1 + 2A_2 x$, $y_p'' = 2A_2$,
    $2A_2 + A_1 + 2A_2 x - 12A_0 - 12A_1 x - 12A_2 x^2$
    $\quad = (-12A_0 + A_1 + 2A_2) + 2(-6A_1 + A_2)x - 12A_2 x^2 = 4x^2$;
    solve the system $-12A_0 + A_1 + 2A_2 = 0$, $-6A_1 + A_2 = 0$, $-12A_2 = 4$ to get
    $A_0 = -13/216$, $A_1 = -1/18$, $A_2 = -1/3$ so $y = c_1 e^{3x} + c_2 e^{-4x} - \frac{13}{216} - \frac{1}{18}x - \frac{1}{3}x^2$.

9. $m^2 - 6m = 0$; $m = 0, 6$ so $y_c = c_1 + c_2 e^{6x}$. Let $y_p = A_0 x + A_1 x^2$, then $y_p' = A_0 + 2A_1 x$, $y_p'' = 2A_1$,
    $2A_1 - 6A_0 - 12A_1 x = (-6A_0 + 2A_1) - 12A_1 x = x - 1$; solve the system $-6A_0 + 2A_1 = -1$,
    $-12A_1 = 1$ to get $A_0 = 5/36$, $A_1 = -1/12$, so $y = c_1 + c_2 e^{6x} + \frac{5}{36}x - \frac{1}{12}x^2$.

11. $m^2 = 0$, $m = 0$ so $y_c = c_1 + c_2 x$. Let $y_p = A_0 x^2 + A_1 x^3 + A_2 x^4 + A_3 x^5$,
    then $y_p' = 2A_0 x + 3A_1 x^2 + 4A_2 x^3 + 5A_3 x^4$, $y_p'' = 2A_0 + 6A_1 x + 12A_2 x^2 + 20A_3 x^3$;
    $A_0 = -1/2$, $A_1 = 0$, $A_2 = 0$, $A_3 = 1/20$ so $y = c_1 + c_2 x - x^2/2 + x^5/20$.

13. $m^2 - m - 2 = 0$, $(m+1)(m-2) = 0$; $m = -1, 2$ so $y_c = c_1 e^{-x} + c_2 e^{2x}$. Let
    $y_p = A_1 \cos x + A_2 \sin x$, then $y_p' = -A_1 \sin x + A_2 \cos x$, $y_p'' = -A_1 \cos x - A_2 \sin x$,
    $-A_1 \cos x - A_2 \sin x + A_1 \sin x - A_2 \cos x - 2A_1 \cos x - 2A_2 \sin x$
    $\quad = (-3A_1 - A_2)\cos x + (A_1 - 3A_2)\sin x = 10 \cos x$; solve the system $-3A_1 - A_2 = 10$,
    $A_1 - 3A_2 = 0$ to get $A_1 = -3$, $A_2 = -1$ so $y = c_1 e^{-x} + c_2 e^{2x} - 3 \cos x - \sin x$.

15. $m^2 - 4 = 0$; $m = \pm 2$ so $y_c = c_1 e^{-2x} + c_2 e^{2x}$. Let $y_p = A_1 \cos 2x + A_2 \sin 2x$,
    then $y_p' = -2A_1 \sin 2x + 2A_2 \cos 2x$, $y_p'' = -4A_1 \cos 2x - 4A_2 \sin 2x$,
    $-8A_1 \cos 2x - 8A_2 \sin 2x = 2 \sin 2x + 3 \cos 2x$, $A_1 = -3/8$, $A_2 = -1/4$;
    $y = c_1 e^{-2x} + c_2 e^{2x} - \frac{3}{8} \cos 2x - \frac{1}{4} \sin 2x$.

17. $m^2 + 1 = 0$; $m = \pm i$ so $y_c = c_1 \cos x + c_2 \sin x$. Let $y_p = A_1 x \cos x + A_2 x \sin x$,
    then $y_p' = (A_1 + A_2 x)\cos x + (A_2 - A_1 x)\sin x$, $y_p'' = (2A_2 - A_1 x)\cos x - (2A_1 + A_2 x)\sin x$,
    $2A_2 \cos x - 2A_1 \sin x = \sin x$, $A_1 = -1/2$, $A_2 = 0$; $y = c_1 \cos x + c_2 \sin x - \frac{1}{2}x \cos x$.

19. $m^2 - 3m + 2 = 0$, $(m-1)(m-2) = 0$; $m = 1, 2$ so $y_c = c_1 e^x + c_2 e^{2x}$. Let $y_p = A_0 + A_1 x$,
    then $y_p' = A_1$, $y_p'' = 0$; solve the system $2A_0 - 3A_1 = 0$, $2A_1 = 1$ to get $A_0 = 3/4$, $A_1 = 1/2$;
    $y = c_1 e^x + c_2 e^{2x} + 3/4 + x/2$.

21. $m^2 + 4m + 9 = 0$; $m = -2 \pm \sqrt{5}\, i$ so $y_c = e^{-2x}(c_1 \cos \sqrt{5}\, x + c_2 \sin \sqrt{5}\, x)$.
    Let $y_p = A_0 + A_1 x + A_2 x^2$, then $y_p' = A_1 + 2A_2 x$, $y_p'' = 2A_2$; solve the system
    $9A_0 + 4A_1 + 2A_2 = 0$, $9A_1 + 8A_2 = 3$, $9A_2 = 1$, to get $A_0 = -94/729$, $A_1 = 19/81$, $A_2 = 1/9$
    so $y = e^{-2x}(c_1 \cos \sqrt{5}\, x + c_2 \sin \sqrt{5}\, x) - 94/729 + 19x/81 + x^2/9$.

23. $m^2 + 4 = 0$; $m = \pm 2i$ so $y_c = c_1 \cos 2x + c_2 \sin 2x$; $\sin x \cos x = \frac{1}{2} \sin 2x$,

let $y_p = A_1 x \cos 2x + A_2 x \sin 2x$, then $y_p' = (A_1 + 2A_2 x) \cos 2x + (A_2 - 2A_1 x) \sin 2x$,

$y_p'' = (4A_2 - 4A_1 x) \cos 2x - (4A_1 + 4A_2 x) \sin 2x$, $4A_2 \cos 2x - 4A_1 \sin 2x = \frac{1}{2} \sin 2x$,

$A_1 = -1/8$, $A_2 = 0$; $y = c_1 \cos 2x + c_2 \sin 2x - \frac{1}{8} x \cos 2x$.

25. **(a)** Let $y = y_1 + y_2$, then $y' = y_1' + y_2'$, $y'' = y_1'' + y_2''$ so

$y'' + p(x)y' + q(x)y = [y_1'' + p(x)y_1' + q(x)y_1] + [y_2'' + p(x)y_2' + q(x)y_2] = r_1(x) + r_2(x)$.

(b) $m^2 + 3m - 4 = 0$, $(m-1)(m+4) = 0$; $m = 1, -4$ so $y_c = c_1 e^x + c_2 e^{-4x}$. For $y'' + 3y' - 4y = x$

let $y_1 = A_0 + A_1 x$, then $y_1' = A_1$, $y_1'' = 0$; $-4A_0 + 3A_1 = 0$ and $-4A_1 = 1$ so $A_0 = -3/16$,

$A_1 = -1/4$; $y_1 = -3/16 - x/4$, for $y'' + 3y' - 4y = e^x$ let $y_2 = Axe^x$, then $y_2' = (Ax + A)e^x$,

$y_2'' = (Ax + 2A)e^x$, $5A = 1$, $A = 1/5$; $y_2 = \frac{1}{5} xe^x$ thus $y_1 + y_2 = -\frac{3}{16} - \frac{1}{4} x + \frac{1}{5} xe^x$ is a

particular solution.

(c) If $y_i(x)$ is a solution of $y'' + p(x)y' + q(x)y = r_i(x)$ for $i = 1, 2, \cdots, n$ then

$y_1(x) + y_2(x) + \cdots + y_n(x)$ is a solution of

$y'' + p(x)y' + q(x)y = r_1(x) + r_2(x) + \cdots + r_n(x)$.

27. $m^2 - 1 = 0$; $m = \pm 1$ so $y_c = c_1 e^{-x} + c_2 e^x$. Let $r_1(x) = 1$ and $r_2(x) = e^x$, then $y_1 = A_0$,

$y_1' = y_1'' = 0$, $A_0 = -1$; $y_2 = Axe^x$, $y_2' = (Ax + A)e^x$, $y_2'' = (Ax + 2A)e^x$, $2A = 1$, $A = 1/2$ so

$y = c_1 e^{-x} + c_2 e^x - 1 + \frac{1}{2} xe^x$.

29. $m^2 + 4 = 0$; $m = \pm 2i$ so $y_c = c_1 \cos 2x + c_2 \sin 2x$. Let $r_1(x) = 1 + x$ and $r_2(x) = \sin x$, then

$y_1 = A_0 + A_1 x$, $y_1' = A_1$, $y_1'' = 0$, $4A_0 = 1$, and $4A_1 = 1$ so $A_0 = A_1 = 1/4$;

$y_2 = B_1 \cos x + B_2 \sin x$, $y_2' = -B_1 \sin x + B_2 \cos x$, $y_2'' = -B_1 \cos x - B_2 \sin x$,

$3B_1 = 0$ and $3B_2 = 1$ so $B_1 = 0$, $B_2 = 1/3$; $y = c_1 \cos 2x + c_2 \sin 2x + \frac{1}{4} + \frac{1}{4} x + \frac{1}{3} \sin x$.

31. $m^2 - 2m + 1 = 0$, $(m-1)^2 = 0$; $m = 1$ so $y_c = (c_1 + c_2 x)e^x$. Let

$r_1(x) = \frac{1}{2} e^x$ and $r_2(x) = -\frac{1}{2} e^{-x}$, then $y_1 = Ax^2 e^x$, $y_1' = (Ax^2 + 2Ax)e^x$

$y_1'' = (Ax^2 + 4Ax + 2A)e^x$, $2A = 1/2$, $A = 1/4$; $y_2 = Be^{-x}$, $y_2' = -Be^{-x}$,

$y_2'' = Be^{-x}$, $4B = 1/2$, $B = 1/8$; $y = (c_1 + c_2 x)e^x + \frac{1}{4} x^2 e^x + \frac{1}{8} e^{-x}$.

33. $m^2 + 1 = 0$; $m = \pm i$ so $y_c = c_1 \cos x + c_2 \sin x$. Let $r_1(x) = 6$ and

$r_2(x) = 6 \cos 2x$, then $y_1 = A_0$, $y_1' = y_1'' = 0$, $A_0 = 6$; $y_2 = A_1 \cos 2x + A_2 \sin 2x$,

$y_2' = -2A_1 \sin 2x + 2A_2 \cos 2x$, $y_2'' = -4A_1 \cos 2x - 4A_2 \sin 2x$, $-3A_1 = 6$ and $-3A_2 = 0$

so $A_1 = -2$, $A_2 = 0$; $y = c_1 \cos x + c_2 \sin x + 6 - 2 \cos 2x$.

35. **(a)** $m^2 + \mu^2 = 0$; $m = \pm \mu i$ so $y_c = c_1 \cos \mu x + c_2 \sin \mu x$. Let $y_p = A_1 \cos bx + A_2 \sin bx$,

then $y_p' = -bA_1 \sin bx + bA_2 \cos bx$, $y_p'' = -b^2 A_1 \cos bx - b^2 A_2 \sin bx$,

$A_1 = 0$, $A_2 = a/(\mu^2 - b^2)$; $y = c_1 \cos \mu x + c_2 \sin \mu x + \dfrac{a}{\mu^2 - b^2} \sin bx$.

(b) $y = c_1 \cos \mu x + c_2 \sin \mu x + \sum\limits_{k=1}^{n} \dfrac{a_k}{\mu^2 - k^2 \pi^2} \sin k\pi x.$

37. $m^2 - m = 0$; $m = 0, 1$ so $y_c = c_1 + c_2 e^x$. Let $y_p = A_0 x + A_1 x^2$, then

$y'_p = A_0 + 2A_1 x$, $y''_p = 2A_1$; $A_0 = 0$, $A_1 = 2$, $y = c_1 + c_2 e^x + 2x^2$. If $y'(x_0) = y''(x_0) = 0$,

then $4 - 4x_0 = 0$, $x_0 = 1$ so $y'(1) = 0$ and $y''(1) = 0$; $y' = c_2 e^x + 4x$, $y'' = c_2 e^x + 4$,

$y'(1) = c_2 e + 4 = y''(1) = 0$ if $c_2 = -4/e$ so $y = c_1 - 4e^{x-1} + 2x^2$, or simply

$y = c - 4e^{x-1} + 2x^2$ where c is an arbitrary constant.

EXERCISE SET 19.3

1. $m^2 + 1 = 0$; $m = \pm i$ so $y_c = c_1 \cos x + c_2 \sin x$. $u' \cos x + v' \sin x = 0$ and $-u' \sin x + v' \cos x = x^2$;
    $u' = -x^2 \sin x$, $v' = x^2 \cos x$ so $u = x^2 \cos x - 2x \sin x - 2 \cos x$, $v = x^2 \sin x + 2x \cos x - 2 \sin x$,
    $y_p = u \cos x + v \sin x = x^2 - 2$, $y = c_1 \cos x + c_2 \sin x + x^2 - 2$.

3. $m^2 + m - 2 = 0$, $(m-1)(m+2) = 0$; $m = 1, -2$ so $y_c = c_1 e^x + c_2 e^{-2x}$.
    $u'e^x + v'e^{-2x} = 0$ and $u'e^x - 2v'e^{-2x} = 2e^x$;
    $u' = \frac{2}{3}$, $v' = -\frac{2}{3}e^{3x}$ so $u = \frac{2}{3}x$, $v = -\frac{2}{9}e^{3x}$, $y_p = ue^x + ve^{-2x} = \frac{2}{3}xe^x - \frac{2}{9}e^x$.
    But $-\frac{2}{9}e^x$ satisfies the complementary equation so $y = c_1 e^x + c_2 e^{-2x} + \frac{2}{3}xe^x$.

5. $m^2 + 4 = 0$; $m = \pm 2i$ so $y_c = c_1 \cos 2x + c_2 \sin 2x$. $u' \cos 2x + v' \sin 2x = 0$ and
    $-2u' \sin 2x + v' \cos 2x = \sin 2x$; $u' = -\frac{1}{2} \sin^2 2x$, $v' = \frac{1}{2} \sin 2x \cos 2x$ so
    $u = -\frac{1}{4}x + \frac{1}{16} \sin 4x = -\frac{1}{4}x + \frac{1}{8} \sin 2x \cos 2x$, $v = -\frac{1}{8} \cos^2 2x$,
    $y_p = u \cos 2x + v \sin 2x = -\frac{1}{4}x \cos 2x$, $y = c_1 \cos 2x + c_2 \sin 2x - \frac{1}{4}x \cos 2x$.

7. $m^2 + 1 = 0$; $m = \pm i$ so $y_c = c_1 \cos x + c_2 \sin x$. $u' \cos x + v' \sin x = 0$ and
    $-u' \sin x + v' \cos x = \tan x$; $u' = -\tan x \sin x = \cos x - \sec x$, $v' = \tan x \cos x = \sin x$
    so $u = \sin x - \ln|\sec x + \tan x|$, $v = -\cos x$, $y_p = u \cos x + v \sin x = -\cos x \ln|\sec x + \tan x|$,
    $y = c_1 \cos x + c_2 \sin x - \cos x \ln|\sec x + \tan x|$.

9. $m^2 - 2m + 1 = 0$, $(m-1)^2 = 0$; $m = 1$ so $y_c = c_1 e^x + c_2 x e^x$. $u'e^x + v'xe^x = 0$ and
    $u'e^x + v'(x+1)e^x = e^x/x$; $u' = -1$, $v' = 1/x$ so $u = -x$, $v = \ln|x|$,
    $y_p = ue^x + vxe^x = -xe^x + xe^x \ln|x|$. But $-xe^x$ satisfies the complementary equation so
    $y = c_1 e^x + c_2 x e^x + xe^x \ln|x|$.

11. $m^2 + 1 = 0$; $m = \pm i$ so $y_c = c_1 \cos x + c_2 \sin x$. $u' \cos x + v' \sin x = 0$ and
$-u' \sin x + v' \cos x = 3 \sin^2 x$; $u' = -3 \sin^3 x$, $v' = 3 \sin^2 x \cos x$ so $u = 3 \cos x - \cos^3 x$,
$v = \sin^3 x$, $y_p = u \cos x + v \sin x = 3 \cos^2 x - \cos^4 x + \sin^4 x = 3 \cos^2 x - (\cos^4 x - \sin^4 x)$
$\qquad = \frac{3}{2}(1 + \cos 2x) - (\cos^2 x - \sin^2 x)(\cos^2 x + \sin^2 x) = \frac{3}{2} + \frac{1}{2} \cos 2x$,
$y = c_1 \cos x + c_2 \sin x + \frac{3}{2} + \frac{1}{2} \cos 2x$.

13. $m^2 + 1 = 0$; $m = \pm i$ so $y_c = c_1 \cos x + c_2 \sin x$. $u' \cos x + v' \sin x = 0$ and
$-u' \sin x + v' \cos x = \csc x$; $u' = -1$, $v' = \cos x \csc x = \cot x$ so $u = -x$,
$v = \ln|\sin x|$, $y_p = u \cos x + v \sin x = -x \cos x + \sin x \ln|\sin x|$,
$y = c_1 \cos x + c_2 \sin x - x \cos x + \sin x \ln|\sin x|$.

15. $m^2 + 1 = 0$; $m = \pm i$ so $y_c = c_1 \cos x + c_2 \sin x$. $u' \cos x + v' \sin x = 0$ and
$-u' \sin x + v' \cos x = \sec x \tan x$; $u' = -\tan^2 x$, $v' = \tan x$ so $u = x - \tan x$,
$v = -\ln|\cos x|$, $y_p = u \cos x + v \sin x = x \cos x - \sin x - \sin x \ln|\cos x|$. But $-\sin x$ satisfies
the complementary equation so $y = c_1 \cos x + c_2 \sin x + x \cos x - \sin x \ln|\cos x|$.

17. $m^2 + 2m + 1 = 0$, $(m + 1)^2 = 0$; $m = -1$ so $y_c = c_1 e^{-x} + c_2 x e^{-x}$. $u' e^{-x} + v' x e^{-x} = 0$
and $-u' e^{-x} + v'(1 - x)e^{-x} = e^{-x}/x^2$; $u' = -1/x$, $v' = 1/x^2$ so $u = -\ln|x|$, $v = -1/x$,
$y_p = u e^{-x} + v x e^{-x} = -e^{-x} \ln|x| - e^{-x}$. But $-e^{-x}$ satisfies the complementary equation so
$y = c_1 e^{-x} + c_2 x e^{-x} - e^{-x} \ln|x|$.

19. $m^2 + 4m + 4 = 0$, $(m + 2)^2 = 0$; $m = -2$ so $y_c = c_1 e^{-2x} + c_2 x e^{-2x}$. $u' e^{-2x} + v' x e^{-2x} = 0$
and $-2u' e^{-2x} + v'(1 - 2x)e^{-2x} = x e^{-x}$; $u' = -x^2 e^x$, $v' = x e^x$ so $u = (-x^2 + 2x - 2)e^x$,
$v = (x - 1)e^x$, $y_p = u e^{-2x} + v x e^{-2x} = (x - 2)e^{-x}$, $y = c_1 e^{-2x} + c_2 x e^{-2x} + (x - 2)e^{-x}$.

21. $m^2 + 1 = 0$; $m = \pm i$ so $y_c = c_1 \cos x + c_2 \sin x$. $u' \cos x + v' \sin x = 0$ and
$-u' \sin x + v' \cos x = \sec^2 x$; $u' = -\sec x \tan x$, $v' = \sec x$ so $u = -\sec x$,
$v = \ln|\sec x + \tan x|$, $y_p = u \cos x + v \sin x = -1 + \sin x \ln|\sec x + \tan x|$,
$y = c_1 \cos x + c_2 \sin x - 1 + \sin x \ln|\sec x + \tan x|$.

23. $m^2 - 2m + 1 = 0$, $(m - 1)^2 = 0$; $m = 1$ so $y_c = c_1 e^x + c_2 x e^x$. $u' e^x + v' x e^x = 0$ and
$u' e^x + v'(x + 1)e^x = e^x/x^2$; $u' = -1/x$, $v' = 1/x^2$ so $u = -\ln|x|$, $v = -1/x$,
$y_p = u e^x + v x e^x = -e^x \ln|x| - e^x$. But $-e^x$ satisfies the complementary equation so
$y = c_1 e^x + c_2 x e^x - e^x \ln|x|$.

25. $m^2 - 1 = 0$; $m = \pm 1$ so $y_c = c_1 e^x + c_2 e^{-x}$. $u' e^x + v' e^{-x} = 0$ and $u' e^x - v' e^{-x} = e^x \cos x$;
$u' = \frac{1}{2} \cos x$, $v' = -\frac{1}{2} e^{2x} \cos x$ so $u = \frac{1}{2} \sin x$, $v = -\frac{1}{10} e^{2x}(2 \cos x + \sin x)$,
$y_p = u e^x + v e^{-x} = \frac{1}{2} e^x \sin x - \frac{1}{10} e^x(2 \cos x + \sin x) = \frac{1}{5} e^x(2 \sin x - \cos x)$,
$y = c_1 e^x + c_2 e^{-x} + \frac{1}{5} e^x(2 \sin x - \cos x)$.

27.　$m^2 + 2m + 1 = 0$, $(m+1)^2 = 0$; $m = -1$ so $y_c = c_1 e^{-x} + c_2 x e^{-x}$. $u' e^{-x} + v' x e^{-x} = 0$

and $-u' e^{-x} + v'(1-x)e^{-x} = e^{-x} \ln|x|$; $u' = -x \ln|x|$, $v' = \ln|x|$ so $u = -\frac{1}{2} x^2 \ln|x| + \frac{1}{4} x^2$,

$v = x \ln|x| - x$,

$y_p = u e^{-x} + v x e^{-x} = -\frac{1}{2} x^2 e^{-x} \ln|x| + \frac{1}{4} x^2 e^{-x} + x^2 e^{-x} \ln|x| - x^2 e^{-x} = \frac{1}{2} x^2 e^{-x} \ln|x| - \frac{3}{4} x^2 e^{-x}$,

$y = c_1 e^{-x} + c_2 x e^{-x} + \frac{1}{2} x^2 e^{-x} \ln|x| - \frac{3}{4} x^2 e^{-x}$.

29.　$m^2 + 1 = 0$; $m = \pm i$ so $y_c = c_1 \cos x + c_2 \sin x$. $u' \cos x + v' \sin x = 0$, $-u' \sin x + v' \cos x = r(x)$;

$u' = -r(x) \sin x$, $v' = r(x) \cos x$ so $u = -\int r(x) \sin x \, dx$,

$v = \int r(x) \cos x \, dx$, $y = c_1 \cos x + c_2 \sin x - \left[\int r(x) \sin x \, dx \right] \cos x + \left[\int r(x) \cos x \, dx \right] \sin x$.

EXERCISE SET 19.4

1.　(a)　$M = w/g = 64/32 = 2$, $k/M = 8/2 = 4$; $y'' + 4y = 0$, $y(0) = 1$, $y'(0) = 0$.

　　(b)　$m^2 + 4 = 0$; $m = \pm 2i$ so $y = c_1 \cos 2t + c_2 \sin 2t$, $y' = -2c_1 \sin 2t + 2c_2 \cos 2t$, $c_1 = 1$ and $c_2 = 0$, $y = \cos 2t$.

3.　(a)　$k/M = g/\ell = 980/5 = 196$; $y'' + 196y = 0$, $y(0) = -10$, $y'(0) = 0$.

　　(b)　$m = \pm 14i$ so $y = c_1 \cos 14t + c_2 \sin 14t$, $y' = -14c_1 \sin 14t + 14c_2 \cos 14t$, $c_1 = -10$ and $c_2 = 0$, $y = -10 \cos 14t$.

5.　(a)　$M = w/g = (1/2)/32 = 1/64$, $k/M = 64$, $y_0 = 2$; $y = 2 \cos 8t$.

　　(b)　$|y_0| = 2$　　　　　　　　　　(c)　$T = \pi/4$　　　　　　　　　　(d)　$f = 4/\pi$

7.　(a)　$y = -\dfrac{1}{4} \cos 8\sqrt{6}\, t$.

　　(b)　$|y_0| = 1/4$ ft.　　　　　　　(c)　$T = \pi/(4\sqrt{6})$　　　　　　(d)　$f = 4\sqrt{6}/\pi$

9.　(a)　$M = w/g = 32/32 = 1$, $k/M = 8$; $y'' + 4y' + 8y = 0$, $y(0) = -3$, $y'(0) = 0$.

　　(b)　$m^2 + 4m + 8 = 0$; $m = -2 \pm 2i$ so $y = e^{-2t}(c_1 \cos 2t + c_2 \sin 2t)$,

　　　　$y' = 2e^{-2t}[(c_2 - c_1)\cos 2t - (c_1 + c_2)\sin 2t]$, $c_1 = c_2 = -3$, $y = -3e^{-2t}(\cos 2t + \sin 2t)$.

　　(d)　$\alpha = \beta = 2$, $\omega = \tan^{-1}(\alpha/\beta) = \tan^{-1} 1 = \pi/4$; $y = -3\sqrt{2}\, e^{-2t} \cos(2t - \pi/4)$.

　　(e)　$T = \pi$　　　　　　　　　　(f)　$f = 1/\pi$

11.　(a)　$\alpha = 1/5$, $\beta = \sqrt{2}/5$; $y = \dfrac{5}{2}\sqrt{6}\, e^{-t/5} \cos[\sqrt{2}t/5 - \tan^{-1}(1/\sqrt{2})]$.

(b) $T = 10\pi/\sqrt{2}$ **(c)** $\sqrt{2}/(10\pi)$

13. **(a)** $y'' + 64y = 0$, $y(0) = 0$, $y'(0) = -2$. $y = c_1 \cos 8t + c_2 \sin 8t$, $y' = -8c_1 \sin 8t + 8c_2 \cos 8t$,

$c_1 = 0$ and $c_2 = -1/4$; $y = -\dfrac{1}{4} \sin 8t$.

(b) $|y_0| = 1/4$ **(c)** $T = \pi/4$ **(d)** $f = 4/\pi$

15. $M = w/g = 1/8$, $k/M = 50$; $y'' + 2y' + 50y = 0$, $y(0) = 1/3$, $y'(0) = -5$.

$m^2 + 2m + 50 = 0$; $m = -1 \pm 7i$ so $y = e^{-t}(c_1 \cos 7t + c_2 \sin 7t)$,

$y' = e^{-t}[(7c_2 - c_1)\cos 7t - (7c_1 + c_2)\sin 7t]$, $c_1 = 1/3$ and $c_2 = -2/3$,

$y = \dfrac{1}{3}e^{-t}(\cos 7t - 2\sin 7t)$.

17. $T = 2\pi\sqrt{M/k} = 2\pi\sqrt{w/(kg)}$, $T^2 = 4\pi^2 w/(32k)$, $\pi^2 w = 8T^2 k$; for w and $w + 4$, $\pi^2 w = 72k$ and $\pi^2(w + 4) = 200k$. Solve this system of equations to get

(a) $k = \pi^2/32$ **(b)** $w = 9/4$

19. Let ℓ be the depth to which the cylinder is submerged in the water at equilibrium, then $\rho\pi r^2 \ell = \delta\pi r^2 h$ so $\ell = \delta h/\rho$, $M/k = \ell/g = \delta h/(\rho g)$, $T = 2\pi\sqrt{\delta h/(\rho g)}$.

21. **(a)** $y = y_0\cos(\sqrt{k/M}\, t)$, $y' = -y_0\sqrt{k/M}\sin(\sqrt{k/M}\, t)$; when $\sin(\sqrt{k/M}t)$ is 1 or -1, $\cos(\sqrt{k/M}\, t)$ is 0 so $y = 0$ when $|y'| = |y_0|\sqrt{k/M} = 2\pi|y_0|/T$.

(b) $y'' = -y_0(k/M)\cos(\sqrt{k/M}\, t)$; $|y''|$ is maximum when $\cos(\sqrt{k/M}\, t)$ is 1 or -1, and hence $|y|$ is maximum. The maximum value of $|y''|$ is $|y_0|k/M = 4\pi^2|y_0|/T^2$.

23. Let $\omega = \tan^{-1}(\alpha/\beta)$, then $\cos\omega = \beta/\sqrt{\alpha^2 + \beta^2}$ and $\sin\omega = \alpha/\sqrt{\alpha^2 + \beta^2}$,

$\beta\cos\beta t + \alpha\sin\beta t = \sqrt{\alpha^2 + \beta^2}(\cos\beta t\cos\omega + \sin\beta t\sin\omega) = \sqrt{\alpha^2 + \beta^2}\cos(\beta t - \omega)$

so $y(t) = \dfrac{y_0\sqrt{\alpha^2 + \beta^2}}{\beta}e^{-\alpha t}\cos(\beta t - \omega)$.

25. **(a)** $m = -\dfrac{c}{2M} = -\alpha$ so $y = (c_1 + c_2 t)e^{-\alpha t}$, $y' = (c_2 - \alpha c_1 - \alpha c_2 t)e^{-\alpha t}$, $c_1 = y_0$ and $c_2 = \alpha y_0$; $y = y_0(1 + \alpha t)e^{-\alpha t}$.

(b) $\displaystyle\lim_{t \to +\infty} y_0(1 + \alpha t)e^{-\alpha t} = \lim_{t \to +\infty}\frac{y_0(1 + \alpha t)}{e^{\alpha t}} = \lim_{t \to +\infty}\frac{y_0}{e^{\alpha t}} = 0.$

(c) $\alpha > 0$ so $y_0(1 + \alpha t)e^{-\alpha t} \neq 0$ for $t > 0$.

APPENDIX B

Trigonometry Review

EXERCISES, TRIGONOMETRIC FUNCTIONS AND IDENTITIES

1. (a) $5\pi/12$ (b) $13\pi/6$ (c) $\pi/9$ (d) $23\pi/30$

3. (a) $12°$ (b) $(270/\pi)°$ (c) $288°$ (d) $540°$

5. (a)   (b)

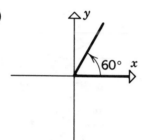
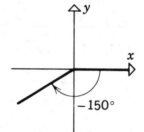

(c)   (d)

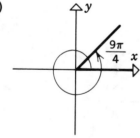
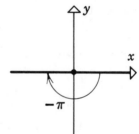

7. $150° = 5\pi/6$ rad so

(a) $s = 4(5\pi/6) = 10\pi/3$ cm   (b) $A = \dfrac{1}{2}(4)^2(5\pi/6) = 20\pi/3$ cm^2

9. (a) $2\pi r = R(2\pi - \theta)$, $r = \dfrac{2\pi - \theta}{2\pi}R$

(b) $h = \sqrt{R^2 - r^2} = \sqrt{R^2 - (2\pi - \theta)^2 R^2/(4\pi^2)} = \dfrac{\sqrt{4\pi\theta - \theta^2}}{2\pi}R$

11. $\sin\theta = \sqrt{21}/5$, $\tan\theta = \sqrt{21}/2$ **13.** $\sin\theta = 5/\sqrt{34}$, $\cos\theta = 3/\sqrt{34}$

15. $\sin\theta = 3/\sqrt{10}$, $\cos\theta = 1/\sqrt{10}$ **17.** $\tan\theta = \sqrt{21}/2$, $\csc\theta = 5/\sqrt{21}$

19. Let x be the length of the side adjacent to θ, then $\cos\theta = x/6 = 0.3$, $x = 1.8$.

21. Let x be the length of the hypotenuse, then $\sin\theta = 2.4/x = 0.8$, $x = 2.4/0.8 = 3$.

23.

	$\sin\theta$	$\cos\theta$	$\tan\theta$	$\csc\theta$	$\sec\theta$	$\cot\theta$
(a)	$a/3$	$\sqrt{9-a^2}/3$	$a/\sqrt{9-a^2}$	$3/a$	$3/\sqrt{9-a^2}$	$\sqrt{9-a^2}/a$
(b)	$a/\sqrt{a^2+25}$	$5/\sqrt{a^2+25}$	$a/5$	$\sqrt{a^2+25}/a$	$\sqrt{a^2+25}/5$	$5/a$
(c)	$\sqrt{a^2-1}/a$	$1/a$	$\sqrt{a^2-1}$	$a/\sqrt{a^2-1}$	a	$1/\sqrt{a^2-1}$

25. Let h be the altitude as shown
in the figure, then $h = 3\sin 60° = 3\sqrt{3}/2$
so $A = \dfrac{1}{2}(3\sqrt{3}/2)(7) = 21\sqrt{3}/4$.

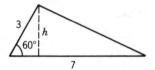

27. Let x be the distance above the ground, then $x = 10\sin 67° \approx 9.2$ ft.

29. From the figure, $h = x - y$
but $x = d\tan\beta$, $y = d\tan\alpha$
so $h = d(\tan\beta - \tan\alpha)$.

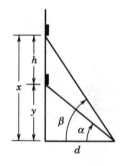

31. From the figure, area $= \dfrac{1}{2}hc$ but $h = b\sin A$

so area $= \dfrac{1}{2}bc\sin A$. The formulas

area $= \dfrac{1}{2}ac\sin B$ and area $= \dfrac{1}{2}ab\sin C$

follow by drawing altitudes from
vertices B and C, respectively.

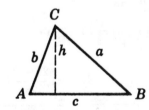

33.

θ	$\sin\theta$	$\cos\theta$	$\tan\theta$	$\csc\theta$	$\sec\theta$	$\cot\theta$
(a) $225°$	$-1/\sqrt{2}$	$-1/\sqrt{2}$	1	$-\sqrt{2}$	$-\sqrt{2}$	1
(b) $-210°$	$1/2$	$-\sqrt{3}/2$	$-1/\sqrt{3}$	2	$-2/\sqrt{3}$	$-\sqrt{3}$
(c) $5\pi/3$	$-\sqrt{3}/2$	$1/2$	$-\sqrt{3}$	$-2/\sqrt{3}$	2	$-1/\sqrt{3}$
(d) $-3\pi/2$	1	0	$-$	1	$-$	0

35. $\sin\theta = -3/5$, $\cos\theta = -4/5$, $\tan\theta = 3/4$, $\csc\theta = -5/3$, $\sec\theta = -5/4$, $\cot\theta = 4/3$

37. **(a)** from identity (23b) with $\alpha = 45°$ and $\beta = 30°$,

$$\cos 15° = \cos(45° - 30°) = \cos 45° \cos 30° + \sin 45° \sin 30°$$
$$= (1/\sqrt{2})(\sqrt{3}/2) + (1/\sqrt{2})(1/2) = (\sqrt{6} + \sqrt{2})/4, \text{ or from (27a) with } \alpha = 30°,$$
$$\cos^2 15° = \cos^2(30°/2) = (1 + \cos 30°)/2 = (2 + \sqrt{3})/4 \text{ so } \cos 15° = \frac{1}{2}\sqrt{2 + \sqrt{3}}$$

(b) from (27b) with $\alpha = 45°$, $\sin^2 22.5° = \sin^2(45°/2) = (1 - \cos 45°)/2 = (2 - \sqrt{2})/4$

so $\sin 22.5° = \frac{1}{2}\sqrt{2 - \sqrt{2}}$

(c) from (27a) with $\alpha = 150°$, $\cos^2 75° = \cos^2(150°/2) = (1 + \cos 150°)/2 = (2 - \sqrt{3})/4$

so $\cos 75° = \frac{1}{2}\sqrt{2 - \sqrt{3}}$;

from (27b) with $\alpha = 75°$, $\sin^2 37.5° = \sin^2(75°/2) = (1 - \cos 75°)/2 = (2 - \sqrt{2 - \sqrt{3}})/4$

so $\sin 37.5° = \frac{1}{2}\sqrt{2 - \sqrt{2 - \sqrt{3}}}$

39. **(a)** $\sin 2\theta = 2\sin\theta\cos\theta = 2(\sqrt{5}/3)(2/3) = 4\sqrt{5}/9$

(b) $\cos 2\theta = 2\cos^2\theta - 1 = 2(2/3)^2 - 1 = -1/9$

41. **(a)** $\sin(\pi/2 + \theta) = \sin(\pi/2)\cos\theta + \cos(\pi/2)\sin\theta = (1)\cos\theta + (0)\sin\theta = \cos\theta$

(b) $\cos(\pi/2 + \theta) = \cos(\pi/2)\cos\theta - \sin(\pi/2)\sin\theta = (0)\cos\theta - (1)\sin\theta = -\sin\theta$

(c) $\sin(3\pi/2 - \theta) = \sin(3\pi/2)\cos\theta - \cos(3\pi/2)\sin\theta = (-1)\cos\theta - (0)\sin\theta = -\cos\theta$

(d) $\cos(3\pi/2 + \theta) = \cos(3\pi/2)\cos\theta - \sin(3\pi/2)\sin\theta = (0)\cos\theta - (-1)\sin\theta = \sin\theta$

43. **(a)** Add (22a) and (23a) to get $\sin(\alpha - \beta) + \sin(\alpha + \beta) = 2\sin\alpha\cos\beta$ so
$\sin\alpha\cos\beta = (1/2)[\sin(\alpha - \beta) + \sin(\alpha + \beta)]$.

(b) Subtract (22b) from (23b).   **(c)** Add (22b) and (23b).

45. $\sin\alpha + \sin(-\beta) = 2\sin\dfrac{\alpha-\beta}{2}\cos\dfrac{\alpha+\beta}{2}$, but $\sin(-\beta) = -\sin\beta$ so

      $\sin\alpha - \sin\beta = 2\cos\dfrac{\alpha+\beta}{2}\sin\dfrac{\alpha-\beta}{2}$.

47. $\dfrac{\cos\theta\tan\theta + \sin\theta}{\tan\theta} = \dfrac{\cos\theta(\sin\theta/\cos\theta) + \sin\theta}{\sin\theta/\cos\theta} = 2\cos\theta$

49. $\tan\theta + \cot\theta = \dfrac{\sin\theta}{\cos\theta} + \dfrac{\cos\theta}{\sin\theta} = \dfrac{\sin^2\theta + \cos^2\theta}{\sin\theta\cos\theta} = \dfrac{1}{\sin\theta\cos\theta} = \dfrac{2}{2\sin\theta\cos\theta} = \dfrac{2}{\sin 2\theta} = 2\csc 2\theta$

51. $\dfrac{\sin\theta + \cos 2\theta - 1}{\cos\theta - \sin 2\theta} = \dfrac{\sin\theta + (1 - 2\sin^2\theta) - 1}{\cos\theta - 2\sin\theta\cos\theta} = \dfrac{\sin\theta(1 - 2\sin\theta)}{\cos\theta(1 - 2\sin\theta)} = \tan\theta$

53.   Using (28a), $2\cos 2\theta\sin\theta = 2(1/2)[\sin(-\theta) + \sin 3\theta] = \sin 3\theta - \sin\theta$

55. $\tan(\theta/2) = \dfrac{\sin(\theta/2)}{\cos(\theta/2)} = \dfrac{2\sin(\theta/2)\cos(\theta/2)}{2\cos^2(\theta/2)} = \dfrac{\sin\theta}{1 + \cos\theta}$

57. $\sin 3\theta = \sin(2\theta + \theta) = \sin 2\theta\cos\theta + \cos 2\theta\sin\theta = (2\sin\theta\cos\theta)\cos\theta + (\cos^2\theta - \sin^2\theta)\sin\theta$

         $= 2\sin\theta\cos^2\theta + \sin\theta\cos^2\theta - \sin^3\theta = 3\sin\theta\cos^2\theta - \sin^3\theta$;

      similarly, $\cos 3\theta = \cos^3\theta - 3\sin^2\theta\cos\theta$

59.   Consider the triangle having a, b, and d as sides. The angle formed by sides a and b is $\pi - \theta$ so from the law of cosines, $d^2 = a^2 + b^2 - 2ab\cos(\pi - \theta) = a^2 + b^2 + 2ab\cos\theta$,

      $d = \sqrt{a^2 + b^2 + 2ab\cos\theta}$.

61.   **(a)** $\theta = \pm n\pi,\ n = 0, 1, 2, \ldots$ **(b)** $\theta = \pi/2 \pm n\pi,\ n = 0, 1, 2, \ldots$

      (c) $\theta = \pm n\pi,\ n = 0, 1, 2, \ldots$ **(d)** $\theta = \pm n\pi,\ n = 0, 1, 2, \ldots$

      (e) $\theta = \pi/2 \pm n\pi,\ n = 0, 1, 2, \ldots$ **(f)** $\theta = \pm n\pi,\ n = 0, 1, 2, \ldots$

63. $\theta = 5\pi/4 \pm 2n\pi$ and $\theta = 7\pi/4 \pm 2n\pi,\ n = 0, 1, 2, \ldots$

65. $\theta = \pi/3 \pm 2n\pi$ and $\theta = 5\pi/3 \pm 2n\pi,\ n = 0, 1, 2, \ldots$

67. $\theta = \pi/3 \pm n\pi,\ n = 0, 1, 2, \ldots$

69. $\theta = 4\pi/3 \pm 2n\pi$ and $\theta = 5\pi/3 \pm 2n\pi,\ n = 0, 1, 2, \ldots$

71. $\theta = \pi \pm 2n\pi,\ n = 0, 1, 2, \ldots$ **73.** $\theta = \pi/6 \pm n\pi,\ n = 0, 1, 2, \ldots$

75. $\theta = 7\pi/6 \pm 2n\pi$ and $\theta = 11\pi/6 \pm 2n\pi$, $n = 0, 1, 2, \ldots$

77. $\theta = \pi/6 \pm 2n\pi$ and $\theta = 11\pi/6 \pm 2n\pi$, $n = 0, 1, 2, \ldots$

EXERCISES, GRAPHS OF TRIGONOMETRIC FUNCTIONS

1. (a) $2\pi/5$ (b) 6π (c) 8π (d) $\pi/7$

     (e) 2   (f) 10   (g) 1   (h) $2k\pi$

3. (a) 5   (b) 1/3   (c) 1/2   (d) 1

5.

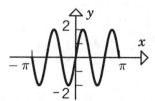

7.

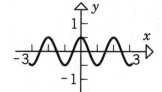

9.

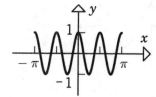

11.

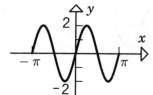

13. (a)

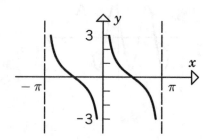

(b)

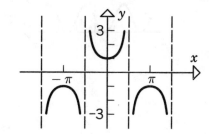

(c)

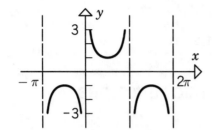

15.

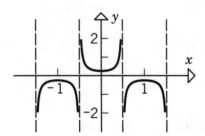

17.

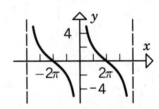

19.

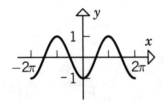

21.

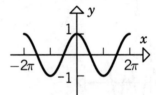

23.

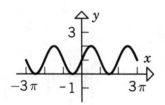

25.

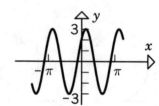

27.

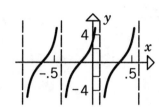

29.

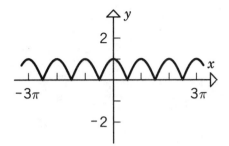

31.

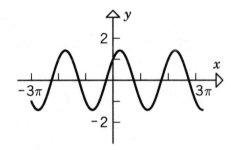

33. (a) odd   (b) even   (c) odd   (d) even

     (e) even   (f) even   (g) odd   (h) even

35. If $|x| \geq \pi/2$ then $|\sin x| \leq |x|$ because $|\sin x| \leq 1 < \pi/2$ for all x. If $x = 0$ then $\sin x = 0$ so $|\sin x| \leq |x|$. If $0 < |x| < \pi/2$ then, from Exercise 34, $\sin |x| \leq |x|$. But $\sin |x| = |\sin x|$ if $0 < |x| < \pi/2$ so $|\sin x| \leq |x|$.

37. $\tan bx$ repeats when bx changes by π or equivalently when x changes by π/b so the fundamental period is $\pi/|b|$.

APPENDIX C

Supplementary Material

EXERCISES, ONE-SIDED AND INFINITE LIMITS

1. If $x > 2$ then $|(x + 1) - 3| = |x - 2| = x - 2 < \epsilon$ when $x < 2 + \epsilon$; take $\delta = \epsilon$.

3. If $x > 4$ then $|\sqrt{x - 4} - 0| = \sqrt{x - 4} < \epsilon$ when $x - 4 < \epsilon^2$, $x < 4 + \epsilon^2$; take $\delta = \epsilon^2$.

5. If $x > 2$ then $|f(x) - 2| = |x - 2| = x - 2 < \epsilon$ when $x < 2 + \epsilon$; take $\delta = \epsilon$.

7. If $x > 0$ then $|1/x^2 - 0| = 1/x^2 < \epsilon$ when $x^2 > 1/\epsilon$, $x > 1/\sqrt{\epsilon}$; take $N = 1/\sqrt{\epsilon}$.

9. If $x < -2$ then $|1/(x + 2) - 0| = 1/|x + 2| = -1/(x + 2) < \epsilon$ when $x + 2 < -1/\epsilon$, $x < -2 - 1/\epsilon$; take $N = -2 - 1/\epsilon$.

11. If $x > -1$ then $|x/(x + 1) - 1| = |-1/(x + 1)| = 1/(x + 1) < \epsilon$ when $x + 1 > 1/\epsilon$, $x > -1 + 1/\epsilon$; take $N = 1/\epsilon$.

13. If $x < -5/2$ then $\left|\dfrac{4x - 1}{2x + 5} - 2\right| = \left|\dfrac{-11}{2x + 5}\right| = \dfrac{11}{|2x + 5|} = -\dfrac{11}{2x + 5} < \epsilon$ when $2x + 5 < -\dfrac{11}{\epsilon}$, $x < -\dfrac{5}{2} - \dfrac{11}{2\epsilon}$; take $N = -\dfrac{5}{2} - \dfrac{11}{2\epsilon}$.

15. If $x \neq 3$ and $N > 0$ then $1/(x - 3)^2 > N$ when $(x - 3)^2 < 1/N$, $|x - 3| < 1/\sqrt{N}$; take $\delta = 1/\sqrt{N}$.

17. If $x \neq 0$ and $N > 0$ then $1/|x| > N$ when $|x| < 1/N$; take $\delta = 1/N$.

19. If $x \neq 0$ and $N < 0$ then $-1/x^4 < N$ when $x^4 < -1/N = 1/|N|$, $x < 1/\sqrt[4]{|N|}$; take $\delta = 1/\sqrt[4]{|N|}$.

21. (15): Let f be defined on some open interval extending to the right from a. We will write $\lim\limits_{x \to a^+} f(x) = +\infty \ (-\infty)$ if given any positive (negative) number N we can find a number $\delta > 0$ such that $f(x)$ satisfies $f(x) > N \ (f(x) < N)$ whenever x satisfies $a < x < a + \delta$.

(16): Similar to (15) with "right" replaced by "left", a^+ replaced by a^-, and $a < x < a + \delta$ replaced by $a - \delta < x < a$.

23. (a) If $x > 1$ and $N < 0$ then $1/(1 - x) < N$ when $1 - x > 1/N = -1/|N|$, $x < 1 + 1/|N|$; take $\delta = 1/|N|$.

(b) If $x < 1$ and $N > 0$ then $1/(1-x) > N$ when $1 - x < 1/N$, $x > 1 - 1/N$; take $\delta = 1/N$.

25. (a) If $M > 0$ then $x + 1 > M$ when $x > M - 1$; take $N = M$

(b) If $M < 0$ then $x + 1 < M$ when $x < M - 1$; take $N = M - 1$.

EXERCISES, PROOFS OF LIMIT THEOREMS

1. $|k - k| = |0| = 0 < \epsilon$ when $x > N$ for any $N > 0$.

3. $|[f(x) + g(x)] - [L_1 + L_2]| = |[f(x) - L_1] + [g(x) - L_2]| \le |f(x) - L_1| + |g(x) - L_2|$.

Given $\epsilon > 0$, there exist negative numbers N_1 and N_2 such that $|f(x) - L_1| < \epsilon/2$ and $|g(x) - L_2| < \epsilon/2$ whenever $x < N_1$ and $x < N_2$, respectively. Let $N = \min(N_1, N_2)$, if $x < N$ then $|f(x) - L_1| + |g(x) - L_2| < \epsilon/2 + \epsilon/2 = \epsilon$ so $|[f(x) + g(x)] - [L_1 + L_2]| < \epsilon$.

5. From Theorem 8, $\lim_{x \to a}(-1) = -1$ so from Theorem 10,

$$\lim_{x \to a}[(-1)g(x)] = \lim_{x \to a}(-1) \lim_{x \to a} g(x) = (-1)L_2 = -L_2 \text{ thus, using Theorem 9,}$$

$$\lim_{x \to a}[f(x) - g(x)] = \lim_{x \to a}[f(x) + (-g(x))] = \lim_{x \to a} f(x) + \lim_{x \to a}[-g(x)] = L_1 + (-L_2) = L_1 - L_2$$

7. (a) Given $N < 0$, there exist numbers $\delta_1 > 0$ and $\delta_2 > 0$ such that $f(x) < N/2$ and $g(x) > -N/2$ whenever $0 < |x - a| < \delta_1$ and $0 < |x - a| < \delta_2$, respectively. Let $\delta = \min(\delta_1, \delta_2)$ then $f(x) < N/2$ and $-g(x) < N/2$ whenever $0 < |x - a| < \delta$ so $f(x) - g(x) < N/2 + N/2 = N$.

(b) No, for example $\lim_{x \to 0}[(-1/x^2) + (1/x^4)] = \lim_{x \to 0}[(1 - x^2)/x^4] = +\infty$

9. If $\lim_{x \to a} f(x) = L$ then given $\epsilon > 0$, there exists a $\delta > 0$ such that $|f(x) - L| < \epsilon$ whenever $0 < |x - a| < \delta$. But $|f(x) - L| = |[f(x) - L] - 0| < \epsilon$ whenever $0 < |x - a| < \delta$ so $\lim_{x \to a}[f(x) - L] = 0$. If $\lim_{x \to a}[f(x) - L] = 0$ then given $\epsilon > 0$, there exists a $\delta > 0$ such that $|[f(x) - L] - 0| < \epsilon$ whenever $0 < |x - a| < \delta$, but $|[f(x) - L] - 0| = |f(x) - L| < \epsilon$ whenever $0 < |x - a| < \delta$ so $\lim_{x \to a} f(x) = L$.

11. If $\lim_{x \to a} f(x) = L$ then given $\epsilon > 0$ there exists a $\delta > 0$ such that $|f(x) - L| < \epsilon$ whenever $0 < |x - a| < \delta$ or equivalently whenever $a < x < a + \delta$ or $a - \delta < x < a$ so

$$\lim_{x \to a^+} f(x) = \lim_{x \to a^-} f(x) = L.$$

EXERCISES, PROOFS OF KEY RESULTS

1. **(a)** In the proof of part (a), interchange the words "increasing" and "decreasing", and reverse the order of the inequality symbols.

   (b) If $f'(x)$ has the same sign on the intervals (a, x_0) and (x_0, b), then f is increasing on (a, b) or decreasing on (a, b) so f does not have a relative extremum at x_0.

3. **(a)** 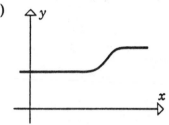   **(b)**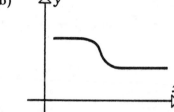

5. Let $h(x) = f(x) - g(x)$, then $h'(x) = f'(x) - g'(x) < 0$ thus h is decreasing on (a, b), that is $h(x_2) < h(x_1)$ if $x_2 > x_1$ so $f(x_2) - g(x_2) < f(x_1) - g(x_1)$, $f(x_2) - f(x_1) < g(x_2) - g(x_1)$.

7. **(a)** From Exercise 6(a) it follows that $f'(x)$ is increasing on some open interval I containing x_0 so f is concave up on I.

   (b) Similar to the proof in part (a).

EXERCISES, CRAMER'S RULE

1. $\begin{vmatrix} 3 & -4 \\ 2 & 1 \end{vmatrix} = 11$, $\begin{vmatrix} -5 & -4 \\ 4 & 1 \end{vmatrix} = 11$, $\begin{vmatrix} 3 & -5 \\ 2 & 4 \end{vmatrix} = 22$; $x = 1$, $y = 2$

3. $\begin{vmatrix} 2 & -5 \\ 4 & 6 \end{vmatrix} = 32$, $\begin{vmatrix} -2 & -5 \\ 1 & 6 \end{vmatrix} = -7$, $\begin{vmatrix} 2 & -2 \\ 4 & 1 \end{vmatrix} = 10$; $x_1 = -7/32$, $x_2 = 5/16$

5. $\begin{vmatrix} 1 & 2 & 1 \\ 2 & 1 & -1 \\ 1 & -1 & 1 \end{vmatrix} = -9$, $\begin{vmatrix} 3 & 2 & 1 \\ 0 & 1 & -1 \\ 6 & -1 & 1 \end{vmatrix} = -18$, $\begin{vmatrix} 1 & 3 & 1 \\ 2 & 0 & -1 \\ 1 & 6 & 1 \end{vmatrix} = 9$, $\begin{vmatrix} 1 & 2 & 3 \\ 2 & 1 & 0 \\ 1 & -1 & 6 \end{vmatrix} = -27$;

   $x = 2$, $y = -1$, $z = 3$

7. $\begin{vmatrix} 1 & 1 & -2 \\ 2 & -1 & 1 \\ 1 & -2 & -4 \end{vmatrix} = 21, \quad \begin{vmatrix} 1 & 1 & -2 \\ 2 & -1 & 1 \\ -4 & -2 & -4 \end{vmatrix} = 26, \quad \begin{vmatrix} 1 & 1 & -2 \\ 2 & 2 & 1 \\ 1 & -4 & -4 \end{vmatrix} = 25, \quad \begin{vmatrix} 1 & 1 & 1 \\ 2 & -1 & 2 \\ 1 & -2 & -4 \end{vmatrix} = 15;$

$x_1 = 26/21, \; x_2 = 25/21, \; x_3 = 5/7$

9. $\begin{vmatrix} \cos\theta & -\sin\theta \\ \sin\theta & \cos\theta \end{vmatrix} = 1, \quad \begin{vmatrix} x & -\sin\theta \\ y & \cos\theta \end{vmatrix} = x\cos\theta + y\sin\theta, \quad \begin{vmatrix} \cos\theta & x \\ \sin\theta & y \end{vmatrix} = y\cos\theta - x\sin\theta;$

$x' = x\cos\theta + y\sin\theta, \; y' = -x\sin\theta + y\cos\theta$